1979

Atomic Symbols of Atoms in Alphabetical Order of N

Element	Symbol	Element	Symbol
Actinium	Ac	Mendelevium	
Aluminum	Al	Mercury	
Americium	Am	Molybdenum	Mo
Antimony	Sb	Neodymium	Nd
Argon	Ar	Neon	Ne
Arsenic	As	Neptunium	Np
Astatine	At	Nickel	Ni
Barium	Ba	Niobium	Nb
Berkelium	Bk	Nitrogen	N
Beryllium	Be	Nobelium	No
Bismuth	Bi	Osmium	Os
Boron	B	Oxygen	O
Bromine	Br	Palladium	Pd
Cadmium	Cd	Phosphorus	P
Calcium	Ca	Platinum	Pt
Californium	Cf	Plutonium	Pu
Carbon	C	Polonium	Po
Cerium	Ce	Potassium	K
Cesium	Cs	Praseodymium	Pr
Chlorine	Cl	Promethium	Pm
Chromium	Cr	Protactinium	Pa
Cobalt	Co	Radium	Ra
Copper	Cu	Radon	Rn
Curium	Cm	Rhenium	Re
Dysprosium	Dy	Rhodium	Rh
Einsteinium	Es	Rubidium	Rb
Erbium	Er	Ruthenium	Ru
Europium	Eu	Samarium	Sm
Fermium	Fm	Scandium	Sc
Fluorine	F	Selenium	Se
Francium	Fr	Silicon	Si
Gadolinium	Gd	Silver	Ag
Gallium	Ga	Sodium	Na
Germanium	Ge	Strontium	Sr
Gold	Au	Sulfur	S
Hafnium	Hf	Tantalum	Ta
Hahnium	Ha	Technetium	Tc
Helium	He	Tellurium	Te
Holmium	Ho	Terbium	Tb
Hydrogen	H	Thallium	Tl
Indium	In	Thorium	Th
Iodine	I	Thulium	Tm
Iridium	Ir	Tin	Sn
Iron	Fe	Titanium	Ti
Krypton	Kr	Tungsten	W
Kurchatovium	Ku	Uranium	U
Lanthanum	La	Vanadium	V
Lawrencium	Lr	Xenon	Xe
Lead	Pb	Ytterbium	Yb
Lithium	Li	Yttrium	Y
Lutetium	Lu	Zinc	Zn
Magnesium	Mg	Zirconium	Zr
Manganese	Mn		

INSIDE CHEMISTRY

McGraw-Hill Book Company

New York St. Louis San Francisco Auckland Bogotá Düsseldorf
Johannesburg London Madrid Mexico Montreal New Delhi Panama
Paris São Paulo Singapore Sydney Tokyo Toronto

CHARLES COMPTON

Williams College

INSIDE CHEMISTRY

The cover photograph shows hot-air balloons made of synthetic
polymers soaring aloft due to the heat generated by the
combustion of propane (Sections 6-7 and 14-3).

INSIDE CHEMISTRY

1 2 3 4 5 6 7 8 9 0   VHVH   7 8 3 2 1 0 9

This book was set in Clarendon Light by Monotype Composition Company.
The editors were Donald C. Jackson, Michael Gardner, and Sibyl Golden;
the designer was Anne Canevari Green;
the production supervisor was Leroy A. Young.
The drawings were done by J & R Services, Inc.
The cover photograph was taken by Donald D. Woodman.
Von Hoffmann Press, Inc., was printer and binder.

Library of Congress Cataloging in Publication Data

Compton, Charles, date
    Inside chemistry.

    Bibliography: p.
    Includes index.
    1. Chemistry.   I.   Title.
QD31.2.C65   540   78-13379
ISBN 0-07-012350-0

TO MARY AND GORDON TWEEDY

CONTENTS IN BRIEF

CONTENTS

CHAPTER 3   **HOW IS THE STRUCTURE OF MOLECULES AND IONS EXPRESSED?**   **43**

* Items marked by an asterisk are optional and can be deleted without loss of continuity to text.

CHAPTER 7   **HOW ARE CHEMICAL REACTIONS CLASSIFIED AND
            ORGANIZED?**   151

SUPPLEMENTS

PREFACE

This text has been written to meet the special needs and interests of nonscience students. It emphasizes topics in the fields of energy, food, medicine, environment, industrial production, nuclear reactions, and the chemistry of living organisms. No previous study of science is assumed. The text presents an informative view of contemporary chemistry and includes sufficient coverage of fundamental principles to provide understanding, without being excessively technical.

Recognizing that there are many different approaches to the organization of chemistry courses for nonscience students, the plan of **the text provides great flexibility.** After a core of eight chapters (about a third of the text) each of the remaining chapters is completely freestanding and **may be studied in any order.** Moreover, if an expansion of the basic material is desired, most chapters contain optional sections at the end, clearly labeled as such, for which there are separate practice exercises. These sections may be omitted, however, with **no loss of continuity.** In addition, there are five supplements which provide an optional opportunity to introduce quantitative topics, and one which expands on nomenclature.

The approach in the body of the text is essentially nonquantitative. While the later chapters are devoted principally to topics of current interest, applications are also introduced in the core chapters. Over 25 short essays on wide-ranging topics are distributed throughout. These stand apart from the body of the text and have been included to increase interest in the study of chemistry.

SOME POSSIBLE CHAPTER SEQUENCES

The material of the text can easily be organized to provide the basis for courses with different emphases, levels, and lengths. Consider the following examples:

A Biochemical or Life Science Emphasis

Chaps. 1 to 8   Core of cumulative material
Chap. 10   What Are the Substances of Living Organisms?
Chap. 11   What Is the Relationship between Chemistry and Food?
Chap. 12   The Relationship between Chemistry and Medicine

Note: The supplements may be introduced at any point after Chapter 6 whatever the special emphasis of the course, for example,

Supplement 4   The Weight Relationships of Chemical Reactions
Supplement 5   The Concentration of Solutions

An Environmental Emphasis

Chaps. 1 to 8   Core of cumulative material
Chap. 9   What Are Nuclear Reactions?
Chap. 11   What Is the Relationship between Chemistry and Food?
Chap. 14   What Will Be the Energy Sources in the Future?
Chap. 15   Chemistry and Environmental Problems

An Emphasis on Economic and Practical Aspects

Chaps. 1 to 8   Core of cumulative material
Chap. 9   What Are Nuclear Reactions?
Chap. 11   What Is the Relationship between Chemistry and Food?
Chap. 13   What Is the Role of the Chemical Industry?
Chap. 14   What Will Be the Energy Sources in the Future?

Note: The optional sections of chapters may be included but need not be; thus the level and length of courses may be varied.

The **core material** itself includes: Chapter 1, What Is Chemistry?; Chapter 2, What Is the Structure of Atoms?; Chapter 3, How Is the Structure of Molecules and Ions Expressed?; Chapter 4, What Is the Nature of Solids, Liquids, and Gases?; Chapter 5, What Is the Nature of Solutions?; Chapter 6, What Is the Nature of Chemical Reactions?; Chapter 7, How Are Chemical Reactions Classified and Organized?; and Chapter 8, What Is Organic Chemistry?

AIDS FOR THE STUDENT

1. The principal points of each chapter are comprehensively summarized at the end in a series of Summarizing Questions for Self-Study. A succinct and explicit answer immediately follows each question.

2. The special terms introduced in each chapter are listed at the end together with cross-references to the sections where the terms are defined.

3. The glossary of all terms defined in text (with cross-references) together with many other terms provides a quick source of definitions of all common terms.

4. In almost all equations the names of all substances are given immediately below the formulas and the equation has an explanatory caption.

5. Numerous Notes to Students serve as instructional aids, especially in early chapters.

6. Detailed structural formulas have been used throughout, and a special effort has been made to make such formulas understandable and useful.

7. Numerous practice exercises cover chapter content comprehensively. Practice exercises related to optional sections are clearly labeled.

8. Special care has been taken to define terms explicitly.

ACKNOWLEDGMENTS

It is a pleasure to acknowledge the many useful comments, corrections, and suggestions of the reviewers, David L. Adams of North Shore Community College, Keith P. Anderson of Brigham Young University, Ronald Backus of American River College, Paul F. Endres of Bowling Green State University, and Dennis D. Scott of San Bernadino Valley College. I am also very grateful for the assistance of my colleagues at Williams, especially Danier Kleier, Hodge Markgraf, Paul McFarland, Thomas Newton, John Ricci, Allen Taylor, and Harold Warren. I am particularly indebted to Raymond Chang for his encouragement and helpful suggestions from the beginning of the project. Special thanks are due Ms. Nancy Barber, whose care and skill managed to convert unbelievably disordered copy into a useful manuscript; and Hans Oettgen for his interest and patience in taking many of the photographs. I would also like to express my gratitude to the staff at McGraw-Hill, particularly to Donald C. Jackson who initiated and guided the project. Finally, and by no means least, I am very grateful to my wife for her great help and understanding patience.

The author welcomes comments, corrections, and other suggestions for improvement.

Charles Compton

1

WHAT IS CHEMISTRY?

1-1 INTRODUCTION

Chemistry is concerned with the materials of the universe. It may seem, at first thought, to be far removed from human experience. The investigation of matter and of the transformations matter undergoes suggest activities which might be interesting only to steelmakers or possibly paint manufacturers. On reflection you will probably grant that the structure and behavior of matter is of concern to the development of such things as medicines, motor fuels, and new textiles. But only a fairly detailed study would reveal how the work of chemists has changed fields like psychiatry and has supplied evidence about the beginnings of the universe and the origin of life.

The special domain of the chemist is matter in all its forms, from acetylene and absinthe to zinc and zircon, from the composition of aardvarks to the materials of zinnias. Chemists seek to determine the fundamental regularities and principles regarding the structure and transformations of all forms of matter. These principles, in turn, are used to interpret newly discovered properties and to develop new materials. The ultimate aim of chemistry is to understand the nature of matter so well that we can predict the behavior of all forms of matter under all circumstances.

1-2 THE FOCUS ON SUBSTANCES

A principal underlying assumption of chemistry is that all forms of matter, no matter how complicated, can be understood in terms of the individual substances of which they are composed. A chemist uses the word **substance** in a special and precise way, to refer to forms of matter which are of definite composition and consistent behavior. The colorless gas carbon dioxide, for example, qualifies as a substance, but ginger ale does not. Ginger ale is a mixture of several substances, two of which are water and carbon dioxide (the bubbles). Moreover, the mixture of substances in one kind of ginger ale may differ from the mixture in another, whereas carbon dioxide is the same wherever it appears—Guatemala, Mars, or Cross Corners, Wisconsin.

Ordinary table salt (the chemical name is sodium chloride) also has the attributes of a substance. It has the same fundamental structure and properties no matter where it is found or how it is obtained. Seawater, on the other hand, is not a substance but a mixture of many substances, two of which are table salt and water. The composition of seawater varies from place to place. The seawater of the Red Sea, for example, has a different proportion of some substances than the seawater of the North Atlantic Ocean.

Table sugar, called sucrose by the chemist, also qualifies as a substance. It has the same structure and properties whether it is isolated from sugar cane or from the sap of the maple tree. Sugar cane and maple tree sap, on the other hand, are complex mixtures of many substances, one of which is sucrose.

Ask chemists to characterize any form of matter and they will first think of it in terms of its component substances. Ask them, for example, to consider the air we breathe, and they will think in terms of the oxygen, nitrogen, carbon dioxide, and other substances it contains. Confront them with a problem involving a chimpanzee and they will turn first to a consideration of the proteins and all the other substances which go to make up a chimpanzee, although they will be the first to recognize that this may not tell you all you would like to know about a chimpanzee.

1-3 HOW DO CHEMISTS ATTEMPT TO UNDERSTAND SUBSTANCES?

Chemists, then, attempt to understand the behavior of materials by determining the properties of the substances of which they are composed. To understand substances, we assume that they are made up of invisibly small fundamental units called **molecules.** All the molecules of a given substance are identical, except for some relatively subtle differences which we can ignore for now. Each molecule of water is like every other molecule of water. Each molecule of table sugar is identical to every other molecule of table sugar.† Molecules of water and molecules of table sugar, however, differ. Indeed, it is the difference in structure and composition between the molecules of two substances that is responsible for the difference in their observable properties. Chemists know a great deal about a substance if they

---

† These statements will be refined in later discussions.

know the structural details of its molecules. Chemistry is essentially the molecular approach to understanding the world about us.

Molecules themselves are composed of still smaller units, called atoms. The atoms in turn are of great internal complexity and are made up of even smaller units. Understanding the nature of molecules requires an appreciation of the structural details of atoms, but chemists are principally concerned with structure at the molecular level.

Complex as the structures of molecules are, their sizes are extremely small—so small they are sometimes difficult to comprehend. If 50 million molecules of the gas carbon dioxide were placed end to end and touching, for example, they would extend along a line for less than 1 inch.

1-4 THE SPECIAL VIEWPOINT OF CHEMISTRY

The chemical viewpoint reveals relationships between the various materials in our environment that are apparent in no other way. By means of chemistry, for example, a colorless liquid obtainable from grain (ethyl alcohol) and water can be converted with the aid of a substance obtained from a natural yellow powder (sulfur), into a synthetic colorless liquid (ethyl ether), which was the first effective surgical anesthetic. The use of ethyl ether, which does not occur in nature, changed surgical operations from the nightmares they once were into tremendously useful techniques of modern medical practice. The relationships between the various forms of matter revealed by the application of the chemical point of view have led not only to new methods of combating discomfort and poor health but also to more abundant and nutritious food, better clothing, and more durable shelter. The ideas of chemistry have also contributed to people's understanding of themselves and their relationship to the world about them.

The applications of chemistry, however, like most human activity, are double-edged. They can be used to reduce famine, health problems, and poverty, but they can also be used to create forces of deterioration and destruction. Directing the power of chemistry exclusively toward beneficial ends is one of the important challenges facing organized society.

1-5 CHEMISTRY AND THE RELIEF OF PAIN

As a means of illustrating the chemical viewpoint and how chemists go about their work, we shall review one of the many areas of present chemical research: the search for better agents for the relief of pain. Pain is a common denominator of all poor health and disease. Yet, despite determined efforts over many years, a completely satisfactory means for controlling pain remains to be developed.

The ideal pain-relieving agent would meet a highly demanding set of requirements. It should have a broad margin of safety; it should not impair ordinary activity; it should be easily and comfortably administered; it should be rapidly effective in relieving pain of various origins for long periods; it should not develop resistance to its effectiveness; it should not be addicting; it should be inexpensive and chemically stable for easy storage and distribution. At present no agent completely meets all these specifications.

For many years morphine, the active principle of opium, has been the drug most frequently used by the medical profession for the relief of severe pain. Although it is reasonably effective, it has many serious deficiencies. It has a depressing effect on the respiration process, and dosage must be carefully limited to avoid distressing and even fatal consequences. In even limited amounts it may cause nausea, urinary retention, and constipation. If it is used very long, patients usually develop a tolerance to its analgesic effect, and larger and larger dosages have to be used. Prolonged use generally leads to addiction, characterized by a physiological and psychological dependence, resulting in disagreeable and dangerous symptoms when use of the drug is stopped.

How does one go about finding a better pain-relieving agent? Since morphine is obtained from unripe poppy seed capsules, should one examine other parts of the poppy plants? What about the seeds of other plants? Or their leaves, or stems, or roots? What are the methods used by chemists in tackling such a problem?

Morphine was originally used in the form of crude opium, obtained by drying the sticky exudate from cut unripe poppy seed capsules (Fig. 1-1). By the turn of the eighteenth century

FIGURE 1-1
(*a*) Opium poppy, the source of morphine; (*b*) the opium-containing exudate, which oozes from the cut unripe seed capsule; (*c*) some crystals of pure morphine.

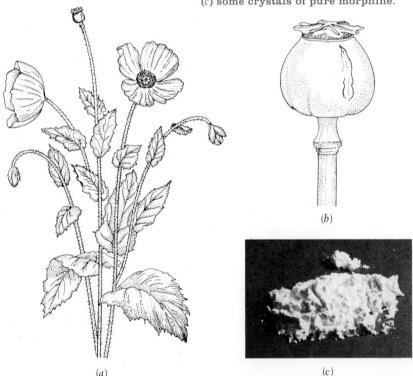

(*a*)

(*b*)

(*c*)

chemists had learned how to isolate substances from naturally occurring mixtures such as opium. Pure citric acid, for example, was isolated from lemons in 1780, cholesterol from animal fat in 1815, and quinine from the bark of the cinchona tree in 1817. About 1803 a young German pharmacist's assistant, Frederich Sertürner, isolated pure morphine from opium. Morphine accounts for only about 10 percent of the total weight of opium, but it is the substance primarily responsible for its pain-relieving properties. The isolation and identification of the pure substance made it possible to give smaller and more accurate doses. But the main drawbacks (toxicity, tolerance development, and addiction) remained.

The isolation of the pure substance, however, made it possible to apply chemical principles to the investigation of the active ingredient in opium. The complete characterization of morphine was not easily achieved. **Characterization** means the determination of the distinctive properties of a substance and the structure of its fundamental components. It was not until 1925 that significant progress was made and only in the 1950s that the structure was completely determined. This achievement required the creative and concentrated efforts of many chemists: the Englishmen Robert Robinson and John Gulland and the Americans Lyndon Small and Marshall Gates come principally to mind.

1-6 HOW IS THE STRUCTURE
OF A MOLECULE EXPRESSED?

When the structure of a molecule is completely understood, it becomes possible to express it by a very useful combination of symbols known as a **structural formula**. In this notation the individual atoms of the molecule are represented by letters of the alphabet, and their sequence is indicated by connecting the atomic symbols by dashes. A molecule of water, for example, contains an atom of oxygen in between two atoms of hydrogen. Each oxygen atom is represented by the symbol O and each hydrogen atom by the symbol H. The structural formula for water is written H—O—H, which indicates the atom sequence in the molecule. (There are more than 100 different kinds of atoms,† and each has been assigned a distinguishing symbol.) The structural formula for the substance ethyl alcohol is shown in the margin; C represents a carbon atom, and the H's and O represent hydrogen and oxygen atoms, as before.

$$\begin{array}{ccc} & H & H \\ & | & | \\ H- & C- & C-O-H \\ & | & | \\ & H & H \end{array}$$

Ethyl alcohol

| Note to Students | If this chapter is to fulfill its objective as a general orientation to the nature of chemistry, we must include some topics which can be discussed fully only after you have studied later chapters. You need have only a very general understanding at this time and appreciate the overall patterns of the formulas used rather than their many details. The summarizing questions for self-study at the end of the chapter will make clear to you what parts of the discussion you should understand after you have read the chapter. |

---

† This statement will later be refined.

A formula is a kind of highly condensed shorthand summarizing a wealth of information about a substance. The symbolism employed is rich in subtle details which will be elaborated in later chapters, but two points should be made here.

1. There are different kinds of links (called *bonds*) between atoms, and distinguishing symbols are associated with each of these different links. The single dash between the atoms in the formula for the water molecules, H—O—H, for example, represents one type of linkage. A double dash between the atoms of the carbon dioxide molecule, O=C=O, represents a second type. A third type is represented by a triple dash, as in the link between the two carbon atoms in a molecule of acetylene, H—C≡C—H. For the present, however, we are interested primarily in how formulas represent the overall atom sequence in a molecule, and we can ignore differences in the bonds between the atoms.

2. Formulas are not pictures of molecules. They are extremely useful representations of the structural features of molecules, but the representation is symbolic not pictorial.

Once the structure of the morphine molecule had been completely solved, it was possible to represent it by the structural formulas given below, where N represents a nitrogen atom, and the other symbols have already been explained.

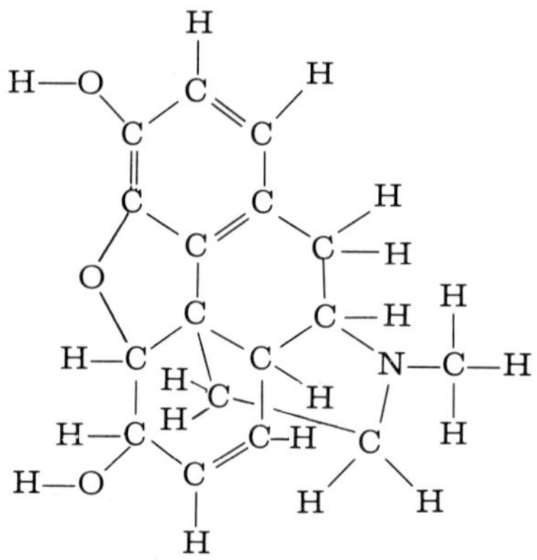

Detailed structural formula for morphine; each atom is represented by a symbol; C = carbon atom, H = hydrogen atom, O = oxygen atom, and N = nitrogen atom

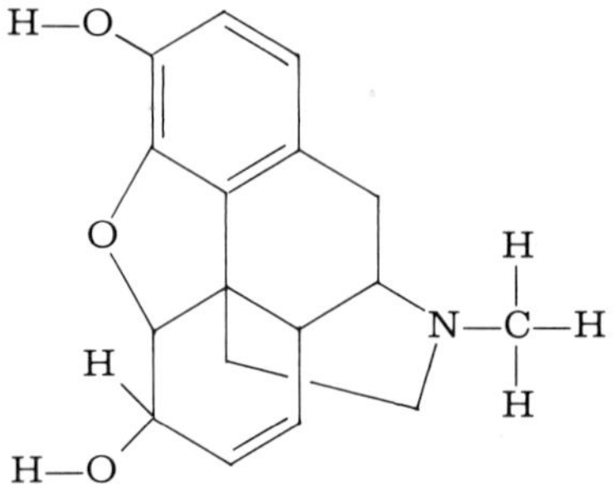

Abbreviated structural formula for morphine; most carbon atoms and hydrogen atoms are not specifically represented by symbols; the broken line in the lower center means that this bond is behind the others

Relationship between the detailed structural formula for morphine and the more convenient abbreviated formula; to a chemist both forms convey the same information

1-7 DETERMINING THE STRUCTURE OF A MOLECULE

How do chemists go about determining the structure of a molecule like morphine? How can they be sure about the identity of individual atoms and their sequence in a molecule that has never been characterized before?

The first step is to isolate and purify the substance. This involves many special techniques. It requires, for example, criteria for determining the purity of a substance. It is essential for chemists to be able to determine whether or not a substance is pure even though they do not know what the substance is.

The second step is to determine such properties as the boiling point, the melting point, and the solubility behavior. At what temperature does the substance melt? At what temperature does it boil? Does it dissolve in water? Does it dissolve in ethyl alcohol?

The third step is to determine the kinds of atoms in the substance. This requires a qualitative analysis, employing highly sophisticated methods which required many years to develop. In the characterization of morphine, it was discovered that the molecule involves a combination of carbon, C, oxygen, O, nitrogen, N, and hydrogen, H, atoms but no other kinds of atoms are present.

The fourth step is to determine the number of each of the different kinds of atoms in the molecule. How many carbon atoms; how many nitrogen, oxygen, and hydrogen atoms? The answers to these questions require the chemist to find out, for example, the percentage by weight of carbon atoms present. In other words, for each 100 grams of pure morphine, how many grams of carbon atoms are present? This is known as a quantitative analysis.

In 1831 the German chemist Justus Liebig became the first to determine the number of each kind of atom in morphine. He was also the first to develop many of the methods used for such determinations. Liebig found that each molecule of morphine contains 17 carbon atoms, 19 hydrogen atoms, 3 oxygen atoms, and 1 nitrogen atom. This information is summarized succintly by writing $C_{17}H_{19}O_3N$, and a chemist calls such a representation a **molecular formula.** It expresses the number of each kind of atom in the molecule.

The fifth step in characterizing a substance is to determine the arrangement of atoms in the molecule and the kinds of linkages that hold them together. Problems multiply quickly at this stage, since usually many different arrangements are possible. If a substance contains only two kinds of atoms, carbon and hydrogen, and only 8 carbon atoms and 18 hydrogen atoms, i.e., the molecular formula is C_8H_{18}, there are 18 different possible arrangements each corresponding to a separate substance. That is, that there are 18 different substances, each with a characteristic set of properties, and each with the same molecular formula, C_8H_{18}. If a molecule is known to contain 20 carbon atoms and 42 hydrogen atoms (molecular formula $C_{20}H_{42}$), there are more than 350,000 possible arrangements; i.e., more than 350,000 different substances have a molecular formula of $C_{20}H_{42}$. (The steps in determining the structure of a substance are summarized in Table 1-1.)

Determining the arrangement of atoms in a molecule as complex as morphine, with a molecular formula of $C_{17}H_{19}O_3N$, is one of the most challenging assignments in the physical sciences. Chemists have developed an impressive storehouse of tech-

Table 1-1
Principal Steps in Determining
the Structure of a Substance

1. Isolate and purify
2. Determinine properties, e.g., boiling point, melting point, and solubility behavior
3. Determine the kinds of atoms in the molecule (qualitative analysis)
4. Determine the number of each of the different kinds of atoms in the molecule (quantitative analysis)
5. Determine the arrangement of atoms within the molecule and the nature of the bonds which hold them
6. (Usually) synthesize the substance from substances of previously determined structure

niques, and they are joined in this work by physicists and mathematicians.

The general approach to ferreting out the sequence of atoms is to search for characteristic groups of atoms within the molecule. In this respect the behavior of the substance as it is allowed to interact with other substances is often revealing. In recent years highly sophisticated methods of determining structure, involving complex instruments, have been developed. The end result of such diagnostic probing is a tentative proposal of a structure for the molecule, usually expressed as a structural formula. Although chemists may be fairly confident about the validity of the proposed structure, they are often not really certain until they have gone to the laboratory and built up the proposed arrangement of atoms from simple molecules whose structures are known. This requires them to draw upon the accumulated knowledge of molecular interactions, i.e., chemical reactions. When the proposed structure has been completely constructed i.e., when it has been synthesized from known substances, and when the end result of this synthesis is found to be a substance which is identical in all respects with the substance being characterized, the structure of the substance is considered to have been established.

For morphine it was not until the 1950s that all the structural details were settled, almost 150 years after the original isolation of the substance. Such structure determinations have frequently extended over many decades, and individual chemists have devoted the better part of an entire career to the characterization of a single substance. In recent years the sophisticated instrumental methods for determining structure involving x-ray-diffraction studies have reduced the time to a few years or even a few months.

1-8 WHAT IS THE POINT OF DETERMINING THE STRUCTURE OF MOLECULES?

The molecular structure of morphine was eventually settled, and a great deal was learned about its nature. But this new information did not immediately lead to any improvement in the use of the substance as a pain-relieving agent. Its structure had been revealed, but the mechanism of its physiological action remained unaccounted for.

Suppose, however, the structure were altered in some slight way. What would be the effect on the physiological activity? Would the pain-relieving properties be increased or decreased? Would the toxicity be reduced? In pursuing the answers to these questions, over 100 such variations on the molecular structure of morphine have been synthesized. The term **structural variant** or **molecular analog** is applied to these synthetic variations of a known molecular structure.

One of the first variations of the morphine structure resulted in the synthesis of heroin by the German chemists Ludwig Knorr and Heinrich Horlein in 1909. The close similarity between the structure of this substance and that of natural morphine can be seen by comparing their formulas:

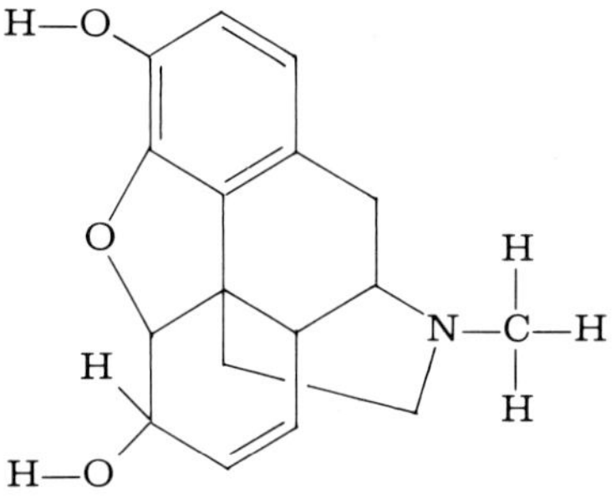

Morphine 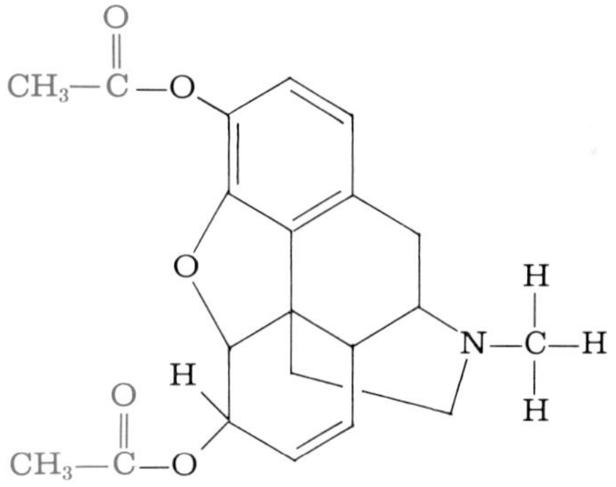   Heroin

The close relationship between the structures of morphine and heroin

---

Note to Students   ·Only the overall patterns suggested by the formulas need be examined at this point.

When first tested on patients, heroin appeared to be a very promising discovery. It was more powerful in pain relief than morphine, and early results suggested that it is nonaddicting. Extended testing, however, revealed that, on the contrary, addiction occurs so readily that heroin was for many years considered too dangerous for physicians to use. It is popular in the illicit drug trade probably because its euphoriant action is so strong.

1-9   THE DEVELOPMENT
OF NARCOTIC ANTAGONISTS

One of the most interesting variations of the morphine structure is found in the substance nalorphine, first synthesized in 1942 by the American chemists John Weijlard and A. E. Ericksen in the laboratories of Merck and Company. The structure of nalorphine differs only slightly from the structure of morphine, as the following comparison reveals:

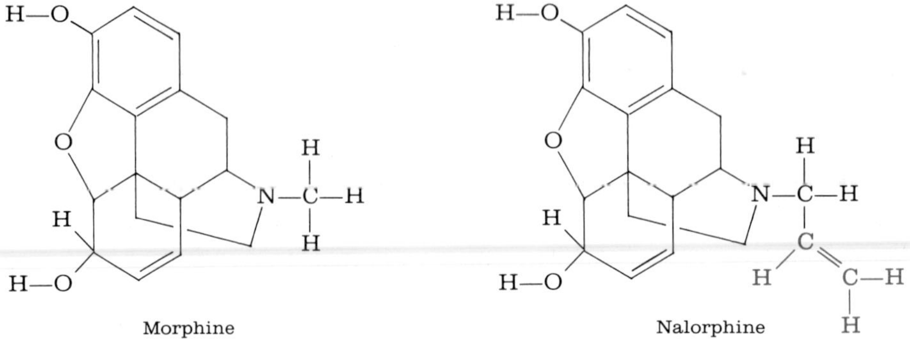

Morphine

Nalorphine

The close relationship between the structure of morphine and that of the narcotic antagonist nalorphine

The modest structural change of replacing one of the hydrogen atoms of the $-CH_3$ group of atoms on the nitrogen, N, atom of morphine with a $-HC=CH_2$ group creates a substance with markedly different physiological action. Nalorphine is actually able to reverse the effects of morphine. A **narcotic antagonist** is a substance which can block the effects of narcotics. Nalorphine and other antagonists are widely used in the treatment of morphine poisoning, brought on, for example, by an overdose.

Nalorphine was examined to determine whether it has any pain-relieving action, but the original tests, using small animals, suggested that it had none. Trials with small animals usually are good indicators of analgesic activity in human beings, but it is one of the many pitfalls of investigations of pain-relieving agents that small-animal tests are not always reliable. Nalorphine's ability to relieve pain in people went unrecognized until 1954 and even then was discovered only by accident.

The American pharmacologists Henry Beecher and Louis Lasagna, then at the Massachusetts General Hospital, were testing their hypothesis that a mixture of nalorphine and morphine would be less toxic than morphine alone. (They hoped that the nalorphine as a narcotic antagonist would counteract the morphine's depression of respiratory function.) They treated a group of patients with a nalorphine-morphine mixture, and to strengthen the validity of their results they treated a comparable group (a control group) with nalorphine alone. To their surprise, nalorphine was as good as the nalorphine-morphine mixture in relieving pain. This unexpected action was confirmed, and it was discovered that nalorphine is nonaddicting.

The presumption that strong analgesia is always accompanied by significant addiction was proved false. Unfortunately, however, nalorphine did not provide the practical means of capitalizing on this discovery since it was found to produce anxiety and hallucinations in patients.

The dramatic shift in properties accompanying the slight change in structure between morphine and nalorphine illustrates the complexities of the relationships between molecular structure and pain-relieving properties.

The name agonist is given to substances like morphine which have a pain-relieving and euphoric effect. Many agonists have been synthesized. Unfortunately most of them have side effects which prevent them from being useful drugs. Two substances, meperidine and methadone, have proved useful, but both are as strongly addicting as morphine. Much current research is being directed toward developing an agent which will be as effective as morphine (or better) in the relief of pain without being addicting.

The extensive physiological testing of the many structural variants of morphine has permitted them to be classified as agonists, antagonists, or mixed agonist-antagonists. Morphine itself is, of course, an agonist. It turns out that nalorphine has some agonist attributes mixed in with its antagonist properties. The fact that nalorphine is pain-relieving and nonaddicting has suggested that substances with mixed agonist-antagonist properties might be profitably explored as possible nonaddicting analgesics.

The synthetic morphine analog naloxone has proved to be a very strong antagonist. This substance was synthesized in 1963 in the private laboratory of Mozes J. Lewenstein and later developed by the Endo Laboratories of the Du Pont Company. The structure is closely related to that of morphine. Its antagonist behavior is mostly due to the replacement of the —CH_3 group of atoms with the —$CH_2CH{=}CH_2$ group, which parallels the situation with nalorphine.

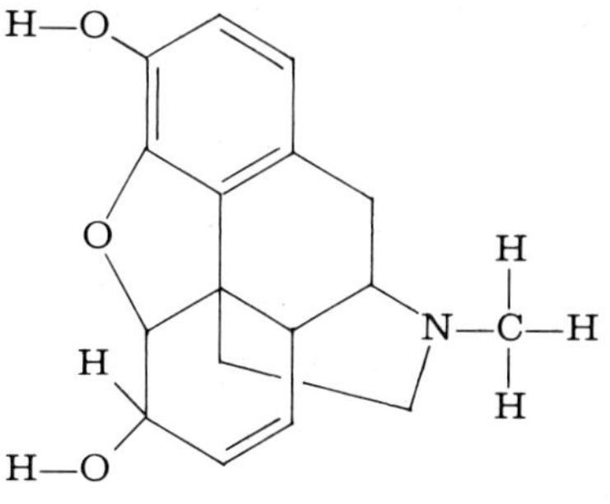

Morphine 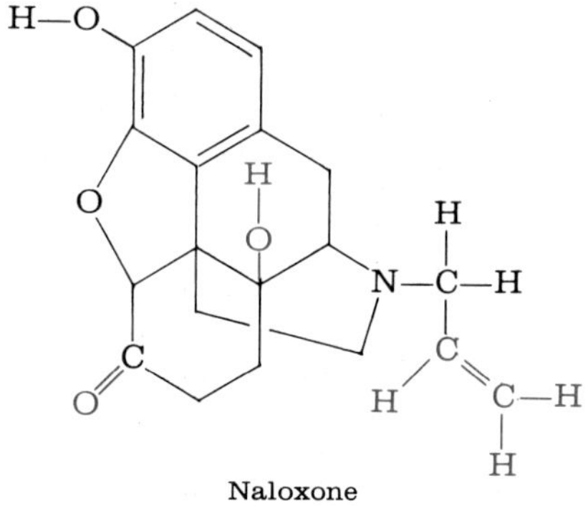   Naloxone

The close relationship between the structure of morphine and that of the narcotic antagonist naloxone

The structural variant phenazocine is a strong agonist. This is somewhat surprising considering that although its structural relationship to morphine remains clear, its structure differs substantially from that of morphine, as the following comparison reveals:

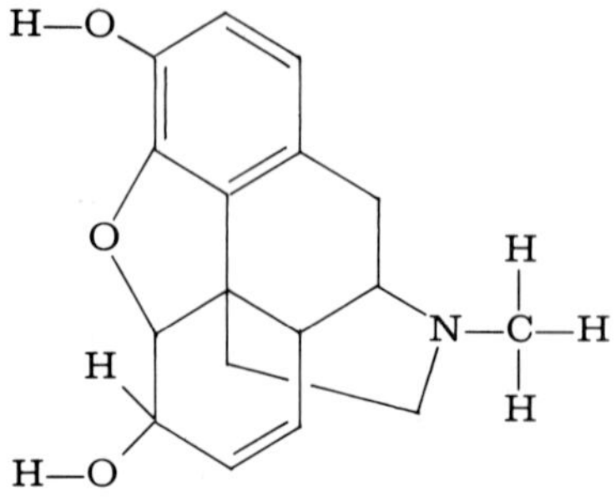

Morphine

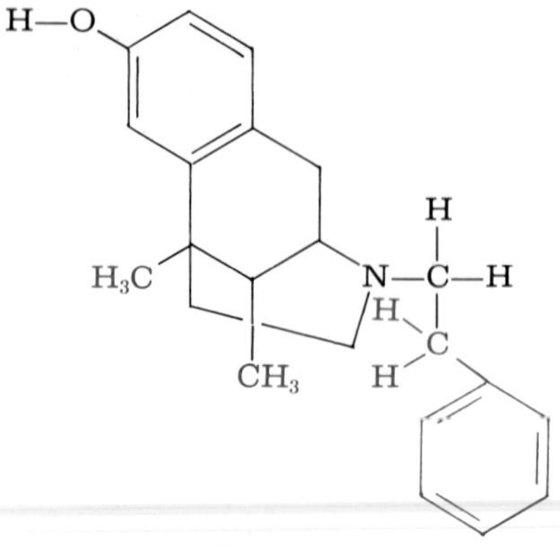

Phenazocine

Comparison of the structures of morphine and phenazocine

Phenazocine was synthesized in 1959 by the Americans Everett May and Nathan B. Eddy, of the National Institutes of Health. It proved to be even more powerful as an analgesic than morphine but unfortunately dangerously addicting.

1-11   THE DEVELOPMENT OF PENTAZOCINE

The pain-relieving action associated with the structure of phenazocine was explored further by a group of chemists at the Sterling-Winthrop Institute for Therapeutic Research headed by Sydney Archer. They synthesized a series of variations on the phenazocine molecule in which the group on the nitrogen, N, was replaced by groups of atoms related to the $-CH_2CH=CH_2$ group of the antagonists nalorphine and naloxone. In one of the most promising of these, pentazocine, the replacing group is $-CH_2CH=C-CH_3$. The following formulas illustrate the structural relationships:

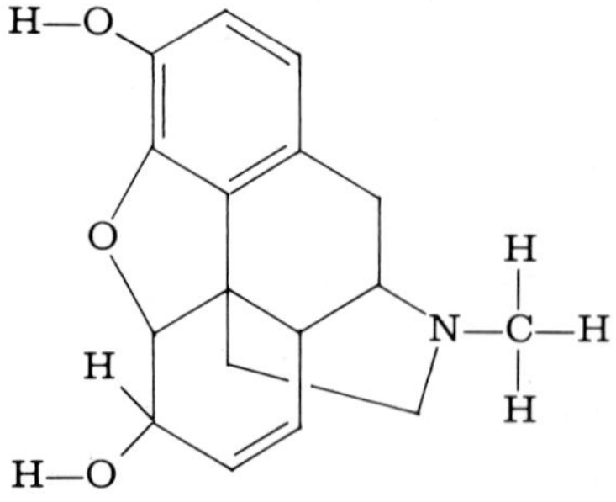

$$\underset{\displaystyle CH_3}{\displaystyle |}$$

Morphine

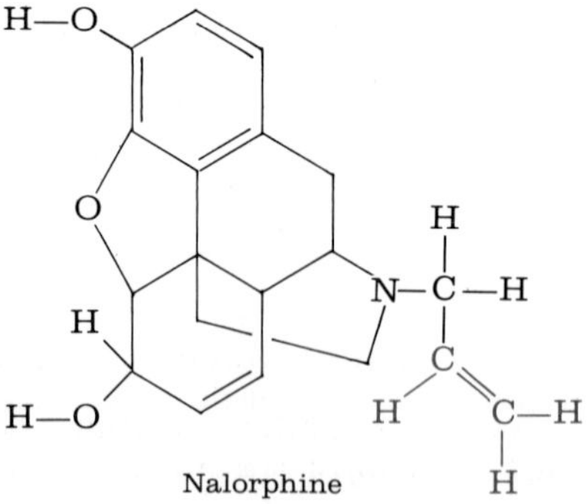

Nalorphine

13

1-13   ARE THERE AGENTS
USEFUL IN THE TREATMENT
OF NARCOTIC ADDICTION?

Comparison of the structures of morphine, nalorphine, phenazocine, and pentazocine

Pentazocine is less powerful in the relief of pain than morphine, but its "abuse potential" is so low that it is not included among drugs which are strictly controlled as being dangerously addicting. Presumably its molecule incorporates a balance of agonist and antagonist properties. It is a useful drug, especially in the control of chronic pain, where low addiction liability is especially important. It is not the ideal pain-relieving agent, however, and the search for safer agents with minimum side effects continues.

1-12   WHAT IS THE NEXT STEP?

Much of the present research is devoted to synthesizing structural variants which will incorporate a balance of agonist and antagonist properties. Although the agonist-antagonist hypothesis is a provocative idea, it would be better to work from a knowledge of the actual mechanism of pain-relieving action. How, for example, does the molecule of morphine interact with the molecules which make up human beings to bring about its analgesic and euphoric effects? What molecular interactions are responsible for its addicting properties?

Recent results are beginning to provide some preliminary answers to such questions. They suggest that both agonists and antagonists act by associating with specific receptor sites on large molecules involved in the transmission of nerve impulses. Perhaps in the not too distant future the molecular mechanism will be well enough understood to enable us to reach the long-sought goal of a strong pain-relieving agent free of addiction liability and other unfavorable side effects.

1-13   ARE THERE AGENTS USEFUL IN THE TREATMENT OF NARCOTIC ADDICTION?

The National Institute of Drug Abuse estimates that there are at least a quarter of a million heroin addicts in the United States who use the drug daily. About 5000 people die of heroin overdoses each year. It is estimated that the crimes committed by heroin addicts cost the nation about $6 billion a year, and the human suffering involved in heroin addiction is appalling.

About 60,000 heroin addicts are currently being treated in programs emphasizing psychological rehabilitation. Another 90,000 are being treated in methadone maintenance clinics. Methadone is a useful analgesic which was synthesized during World War II by chemists at I. G. Farbenindustrie. Since 1964 it has been used in the long-term treatment of heroin addicts. Daily oral doses block the effects of heroin and permit addicts to do without the drug. Although addicting itself, methadone enables addicts to maintain a psychological and physiological balance and helps them to lead useful lives. Since it involves replacing one addicting drug with another, however, the treatment is a controversial one.

Narcotic antagonists are being explored as possible agents for the treatment of heroin addiction. If these substances are taken before heroin is used, the heroin produces no effect. Caution must be exercised in their use, however, since they can cause painful withdrawal symptoms. One of the first narcotic antagonists to be studied was the morphine analog naloxone, the molecular structure of which is given on page 11. Another is cyclazocine, which has a structure related to that of pentazocine (page 13). Encouraging was the finding that if cyclazocine and morphine are administered together, the effects of the morphine are blocked but when the use of morphine is subsequently stopped, no withdrawal symptoms occur. The advantages of using antagonists include not only breaking the addiction habit but protecting the patient, with continuing treatment, from becoming readdicted.

Much of the present research is devoted to synthesizing structural variants of naloxone, cyclazocine, and other known antagonists, as well as investigating various methods of administration in the hope of developing a treatment with increased potency and duration of action. The recent progress in understanding the molecular mechanism responsible for the effects of agonists and antagonists in human beings is also providing useful leads.

Note to Students

If you wish to read more about the current work on the development of pain-relieving agents and the treatment of narcotic addiction, some references are given at the end of the chapter. Most of them, however, will be easier to understand after you have studied a few more chapters. Specific suggestions are given with the references.

1-14 THE UNFINISHED NATURE OF CHEMISTRY

We have discussed the search for improved pain-relieving agents in this chapter because it is an unfinished investigation.

Its incompleteness illustrates the unfinished nature of chemistry, which for all its accomplishments continues to generate more questions than it answers. The unfinished quality of chemistry is one of the most intriguing things about it.

This discussion also illustrates how chemistry provides answers and accomplishments that would be impossible without it. Only through chemical structural principles can the crucial relationships between morphine, pentazocine, cyclazocine, and naloxone be recognized. Only through chemistry could the structural variants have been synthesized. And only through chemistry will the molecular mechanism of the pain-relieving properties of morphine be determined.

Chemistry, of course, ranges far beyond the search for pain-relieving agents. It aims to answer questions of a broader scope than how a molecule of morphine interacts with the molecules of human bodies to relieve pain without loss of sensibility. What, for example, are the key molecular interactions responsible for the effects of any individual substance that is swallowed? And, for that matter, what are the molecular reactions responsible for normal health? Illness? Aging? Death? Birth?

1-15 THE WORK OF CHEMISTS

Chemistry is an activity carried on by thousands of chemists in practically every country of the world. They may be found working in industrial research laboratories and production departments, private and government institutes, and the laboratories of colleges and universities. Chemists and other professionals who are concerned with chemistry are engaged in a remarkably diversified range of activities, like monitoring the quality of a public water supply, flameproofing children's pajamas, developing a treatment for a disease, processing a photographic film, analyzing body fluids to diagnose an illness, and determining the authenticity of a painting. Most chemists work in fields not considered as being "chemical," e.g., processing food, manufacturing electronic equipment, and making metal, glass, paper, and petroleum products. Moreover, the work of chemists overlaps the work of many other scientists, especially biologists, geologists, and physicists, since all are concerned in one way or another with various forms of matter and their transformations.

In academic institutions chemists are involved in research and in bringing students up to the "frontiers" so that they can make their own contribution to the body of knowledge in chemistry or a related science. Many chemists of colleges and universities are also actively engaged in introducing the nature of chemistry to students who do not intend to become scientists, so that, among other things, they can better understand the many issues of public policy related to chemistry. These issues range all the way from environmental pollutants to food supply and space flight.

Organic chemists specialize in the substances containing carbon, which are given special emphasis because they are so widespread in the substances of living organisms and the substances derived from them. Inorganic chemists are concerned primarily with substances which do not contain carbon, the substances associated with the nonliving world. Analytical chemists specialize in the identification and composition of the various forms of matter. Biochemists focus on the changes of matter in living plants and animals. Physical chemists, or theoretical chemists, are concerned primarily with the study of the principles that underlie the structure of substances and the changes they undergo. Other chemists specialize in areas which do not fall readily into any classification, and still others work in more than one field.

1-16 ABOUT THIS TEXT

This text is addressed primarily to students who have little or no background in science. It aims to develop a foundation of chemical concepts and facts sufficient to enable you to understand the current developments in many fields. Chemistry is such a far-ranging subject that we must be highly selective in the choice of subject material, but we have tried to focus on topics of general interest and the background necessary for their understanding.

Special emphasis has been placed on the relationship between chemistry and energy, the food supply, the environment, and human health. Equal prominence is given to the role of industrial chemistry in the use of natural resources, the chemistry of plants and animals, and nuclear chemistry.

The fundamental background required by these topics is presented in the first eight chapters. The chapters on special topics following the first eight can be studied in any order, making it possible to go directly to areas of special interest. The material contained in the supplements expands some of the technical details in the body of the chapters. It is hoped that you will find your study interesting, informative, and worth the effort.

Note to Students

To help you distinguish the most important points of discussions from the less important ones, summarizing questions for self-study are given at the end of each chapter. If you examine these questions with care after first reading the chapter, you will have a useful guide to a second reading.

Many practice exercises are included so that you can test the extent of your understanding of each chapter. The study of chemistry, like the study of most subjects, involves the introduction of many special terms. The new terms introduced in each chapter are listed at the end, and you should be sure that you understand their meaning. These Notes to Students provide further guidance as you progress.

KEY WORDS

1. Substance (Sec. 1-2)
2. Molecule (Sec. 1-3)
3. Characterization (Sec. 1-5)
4. Structural formula (Sec. 1-6)
5. Molecular formula (Sec. 1-7)
6. Structural variant (Sec. 1-8)
7. Molecular analog (Sec. 1-8)
8. Narcotic antagonist (Sec. 1-9)

"The codeine is O.K., and the phenobarbital is O.K., but the Food and Drug
Administration says no to the powdered bat's tooth."

(© 1970 American Scientist. Reprinted by permission of Sidney Harris.)

SUMMARIZING QUESTIONS FOR SELF-STUDY

Sections 1-1 to 1-4

1. Q. What does a chemist mean by the word substance?

A. A form of matter which is of definite composition and consistent behavior, e.g., water, oxygen, carbon dioxide. (To be compared with a mixture, which contains two or more substances.)

2. Q. In what way does chemistry place special emphasis on substances?

A. They are one of the main keys to understanding the behavior of matter. It is a principal underlying assumption of chemistry that all forms of matter can be understood in terms of the individual substances of which they are composed and the interactions between the substances.

3. Q. What is a molecule?

A. The invisibly small fundamental unit of a substance, composed of an atom or more than one atom linked together in a specific and characteristic arrangement.

Section 1-5

4. Q. What are the properties of an ideal pain-relieving agent?

A. It should have a broad margin of safety; it should not impair ordinary activity; it should be easily and comfortably administered; it should be rapidly effective in relieving pain of various origins for long periods; it should not develop resistance to its effectiveness; it should not be addicting; it should be inexpensive and chemically stable for easy storage and distribution.

5. Q. What is morphine and from what is it obtained?

A. Morphine is a substance obtained from the capsules of unripe poppy seeds. It is widely used for the relief of pain.

6. Q. Why is morphine discussed in this chapter?

A. It illustrates the isolation and chemical characterization of substances from living systems (in this case a plant).

7. Q. What are the advantages and disadvantages of morphine as a pain-relieving agent?

A. It is one of the most effective agents for the relief of pain; but it has many serious deficiencies. It has a depressing effect on the respiration process, and dosage must be carefully limited to avoid distressing and even fatal consequences. Even in limited amounts it may cause nausea, urinary retention, and constipation. If it is used over an extended period, patients usually develop a tolerance to its analgesic effect, and this necessitates increasingly larger dosages. Prolonged use generally leads to addiction.

Sections 1-6 and 1-7

8. Q. What is a structural formula of a substance?
A. It is a symbolic representation of the atoms contained in the molecule of a substance and of the arrangement of bonds that hold them together.

9. Q. What is the first step in determining the structure of a molecule?
A. Its isolation and purification.

10. Q. What is the second step?
A. Determining properties of the substance, e.g., the temperature at which the substance boils, the temperature at which it melts, and what common substances it does and does not dissolve in.

11. Q. What is the third step?
A. Determining the kinds of atoms in the substance.

12. Q. What is the fourth step?
A. Determining the number of each of the different kinds of atoms in the molecule.

13. Q. What is the fifth step?
A. Determining the arrangement of atoms within the molecule and the nature of the bonds which hold them together. This step often involves a confirmation of the proposed structure by synthesizing it from substances of previously determined structure.

Section 1-8

14. Q. What is the purpose of synthesizing variations of the structure of a molecule?

A. To attempt to create substances with properties which are more useful than those of the original substance.

15. Q. What was the result of the first significant synthetic variation of the morphine molecule?
A. Heroin, originally thought to be superior to morphine as a pain-relieving agent.

Section 1-9

16. Q. What are narcotic antagonists? Of what use are they?
A. Narcotic antagonists are substances with the ability to reverse the physiological effects of morphine. They are used, e.g., nalorphine, as antidotes for morphine poisoning. They are being investigated, e.g., naloxone and cyclazocine, as possible agents for the treatment of drug addiction.

Sections 1-10 and 1-11

17. Q. What is the special significance of the structural variation known as pentazocine?
A. It is a relatively strong pain-relieving substance with low addiction liability.

General

18. Q. Comment on the statement "chemistry is essentially the molecular approach to understanding the world about us."
A. Chemists attempt to understand all forms of matter by focusing on the individual substances of which they are composed. In attempting to understand the properties of substances they focus on the structure of their fundamental units, molecules.

PRACTICE EXERCISES

1. Which of the following are substances?

   (a) Water
   (b) Sodium chloride
   (c) A solution of sodium chloride in water
   (d) Oxygen
   (e) Nitrogen
   (g) Carbon dioxide
   (f) Air
   (i) Morphine
   (h) Opium
   (k) Heroin
   (j) Ethyl alcohol

2. A white solid x is suspected of being sodium chloride, which is a white solid that is very soluble in water at room temperature. x is only partly soluble in water at room temperature. What can be concluded about the nature of x? Explain.

3. A volatile liquid y is suspected of being ethyl ether, known to be easily evaporated even at room temperature. A small amount of y is heated to well above room temperature. Most of it evaporates, but a white solid residue remains. What can be concluded about the nature of y? Explain.

4. The molecular formula of a substance is known to be $C_8H_{13}O_3N$. What does this formula tell us about the molecules of the substance?

5. Two substances have the molecular formula $C_6H_4O_4N_2$. How must the structures of the two substances differ?

6. By a series of diagnostic tests a team of chemists is able to determine the kinds of atoms in the molecule of a substance, their arrangement, and the nature of the linkages which hold the atoms together. To be certain about the structure of the substance, what further step is often carried out?

7. The molecular formula for morphine is $C_{17}H_{19}O_3N$. With the aid of the structural formulas on pages 9, 10, and 11, determine the molecular formulas of (a) heroin, (b) nalorphine, and (c) naloxone.

8. Morphine is obtained from the unripe seeds of poppy plants. Do heroin and nalorphine occur in nature?

9. What is the principal difference in physiological activity between morphine and heroin?

10. What is the principal difference in physiological activity between morphine and nalorphine?

11. How did chemists happen to synthesize heroin and nalorphine?

12. Both morphine and pentazocine are currently used as pain-relieving agents. What is the principal difference in physiological activity between the two substances?

13. Methadone is currently being used in the long-term treatment of heroin addicts. Narcotic antagonists are also being explored as possible agents in the treatment of heroin addiction. How do the two approaches compare?

14. What is the difference between an organic chemist and an inorganic chemist?

15. What is the difference between a biochemist and a physical chemist?

SUGGESTIONS FOR FURTHER READING

Although the following articles are of an introductory nature, perhaps it would be better to postpone reading them until after you have completed Chap. 3.

Gates, Marshall: Analgesic Drugs, **Scientific American,** November 1966, p. 131.

Hammond, Allen L.: Narcotic Antagonists: New Methods of Treating Heroin Addiction, **Science: 173:** 503 (1971).

Lennard, Henry L., Leon J. Epstein, and Mitchell S. Rosenthal: The Methadone Illusion, **Science, 176:**881 (1972).

Maugh, Thomas H., II: Narcotic Antagonists: The Search Accelerates, **Science, 177:**249 (1972).

Walsh, John: Methadone and Heroin Addiction: Rehabilitation without a "Cure," **Science, 168:** 684 (1970).

The following articles are at a more advanced level; postpone reading them until you have completed Chap. 3.

Drugs for Treating Narcotics Addicts, **Chemical and Engineering News,** Mar. 28, 1977, p. 30.

Eddy, Nathan B., and Everette L. May: The Search for a Better Analgesic, **Science, 181:**407 (1973).

Snyder, Solomon H.: Opiate Receptors and Internal Opiates, **Scientific American,** March 1977, p. 44.

WHAT IS THE STRUCTURE OF ATOMS?

2-1 THE COMMON COMPOSITION OF MATTER

Since the earliest years of recorded history thoughtful people have asked: "What is the stuff of which all things are made?" The Ionian philosopher Thales, of the sixth century B.C., suggested one of the first answers of consequence when he proposed that everything is made up of water in various forms. Although this suggestion no longer seems very reasonable, it proved to be highly significant. It was one of the first proposals that there is a common composition which underlies all forms of matter, despite the extraordinary variety of outward appearances.

How did the idea of the common composition of matter come to be incorporated into modern chemistry? Thales' concept was expanded and modified two centuries later into a more elaborate proposal by the Greek atomists Leucippus and Democritus. These philosophers suggested that the underlying structural elements of all forms of matter are invisibly small solid particles which they called atoms. All the changes in nature, from the evaporation of liquids to the growth of animals, were attributed to alterations in the arrangements of atoms, while the atoms themselves maintained their integrity.

In the eighteenth century the English physicists Isaac Newton and Robert Boyle used the idea of atoms to interpret the behavior of matter. At the turn of the nineteenth century, the English chemist John Dalton (Fig. 2-1) (among others) drew upon the atomic concept to explain the newly discovered quantitative relationships accompanying changes in matter. His proposals led directly to the refinements of the Italians Amedeo Avogadro (Fig. 2-2) and Stanislao Cannizzaro, which brought the atomic hypothesis into the mainstream of scientific thinking.

The atoms of the atomists were believed to vary in size and shape but to be identical in substance. To Dalton, however, the most important distinguishing characteristic of each kind of atom was its mass. The eighteenth-century French chemist Antoine Lavoisier had found it useful to classify matter into elements and compounds on an operational basis. Substances, like water, which were known to be decomposable into simpler substances were called **compounds.** (Water can be decomposed into two different substances called hydrogen and oxygen.) Substances like hydrogen and oxygen, which cannot be decomposed into simpler substances, were called **elements.** Dalton assigned to each element a distinctive kind of atom with a characteristic mass. Hydrogen was composed of atoms peculiar to hydrogen, possessing identical masses. Oxygen was composed of atoms whose characteristic masses differed from those of hydrogen atoms. Dalton held that compounds like water are made up of identical structural units, each containing more than one kind of atom. Each fundamental unit of water, for example, contains atoms of hydrogen and oxygen in a fixed ratio (now known to be 2:1).

The atoms of Dalton, Avogadro, and Cannizzaro were considered to be solid and "fundamental." At the turn of the twentieth century, however, the work of the English physicists Joseph John Thomson and Ernest Rutherford (Fig. 2-3) (among others) suggested that atoms have a porous and complex structure made up of a small, dense nucleus surrounded by widely dispersed arrangements of tiny moving electrons. As a result of further investigations, all atoms were considered to be composed of only three different particles, electrons, protons, and neutrons. Thus the idea of the common composition of matter, first championed by Thales more than 2500 years ago, persists in the concepts of twentieth-century science.

2-2 WHAT IS THE NATURE OF ELECTRONS, PROTONS, AND NEUTRONS?

Electrons are tiny bits of exceedingly small mass and negative electric charge. The mass of each electron is almost 2000 times smaller than the mass of the simplest atom (the hydrogen atom), which in itself is so small that there are

$$602,200,000,000,000,000,000,000$$

of them in 1 gram (about $\frac{1}{500}$ pound). (Units of mass are discussed in Supplement 3, Sec. S3-1.) The mass of a **proton** is very close to the mass of a hydrogen atom, and each proton has a positive

FIGURE 2-1
John Dalton (1766–1844), a teacher of mathematics and chemistry in Manchester, England. He is described by his contemporaries as an inferior experimenter and poor lecturer, but in his work on gases and his speculations on the structure of matter he revealed insight and genius which have had a marked influence on all the physical science which followed him. (*Library of Congress.*)

charge of the same magnitude as the negative charge on an electron. A **neutron** has a mass about the same as a proton, but it has no electric charge. Electrons are considered to be truly elementary in nature, but protons and neutrons have complex structures. Table 2-1 summarizes the properties of these particles.

How are electrons, protons, and neutrons arranged in atoms? How are the atoms arranged in molecules? What structural features characterize a given kind of atom? How does the structure of one kind of atom differ from another kind? Before we can discuss these questions, we must become familiar with some of the basic language used to describe our ideas of the nature and structure of matter.

Table 2-1
Fundamental Particles of Matter

Particle	Relative mass†	Electric charge
Electron	1	− 1
Proton	1836	1
Neutron	1839	0

† Mass relative to that of the electron.

2-3 THE BASIC LANGUAGE OF THE CONCEPTS OF MATTER

The **inertia** of a body of matter refers to its tendency to remain at rest if it is at rest or to keep moving with a constant velocity if it is moving. **Force** is what is exerted on a motionless body of matter to put it in motion or to change its velocity if it is already in motion. **Matter** itself is the stuff of the physical universe; it occupies space and has inertia. The term **mass** refers to a quantity of matter. A **vacuum** is a region which contains no matter.

The **weight** of a quantity of matter is the force exerted on it by the gravity of the earth. If two masses have identical weights, they contain the same quantity of matter; 100 grams of oxygen contains the same mass as 100 grams of books. (Chemists, regrettably, tend to use the terms *mass* and *weight* interchangeably.)

Energy may be defined for our purposes as the capacity to do work, e.g., the capacity of a force to set a body of matter in motion. Energy occurs in many forms: energy associated with moving bodies is called mechanical energy; burning coal produces heat (or thermal) energy and light energy; a dry-cell battery produces electric energy.

Electricity is a kind of force. An **electric charge** is a quantity of electricity. To account for the behavior of electricity, it is assumed that two kinds of charge exist: positive charge and negative charge. Particles of matter may have electric charge. The negative charge on an electron is equal to the smallest unit of charge. An **electric current** can be viewed as a flow of particles possessing electric charges. When a copper wire conducts an electric current, for example, electrons flow along the wire.

A **substance** is any form of matter with a definite composition and a consistent set of properties, for example water, oxygen, table sugar, copper, and gold. (To describe a substance as "pure" is simply another way of labeling it a substance. An impure substance is a substance mixed with one or more other substances which for one reason or another are of less interest.) A **mixture** contains two or more substances. A solution of table sugar and water for example, is a mixture of table sugar and water. In nature most forms of matter are mixtures of more than one substance. Wood, seawater, and air, for example, are all mixtures.

An **element** is a substance which is not readily decomposable into simpler substances. Elements include such substances as carbon, iron, sulfur, oxygen, nitrogen, hydrogen, mercury,

FIGURE 2-2
Amedeo Avogadro (1776–1856), professor of physics at Turin, Italy. His many contributions to physics and chemistry included research on the nature of gases and far-reaching proposals about the structure of atoms and molecules. (*Library of Congress.*)

gold, and iodine. More than 100 elements have now been recognized. **A compound** is a substance which can be decomposed into simpler substances. Water is a compound, since it can be decomposed (with the use of electric energy) into hydrogen and oxygen. Ammonia is a compound since it can be decomposed (by heating) into nitrogen and hydrogen. All substances known, except for the 100 or so elements, are compounds. (We shall return to the difference between an element and a compound in Sec. 2-10.)

2-4 WHAT IS INSIDE ATOMS?

Although the internal structure of atoms is of considerable complexity and is still the subject of intensive investigation, the following list of the principal basic concepts will be sufficient for our immediate purposes.

1. Atoms are composed of electrons, protons, and neutrons held together by strong forces. Although electrons and protons have an electric charge, atoms as a whole are neutral.

2. Most of the volume of an atom is empty space. The dimensions of the component particles are very small compared with the overall dimensions of the atom.

3. Most of the mass of an atom is concentrated in a minute nucleus at the center. This is a tightly packed arrangement of the protons and neutrons. If the diameter of an atom were about 1000 times the width of this page, the diameter of the nucleus would be indicated by the printed dot over the letter i. The **mass number** of an atom is the sum of the number of protons and neutrons in the nucleus. For example, if the nucleus of an atom contains four protons and five neutrons, the mass number of the atom is 9.

4. The nucleus of each atom has one or more positive charges, one for each proton. All the atoms of a specific element have the same number of protons in the nucleus and therefore the same distinctive nuclear charge. The nuclear charge on each hydrogen atom, for example, is 1; for each helium atom 2; for each carbon atom 6, etc. The **atomic number** is the number of positive charges on the nucleus. It corresponds to the number of protons.

FIGURE 2-3
Ernest Rutherford (1871–1937), a British physicist who made extraordinary contributions to our knowledge of radaoactivity and the structure of the atom. "Science goes step by step," he observed," and every man depends on the work of his predecessors. . . . Scientists are not dependent on the ideas of a single man, but on the combined wisdom of thousands. . . ." (*Drawing by Robert Kastor, Library of Congress.*)

Note to Students

This intensive discussion of the structure of atoms is necessitated by the fundamental role that atoms play in the structure of matter and the important relationship between the structure of matter and its properties. Unavoidably it involves a barrage of new terms, which in this condensed discussion are introduced in machine-gun style. Do not try to keep them all in mind during the first reading. Return to this section after you have consulted the summarizing questions for self-study at the end of the chapter, and again as you work through the Practice Exercises, especially 3, 4, and 7 to 9. By then you will have a handy background for Chap. 3.

5. The space around the nucleus is occupied by the electrons. There is one electron for each proton in the nucleus. The number of electrons therefore corresponds to the nuclear charge and is the same for each atom of a given element. Each hydrogen atom, with a nuclear charge of 1, for example, contains one electron. Each carbon atom, with a nuclear charge of 6, contains six electrons. The number and behavior of electrons in an atom are very important in determining the properties of an atom.

6. The electrons move very rapidly about the nucleus, but their nature is such that structural concepts cannot define both their velocity and their position precisely at any particular moment. Instead, the paths of electrons are described in terms of probable positions and probable velocities. The **atomic orbital** is the volume within which an electron is most probably to be found in an atom. An atomic orbital is frequently described as an electron cloud, conveniently approximated by thinking of it as a time-exposure "photograph" of the rapidly moving electron. Where the "photograph" is brightest marks the space where the electron spends most of its time. It is not possible to say, however, exactly when the electron was at any particular spot (see Fig. 2-4).

7. Although the density of the electron cloud falls off very rapidly some distance from the nucleus, there is a small probability of finding an electron very far away from the nucleus. For this reason atomic orbitals have no definite outer boundaries. For convenience, however, pictorial representations usually enclose the volume where the electron is most probably to be found about 95 percent of the time (see Fig. 2-5).

8. Atomic orbitals have various sizes and shapes. Some orbitals are spherical and symmetrically distributed about the nucleus (Fig. 2.4). Other orbitals approximate the shape of dumbells or have more complex shapes.

9. Because of the indefinite paths of the electrons in atoms, the overall dimensions of an atom are not easy to determine. Moreover, an atom's dimensions are greatly influenced by the environment in which it happens to be, what and how many atoms it is linked to, and how it is linked. When a hydrogen atom, for example, is linked to another atom in one of the known ways, its overall radius has been estimated to be in the range of 0.28 to 0.38 angstrom units, depending on the nature of the other atom. [One angstrom unit (Å) is equal to 0.00000001 centimeter (10^{-8} centimeter),† and there are

---

Note to Students

One of the points here is that atoms are very tiny. So small, as a matter of fact, that in a penny there are about

$$29,000,000,000,000,000,000,000 \text{ atoms}$$

---

† The expression of very small numbers is discussed in Supplement 1, Sec. S1-2.

2.54 centimeters to an inch. See Supplement 3, Sec. S3-1, for
a discussion of units of length.] If it is linked in another
way, its radius approximates 2.08 angstrom units.

FIGURE 2-4
Schematic representation of the cross sections of the orbitals of (a)
the electron in the hydrogen atom and (b) one of the electrons in
the carbon atom. The varying concentrations of dots are intended
to represent the relative densities of electric charge brought about
by the moving electron. Note the lack of definite outer boundaries.

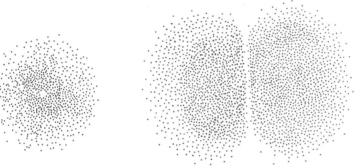

FIGURE 2-5
Three different schematic representations of the structure of a
hydrogen atom. The + symbol represents the positive charge on
the nucleus. The varying concentrations of dots represent the
relative densities of electric charge brought about by the moving
electron. Note that the simplified outline representation (c) does
not enclose all the space occupied by the electric charge.

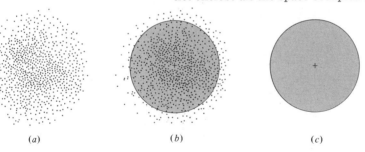

(a)   (b)   (c)

2-5   THE STRUCTURE OF HYDROGEN; ISOTOPES

The atoms and molecules of the element hydrogen have the simplest structure of all the atoms and molecules found in nature;
they therefore provide a logical starting point for discussion of
the structure of matter.

Note to Students   Although this chapter is primarily devoted to the structure of
                          atoms, some discussion of hydrogen molecules is desirable
                          here to clarify the relationship between atoms and molecules.

Each hydrogen molecule contains two hydrogen atoms held tightly together. Before discussing the structure of the molecule, let us consider the structure of an individual hydrogen atom. The nucleus of the hydrogen atom contains one proton and therefore has a single positive charge. One electron travels extremely rapidly about the nucleus, spending most of its time within an orbital which (under normal conditions of energy content) is spherical and symmetrically distributed around the nucleus (see Fig. 2-4). The atomic number of the hydrogen atom is 1 (one positive charge on the nucleus), and its mass number is 1 (one proton, no neutrons).

But this is not the complete description. It happens that there are two other kinds of hydrogen atoms in nature. The three hydrogen atoms differ from each other, however, only in the number of neutrons in the nucleus. Each of them has an atomic number of 1, which means that each has one proton in the nucleus and one electron. Since they differ only in the number of neutrons, they differ only in their masses. An **isotope** is an atom which has the same atomic number as another atom but differs from it in mass number. The atomic number is what defines the number of nuclear charges and the number of electrons. It is the number of electrons, moreover, which determines the chemical properties of an atom. All isotopes of a given element therefore have the same chemical properties. The properties which differ are those which depend on the individual masses of the atoms.

THE ELEMENT HYDROGEN

Hydrogen atoms are the most abundant of all of the different kinds of atoms, constituting as they do at least 90 percent of all the atoms in the universe. They are incorporated into some of the most interesting substances, as well as some of the most important. Water molecules contain hydrogen atoms, as do the molecules of sugar, proteins, and the components of crude petroleum.

Elemental hydrogen is made up of molecules containing two hydrogen atoms. It is a colorless, odorless gas with the lowest density of all forms of matter; that is, 1 liter weighs only about 0.08 gram under ordinary conditions, whereas the same volume of air weighs about 15 times as much, or 1.2 grams. Its low density is responsible for its ability to lift balloons and other lighter-than-air aircraft. Only traces of molecular hydrogen gas occur in the earth's atmosphere since the earth's gravitational field is not strong enough to hold onto the hydrogen released into it from volcanoes, the decomposing action of light on water, and other sources. But more than 7 million tons of the element are produced commercially in the United States each year.

What is all this hydrogen used for? About two-thirds of it is converted into ammonia as a first step toward the synthesis of important fertilizers, among other things. Consequently hydrogen production is a significant factor in the supply of food.

Elemental hydrogen is ordinarily a gas. It tends so strongly to be a gas that a special effort is required to persuade it to condense to a liquid: cooling it at least to $-253°C$.[†] Keeping large quantities at this low temperature presents a special challenge, and its handling is further complicated by hydrogen's famous explosive nature when mixed with oxygen, which is abundant in the atmosphere. Yet as much as 150 tons of liquid hydrogen has been required each day for the space program of the United States. It is used as one of the fuels to propel rockets into outer space (Fig. 2-6).

† Temperature scales are discussed in Supplement 3, Sec. S3-5.

FIGURE 2-6
Storage facility for liquid hydrogen close to where a Saturn rocket is being prepared for launching. The upper stages of this rocket use liquid hydrogen and liquid oxygen as fuels. The first stage uses kerosine and oxygen. (*Courtesy of NASA.*)

All atoms of the same element have the same atomic number, but the atoms of the same element may include more than one isotope, each of which has its own characteristic mass number. The atomic and mass numbers of an isotope are indicated in symbols by either of the two ways shown in the margin. Table 2-2 gives some of the principal characteristics of the three isotopes of hydrogen.

The isotopes of hydrogen have been given distinctive names: protium (or hydrogen), symbol $_1^1H$; deuterium, symbol $_1^2H$ or $_1^2D$; and tritium, symbol $_1^3H$ or $_1^3T$. Isotopes of other elements are

Atomic number $\longrightarrow$ $_1H^1$ $\longleftarrow$ Mass number

Atomic symbol

Mass number $\longrightarrow$ $_1^1H$ $\longleftarrow$ Atomic symbol

Atomic number

Symbols used to represent isotopes; this book uses the second form

Table 2-2
The Isotopes of Hydrogen

	Composition			Atomic number	Mass number	Isotope symbol
	Neutrons	Protons	Electrons			
Protium†	0	1	1	1	1	$_1^1H$
Deuterium‡	1	1	1	1	2	$_1^2H$ or $_1^2D$
Tritium‡	2	1	1	1	3	$_1^3H$ or $_1^3T$

† More commonly called hydrogen.

‡ The symbol for deuterium is either H or D; the symbol for tritium is either H or T.

usually not given special names but are identified only by their mass numbers. The isotopes of helium, for example, are called helium 3 and helium 4, the 3 and 4 representing their mass numbers. The information of Table 2-2 indicates that it is a simple matter to determine the number of protons, electrons, and neutrons in an atom from the isotope symbol. In the isotope symbol for a tritium atom ^3_1H, for example, the atomic number gives the number of protons and electrons directly (one each), and the difference between the mass number and the atomic number gives the number of neutrons (two). The isotope symbol $^{13}_6C$ for carbon 13 indicates that the number of protons in the isotope is 6, the number of electrons is 6, and the number of neutrons is 7. Moreover, if the atomic number and mass number of an isotope are known, it is a simple matter to write the isotope symbol. If the atomic number of an isotope is 15, for example, the atom is an isotope of phosphorus (Table 2-3). If the mass number is known to be 31, the isotope symbol is $^{31}_{15}P$.

The various isotopes of a given element may differ greatly in their abundance in nature. Naturally occurring hydrogen, for example, is made up of 99.985 percent protium atoms, 0.015 percent deuterium atoms, and only very small traces of tritium atoms. Elemental hydrogen is found in nature as molecules, each of which contains two atoms. Because these atoms may be any of the three isotopes of hydrogen (hydrogen, H; deuterium, D; tritium, T), the following different hydrogen molecules are possible: H—H, H—D, H—T, D—D, D—T, T—T. Since the percentage of the ^1_1H isotope in the mixture of hydrogen atoms in nature is more than 99 percent, with less than 0.02 percent deuterium and an extremely small trace of tritium, the most abundant hydrogen molecule in nature is H—H, and T—T is so extremely rare that it probably should not be considered at all.

2-6   HOW IS THE STRUCTURE
OF HYDROGEN REPRESENTED?

Although the names hydrogen atom and hydrogen molecule have a similar sound, they refer to the fundamental particles of two quite different forms of matter, which differ dramatically in their behavior. The form of matter composed of hydrogen atoms, called atomic hydrogen, is a highly reactive gas, combining readily with many substances and existing only under special circumstances. The substance composed of hydrogen molecules is the ordinary elemental hydrogen. Although it is a reasonably reactive gas and will, for example, combine with oxygen to form water, it is not nearly as reactive as atomic hydrogen.

Let us assume that both atoms in a hydrogen molecule are the most common isotope of hydrogen (protium atoms). Since each of these atoms is made up of one proton and one electron, each diatomic hydrogen molecule derived from them is therefore composed of two protons and two electrons. The two protons in a hydrogen molecule remain as two separated nuclei. The paths of the two electrons go around both nuclei within the same orbital, roughly the shape of a stubby sausage. Since the orbital is associated with the entire molecule, it is called a molecular orbital. A **molecular orbital** is the volume within which an electron is most probably to be found within a molecule. Pictorial representations often enclose only the space within which the electrons spend most of their time.

The two nuclei of the hydrogen atoms are held together by the attraction of the positive charges of the nuclei for the negative charges of the electrons, and vice versa. The particles of the molecule are held in a dynamic balance, the forces of attraction between the positive nuclei and the negative electrons being opposed by the repulsive forces between the two positive nuclei and the repulsive forces between the two negative electrons. Although the forces of attraction prevail, the two nuclei vibrate back and forth with respect to each other, and the electrons move rapidly around both of them. In simplified language, the two nuclei are described as being bonded by the two electrons (Fig. 2-7), and the two electrons are called the **bonding pair of electrons.** In a still more simplified fashion, the structure of the molecule is described as two hydrogen atoms bonded together by the two electrons. Actually, of course, the two electrons which bond the atoms are part of the atoms themselves.

FIGURE 2-7
Schematic representation of the bonding of two hydrogen atoms: (*a*) **separate atoms;** (*b*) overlapping atomic orbitals; (*c*) outline representation of hydrogen molecule; and (*d*) a cross section of electron density in a hydrogen molecule.

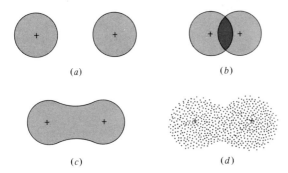

(*a*)   (*b*)

(*c*)   (*d*)

When the structure of the hydrogen molecule is described in these terms, i.e., two nuclei mutually bonded by two electrons within a molecular orbital, it is referred to as the **molecular-orbital description.** For some purposes it is useful to imagine that the molecular orbital has been formed from the individual atomic orbitals of the two separate hydrogen atoms. In so imagining, the atomic orbitals are said to overlap as the molecular orbital is formed. Ordinarily, a hydrogen molecule is not formed in this way, but picturing such a process helps us understand molecular structure.

Note to Students   The nature of the links between atoms and the methods of representing them will be discussed further in Chap. 3. The discussion here is of an introductory nature and focuses only on the hydrogen molecule.

In the electronic-formula notation an individual hydrogen atom is represented by H·, with the dot representing the electron. The completed molecule is represented by the symbol H:H. The two dots represent the bonding pair of electrons. The expression is simplified further in structural-formula notation in which the bonding pair of electrons is represented by a dash, H—H.

Structural formulas are the most convenient method of expressing molecular structure since, among other things, they are easily written with clarity. Electronic formulas are less popular, since writing the dots can be tedious and at times less clear. The molecular-orbital description gives a better "image" of the overall structure than either structural or electronic formulas, and it has many other advantages. The pictorial aspects of molecular orbitals, however, present many difficulties. In practice the chemist employs all representations interchangeably, using in a given instance whichever method serves the purpose most clearly or most conveniently.

2-7   WHAT IS THE STRUCTURE OF HELIUM?

Helium, like hydrogen, is an element; i.e., it cannot be decomposed into simpler substances. It is composed of single atoms which under ordinary conditions have no tendency to join together in pairs or larger groups. The nucleus of a helium atom contains two protons and therefore has a double positive charge, twice that of the nucleus of a hydrogen atom. There are two electrons around the nucleus, which spend most of their time within the same spherical orbital (see Fig. 2-8). Two isotopes exist,

---

THE ELEMENT HELIUM

Helium is a colorless gas which is probably best known for its ability to lift lighter-than-air aircraft (Fig. 2-9). Its lifting ability is not as good as that of hydrogen, but unlike hydrogen it is nonflammable. It is therefore much preferred for safety. As a matter of fact, helium is unusually inert; it does not react with any other substance.

The element is found only in small quantities in the earth's crust and almost not at all in the atmosphere. As with hydrogen, the earth's gravitational field is not strong enough to prevent its escape. Helium is found abundantly in the universe, however, being the second most abundant element after hydrogen. It is a major constituent of the sun and most other stars. Helium was originally discovered when its presence in the sun was detected by analyzing the characteristics of the light energy helium radiates at the sun's high temperatures. For more than 20 years helium was regarded more or less as a hypothetical element, present probably on the sun but not on the earth. It

was eventually found, however, entrapped in certain uranium ores and in underground pockets containing natural gas. Today about 800 million cubic feet of helium is used annually in the United States in welding processes and a variety of special uses, including important techniques of chemical analysis. The United States produces more than 80 percent of the world's helium, most of it found underground along with natural gas. Currently, supply exceeds demand, and much of the helium in natural gas is not isolated and is lost forever to the atmosphere. Since future methods of energy generation, transmission, and storage may require substantial amounts of helium, consideration is now being given to storage for future use. Helium occurs in the atmosphere in very small amounts, and extraction from the air could conceivably furnish an inexhaustible supply; but it would be 200 or 300 times more expensive than the present sources.

helium 4, whose nucleus contains two neutrons and two protons, and helium 3, whose nucleus contains one neutron and two protons. Natural helium is almost entirely helium 4, with only a trace of helium 3.

FIGURE 2-8
Representations of a helium atom: (*a*) a suggestion of the electric charge distribution caused by the motion of the two electrons; (*b*) a more simplified representation, indicating the two positive charges on the nucleus and enclosing the region of about 95 percent of the charge distribution.

(*a*)

(*b*)

FIGURE 2-9
The Goodyear airship "Columbia" over Pasadena, California. Such blimps often hover over sporting events, making it possible to take spectacular television shots from the air. The blimp is kept in the air by the low density of helium. (*Courtesy of The Goodyear Tire and Rubber Company.*)

2-8 THE ELECTRONIC STRUCTURE OF ATOMS

When the details of atomic structure were first appreciated (about 1919 by Rutherford, Bohr, Lewis, and Langmuir), the electrons were imagined as revolving around the nucleus in a series of precisely defined concentric orbits at various fixed distances. Each orbit was considered to have a limited capacity for electrons. Experimental evidence suggested that the orbit closest to the nucleus, identified as the *K* orbit, or orbit 1, has a capacity for only 2 electrons; the electron capacity for orbit L, or orbit 2, is 8 electrons; for the orbit M (orbit 3) 18 electrons, and orbit N (orbit 4) 32 electrons, and so on.

Electrons were believed to group themselves in those orbits closest to the nucleus, but if inner orbits were filled to capacity, the overflow electrons were distributed in the next outer orbit until its capacity was filled, and so on. On the basis of these principles, each atom was assigned an electronic arrangement. The term **electron configuration** refers to the arrangement of electrons in an atom.

As the electronic nature of atoms was investigated further, each electron was assigned a quantum number, a number designating a characteristic of the electron. (The word quantum derives from a concept of energy in which it is considered to be packaged in separate units, each called a quantum. This idea about the nature of energy was used by Bohr in his interpretation of atomic structure.) Electrons with quantum number 1 were in the first orbit (closest to the nucleus). Those with quantum number 2 were in the second orbit, and so forth. All the electrons within a given orbit were considered to have the same quantum number and to be otherwise identical.

A more detailed examination of the structure of atoms revealed, however, that the situation was more complicated; actually each electron in a given atom has a distinctive set of attributes which distinguishes it from all the other electrons in the atom. Moreover, as mentioned in Sec. 2-4, the nature of electrons is such that their motion cannot be described in terms of precisely defined paths or orbits but only in terms of atomic orbitals, volumes within which an electron is most probably to be found in an atom. To characterize the electrons adequately, it was necessary to describe each electron in a given atom by a set of four quantum numbers, rather than one quantum number. For our purposes, however, it will be sufficient to consider the electronic structure of atoms from the more simplified view held about 50 years ago. The simplified configurations for the atoms with the 20 lowest atomic numbers are given in Table 2-3.

2-9 WHAT ARE ELECTRONIC SYMBOLS?

From the standpoint of molecular structure, the most important structural feature of atoms is the disposition of their electrons. In the representation of the bonds between atoms attention is focused on the electrons around the "outside" of an atom. From Table 2-3 the electron configuration of the fluorine atom, atomic symbol F, is two electrons in the first orbit close to the nucleus, and seven electrons in the second orbit, which in fluorine is the

---

Note to Students

This chapter is limited primarily to interpreting electronic symbols since they represent the aspect of the structure of atoms which is used directly in representing molecular structure.

---

outermost orbit. A convenient abbreviated representation of the fluorine atom is the symbol $:\!\overset{..}{\underset{.}{F}}\!\cdot$, in which the seven dots represent the seven outermost electrons and the F represents the nucleus plus the two inner electrons. The **electronic symbol** is the representation of an atom in which (1) the letter atomic symbol represents the nucleus of the atom and all the electrons except those in the outermost orbit (a combination sometimes called the *core* or *kernel* of the atom) and (2) the electrons in the outermost orbit are represented by dots, one dot for each electron. The electronic symbols for the 20 simplest atoms are given in Table 2-3.

Although electronic symbols are rather primitive representations of atoms, they are remarkably effective in providing a basis for useful formulas. For example, the formula for molecular hydrogen can be written H:H. The H's represent the two hydrogen nuclei, and the two dots represent the two electrons involved in the bond which joins them. In a similar way the electronic symbols for hydrogen and fluorine can be used to represent the structures of the molecules of elemental fluorine, $:\!\overset{..}{\underset{..}{F}}\!:\!\overset{..}{\underset{..}{F}}\!:$, and of the substance hydrogen fluoride, $H\!:\!\overset{..}{\underset{..}{F}}\!:$. The use of elec-

Table 2-3
Electronic Structure of the 20 Simplest Atoms

Element	Atomic symbol	Electronic symbol	Atomic number	Electron configuration by orbits			
				Orbit 1 (maximum 2)	Orbit 2 (maximum 8)	Orbit 3 (maximum 18)	Orbit 4 (maximum 32)
Hydrogen	H	H·	1	1			
Helium	He	He:	2	2			
Lithium	Li	Li·	3	2	1		
Beryllium	Be	·Be·	4	2	2		
Boron	B	·Ḃ·	5	2	3		
Carbon	C	·Ċ·	6	2	4		
Nitrogen	N	·Ṅ·	7	2	5		
Oxygen	O	:Ö·	8	2	6		
Fluorine	F	:Ḟ·	9	2	7		
Neon	Ne	:Nė:	10	2	8		
Sodium	Na	Na·	11	2	8	1	
Magnesium	Mg	·Mg·	12	2	8	2	
Aluminum	Al	·Äl·	13	2	8	3	
Silicon	Si	·Ṡi·	14	2	8	4	
Phosphorus	P	·Ṗ·	15	2	8	5	
Sulfur	S	:Ṡ·	16	2	8	6	
Chlorine	Cl	:Ċl·	17	2	8	7	
Argon	Ar	:Är:	18	2	8	8	
Potassium	K	K·	19	2	8	8	1
Calcium	Ca	·Ca·	20	2	8	8	2

tronic symbols to represent atomic bonds will be discussed
further in Chap. 3.

2-10 THE NATURE OF ELEMENTS RECONSIDERED

An element is defined (Sec. 2-3) as a substance which is not decomposable into other substances. The element hydrogen, composed of diatomic molecules, is a gas which under ordinary conditions is very stable. The fact that the molecules of hydrogen
can be decomposed at very high temperatures into individual
atoms of hydrogen does not invalidate the classification of
hydrogen as an element. The form of matter made up of
hydrogen *atoms* is so extremely reactive that it has a very short
existence and is ordinarily not considered a substance.

From a structural point of view an element is a substance
composed of atoms all of which have the same atomic number.
This means that each atom of a given element has the same
number of protons, nuclear charges, and electrons. It is the
number of electrons in an atom which determines its characteristic properties. The atoms of a given element, of course, may
be composed of more than one isotope (Sec. 2-5). The atoms of the
element helium, for example, as already mentioned, include the
isotopes helium 3 and helium 4. The differences in mass between the various isotopes of an element give rise to only minor
differences in properties and have no effect on the bonding behavior of atoms.

Some elements such as helium are composed of molecules
containing only a single atom. Such units are called, appropriately, monoatomic molecules. The molecules of some of the
other elements contain two atoms (diatomic) like hydrogen and
fluorine. Still other elements contain even more atoms in their
fundamental molecules. (For information about selected elements see Table 2-4.)

FIGURE 2-10
Mercury, one of nine elements known to the ancient world, is
the only common metal which is liquid under ordinary conditions. (*Photograph by Hans Oettgen.*)

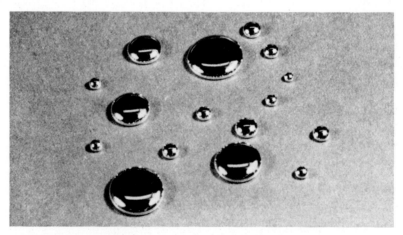

Table 2-4
Some Elemental Substances of Interest

Element name	Atomic symbol†	Comment
Aluminum	Al	A lightweight silvery metal, used in construction and electric transmission lines
Argon	Ar	A nonreactive, colorless gas (one of the so-called noble gases), used to fill electric light bulbs
Bromine	Br‡	A dangerous toxic reddish-brown liquid isolated from seawater; used to make medicinals, sanitizers, photographic chemicals, and other useful substances
Carbon	C	Occurs as diamonds, graphite (black solid), and in an amorphous form (black solid); one of the elements known in ancient times; used, for example, as the "lead" in lead pencils (graphite) and as a reinforcing agent in rubber tires (carbon black); over 2 million compounds of carbon known, many essential to living systems
Chlorine	Cl‡	A toxic greenish-yellow gas, used to disinfect drinking-water supplies
Copper	Cu	A reddish-brown metal used as a conductor of electricity and in coinage; known in ancient times
Gold	Au	A yellowish metal; used for coinage and jewelry; known in ancient times
Helium	He	A colorless, inert gas of low density, used to inflate lighter-than-air aircraft
Hydrogen	H‡	A colorless gas of low density, explosive in air; important in the production of many useful substances
Iodine	I‡	A bluish-black solid; very useful antiseptic
Iron	Fe	A gray, hard, and brittle metal, very reactive; important in the production of steel; known in ancient times
Lead	Pb	A soft, heavy, bluish-white metal, resistant to corrosion; known in ancient times; used for plumbing by the Romans; cumulatively toxic (Sec. 15-4); compounds used to improve performance of gasoline as a motor fuel (tetraethyl lead)
Magnesium	Mg	A light, silvery metal, used in the aerospace industry; isolated from the sea (Sec. 13-7), which serves as an inexhaustible source
Mercury	Hg	A silvery liquid; the only metal liquid under ordinary conditions (Fig. 2-10); cumulatively toxic; known in ancient times; found in Egyptian tombs of 1500 B.C.; used in making thermometers and mercury-vapor lamps
Neon	Ne	A colorless, inert gas; used in neon advertising signs, in which it gives a reddish-orange glow
Nickel	Ni	A silvery metal; used in coinage and in making stainless steel
Nitrogen	N‡	A colorless gas constituting 78% of the atmosphere; used to make fertilizers and explosives

(*Continued on next page*)

Table 2-4
Some Elemental Substances of Interest *(continued)*

Element name	Atomic symbol†	Comment
Oxygen	O‡	A colorless gas constituting 21% of the atmosphere; essential for animal life; used in the production of steel; atoms of oxygen more abundant than those of any other element in crust of earth
Plutonium	Pu	A highly toxic, silvery metal; a synthetic element first made in 1940; used as a nuclear fuel and in atom bombs (Sec. 9-9B)
Silicon	Si	Element not found as such in nature but prepared as a brown powder or a grayish solid; silicon atoms are the second most abundant; silicon compounds are widespread in rock formations and constitute about 26% of the earth's crust
Silver	Ag	A grayish-white metallic solid; used in coinage and jewelry; compounds important in photography; known in ancient times
Sodium	Na	A highly reactive, soft, silvery-looking metal; used in industrial production of many widely used substances
Sulfur	S§	A brittle yellow solid, found in elemental form in underground deposits; known in ancient times as brimstone; used in vulcanization of rubber and in industrial production of many widely used substances
Uranium	U	A heavy, silvery white, radioactive metal; used as a nuclear fuel and in atom bombs (Sec. 9-8A)
Zinc	Zn	A bluish-white lustrous metal; alloyed with copper to make brass

† Although the symbol for many elements is clearly an abbreviation of the name of the element, some symbols seem quite unrelated because they were formed from the Latin name, e.g., Au, from *aurum*, "gold," and Pb, from *plumbum*, "lead."
‡ The element is twice this, i.e., Br_2, Cl_2, H_2, I_2, N_2, and O_2.
§ The element is most commonly S_8.

As found in elements, atoms are frequently referred to (ambiguously) as **atoms in the free state,** even though they actually may be linked to other atoms. Atoms are referred to as being in the **combined state** only when they occur in molecules containing atoms of more than one atomic number, as in compounds. An atom of fluorine in the element fluorine, $:\overset{..}{\underset{..}{F}}:\overset{..}{\underset{..}{F}}:$, is an atom in the free state. An atom of fluorine in the compound hydrogen fluoride, $H:\overset{..}{\underset{..}{F}}:$, is an atom in the combined state.

An atom in the combined state is sometimes referred to as an "element" in the combined state. Hydrogen fluoride may be described, for example, as containing hydrogen and fluorine in the combined state or may be said to be made up of the elements of hydrogen and fluorine. These statements actually refer, of course, to atoms of hydrogen and fluorine, not to the elements hydrogen and fluorine.

In modern usage the term **molecule** refers to the smallest particle of any substance as it ordinarily exists, whether the substance is an element or a compound. An **atom,** on the other hand, is the smallest conceivable unit of an element but not necessarily the smallest unit of the element as it normally is found. The smallest naturally existing unit of the element hydrogen, for example, contains two hydrogen atoms bound together in a molecule.

HOW ARE DIAMONDS SYNTHESIZED?

The way atoms are grouped and bonded together can exert a profound influence on the properties of substances. One of the most striking examples is two substances made up of only carbon atoms, graphite and diamond. Graphite is a soft, slippery black solid, commonly used as the "lead" of pencils. Diamond is a clear, crystalline substance, one of the very hardest materials, highly valued as a gem and used industrially in grindstones and cutting tools. The name allotrope is applied to an elemental substance made up of the same kind of atoms found in another elemental substance. Two or more allotropes, such as graphite and diamond, are frequently referred to as the allotropic forms of an element.

The carbon atoms in graphite are grouped in separate, flat layers of hexagonal rings, each carbon atom bonded to three other carbon atoms (Fig. 2-11a). The carbon atoms in diamond are more compactly arranged in a continuous three-dimensional pattern in which each carbon atom is bonded to four other carbon atoms (Fig. 2-11b). In both graphite and diamond the carbon atoms are held together by bonding electron pairs somewhat similar to the bonds between two hydrogen atoms in a hydrogen molecule.

The intriguing possibility of synthesizing diamonds from more common forms of carbon has attracted considerable interest ever since the French chemist Antoine Lavoisier demonstrated in 1797 that diamond is simply carbon. Curiosity was strengthened when it was later shown that if diamond is heated to between 1500 and 2000°C in a vacuum, it is converted into graphite. But the many attempts made over more than a century to convert graphite into diamond led to no success. It was eventually realized that if the transformation were to be carried out at all, a combination of a very high temperature and a very high pressure would be required.

In the early 1950s a team of research workers at the General Electric Company's laboratories in Schenectady, New York, embarked upon a thoroughgoing experimental inquiry into the nature of the forms of carbon and the means of achieving a suitable high-pressure–high-temperature combination. Success came in 1955. Using temperatures in the neighborhood of 2000°C, pressures in the range of 100,000 atmospheres, or 1.5 million pounds per square inch, and catalysts such as metallic nickel, they were able to carry out a reproducible synthesis of diamond from more common forms of carbon such as graphite.

The first synthetic diamonds were of only industrial grade, but they found a ready use in such applications as grinding wheels, in which they are in some respects superior to natural diamonds and competitive in cost. In 1970 a way was found to modify the process to produce flawless 1-carat diamonds of very high quality. (A carat, equivalent to 200 milligrams, is a unit of weight used for gems.) Moreover, these diamonds do not have to be cut into gems as natural diamonds do. The synthetic diamonds are produced with crystal planes similar to the facets of a carefully cut gem (Fig. 2-12). At present, unfortunately, the cost is much higher than that of mining natural diamonds.

KEY WORDS

1. Electron (Sec. 2-2)	12. Electric charge (Sec. 2-3)	23. Deuterium (Sec. 2-5)
2. Proton (Sec. 2-2)	13. Electric current (Sec. 2-3)	24. Tritium (Sec. 2-5)
3. Neutron (Sec. 2-2)	14. Substance (Sec. 2-3)	25. Molecular orbital (Sec. 2-6)
4. Inertia (Sec. 2-3)	15. Mixture (Sec. 2-3)	26. Molecular-orbital description (Sec. 2-6)
5. Force (Sec. 2-3)	16. Element (Sec. 2-3)	27. Bonding pair of electrons (Sec. 2-6)
6. Matter (Sec. 2-3)	17. Compound (Sec. 2-3)	28. Electron configuration (Sec. 2-8)
7. Mass (Sec. 2-3)	18. Atomic number (Sec. 2-4)	29. Electronic symbol (Sec. 2-9)
8. Vacuum (Sec. 2-3)	19. Mass number (Sec. 2-4)	30. Molecule (Sec. 2-10)
9. Weight (Sec. 2-3)	20. Atomic orbital (Sec. 2-4)	31. Atom (Sec. 2-10)
10. Energy (Sec. 2-3)	21. Isotope (Sec. 2-5)	32. Atom in free state (Sec. 2-10)
11. Electricity (Sec. 2-3)	22. Protium (Sec. 2-5)	33. Atom in combined state (Sec. 2-10)

ic representations of the structure of (*a*) **graphite** and (*b*)
ond. In graphite each carbon atom is joined to three other
oon atoms in the layers of carbon rings. In the more compact
arrangement of diamond each carbon atom is joined to four other
carbon atoms.

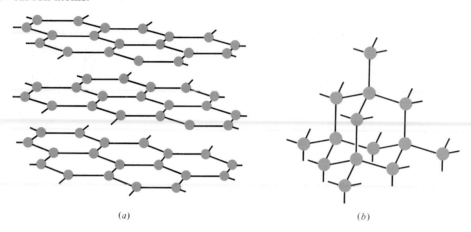

(*a*) (*b*)

FIGURE 2-12
Synthetic diamonds of gem quality and the powdered graphite
from which they are made. The larger crystals are approximately
1 carat in weight. Although these diamonds have undergone
slight polishing, they have not been cut and retain the shape in
which they were made. (*Courtesy of General Electric Research and Development Center.*)

HOWEVER, THE CHANGE IS SPECTACULAR ON THE MOLECULAR LEVEL.

(© 1979 by American Scientist. Reprinted by permission of Sidney Harris.)

SUMMARIZING QUESTIONS FOR SELF-STUDY

Sections 2-1 to 2-3

1. Q. About 500 B.C. Heraclitus proposed that fire was the primordial element from which all forms of matter have arisen. How does this suggestion relate to the current ideas about the structure of matter?

A. This is a variation of the earlier suggestion of Thales of a common composition of matter. The concept of a common composition remains a part of present-day thinking, i.e., that all atoms are composed of electrons, protons, and neutrons.

2. Q. If a particle of matter contains one proton, one neutron, and one electron, (*a*) how will its mass compare with that of a proton alone and (*b*) what overall electric charge will it have?

A. (*a*) Since a proton and a neutron have masses that are about the same, and since the mass of an electron is negligibly small, the particle will have a mass about twice as great as the mass of a proton alone. (*b*) Since a neutron has no charge, an electron has a unit negative charge, and a proton has a unit positive charge, the particle will have no charge.

3. Q. What is the difference between the mass and weight of a neutron?

A. The mass of the neutron refers to the quantity of matter contained in it. The weight of a neutron is a measure of the force exerted on that matter by the gravity of the earth.

4. Q. If ammonia can be decomposed into hydrogen and nitrogen, is it an element or a compound?

A. It is a compound, since all substances which are decomposable into simpler substances are compounds.

5. Q. On what basis is nitrogen described as an element?

A. Nitrogen is described as an element since it cannot be decomposed into simpler substances.

Section 2-4

6. Q. If an atom contains three protons and four neutrons, what will be: (a) the charge on its nucleus, (b) its atomic number, and (c) its mass number? (d) How many electrons will it have?

A. (a) Three units of positive charge. (Neutrons have no charge, but each proton has a unit positive charge and all the protons of an atom are found in its nucleus.) (b) Three. (The atomic number is the number of charges on the nucleus of an atom.) (c) Seven. (The mass number is equal to the sum of the number of neutrons and protons in an atom.) (d) Three. [Atoms as a whole are neutral and the number of electrons (each with a unit negative charge) is equal to the number of protons (each with a unit positive charge).]

7. Q. If an atom contains 11 protons and 12 neutrons, what will be: (a) the charge on its nucleus, (b) its atomic number, and (c) its mass number? (d) How many electrons will it have?

A. (a) Eleven units of positive charge (same explanation as in 6a). (b) Eleven (same explanation as in 6b). (c) Twenty-three (same explanation as in 6c). (d) Eleven (same explanation as in 6d).

8. Q. Why is the position of an electron described as being within an atomic orbital rather than specified, say, as being 1 angstrom from the nucleus of an atom?

A. The elusive nature of electrons is such that their velocity and their position cannot both be specified at a given moment. Instead, the paths of electrons are described in terms of probable positions and probable velocities. Moreover, there is a small probability of finding an electron at a given instant very far away from the nucleus. For these reasons the position of an electron is described in terms of an atomic orbital, a volume within which an electron is most probably to be found in an atom.

9. Q. Why are some diagrammatic representations of atomic orbitals vague about the outer boundaries?

A. See the answer for Question 8.

Sections 2-5 to 2-7

10. Q. The molecules of element X are known to contain two atoms each. If two naturally occurring isotopes are known, how many different molecules are present in the natural element?

A. If the two isotopes are labeled X_1 and X_2, the following three different molecules are possible: X_1X_1, X_2X_2, X_1X_2.

11. Q. The molecules of element Z are known to contain two atoms each. If four naturally occurring isotopes are known, how many different molecules are present in the natural element?

A. If the isotopes are labeled Z_1, Z_2, Z_3, and Z_4, the following 10 different molecules are possible: Z_1Z_1, Z_1Z_2, Z_1Z_3, Z_1Z_4, Z_2Z_2, Z_2Z_3, Z_2Z_4, Z_3Z_3, Z_3Z_4, Z_4Z_4.

12. Q. Why doesn't the presence of different isotopes alter the bonding behavior of atoms?

A. Isotopes are atoms which differ only in the number of neutrons. Since isotopes have the same number of electrons and the bonds with other atoms are determined by the number of electrons, the bonding behavior of isotopes does not differ.

13. Q. What is the difference between an atomic orbital and a molecular orbital?

A. An atomic orbital is the volume within which an electron is most probably to be found in an individual atom. A molecular orbital is the volume in which an electron is most probably to be found in a molecule.

14. Q. How many atoms are present in each molecule of helium?

A. One. (Molecules containing one atom are called monoatomic molecules.)

15. Q. What is a bonding pair of electrons?

A. A bonding pair of electrons is a pair of electrons shared by two atoms in a molecule. It contributes to the forces of attraction that serve to hold the nuclei together.

Sections 2-8 to 2-10

16. Q. What is meant by the electron configuration of an atom?

A. A representation of the arrangement of electrons in an isolated atom.

17. Q. What is the electronic symbol of an atom?

A. A type of representation of the structure of an atom in which the nucleus and inner electrons of an atom are represented by the atomic symbol and each of the outermost electrons is represented by a dot, for example, $:\ddot{F}\cdot$, H$\cdot$

18. Q. What is the electron configuration of the fluorine atom by orbits?

A. 2, 7.

19. Q. What is the electronic symbol of the fluorine atom?

A. $:\ddot{F}\cdot$.

20. Q. What are the electron configurations by orbit and electronic symbols of the atoms lithium, nitrogen, magnesium, sulfur, and calcium?

A.

	Electron configuration by orbits	Electronic symbol
Lithium	2, 1	Li$\cdot$
Nitrogen	2, 5	$\cdot\ddot{N}\cdot$
Magnesium	2, 8, 2	$\cdot$Mg$\cdot$
Sulfur	2, 8, 6	$:\ddot{S}\cdot$
Calcium	2, 8, 8, 2	$\cdot$Ca$\cdot$

21. Q. What is the difference between an atom and a molecule? Give two examples of each.

A. A molecule is the smallest particle of a substance as it ordinarily exists, e.g., hydrogen molecules, H∶H; helium molecules (monoatomic), He∶. An atom is the smallest conceivable unit of an element, e.g., hydrogen atoms, H·; helium atoms, He∶.

PRACTICE EXERCISES

1. What is the difference between the mass and the weight of a rock?

2. If a given volume of air is found to weigh 50.00 grams and a group of marbles is also found to weigh 50.00 grams, what can be said about the quantity of matter contained in the volume of air and group of marbles?

3. What is the difference between (a) an electric charge and an electric current; (b) a substance and a mixture; (c) an element and a compound?

4. What is the difference between (a) an atom of hydrogen and a molecule of hydrogen; (b) an atom and an element; (c) an atom in the free state and an atom in the combined state; (d) the mass number of an atom and the atomic number of an atom; (e) an isotope and an atom?

5. What is the difference between (a) an atomic orbital and a molecular orbital; (b) an orbit of an electron and an atomic orbital; (c) an electron configuration of an atom and an electronic symbol of an atom?

6. Is it correct to say that hydrogen fluoride, HF, contains the elements hydrogen and fluorine?

7. Give names for atoms with the following atomic number: (a) 14; (b) 4; (c) 16; (d) 9 (use Table 2-3).

8. For each of the isotopes in the table for Exercise 8 give the number of protons, number of neutrons, number of electrons, atomic symbol, and isotope symbol.

9. For each of the isotopes in the table for Exercise 9 give the number of neutrons, number of electrons, the electron configuration by orbits, and the electronic symbol (use Table 2-3).

10. Fill in the electron configurations by orbits and electronic symbols in the table for Exercise 10 (use Table 2-3).

11. A molecule of hydrogen has the formula H_2. Is it correct to say, as implied in Sec. 1-3, that each molecule of hydrogen is like every other molecule of hydrogen?

12. Complete the table for Exercise 12 (use Table 2-3).

Table for Exercise 8

Isotope	Atomic number	Mass number	Number of protons	Number of neutrons	Number of electrons	Atomic symbol	Isotope symbol
(a)	11	23	_____	_____	_____	_____	_____
(b)	3	7	_____	_____	_____	_____	_____
(c)	13	27	_____	_____	_____	_____	_____
(d)	19	39	_____	_____	_____	_____	_____

Table for Exercise 9

Isotope	Isotope symbol	Number of neutrons	Number of electrons	Electron configuration by orbits	Electronic symbol
(a)	$^{14}_{6}C$	_____	_____	_____	_____
(b)	$^{3}_{1}H$	_____	_____	_____	_____
(c)	$^{19}_{9}F$	_____	_____	_____	_____
(d)	$^{40}_{18}Ar$	_____	_____	_____	_____
(e)	$^{17}_{8}O$	_____	_____	_____	_____
(f)	$^{25}_{12}Mg$	_____	_____	_____	_____

Table for Exercise 10

	Electron configuration by orbit	Electronic symbol
Helium 3	_____	_____
Boron 11	_____	_____
Calcium 43	_____	_____
Neon 22	_____	_____

Table for Exercise 12

Isotope name	Mass number	Atomic number	Number of neutrons	Isotope symbol	Electron configuration by orbits	Electronic symbol
Boron 11	_____	_____	_____		_____	_____
_____	_____	_____	_____	$^{32}_{15}P$	_____	_____
_____	20	8	_____	_____	_____	_____
_____	_____	19	24	_____	_____	_____
_____	_____	_____	16	_____	2, 8, 4	_____

SUGGESTIONS FOR FURTHER READING

The following articles are of an introductory nature.

Dinga, Gustav P.: The Elements and the Derivation of Their Names and Symbols, **Chemistry**, February 1968, p. 10.

Ihde, Aaron J.: The Karlruhe Congress: A Centennial Retrospect, **Journal of Chemical Education, 38**:83 (1961). An important step in the establishment of the atomic theory.

Keller, Eugenia: Hydrogen: The Simplest Element, **Chemistry**, November 1969, p. 19.

———: Origin of the Elements, **Chemistry**, July-August 1972, p. 17.

Milliken, Robert: Electrons: What They Are and What They Do, **Chemistry**, April 1967, p. 13.

Pratt, Christopher J.: Sulfur, **Scientific American**, May 1970, p. 63.

Szabadvary, Ferenc: Great Moments in Chemistry, II: From Thales to Bohr, **Chemistry**, November 1969, p. 6.

The following articles are at a more advanced level.

Bamberger, C. E., and J. Braustein: Hydrogen: A Versatile Element, **American Scientist, 63**:438 (1975).

Boranger, Michel, and Raymond A. Sorenson: The Size and Shape of Atomic Nuclei, **Scientific American**, August 1969, p. 58.

Kendall, Henry W., and Wolfgang K. H. Panofsky: The Structure of the Proton and the Neutron, **Scientific American**, July 1971, p. 61.

Mermin, N. David, and David M. Lee: Superfluid Helium 3, **Scientific American**, December 1976, p. 56.

Schrödinger, Erwin: What Is Matter?, **Scientific American**, September 1953, p. 52.

Trower, W. Peter: Matter and Antimatter, **Chemistry**, October 1969, p. 8.

HOW IS THE STRUCTURE OF MOLECULES AND IONS EXPRESSED?

3

3-1 INTRODUCTION

The discussion of Chap. 1 showed how structural formulas express the molecular structure of substances. This chapter explains the various symbols used in structural formulas and discusses the nature of chemical formulas. We shall find that structural formulas concisely express a great deal of information about a substance which in words would occupy many paragraphs. Structural formulas are therefore a distinctive characteristic of chemical literature.

Let us consider the structural formula used to represent the molecule of novocain, the first significant local anesthetic:

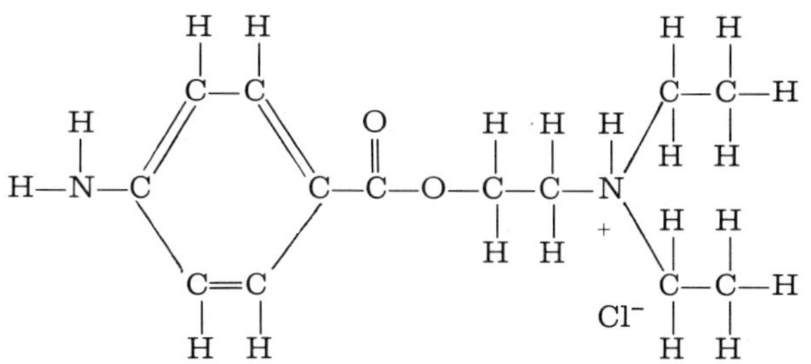

Novocain, the first significant local anesthetic; this chapter explains the symbols used in this formula

In the rest of this chapter we shall learn what the various letters, dashes, and the plus and minus signs of this formula signify.

Much of the symbolism used for structural formulas, the letters and the dashes, was introduced more than a century ago, long before the electronic nature of interatomic linkages was recognized. Although the interpretation of the symbolism has changed, it is a testimony to the insight of the pioneers in this area that their work has had such lasting usefulness. It is characteristic of science that outmoded concepts are usually not abandoned completely but form the basis of the ideas that replace them.

3-2 WHAT ARE VALENCE ELECTRONS?

Atoms differ in their capacity to form bonds to other atoms. A chlorine atom, Cl, for example, combines with only one hydrogen atom, H, in the molecule of the compound hydrogen chloride, HCl. An atom of oxygen, O, however, combines with two atoms of hydrogen in the molecule of water, H_2O. An atom of nitrogen, N, combines with three atoms of hydrogen in the molecule of the substance ammonia, NH_3. But helium, He, does not combine with hydrogen or any other atoms at all.

Valence is the capacity of an atom to form bonds with other atoms. In some methods of specifying combining behavior, each atom is assigned a number or a series of numbers to indicate its valence, but such summaries are too limited to be of general interest.

Note to Students

A summary of the combining behavior of atoms more generally applicable than valence numbers is presented in Sec. 3-18.

The capacity of an atom to combine with other atoms is determined by the number and nature of the electrons in the atom. The name **valence electron** is applied to those electrons in an atom which have the greatest influence on the combining behavior of the atom; in the atoms under consideration these are the outermost electrons, those specifically represented in the electronic symbols.

The electronic symbols of several different atoms may have the same number of valence electrons; what does this mean? Consider, for example, the atoms and their electronic symbols shown in the margin. The seven valence electrons in each of these electronic symbols mean that all four atoms will have similar combining behavior. Each of these atoms has a strong tendency, for example, to combine with one hydrogen atom, as illustrated by the formulas HF, HCl, HBr, and HI.

Thus the valence electrons are of special significance in correlating the combining behavior of many atoms. Since the electronic symbols specify the number of valence electrons for an atom, they are very useful in representing the combinations of atoms found in molecules.

Electronic symbols of four atoms with the same number of valence electrons

Atom	Electronic symbol
Fluorine	:Ḟ·
Chlorine	:C̈l·
Bromine	:B̈r·
Iodine	:Ï·

We can obtain the electronic symbols of the 20 simplest atoms from Table 2-3. Let us set down these symbols in the following way:

H·   He:

Li·   ·Be·   ·Ḃ·   ·Ċ·   ·Ṅ·   :Ö·   :Ḟ·   :N̈e·

Na·   ·Mg·   ·Äl·   ·Si·   ·Ṗ·   :S̈·   :Cl·   :Är:

K·   ·Ca·

It is possible to arrange symbols with the same number of valence electrons in vertical columns in a repeating sequence. Most of the atoms in a vertical group have similar combining behavior.

Long before the electronic nature of matter was recognized, attempts were made to organize atoms according to their combining behavior and other properties. In these early attempts attention was focused on the weights of atoms. We now realize that the total number of electrons in an atom is a better basis for organization than the weights of atoms. The number of electrons in an atom corresponds to the atomic number (Sec. 2-4), and correlations of the combining behavior of atoms now focus on the atomic numbers of atoms. Table 3-1 gives the atomic numbers for all atoms arranged according to increasing number, and Table 3-2 gives them arranged alphabetically by atom name.

It is very hard to figure out completely satisfactory ways of arranging atoms according to their combining behavior. One common arrangement (Table 3-3) is derived from the so-called long form of the periodic table of elements. The table is called a periodic table because the arrangement reflects the periodic recurrence of similar combining properties when the atoms are arranged in order of their atomic numbers. The atoms are organized into groups and periods. The **groups** are the vertical columns of atoms identified by roman numerals at the top. The **periods** are the horizontal rows of atoms identified by the arabic numerals at the left.

To focus on atoms of special interest for our present purposes we can use a separated arrangement of atoms given in Table 3-4; the principal focus is on a group of representative atoms, and the others are arranged in a supplementary manner. Our discussion of the combining behavior of atoms will be confined to the representative atoms whose electronic symbols are shown in color in Table 3-5.

Note to Students

In the discussions of formulas that follow attention will be focused on the atoms in color in Table 3-5. It would be a good idea to examine their electronic symbols carefully and to try to fix them in mind, because they are going to be used repeatedly.

Table 3-1
Atomic Numbers of Atoms in Numerical Order

Atom	Atomic symbol	Atomic number	Atom	Atomic symbol	Atomic number	Atom	Atomic symbol	Atomic number
Hydrogen	H	1	Krypton	Kr	36	Lutetium	Lu	71
Helium	He	2	Rubidium	Rb	37	Hafnium	Hf	72
Lithium	Li	3	Strontium	Sr	38	Tantalum	Ta	73
Beryllium	Be	4	Yttrium	Y	39	Tungsten	W	74
Boron	B	5	Zirconium	Zr	40	Rhenium	Re	75
Carbon	C	6	Niobium	Nb	41	Osmium	Os	76
Nitrogen	N	7	Molybdenum	Mo	42	Iridium	Ir	77
Oxygen	O	8	Technetium	Tc	43	Platinum	Pt	78
Fluorine	F	9	Ruthenium	Ru	44	Gold	Au	79
Neon	Ne	10	Rhodium	Rh	45	Mercury	Hg	80
Sodium	Na	11	Palladium	Pd	46	Thallium	Tl	81
Magnesium	Mg	12	Silver	Ag	47	Lead	Pb	82
Aluminum	Al	13	Cadmium	Cd	48	Bismuth	Bi	83
Silicon	Si	14	Indium	In	49	Polonium	Po	84
Phosphorus	P	15	Tin	Sn	50	Astatine	At	85
Sulfur	S	16	Antimony	Sb	51	Radon	Rn	86
Chlorine	Cl	17	Tellurium	Te	52	Francium	Fr	87
Argon	Ar	18	Iodine	I	53	Radium	Ra	88
Potassium	K	19	Xenon	Xe	54	Actinium	Ac	89
Calcium	Ca	20	Cesium	Cs	55	Thorium	Th	90
Scandium	Sc	21	Barium	Ba	56	Protactinium	Pa	91
Titanium	Ti	22	Lanthanum	La	57	Uranium	U	92
Vanadium	V	23	Cerium	Ce	58	Neptunium	Np	93
Chromium	Cr	24	Praseodymium	Pr	59	Plutonium	Pu	94
Manganese	Mn	25	Neodymium	Nd	60	Americium	Am	95
Iron	Fe	26	Promethium	Pm	61	Curium	Cm	96
Cobalt	Co	27	Samarium	Sm	62	Berkelium	Bk	97
Nickel	Ni	28	Europium	Eu	63	Californium	Cf	98
Copper	Cu	29	Gadolinium	Gd	64	Einsteinium	Es	99
Zinc	Zn	30	Terbium	Tb	65	Fermium	Fm	100
Gallium	Ga	31	Dysprosium	Dy	66	Mendelevium	Md	101
Germanium	Ge	32	Holmium	Ho	67	Nobelium	No	102
Arsenic	As	33	Erbium	Er	68	Lawrencium	Lr	103
Selenium	Se	34	Thulium	Tm	69	Kurchatovium	Ku	104
Bromine	Br	35	Ytterbium	Yb	70	Hahnium	Ha	105

Table 3-2
Atomic Numbers of Atoms in Alphabetical Order of Name

	Symbol	Atomic number		Symbol	Atomic number		Symbol	Atomic number
Actinium	Ac	89	Calcium	Ca	20	Fermium	Fm	100
Aluminum	Al	13	Californium	Cf	98	Fluorine	F	9
Americium	Am	95	Carbon	C	6	Francium	Fr	87
Antimony	Sb	51	Cerium	Ce	58	Gadolinium	Gd	64
Argon	Ar	18	Cesium	Cs	55	Gallium	Ga	31
Arsenic	As	33	Chlorine	Cl	17	Germanium	Ge	32
Astatine	At	85	Chromium	Cr	24	Gold	Au	79
Barium	Ba	56	Cobalt	Co	27	Hafnium	Hf	72
Berkelium	Bk	97	Copper	Cu	29	Hahnium	Ha	105
Beryllium	Be	4	Curium	Cm	96	Helium	He	2
Bismuth	Bi	83	Dysprosium	Dy	66	Holmium	Ho	67
Boron	B	5	Einsteinium	Es	99	Hydrogen	H	1
Bromine	Br	35	Erbium	Er	68	Indium	In	49
Cadmium	Cd	48	Europium	Eu	63	Iodine	I	53

Table 3-2
Atomic Numbers of Atoms in Alphabetical Order of Name (*Continued*)

	Symbol	Atomic number		Symbol	Atomic number		Symbol	Atomic number
Iridium	Ir	77	Osmium	Os	76	Silver	Ag	47
Iron	Fe	26	Oxygen	O	8	Sodium	Na	11
Krypton	Kr	36	Palladium	Pd	46	Strontium	Sr	38
Kurchatovium	Ku	104	Phosphorus	P	15	Sulfur	S	16
Lanthanum	La	57	Platinum	Pt	78	Tantalum	Ta	73
Lawrencium	Lr	103	Plutonium	Pu	94	Technetium	Tc	43
Lead	Pb	82	Polonium	Po	84	Tellurium	Te	52
Lithium	Li	3	Potassium	K	19	Terbium	Tb	65
Lutetium	Lu	71	Praseodymium	Pr	59	Thallium	Tl	81
Magnesium	Mg	12	Promethium	Pm	61	Thorium	Th	90
Manganese	Mn	25	Protactinium	Pa	91	Thulium	Tm	69
Mendelevium	Md	101	Radium	Ra	88	Tin	Sn	50
Mercury	Hg	80	Radon	Rn	86	Titanium	Ti	22
Molybdenum	Mo	42	Rhenium	Re	75	Tungsten	W	74
Neodymium	Nd	60	Rhodium	Rh	45	Uranium	U	92
Neon	Ne	10	Rubidium	Rb	37	Vanadium	V	23
Neptunium	Np	93	Ruthenium	Ru	44	Xenon	Xe	54
Nickel	Ni	28	Samarium	Sm	62	Ytterbium	Yb	70
Niobium	Nb	41	Scandium	Sc	21	Yttrium	Y	39
Nitrogen	N	7	Selenium	Se	34	Zinc	Zn	30
Nobelium	No	102	Silicon	Si	14	Zirconium	Zr	40

Table 3-3
Attempted Organization of Atoms into Groups According to Combining Behavior

Group	I	II											III	IV	V	VI	VII	0
Period 1	H 1																	He 2
2	Li 3	Be 4											B 5	C 6	N 7	O 8	F 9	Ne 10
3	Na 11	Mg 12											Al 13	Si 14	P 15	S 16	Cl 17	Ar 18
4	K 19	Ca 20	Sc 21	Ti 22	V 23	Cr 24	Mn 25	Fe 26	Co 27	Ni 28	Cu 29	Zn 30	Ga 31	Ge 32	As 33	Se 34	Br 35	Kr 36
5	Rb 37	Sr 38	Y 39	Zr 40	Nb 41	Mo 42	Tc 43	Ru 44	Rh 45	Pd 46	Ag 47	Cd 48	In 49	Sn 50	Sb 51	Te 52	I 53	Xe 54
6	Cs 55	Ba 56	* 57–71	Hf 72	Ta 73	W 74	Re 75	Os 76	Ir 77	Pt 78	Au 79	Hg 80	Tl 81	Pb 82	Bi 83	Po 84	At 85	Rn 86
7	Fr 87	Ra 88	† 89–103	Ku 104	Ha 105													

*	La 57	Ce 58	Pr 59	Nd 60	Pm 61	Sm 62	Eu 63	Gd 64	Tb 65	Dy 66	Ho 67	Er 68	Tm 69	Yb 70	Lu 71
†	Ac 89	Th 90	Pa 91	U 92	Np 93	Pu 94	Am 95	Cm 96	Bk 97	Cf 98	Es 99	Fm 100	Md 101	No 102	Lr 103

Table 3-4
Relationship of the Representative Atoms to the Other Atoms†

Representative atoms

Period	Group							
	I	II	III	IV	V	VI	VII	0
1	H 1							He 2
2	Li 3	Be 4	B 5	C 6	N 7	O 8	F 9	Ne 10
3	Na 11	Mg 12	Al 13	Si 14	P 15	S 16	Cl 17	Ar 18
4	K 19	Ca 20	Ga 31	Ge 32	As 33	Se 34	Br 35	Kr 36
5	Rb 37	Sr 38	In 49	Sn 50	Sb 51	Te 52	I 53	Xe 54
6	Cs 55	Ba 56	Tl 81	Pb 82	Bi 83	Po 84	At 85	Rn 86
7	Fr 87	Ra 88						

Period										
4	Sc 21	Ti 22	V 23	Cr 24	Mn 25	Fe 26	Co 27	Ni 28	Cu 29	Zn 30
5	Y 39	Zr 40	Nb 41	Mo 42	Tc 43	Ru 44	Rh 45	Pd 46	Ag 47	Cd 48
6	* 57–71	Hf 72	Ta 73	W 74	Re 75	Os 76	Ir 77	Pt 78	Au 79	Hg 80
7	† 89–103	Ku 104	Ha 105							

*	La 57	Ce 58	Pr 59	Nd 60	Pm 61	Sm 62	Eu 63	Gd 64	Tb 65	Dy 66	Ho 67	Er 68	Tm 69	Yb 70	Lu 71
†	Ac 89	Th 90	Pa 91	U 92	Np 93	Pu 94	Am 95	Cm 96	Bk 97	Cf 98	Es 99	Fm 100	Md 101	No 102	Lr 103

† Adapted from Frank Brescia, John Arents, Herbert Meislich, and Amos Turk, "Fundamentals of Chemistry," 3d ed., p. 107, Academic Press, Inc., 1975, with permission.

Table 3-5
Electronic Symbols of the Representative Atoms

Period	Group							
	I	II	III	IV	V	VI	VII	0
1	H·							He:
2	Li·	·Be·	·B·	·C·	·N·	:O·	:F·	:Ne:
3	Na·	·Mg·	·Al·	·Si·	·P·	:S·	:Cl·	:Ar:
4	K·	·Ca·	·Ga·	·Ge·	·As·	:Se·	:Br·	:Kr:
5	Rb·	·Sr·	·In·	·Sn·	·Sb·	:Te·	:I·	:Xe:
6	Cs·	·Ba·	·Tl·	·Pb·	·Bi·	:Po·	:At·	:Rn:
7	Fr·	·Ra·						

THE DEVELOPMENT OF THE PERIODIC LAW: A STRIKING SIMULTANEOUS DISCOVERY†

Nine of the chemical elements have been known since ancient times: gold, silver, copper, iron, lead, tin, mercury, sulfur, and carbon. By the end of the nineteenth century discoveries had added another 14. By 1865 the number of recognized elements had risen to 64. As the number of known elements grew, many attempts were made to classify them according to their properties. At first they were grouped into families; chlorine, bromine and iodine, for example, were recognized as having similar properties. Most significantly, in 1865 the English chemist John Newlands called attention to the interesting relationship which results from placing the elements in a sequence according to their atomic weights. Every eighth element seemed to fall into a family with similar properties, although the correlation was not very convincing beyond the first 15 or so elements.

The efforts to classify the elements culminated in the proposals of the Russian chemist Dmitri Mendeleev (Fig. 3-1) and the German chemist Lothar Meyer. Working independently of each other, these men studied what was known about the properties of the elements and their compounds, paying particular attention to the combining behavior of atoms, atomic weights, and such properties as melting points, boiling points, and densities. They arrived at the remarkable conclusion that if the elements are arranged in order of increasing atomic weight, they fall into columns containing elements with similar properties. Meyer developed his table a little before Mendeleev but did not publish his work until after Mendeleev's textbook appeared in 1869.

For a table to include all 64 elements known at the time, it was necessary to leave blank spaces in some cases and to rearrange the order somewhat. Mendeleev speculated more thoroughly and confidently than Meyer and proclaimed that where the order had to be rearranged the assigned atomic weights were wrong and that where blank spaces had to be provided missing elements would be discovered. When some of the atomic weights were indeed found to be wrong and some of the missing elements were actually discovered, the validity of the arrangement became accepted. Indeed the order soon became recognized as one of the most useful and far-reaching generalizations in physical science. Although Mendeleev and Meyer formulated their original proposals independently, each man was influenced by the other in revising his ideas, and both acknowledged that they shared equally in the discovery.

The original periodic law stated that the properties of the elements vary periodically if the elements are arranged in order of their atomic weights. When the discovery of the electron and the nuclei of atoms at the turn of the century led to the development of new concepts of atomic structure, it became apparent that the chemical properties of an atom depend most directly on the number of electrons in the atom, which in turn is related to the atomic number. The periodic law was revised to state that the properties of the elements vary periodically if the elements are arranged in order of their atomic numbers. The original law was extremely provocative, however, and led, among other developments, to the discovery of 20 additional elements in about 30 years.

† Adapted, in part, from Charles Compton "An Introduction to Chemistry," pp. 69–74, Van Nostrand Reinhold Company, New York, 1958.

3-4 WHAT IS THE NATURE OF IONS?

Each atom contains equal numbers of protons and electrons and is neutral as a whole. Under certain circumstances, however, some neutral atoms can gain one or more electrons to produce particles with one or more negative charges called **negative ions.** Under other circumstances some neutral atoms can lose one or more electrons to become particles with one or more positive charges called **positive ions.**

The fluorine atom, $:\overset{..}{\underset{..}{F}}\cdot$, for example, can acquire an electron (see margin) to form a negative ion with a single negative charge, called a *fluoride ion;* the symbol for it is $[:\overset{..}{\underset{..}{F}}:]^-$ or simply F^-. The arrow used in a chemical equation is read "becomes" or "yields." The fluoride ion has a single negative charge since it has one more electron than it has protons.

In another example, a sodium atom can lose an electron to become a positive ion with a single positive charge (margin). The sodium ion has a single positive charge, since it has one less electron than it has protons. The term **ion** then refers to a charged particle formed from an atom through the gain or the loss of one or more electrons. The process of ion formation is called **ionization.**

$:\overset{..}{\underset{.}{F}}\cdot$

Neutral
fluorine
atom,
9 protons and
9 electrons

$+$

e^-

Electron

$[:\overset{..}{\underset{..}{F}}:]^-$

Negative
fluoride
ion,
9 protons and
10 electrons

The relationship between the
neutral fluorine atom and the
negative fluoride ion

Na·

Neutral
sodium
atom,
11 protons and
11 electrons

$\downarrow$

Na$^+$

Positive
sodium
ion,
11 protons and
10 electrons

$+$

e^-

Electron

The relationship between the
neutral sodium atom and the
positive sodium ion

3-5 THE OCTET RULE

The American chemist Gilbert N. Lewis (Fig. 3-2) first used electronic symbols in 1917 to correlate the combining behavior of atoms. In generalizing this behavior in terms of electronic symbols, Lewis and his American colleague, Irving Langmuir (Fig. 3-3), paid special attention to the elements known as noble gases. These are the elements listed in the last column on the right of the periodic table, known as group 0: (reading downward) helium, He; neon, Ne; argon, Ar; krypton, Kr; xenon, Xe; and radon, Rn. Originally these elements were thought to be completely unreactive and their atoms to have no tendency whatsoever to combine with any other atoms. (It was probably fortunate that the chemical reactivity of krypton, xenon, and radon was not discovered until much later.) Lewis and Langmuir noted that the atoms of each of these (presumed) unreactive elements have eight valence electrons (or two for the simplest, helium), and they associated with the numbers 2 and 8 an unusual stability.

The cornerstone of this approach was called the **octet rule:** atoms with fewer than eight valence electrons have a strong tendency to achieve a total of eight valence electrons (or two for very simple atoms), and this tendency underlies the combining behavior of atoms. For example, the chlorine atom, with seven valence electrons, $:\overset{..}{\underset{.}{Cl}}\cdot$ (electron distribution in shells 2, 8, 7), forms a more stable species by gaining an electron to attain eight valence electrons and in the process becomes a chloride ion with a negative charge, $[:\overset{..}{\underset{..}{Cl}}:]^-$ (electron distribution by shells 2, 8, 8):

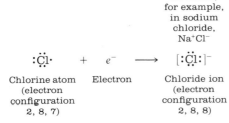

:C̈l· + e^- ⟶ [:C̈l:]⁻

| Chlorine atom (electron configuration 2, 8, 7) | Electron | Chloride ion (electron configuration 2, 8, 8) |

As found, for example, in sodium chloride, Na⁺Cl⁻

Conversion of a chlorine atom into a chloride ion; note that there are eight valence electrons in the chloride ion

The sodium atom, Na·, with one valence electron (electron distribution in shells 2, 8, 1), forms a more stable species by somehow losing an electron to attain eight electrons in the outermost shell and in the process becomes a sodium ion with a positive charge, Na⁺ (electron distribution in shells 2, 8).

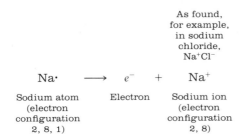

As found, for example, in sodium chloride, Na⁺Cl⁻

Na· ⟶ e^- + Na⁺

| Sodium atom (electron configuration 2, 8, 1) | Electron | Sodium ion (electron configuration 2, 8) |

Conversion of a sodium atom into a sodium ion; there are eight valence electrons in the sodium ion, even though (by convention) they are not represented explicitly

The oxygen atom, with six valence electrons, :Ö· (electron distribution in shells 2, 8, 6) forms a more stable species by gaining two electrons to attain eight electrons and in the process becomes an oxide ion with two negative charges, [:Ö:]²⁻ (electron distribution in shells 2, 8, 8):

As found, for example, in calcium oxide, Ca²⁺O²⁻

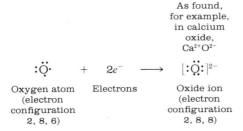

:Ö· + $2e^-$ ⟶ [:Ö:]²⁻

| Oxygen atom (electron configuration 2, 8, 6) | Electrons | Oxide ion (electron configuration 2, 8, 8) |

Conversion of an oxygen atom into an oxide ion; note the eight valence electrons in the ion

The calcium atom, with two valence electrons, ·Ca· (electron distribution in shells 2, 8, 8, 2), forms a more stable species by losing two electrons to attain eight electrons in the outermost orbit

FIGURE 3-1
Dmitri Mendeleev (1834–1907), a Russian chemist best known for his development of a useful periodic table and the periodic law. His technological interests ranged widely and included speculations on the origin of petroleum, the behavior of gases, and industrial chemical production. (*Smithsonian Institution.*)

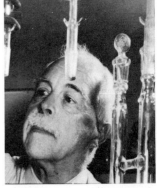

FIGURE 3-2
Gilbert N. Lewis (1875–1946), an American chemist whose contributions to our knowledge of atomic structure and thermodynamics are landmarks in the development of chemistry. (*Courtesy of Bancroft Library, University of California, Berkeley.*)

and in the process becomes a calcium ion with two positive charges, Ca^{2+} (electron distribution in shells 2, 8, 8):

As found, for example, in calcium oxide, $Ca^{2+}O^{2-}$

$$\cdot Ca \cdot \longrightarrow 2e^{-} + Ca^{2+}$$

| Calcium atom (electron configuration 2, 8, 8, 2) | Electron | Calcium ion (electron configuration 2, 8, 8) |

Conversion of a calcium atom into a calcium ion; there are eight valence electrons in the ion even though (by convention) they are not represented explicitly

$$H \cdot + H \cdot \longrightarrow H:H$$

Hydrogen atoms   Hydrogen molecule

Relationship between the structures of two isolated hydrogen atoms and a hydrogen molecule

The hydrogen atom, $H \cdot$, with one valence electron, forms a more stable species by combining with another hydrogen atom to form a molecule in which the hydrogen atoms, by sharing electrons, are both surrounded by the stable number of two electrons (margin).

The nitrogen atom with five valence electrons, $\cdot \ddot{N} \cdot$ (electron distribution in shells 2, 5), forms a more stable species by combining with three hydrogen atoms to form a molecule of ammonia, in which the nitrogen atom, by sharing electrons, is sur-

---

WHAT IS WATER FLUORIDATION?

In the early part of this century dentists in Colorado succeeded in tracing the cause of mottled, stained teeth prevalent among people in certain Colorado areas to a factor in drinking water. It was also observed that people with such stained teeth were less susceptible to tooth decay. By 1931 the cause of the discolored teeth and the resistance to tooth decay was found to be a very small concentration of fluoride ion, F^-, in the drinking water. Further investigation showed that the maximum benefit of the fluoride ion in the prevention of tooth decay was obtained at a concentration of about 1 gram of fluoride ion in 1 million grams of water. If the concentration becomes 2 or 3 times greater than this, teeth become mottled. In the western part of the United States some drinking water was found to have a fluoride ion concentration 6 times greater without any known harm other than the brown tooth stains. In certain regions of India, where the fluoride ion concentration in drinking water has been found to be 10 grams in 1 million grams, some people suffer from abnormal bone growth.

In the late 1930s fluoride ion at a concentration of 1 gram in 1 million grams was added to the drinking water in a few communities on a trial basis, notably in Grand Rapids, Michigan. After 7 or 8 years these experiments showed that tooth decay among children, especially those around the ages of 6 or 7 whose teeth are still being formed, was reduced significantly, up to 60 percent. As a result the fluoridation of public water supplies to this level was officially endorsed by the United States Public Health Service. The fluoridation is achieved by adding sources of fluoride ion such as sodium fluoride, NaF, calcium fluoride, CaF_2, and fluorosilicic acid, H_2SiF_6. Very careful control of the concentration is easily maintained because of the perfection of analytical procedures. The drinking water of over 80 million people in the United States alone is being fluoridated.

Fluoridation of water supplies has often become a controversial issue because evidence of its safety has not satisfied everyone. Concern has been expressed particularly about the effect of even a very small fluoride ion concentration on those who suffer from certain illnesses. Fluoridation has been approved and recommended, however, by the American Dental Association and the American Medical Society, as well as the United States Public Health Service. These authorities consider it to be safe for everyone, regardless of age, health, or the amount of water consumed.

rounded by eight electrons (margin). Note that in the ammonia molecule each hydrogen atom shares electrons, so that it is surrounded by the stable number of two electrons.

The tendency for atoms to achieve electron configurations of the noble gas atoms was considered one of the principal factors in determining the combining properties of atoms. Stable arrangements are attained by giving up, gaining, or sharing electrons. Thus atoms may be linked together by sharing electrons or converted into ions whose positive and negative charges attract each other.

It was soon realized, however, that arrangements other than eight electrons correspond to stable structural situations for many atoms. Atoms of periods 3 and 4 (horizontal rows of the periodic table) can acquire up to 12 electrons in their outermost orbit, and atoms of the fifth period and beyond can acquire up to 16. The octet rule holds, however, for atoms of the first two periods. Accordingly, the maximum number of electrons possible around any of these atoms in stable arrangements is 8. (For the simplest atoms, hydrogen and helium, it is 2.) Remember that these are *maximum* numbers; stable arrangements with fewer electrons are possible.

Although the process of gaining and losing electrons involves factors besides attaining stable numbers of valence electrons, and although the process of sharing electrons is rather complex, the ideas of chemical bonding and formula notation proposed by Lewis and Langmuir led to important clarifications. With some modification they are still useful today.

3-6 WHAT IS THE NATURE OF IONIC BONDS?

Many substances occur naturally in the form of ions. Common table salt (sodium chloride) has a structure containing a three-dimensional repeating pattern of positive sodium ions and negative chloride ions in equal numbers (Fig. 3-4). The bonding which holds the arrangement together is called ionic (or electrovalent) bonding. An **ionic bond** operates between ions and arises from the forces of attraction between positive and negative ions, known as electrostatic attraction. The electrostatic forces of the ionic bonds operate between all the neighboring ions of opposite charge. A diagram of the sodium chloride structure reveals that each positive sodium ion is surrounded by, and attracted to, six negative chloride ions and each negative chloride ion is surrounded by, and attracted to, six positive sodium ions. No single positive ion can be thought of as being bonded to any single negative ion. The term **salt** is applied to substances made up of ions. The white crystalline substance calcium oxide is another example. It is composed of positive calcium ions alternating with negative oxide ions in a three-dimensional lattice arrangement similar to that in sodium chloride. Thousands of different salts are known.

Adding or subtracting a single electron from the fundamental particle of a substance has a profound effect on its properties. A sodium atom, for example, differs from a sodium ion only in having one more electron. Sodium *atoms*, as found in solid elemental sodium, will react with water and many other forms of matter with great violence. Sodium *ions*, as found in table salt,

·N̈·

Nitrogen atom
(electron
configuration
2,5)

+

H· + H· + H·

Hydrogen atoms

↓

H:N̈:H
  H

Ammonia
molecule

Relationship between the structures of an isolated nitrogen atom and three isolated hydrogen atoms and an ammonia molecule

FIGURE 3-3
Irving Langmuir (1881–1957), an American chemist who made basic contributions to many fields of chemistry. He did pioneering work on the structure of atoms and molecules, the fundamental properties of liquid and solid surfaces, the behavior of catalysts, and the seeding of clouds to produce rainfall. (*Courtesy of General Electric Company.*)

FIGURE 3-4
Schematic representation of a portion of a sodium chloride crystal. The larger spheres represent chloride ions, Cl⁻; the smaller spheres represent sodium ions, Na⁺. Each sodium ion is surrounded by (and attracted to) six chloride ions, and each chloride ion is surrounded by (and attracted to) six sodium ions.

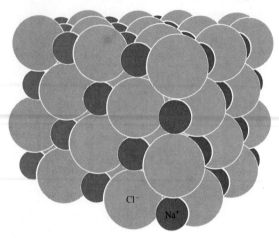

Cl⁻

Na⁺

on the other hand, can be swallowed in complete safety. Indeed, they are a necessary component of the human bloodstream.

H:H

Representation of covalent bond between the two hydrogen atoms in a molecule of hydrogen; note that two electrons are involved

H:N̈:H
H

Representation of the three covalent bonds between the nitrogen atom and each of the three hydrogen atoms in an ammonia molecule; two electrons are involved in each bond

:Ö::C::Ö:

Representation of the two covalent bonds between the carbon atom and each of the two oxygen atoms in a molecule of carbon dioxide; four electrons are involved in each bond

3-7 WHAT IS THE NATURE OF COVALENT BONDS?

A **covalent bond** is a link between two atoms that results from their mutual attraction for electrons they share between them. The covalent bond between the two atoms of a hydrogen molecule has already been discussed (Sec. 2-6). An electronic formula represents this bond as shown in the margin.

Brief mention of the electron-sharing bonds in the ammonia molecule was made in Sec. 3-5. An electronic formula represents these three hydrogen atom-nitrogen atom bonds as shown in the margin. In each of these examples the covalent bonds involve only two electrons. Covalent bonds may also involve four and six electrons. Consider the bonds between the oxygen and carbon atoms in a molecule of the gas carbon dioxide, each of which involves four electrons (margin). The covalent bond between the two nitrogen atoms in the nitrogen molecule involves six electrons (opposite margin).

The name **nonbonding electron** is applied to valence electrons in a molecule which are not involved in interatomic bonds. The formula for the carbon dioxide molecule, for example, indicates that there are eight nonbonding electrons:

Nonbonding electrons ⟶ :Ö: :C: :Ö: ⟵ Nonbonding electrons

Electronic formula of carbon dioxide, showing nonbonding electrons

Interesting examples of covalent bonds are found in the molecule of water, H_2O. The oxygen atom is bonded to both hydrogen atoms through electron sharing, as the electronic formula in the margin suggests.

We have seen (Sec. 3-6) that in the ionic bonds of a salt the positive and negative ions are held together by the mutual attraction of their opposite charges, but each positive ion is attracted to all the negative ions around it and each negative ion is attracted to all the positive ions around it. The bonds formed by covalent bonds, on the other hand, link two specific atoms. Furthermore, covalent bonds have a definite length and a definite direction. The two covalent bonds in the water molecule give the molecule a definite size and shape. The center of each hydrogen atom is 0.96 angstrom unit (Sec. 2-4) from the center of the oxygen atom, and the molecule has a V shape, with an angle of 104.5° between the bonds (Fig. 3-5).

3-8 A SUMMARY OF INTERATOMIC BONDS

A. Ionic Bond

This type of bond results from the mutual attraction between positive and negative ions, as found in the three-dimensional lattice of salts. In formulas, however, only the overall ratio of positive and negative ions is given. The ratio of negative to positive ions is always such that the total positive and negative *charges* are equal. This fact can be used to determine the ratio of the individual positive and negative ions in a salt.

B. Covalent Bonds

A **single covalent bond** involves only one pair of electrons (Table 3-6). Although electrons in formulas are attributed to specific atoms for clarity here, remember that electrons in actual molecules cannot be identified this way. A **double covalent bond** involves two pairs of electrons, and a **triple covalent bond** involves three pairs of electrons (Table 3-6).

3-9 WHAT ARE THE CHARACTERISTICS OF ELECTRONIC FORMULAS OF MOLECULES?

The general features of electronic formulas can be shown if we examine the formula for a relatively simple substance, ethyl alcohol. This compound is a colorless liquid which is responsible for the physiological effect of alcoholic beverages. Its molecule contains two carbon atoms, an oxygen atom, and six hydrogen atoms, joined by single covalent bonds in the sequence shown in the margin. The following features of this formula apply to the electronic formulas of many molecules:

1. There are eight valence electrons around the oxygen atom and each carbon atom, and there are two valence electrons around each hydrogen atom. In arriving at these totals, the valence electrons involved in the bonds are counted twice,

55

3-9 WHAT ARE THE CHARACTERISTICS OF ELECTRONIC FORMULAS OF MOLECULES?

:N:::N:

Representation of the covalent bond between the two nitrogen atoms in a nitrogen molecule; six electrons are involved

H:Ö:H

Electronic formula of the water molecule

Na^+Cl^- or   NaCl

Sodium chloride

Ba^{2+} F^- F^- or BaF_2

Barium fluoride

Examples of ionic-bond representation

  H H
H:C̈:C̈:Ö:H
  H H

Electronic formula of a molecule of ethyl alcohol

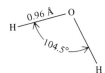

FIGURE 3-5
The shape of a molecule of water. Covalent bonds have a definite length and a definite direction.

once for each of the atoms which they join. These numbers of electrons must not be exceeded: two electrons for hydrogen and eight for each of the other atoms.

2. All the valence electrons of the component electronic symbols have been used but no more than these. The component symbols for six hydrogen atoms, two carbon atoms, and one oxygen atom give a total of 20 electrons. There are 20, no more and no less, in the formula. This is an essential feature of a correct electronic formula, i.e., one that corresponds to the actual structure of a molecule.

3. All the valence electrons in the formula are involved in interatomic bonds except two of the pairs around the oxygen. The electrons involved in bonds are called bonding electrons. The two pairs around the oxygen not involved in the interatomic bonds are called nonbonding electrons (Sec. 3-7).

4. There are an even number of valence electrons in the formula, and they are represented in pairs. Most molecules contain an even number of total electrons and an even number of valence electrons, but this does not have to be the case. (See Example 7, Sec. 3-11.)

3-10 WHAT ARE THE CHARACTERISTICS OF STRUCTURAL FORMULAS?

Since electronic formulas involve writing so many dots, they are usually replaced by **structural formulas,** in which bonding pairs of electrons are represented by dashes, but electronic formulas should always be considered to be the parent formulas from which the structural formulas are derived.

In structural formulas each bonding pair of electrons is replaced by a dash. To keep the notation as simple as possible, nonbonding electrons are often not represented. (There is no

Table 3-6
Examples of Covalent Bonds

Example	Electronic symbols of separated atoms	Notation relating electrons to electronic symbols	Conventional electronic notation
Single: Fluorine	:F· °F°	:F: F:	:F:F:
Hydrogen fluoride	H· °F°	H:F:	H:F:
Water	H· ·O° ×H	H:O ×H	H:O:H
Double: C=C bond in ethylene	·C· ×C× H° H° H° H°	H:C:×C×H H H	H:C::C:H H H
Triple: C≡C bond in acetylene	·C· ×C× H° H°	H° C:×××C×H	H:C:::C:H

general rule on this point, and for many purposes it is helpful to show them.) When they are shown, each nonbonding electron is represented by a dot. The examples in Table 3-7 compare structural formulas with electronic formulas.

3-11 CONVENTIONS OF FORMULA WRITING; FORMULAS OF MOLECULES

EXAMPLE 1 Consider the electronic formula for the gas hydrogen bromide, H:B̈r:.

Comments The number of valence electrons is correct (one from the hydrogen atom, seven from the bromine atom; total, eight). The octet rule is obeyed. The bromine atom is surrounded by its maximum, eight electrons; the hydrogen atom is surrounded by its maximum, two electrons. Conclusion: this is an acceptable formula.

Note to Students	The conventions of formula writing have become established because they lead to symbolic representations which correspond to the actual structure of molecules. Although we have tried to keep formulas as simple as possible, the structures of molecules are inherently complex.

Table 3-7
Comparisons of Electronic and Structural Formulas

		Structural formula	
Substance	Electronic formula	Nonbonding electrons represented	Nonbonding electrons not represented
Hydrogen	H:H	H—H	H—H
Hydrogen fluoride	H:F̈:	H—F̈:	H—F
Fluorine	:F̈:F̈:	:F̈—F̈:	F—F
Water	H:Ö:H	H—Ö—H	H—O—H
Carbon dioxide	:Ö::C::Ö:	:Ö=C=Ö:	O=C=O
Nitrogen	:N:::N:	:N≡N:	N≡N
Ethyl alcohol	H H H:C̈:C̈:Ö:H H H	H—C—C—Ö—H (with H's)	H—C—C—O—H (with H's)
Ethylene	H:C::C:H Ḧ Ḧ	H—C=C—H (with H's)	H—C=C—H (with H's)
Acetylene	H:C:::C:H	H—C≡C—H	H—C≡C—H

EXAMPLE 2 Consider the structural formula for the gas hydrogen bromide, H—Br.

Comments Relating this formula to the corresponding electronic formula given in Example 1 above, we see that the bonding pair of electrons has been correctly represented by a dash. Conclusion: this is an acceptable formula.

EXAMPLE 3 Consider the following electronic formula for the gas hydrogen sulfide, H:$\ddot{\text{S}}$:H.

Comments The number of valence electrons is correct (one from each of the hydrogen atoms and six from the sulfur atom; total, eight). The octet rule is obeyed. The sulfur atom is surrounded by its maximum, eight electrons; the hydrogen atoms are each surrounded by the maximum, two electrons. Conclusion: this is an acceptable formula. The equivalent structural formula is H—S—H.

EXAMPLE 4 Consider the following structural formula for the gaseous element chlorine, Cl—Cl.

Comments In order to evaluate a structural formula it must be translated into the corresponding electronic formula, :$\ddot{\text{Cl}}$:$\ddot{\text{Cl}}$:. The number of valence electrons is correct (seven from each of the chlorine atoms; total 14). The octet rule is obeyed. Each chlorine atom is surrounded by its maximum, eight electrons. Conclusion: the original structural formula is acceptable.

EXAMPLE 5 Consider the following electronic formula for the gas cyanogen fluoride, :$\ddot{\text{F}}$:C:::N:.

Comments The number of valence electrons is correct (seven from the fluorine atom, four from the carbon atom, five from the nitrogen atom; total, sixteen). The octet rule is obeyed. Each atom is surrounded by its maximum, eight. Conclusion: this formula is acceptable. The corresponding structural formula is F—C≡N.

Note to Students

Formulas are fundamental to any discussion of chemistry but it takes some practice to become at ease with the notational details. These examples and additional ones in end-of-chapter exercises will give you some of the necessary practice.

EXAMPLE 6 Consider the structural formula for the gas phosgene,

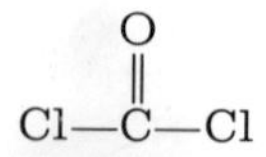

$$\begin{array}{c} \text{O} \\ \parallel \\ \text{Cl—C—Cl} \end{array}$$

Comments In order to evaluate this formula it must be translated into the corresponding electronic formula:

$$:\overset{..}{O}:$$
$$:\overset{..}{\underset{..}{Cl}}:\overset{..}{\underset{..}{C}}:\overset{..}{\underset{..}{Cl}}:$$

59

3-12 WHAT ARE THE
CHARACTERISTICS OF FORMULAS
FOR IONIC SUBSTANCES?

The number of valence electrons is correct (seven from each chlorine atom, four from the carbon atom, and six from the oxygen atom; total, twenty-four). The octet rule is obeyed. Each atom is surrounded by its maximum, eight. Conclusion: the original structural formula is acceptable.

EXAMPLE 7   Consider the following electronic formula for the gas nitric oxide $:N::\overset{..}{O}:$.

Comments   The most striking thing about this formula is that it contains an odd number of electrons (eleven). This is unusual but by no means impossible. The number of valence electrons is correct (six from the oxygen atom, five from the nitrogen atom; total, eleven). The number of electrons around the oxygen atom obeys the octet rule. The seven electrons around the nitrogen atom are one fewer than the octet rule, but this is acceptable. Although our discussions have not emphasized the point, the octet of electrons is a maximum not a minimum. Conclusion: the formula, while unusual, is acceptable. The corresponding structural formula is N=O.

3-12 WHAT ARE THE CHARACTERISTICS OF FORMULAS FOR IONIC SUBSTANCES?

Many ions contain only one atomic nucleus, e.g., the chloride ion, Cl^-, the oxide ion, O^{2-}, the sodium ion, Na^+, and the calcium ion, Ca^{2+}. The structure of ions, however, may be much more complex. Consider, for example, the nitrate ion found in sodium nitrate, among other substances. (Sodium nitrate, a white crystalline salt, is widely used as an agricultural fertilizer. It is one of the substances used to renew the nitrogen content of the soil.) The nitrate ion has a single negative ionic charge and is composed of a nitrogen and three oxygen atoms. The electronic formula is shown in the margin. Note the following features:

Electronic formula of the nitrate ion; see text for an analysis of the symbols

1. The number of valence electrons around the atoms in no case exceeds eight.

2. The electronic symbols of the component atoms are

$$\cdot\overset{..}{N}\cdot \qquad :\overset{..}{\underset{.}{O}}\cdot \qquad :\overset{..}{\underset{.}{O}}\cdot \qquad :\overset{..}{\underset{.}{O}}\cdot$$

The sum of the valence electrons is twenty-three. The total number of valence electrons in the formula is equal to the sum of the valence electrons of the component atoms *plus* the one electron which is responsible for the overall ionic charge of −1. This electron did not "belong" to any of the component atoms originally but was somehow acquired by the group of atoms as a whole. All these electrons appear in the electronic formula.

3. The brackets placed around the ion are an important part of the notation. The ionic charge (in this case, −1) is shown

outside the bracket to the upper right. Ionic charges refer to electrons gained or lost by the atom or group of atoms as a whole.

Let us examine the electronic formula for the salt sodium nitrate. The following features should be noted:

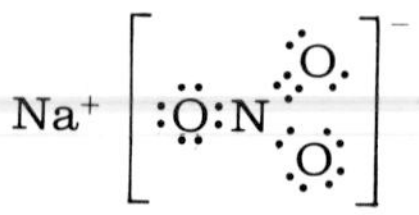

1. Since salts exist as three-dimensional arrays of positive and negative ions, no discrete molecules are found. The formulas therefore indicate only the ratio of positive and negative ions.

2. The number of positive ionic charges equals the number of negative ionic charges. The ionic charge of the ion is shown outside the brackets, as before.

3. The total number of valence electrons in the formula equals the total number in the component electronic symbols. The valence electrons of the component electronic symbols add up to 24, and 24 electrons appear in the electronic formula. From a bookkeeping point of view, the valence electron implied by the negative charge on the ion may be considered to be an electron which is "transferred" from the sodium atom, Na·, to the atoms of the nitrate ion to supply the "extra" electron there. Sodium nitrate, however, may actually be formed by the rather simple coming together of sodium and nitrate ions.

4. A bond involving the Na^+ ion should never be indicated by a dash, since a dash always indicates a covalent bond.

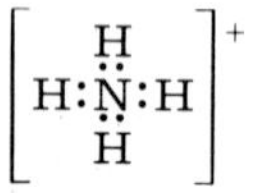

An example of a positive ion containing more than one atom is the ammonium ion, found in ammonium nitrate, a white crystalline salt widely used as a fertilizer. Note the following features:

1. The electronic symbols of the component atoms are

$$H· \quad H· \quad H· \quad H· \quad ·\ddot{N}·$$

The sum of the valence electrons is nine.

2. The total valence electrons represented in the electronic formula (eight) is equal to the sum of the valence electrons in the component electronic symbols (nine), *minus* the one electron responsible for the overall positive ionic charge.

3. The nitrogen atom is surrounded by the maximum of eight electrons. Each hydrogen atom is surrounded by the maximum of two electrons.

4. Using the electronic formulas for the ammonium and nitrate ions the formula for the salt ammonium nitrate is

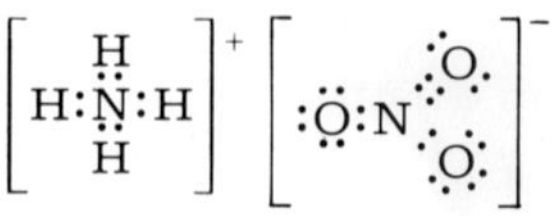

Electronic formula for sodium nitrate; see text for an analysis of the symbols

Electronic formula for the ammonium ion; see text for an analysis of the symbols

Electronic formula for ammonium nitrate; see text for an analysis of the symbols

As discussed in Sec. 3-10, structural formulas are derived from the corresponding electronic formulas. In devising a structural formula for the nitrate ion we therefore begin with the electronic formula discussed in Sec. 3-12. If we follow the conventions for writing structural formulas (Sec. 3-10), i.e., replacing bonding electron pairs with dashes, we arrive quickly at the structural formula shown.

The structural formula for the ammonium ion can be derived from the electronic formula, discussed in the preceding section. Again replacing the bonding pairs of electrons by dashes gives the structural formula shown. Putting together the structural formulas for the ammonium and nitrate ions gives the structural formula for the salt, ammonium nitrate:

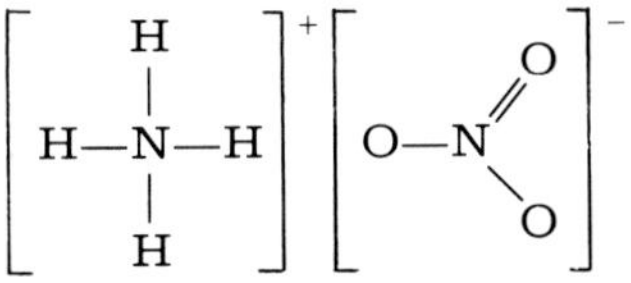

The structural formula for ammonium nitrate

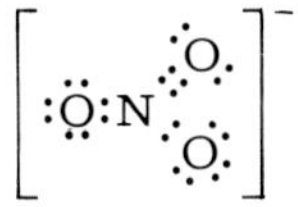

3-14 THE KINDS OF FORMULA NOTATION

The nature of electronic and structural formulas has occupied our attention to this point. While these formula notations are the most useful, they take up a lot of space, so simplified notations are often used. Compare the five common types of formulas shown for the molecular substance hydrazine, N_2H_4, in Table 3-8.

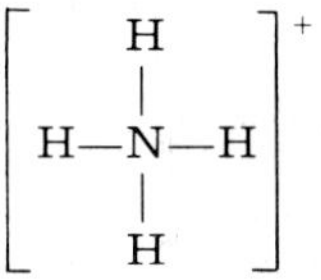

Electronic formula of the
nitrate ion

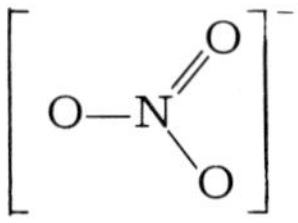

Structural formula for the
nitrate ion

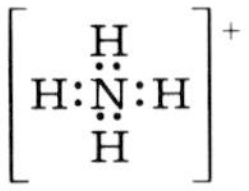

Electronic formula for the
ammonium ion

Structural formula for the
ammonium ion

Table 3-8
Kinds of Formulas

Type of formula	Formula	Comment
Electronic	H H H:N:N:H	Interatomic bonds represented by pairs of dots between adjacent atoms; all valence electrons shown
Structural	H H | | H—N—N—H	Simplification of the electronic formula; interatomic bonds represented by dashes
Condensed	H_2NNH_2	Condensation of the structural formula; interatomic bonds not specifically represented, but some indications of atom sequence retained
Molecular	N_2H_4	Only the numbers of each kind of atom in the molecules indicated
Empirical	NH_2	Only the ratio of the kinds of atoms in the molecule indicated

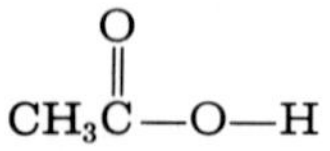

The convenience of the space-saving condensed formulas is apparent, but they sacrifice much important detail. Chemists often condense only part of a structure, retaining the full structural details for parts of the formula requiring special emphasis. Compare the two formulas for acetic acid shown in the margin. The partially condensed formula is frequently used.

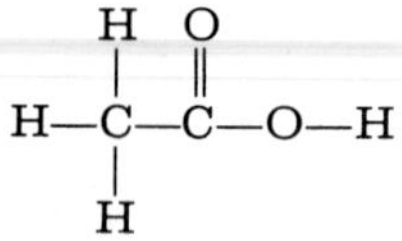

For salts only empirical formulas are ordinarily written. Remember (Sec. 3-6) that in a salt no individual positive ion can be thought of as being bonded to only one negative ion. Each positive ion is attracted to all of the negative ions which surround it, and each negative ion is attracted to all of the positive ions which surround it. The structure of a salt therefore consists of a three-dimensional pattern of positive and negative ions. It contains no individual molecules.

Since the salt contains no molecules, molecular formulas cannot be written. Formulas are limited to showing the relative numbers of positive and negative ions in the salt. The formula of the salt sodium bromide, Na^+Br^-, indicates only that the ratio of sodium ions to bromide ions in the salt is 1:1. The formula of the salt barium fluoride, $Ba^{2+}F^-F^-$, indicates that the salt contains two fluoride ions for each barium ion.

The formulas for salts can be represented in any of the notations Table 3-9 shows for the substances sodium chloride and barium nitrate.

$$CH_3\overset{\displaystyle O}{\overset{\displaystyle \|}{C}}{-}O{-}H$$

Partially condensed structural formula for acetic acid

$$H{-}\overset{\displaystyle H}{\underset{\displaystyle H}{\overset{\displaystyle |}{\underset{\displaystyle |}{C}}}}{-}\overset{\displaystyle O}{\overset{\displaystyle \|}{C}}{-}O{-}H$$

Structural formula for acetic acid

3-15 ADDITIONAL EXAMPLES OF THE FORMULAS OF SALTS

EXAMPLE 1 Consider the electronic formula for the salt potassium bromide, $K^+[:\overset{..}{\underset{..}{Br}}:]^-$.

Comments The number of valence electrons around the bromide ion is eight. The number of electrons around the potassium

Table 3-9
Formula Notations for Salts

	Sodium chloride	Barium nitrate	
Electronic formula	$Na^+[:\overset{..}{\underset{..}{Cl}}:]^-$	Ba^{2+} 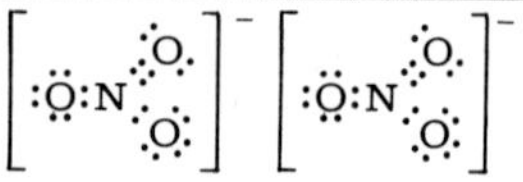	
Structural formula	$Na^+[Cl]^-$	Ba^{2+} 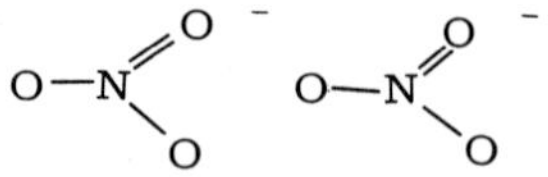	
Empirical formulating indicating ionic charges	Na^+Cl^-	$Ba^{2+}(NO_3^-)_2$†	
Empirical formula omitting ionic charges	NaCl	$Ba(NO_3)_2$†	

† The parentheses are used to enclose ions in empirical formulas only when the ion contains more than one atom.

ion is also eight, but by convention they are not represented. [When the potassium atom (electron configuration 2, 8, 8, 1) loses an electron to form a positive ion, the ion has an electron configuration of 2, 8, 8.] The number of positive ionic charges equals the number of negative ionic charges. The electronic symbols of the component atoms are $K\cdot$ and $:\ddot{B}r\cdot$. The sum of the valence electrons is therefore eight, and they all appear in the electronic formula. Note that brackets are placed around the bromide ion and the ionic charge is indicated outside the brackets. By convention brackets are not placed around a positive ion with only one atom. The structural formula for potassium bromide is $K^+[Br]^-$, and the empirical formula is K^+Br^-, or, more commonly, KBr.

EXAMPLE 2 The electronic formula of lithium hydroxide, $Li^+[:\ddot{O}:H]^-$ is of interest.

Comments The number of valence electrons around the oxygen atom in the hydroxide ion is eight. The hydrogen atom is surrounded by its maximum, two. The lithium ion is also surrounded by two electrons, although they are not represented. [When the lithium atom (electron configuration 2, 1) loses an electron to become a positive ion, the ion has an electron configuration of 2]. The number of positive ionic charges equals the number of negative ionic charges. The electronic symbols of the component atoms are $Li\cdot$, $:\ddot{O}\cdot$, and $H\cdot$. The sum of the valence electrons is therefore eight. All eight appear in the electronic formula. Note that brackets are placed around the atoms of the hydroxide ion and the ionic charge is indicated outside the brackets. By convention no brackets are placed around a positive ion containing only one atom. The structural formula of lithium hydroxide is $Li^+[O—H]^-$. The empirical formula is Li^+OH^- or, more commonly, LiOH.

EXAMPLE 3 The electronic formula for the substance ammonium fluoride provides an example of a positive ion which contains more than one atom.

$$\left[\begin{array}{c} H \\ H:\ddot{N}:H \\ H \end{array}\right]^+ \left[\;:\ddot{F}:\;\right]^-$$

Comments The number of valence electrons around the nitrogen atom and the fluoride ion is eight. The number around the hydrogen atoms is two. The number of positive ionic charges equals the number of negative ionic charges. The electronic symbols of the component atoms are $H\cdot$, $H\cdot$, $H\cdot$, $H\cdot$, $\cdot\ddot{N}\cdot$, and $:\ddot{F}\cdot$. The sum of the valence electrons is therefore sixteen. All these appear in the electronic formula. Note that brackets are placed around both the atoms in the ammonium ion and the fluoride ion and the ionic charges are indicated outside of the brackets. The structural formula of ammonium fluoride is

$$\left[\begin{array}{c} H \\ | \\ H—N—H \\ | \\ H \end{array}\right]^+ \left[\; F \;\right]^-$$

The empirical formula is $NH_4^+F^-$ or, more commonly, NH_4F.

EXAMPLE 4 The electronic formula for the substance sodium carbonate provides an example of a salt in which the numbers of positive and negative ions are not equal:

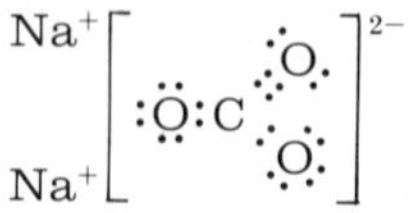

$$
\text{Na}^+ \left[\begin{array}{c} :\overset{..}{\underset{..}{O}}:C \overset{\overset{..}{O}..}{\underset{\underset{..}{O}:}{}} \end{array} \right]^{2-} \\
\text{Na}^+
$$

Comments The number of valence electrons around the carbon atom and each of the oxygen atoms is eight. The number around the sodium ion is also eight, but they are not represented. [When the sodium atom (electron configuration 2, 8, 1) loses an electron to form the sodium ion, the ion has an electron configuration of 2, 8.] The number of positive ionic charges equals the number of negative ionic charges. This requires two sodium ions for each carbonate ion. The electronic symbols of the component atoms are Na·, Na·, $:\overset{..}{\underset{.}{O}}$·, $:\overset{..}{\underset{.}{O}}$·, $:\overset{..}{\underset{.}{O}}$·, and ·$\overset{.}{\underset{.}{C}}$·. The sum of the valence electrons is therefore twenty-four. All these appear in the electronic formula. Brackets are placed around the atoms of the carbonate ion, and the ionic charge is indicated outside the brackets. By convention no brackets are placed around a positive ion containing only one atom. The structural formula of sodium carbonate is

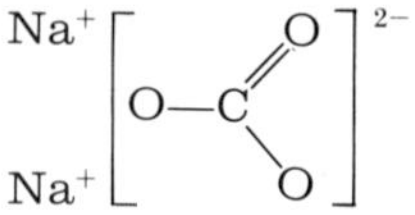

$$
\text{Na}^+ \left[\begin{array}{c} O \\ \| \\ O-C \\ \backslash \\ O \end{array} \right]^{2-} \\
\text{Na}^+
$$

The empirical formula is $\text{Na}^+\text{Na}^+\text{CO}_3{}^{2-}$ or, more commonly, Na_2CO_3.

The names of salts are discussed in Supplement 6.

3-16 THE VALENCE-BOND RESONANCE APPROACH TO REPRESENTING MOLECULAR STRUCTURE

Optional One of the severe limitations of electronic formulas arises from the fact that letters and dots in fixed positions are employed to represent the moving components of molecules and ions, i.e., vibrating nuclei and swiftly moving electrons, some related to more than one nucleus. It is not surprising, therefore, that it is frequently necessary to represent the structure of a given molecule by more than one arrangement of dots. These multiple arrangements do not represent different molecules but result from the inadequacies of formula notation in representing the complex structure of a given molecule or ion.

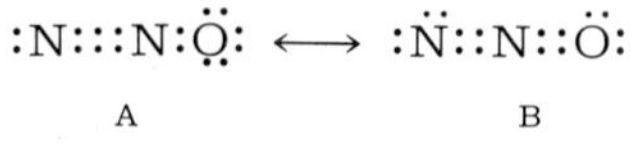

Let us consider, for example, the electronic representation of the gas nitrous oxide (Fig. 3-6). Both formulas A and B are in accordance with the rules of formula writing: each uses all the available electrons, and electron octets are not exceeded around any of the atoms. Which, then, is to be regarded as the correct formula for nitrous oxide? The rather unexpected answer is that while both formulas are useful, neither is the correct formula. This paradox is a direct consequence of the fact that the behavior of valence electrons in nitrous oxide defies adequate representation by a single formula. The best that can be done is

$$:\text{N}:::\text{N}:\overset{..}{\underset{..}{O}}: \longleftrightarrow :\overset{..}{\text{N}}::\text{N}::\overset{..}{\underset{..}{O}}:$$

A   B

FIGURE 3-6
The electronic formulas for two representations of the gas nitrous oxide

to suggest the electronic bonding by more than one approximate representation. This is not surprising, considering the dynamic nature of the electrons. It means, of course, that any given formula must be interpreted with great care.

It should be emphasized that this multiple-formula problem is concerned mostly with the nature of the bonds between atoms. This is demonstrated by the two formulas for nitrous oxide. Both formulas represent the same N, N, O sequence of atoms, but, according to formula A, the bond between the two nitrogen atoms is a triple covalent bond and the bond between the nitrogen atom and the oxygen atom is a single bond. In formula B, however, both bonds are represented as being double covalent. The truth of the matter is that the actual bond between the two nitrogen atoms is "in between" the triple bond of formula A and the double bond of formula B. This is a bond that cannot be represented satisfactorily by a single formula. Similarly, the bond between N and O is actually "in between" the single bond of formula A and the double bond of formula B.

Resonance refers to the discrepancy between the actual bonding in a molecule (or ionic radical) and the formulas used to represent that bonding. The word is an awkward choice, since it may suggest a vibration or other physical phenomenon with which it is not actually associated. The terms **resonance form, resonance structure,** or **resonance approximation** all refer to any approximate formula used as one of a series to represent an actual molecular structure. Thus, both formulas A and B for nitrous oxide are resonance forms. A double-headed arrow is generally used to indicate the relationship between resonance forms. The use of more than one resonance form to represent the structures of molecules and ionic radicals is called the **valence-bond resonance representation.**

George Wheland[†] uses the following analogy:

> We may imagine that a medieval traveler, in the course of his wanderings, saw a rhinoceros; and that, after his return to his home, he attempted to describe this strange beast to his friends. A convenient way for him to convey an approximately correct idea of the animal's appearance would be to say that the rhinoceros is intermediate between a dragon and a unicorn; for, as we may assume, the people to whom he was talking would have fairly clear ideas of what these two latter purely mythical creatures were supposed to look like.

In so doing the traveler would not mean that some rhinoceroses are dragons and the others are unicorns, nor would he mean that a given rhinoceros is a dragon part of the time and a unicorn the rest of the time. Instead, he would mean that the rhinoceros is a new kind of animal, neither dragon nor unicorn, but intermediate between these two mythical animals and partaking to some extent of the character of each. "Similarly the . . . resonance picture [of nitrous oxide describes it as] a hybrid structure, not identical with either of the [mythical resonance] structures, but intermediate between them."

[†] George W. Wheland, "Resonance in Organic Chemistry," p. 4, John Wiley & Sons, Inc., New York 1955; reprinted with permission.

3-16 THE VALENCE-BOND RESONANCE APPROACH TO REPRESENTING MOLECULAR STRUCTURE

FIGURE 3-7
Linus Pauling (born 1901), an American chemist whose work on the structure of matter has had a great influence. He has been interested in a broad range of the fundamental problems of chemistry, especially the nature of the various bonds operating between the particles of matter. With his colleagues he established the helical structure of certain protein molecules. (*Courtesy of Linus Pauling.*)

It is frequently necessary to use more than one electronic or structural formula to represent the structure of a molecule or ion, particularly if information regarding the nature of the bonds between atoms is essential. As a short cut, however, chemists often explicitly represent only one of the formulas, and then with their experienced eyes "read into" this single structure the others that are relevant. Accordingly, the structure of nitrous oxide is frequently represented by either $:N:::N:\ddot{O}:$ or $:N::N::\ddot{O}:$.

3-17 MOLECULAR STRUCTURE IN THREE DIMENSIONS

Optional

The structure of a molecule extends into three dimensions, but formulas are ordinarily written only in two. The molecule of the gas methane, for example, has the overall shape of a regular tetrahedron. The carbon atom is at the center and the hydrogen atoms are at the four corners. This spatial arrangement is represented in the margin together with electronic and structural formulas for methane.

The three-dimensionality of molecules is of great importance to the behavior of matter. We need to know not only which atoms of a molecule are bonded to which other atoms and by what type of links but also their arrangement in three dimensions. **Stereochemistry** refers to that part of chemistry which is concerned with the spatial arrangement of the structural units of matter.

H
..
H:C:H
..
H

Electronic
formula

H
|
H—C—H
|
H

Structural
formula

H
|
..C..
H⟋ ⟍H
H

Actual spatial
arrangement
of methane

3-18 THE COMBINING BEHAVIOR
OF SOME ATOMS OF SPECIAL SIGNIFICANCE

Optional

The combining behavior of individual atoms can vary considerably from substance to substance, and for this reason summarizations that are both comprehensive and succinct become difficult. It is possible, however, to use electronic symbols, the octet rule, the various types of bond formation, and the facts of combining behavior to formulate a set of useful generalizations about the combining behavior of many common atoms. It is important to emphasize at the outset that there are exceptions to these generalizations. You should also realize that truly satisfactory explanations for observed combining behavior involve many intricate details of atomic structure which must await further study of the structure of matter.

For convenient reference the generalizations are presented in Table 3-10, and the atoms will be identified by their electronic symbols.

Table 3-10

Atom	Combining behavior
H·	Shares its one electron to form a single covalent bond, as in elemental hydrogen H:H, or acquires one electron, becoming a negative ion with one charge, hydride ion, H^-; this forms ionic bonds with positive ions as in lithium hydride, Li^+H^-

Atom	Combining behavior
Li·, Na·, K·, Rb·, Cs·	Donate one electron per atom to become a positive ion with one charge, as in the sodium ion, Na^+; these form ionic bonds with negative ions, as in sodium chloride, Na^+Cl^-; another example is cesium ion, Cs^+, as in cesium fluoride, Cs^+F^-
·Be·	Shares its two electrons to form covalent bonds, as in beryllium chloride, $:\ddot{C}l:Be:\ddot{C}l:$
·Mg·, ·Ca·, ·Sr·, ·Ba·, ·Ra·	Donate two electrons per atom to become a positive ion with two charges, as in the calcium ion, Ca^{2+}; these form ionic bonds with negative ions, as in calcium oxide, $Ca^{2+}O^{2-}$
·$\dot{B}$·	Shares its three electrons to form covalent bonds, as in boron trifluoride,

$$\begin{array}{ccc} & :\ddot{F}: & :\ddot{F}: \\ & B & \\ & :\ddot{F}: & \end{array}$$

·$\dot{Al}$·	Shares its three electrons to form covalent bonds, as in aluminum bromide,

$$\begin{array}{ccc} :\ddot{Br}: & :\ddot{Br}: & :\ddot{Br}: \\ :Al: & & :Al: \\ :\ddot{Br}: & :\ddot{Br}: & :\ddot{Br}: \end{array}$$

	or donates three electrons to become a positive ion with three charges, the aluminum ion, Al^{3+}; this forms ionic bonds with negative ions, as in aluminum phosphate, $Al^{3+}PO_4^{3-}$.
·$\dot{C}$·, ·$\dot{Si}$·	Share electrons to form covalent bonds, as in ethane,

$$\begin{array}{cc} H\ H & H \\ H:\ddot{C}:\ddot{C}:H \text{ and silane, } & H:\ddot{Si}:H \\ H\ H & H \end{array}$$

·$\ddot{N}$·, ·$\dot{P}$·	Share electrons to form covalent bonds, as in ammonia, $H:\ddot{N}:H$, and phosphine, $H:\ddot{P}:H$

$$\begin{array}{cc} H & H \end{array}$$

:$\ddot{O}$·, :$\ddot{S}$·	Share electrons to form covalent bonds as in water, $H:\ddot{O}:H$, and hydrogen sulfide, $H:\ddot{S}:H$, *or* acquire two electrons per atom to become a negative ion with two charges: the oxide ion, $[:\ddot{O}:]^{2-}$, and the sulfide ion, $[:\ddot{S}:]^{2-}$; these form ionic bonds with positive ions, as in magnesium oxide, $Mg^{2+}O^{2-}$, and calcium sulfide, $Ca^{2+}S^{2-}$
:$\ddot{F}$·, :$\ddot{C}l$·, :$\ddot{B}r$·, :$\ddot{I}$·	Share electrons to form covalent bonds, as in hydrogen fluoride, $H:\ddot{F}:$, and elemental chlorine, $:\ddot{C}l:\ddot{C}l:$, *or* acquire one electron per atom to become a negative ion with one charge, as in the fluoride ion $[:\ddot{F}:]^-$; these form ionic bonds with positive ions, as in sodium bromide, $Na^+[:\ddot{B}r:]^-$
He:, :$\ddot{N}e$:, :$\ddot{A}r$:	May be assumed to have no tendency to combine with other atoms
:$\ddot{K}r$:, :$\ddot{X}e$:, :$\ddot{R}n$:	Share electrons to form covalent bonds, as in xenon difluoride, $:\ddot{F}:\ddot{X}e:\ddot{F}:$

3-19 CORRELATION OF PRINCIPAL
COMBINING BEHAVIOR WITH ELECTRONIC SYMBOLS

Optional

The combining behavior of the atoms discussed in the preceding section can be summarized (Table 3-11) by relating it to the arrangement of electron symbols discussed in Sec. 3-3. This arrangement is in turn related to the periodic arrangement of the elements (Sec. 3-3). Like the generalizations discussed in the preceding section, these are correlations of principal combining behavior to which there are exceptions. The atoms included have been chosen on the basis of the relative importance of substances containing them and on the simplicity of combining behavior.

When the number of valence electrons in an atom is low, the predominant tendency is to give up the electrons and form positive ions. Except for hydrogen, the atoms of group I, each with one valence electron, have such behavior, and so do the atoms of group II (except for beryllium), each with two valence electrons.

Atoms with six or more valence electrons (except the atoms of group 0) have a tendency to accept electrons to form negative ions. These atoms (including krypton, xenon, and radon) also share electrons in covalent bonds. Atoms with an intermediate number of valence electrons tend only to share electrons in covalent bonds.

3-20 EXAMPLES OF THE
COMBINING BEHAVIOR OF ATOMS

Optional

EXAMPLE 1 Hydrogen fluoride, H:F̈:. A fluorine atom may form either a covalent bond or become a negative ion in an ionic bond. Likewise, a hydrogen atom may form either a covalent bond or

Table 3-11
Correlations of the Principal Combining
Behavior of Selected Atoms with Their Electronic Symbols

Period	Group							
	I	II	III	IV	V	VI	VII	0
1	H·							He:
2	Li·	·Be·	·B·	·Ċ·	·N̈·	:Ö·	:F̈·	:N̈e:
3	Na·	·Mg·	·Al·	·Si·	·P̈·	:S̈·	:C̈l·	:Är:
4	K·	·Ca·					:B̈r·	:K̈r:
5	Rb·	·Sr·					:Ï·	:Ẍe:
6	Cs·	·Ba·						:R̈n:
7		·Ra·						

Code:

▓▓▓ = tendency to lose electrons and form positive ions in ionic bonds.

▒▒▒ = tendency to lose electrons and form positive ions in ionic bonds or to share electrons in covalent bonds.

▓▓▓ = tendency to share electrons in covalent bonds.

▒▒▒ = tendency to share electrons in covalent bonds or form negative ions in ionic bonds through a gain of electrons.

become a negative ion in an ionic bond. Since an ionic bond must involve a positive and a negative ion, the covalent bond represented here is the only possibility.

EXAMPLE 2 Potassium fluoride, $K^+[:\ddot{F}:]^-$. Since a potassium atom forms only ionic bonds in which it is the positive ion and a fluorine atom can become a negative ion, the bond is ionic.

EXAMPLE 3 Methane, $H:\overset{H}{\underset{H}{\ddot{C}}}:H$.

Since a carbon atom forms only covalent bonds and a hydrogen atom is also able to form covalent bonds, the bonding is covalent.

EXAMPLE 4 Sodium hydride $Na^+[H:]^-$. Since a sodium atom forms only ionic bonds in which it is the positive ion and it is possible for a hydrogen atom to become a negative ion, the bonding is ionic.

"I think you should be more explicit here in step two."

(© 1977 American Scientist. Reprinted by permission of Sidney Harris.)

KEY WORDS

1. Valence (Sec. 3-2)
2. Valence electron (Sec. 3-2)
3. Group (in a periodic table) (Sec. 3-3)
4. Period (in a periodic table) (Sec. 3-3)

5. Ion (Sec. 3-4)
6. Positive ion (Sec. 3-4)
7. Negative ion (Sec. 3-4)
8. Ionization (Sec. 3-4)

9. Octet rule (Sec. 3-5)
10. Ionic bond (Sec. 3-6)
11. Salt (Sec. 3-6)
12. Covalent bond (Sec. 3-7)

13. Nonbonding electron (Sec. 3-7)
14. Single covalent bond (Sec. 3-8)
15. Double covalent bond (Sec. 3-8)
16. Triple covalent bond (Sec. 3-8)

17. Structural formula (Sec. 3-10)
18. Condensed formula (Sec. 3-14)
19. Molecular formula (Sec. 3-14)
20. Empirical formula (Sec. 3-14)

Optional:

21. Valence-bond-resonance representation (Sec. 3-16)
22. Resonance (Sec. 3-16)
23. Resonance form (Sec. 3-16)

24. Resonance structure (Sec. 3-16)
25. Resonance approximation (Sec. 3-16)
26. Stereochemistry (Sec. 3-17)

SUMMARIZING QUESTIONS FOR SELF-STUDY

Sections 3-1 to 3-3

1. Q. What is meant by the valence of an atom?

A. The capacity of an atom to combine with other atoms.

2. Q. What are the valence electrons of an atom?

A. Electrons in an atom which are in the outermost orbit; e.g., since the electron configuration of the fluorine atom by orbits is 2, 7 the fluorine atom has seven valence electrons. They are of the greatest influence on the combining behavior of an atom.

3. Q. What is a period of elements?

A. Elements in the same horizontal row of a periodic table.

4. Q. What is a group of elements?

A. Elements in the same vertical column in a periodic table.

5. Q. To what period and group do the following elements belong? (Consult Table 3-2 for the atomic symbols.) (a) Sodium, (b) oxygen, (c) phosphorus, and (d) carbon.

A. (a) Third period, group I; (b) second period, group VI; (c) third period, group V; (d) second period, group IV.

Section 3-4

6. Q. What is the difference between an atom and and ion?

A. An atom is a neutral particle in which the number of protons equals the number of electrons. An ion is a particle with an electric charge in which the number of protons does not equal the number of electrons.

7. Q. What is the difference between a positive ion and a negative ion?

A. A positive ion has one or more unit positive charges, and the number of protons exceeds the number of electrons. A negative ion has one or more unit negative charges, and the number of electrons exceeds the number of protons.

8. Q. If a particle contains three protons, four neutrons, and two electrons, what charge, if any does it have?

A. One positive charge. The protons exceed the electrons by 1.

9. Q. If a particle contains 16 protons, 17 neutrons, and 18 electrons, what charge if any, does it have?

A. Two negative charges. The electrons exceed the protons by 2.

Section 3-5

10. Q. What is the octet rule?

A. It is a useful generalization regarding an underlying factor in the combining behavior of atoms; specifically, atoms with fewer than eight valence electrons have a strong tendency to achieve a total of eight valence electrons (or two for very simple atoms).

11. Q. For what atoms does the octet rule hold according to present concepts?

A. Atoms of the first and second periods of the periodic table.

12. Q. In what three ways can stable octets be achieved as atoms combine?

A. (1) Sharing electrons to form covalent bonds; for example, $H \cdot + :\ddot{F} \cdot \rightarrow H : \ddot{F} :$; (2) gaining electrons to form ions with a negative charge; for example, $:\ddot{O} \cdot + \text{two electrons} \rightarrow [:\ddot{O}:]^{2-}$; (3) losing electrons to form ions with a positive charge; for example, $\cdot Ca \cdot \rightarrow \text{two electrons} + Ca^{2+}$.

Sections 3-6 to 3-8.

13. Q. Explain what is meant by each of the following, and give an example of each: (a) ionic bond, (b) single covalent bond, (c) double covalent bond, and (d) triple covalent bond.

A. (a) An interatomic bond resulting from the mutual attraction between positive and negative ions as found in the three-dimensional lattice of salt crystals, for example, Na^+Cl^-. (b) An interatomic bond resulting from the sharing of one pair of electrons, the H—F bond in hydrogen fluoride, $H : \ddot{F} :$. (c) An interatomic bond resulting from the sharing of two pairs of electrons, e.g., the C=O bond in carbon dioxide, $:\ddot{O} :: C :: \ddot{O} :$. (d) An interatomic bond resulting from the sharing of three pairs of electrons, e.g., the C≡N bond in hydrogen cyanide, H:C:::N:.

Sections 3-9 to 3-15

14. Q. Examine the following formulas and analyze to what extent they are in accordance with the conventions of formula writing.
(a) Hydrogen chloride, H:C̈l:;
   A. Formula acceptable; correct total valence electrons (eight), one from hydrogen H· and seven from chlorine atom :C̈l·; in accordance with the octet rule. This is the electronic formula; the corre-

mula acceptable; correct total valence electrons (eight), one from each of the two hydrogen atoms H·, six from oxygen atom :Ö·; in accordance with the octet rule. (c) Hydrogen cyanide, H:C:::N: **A.** Formula acceptable; correct total valence electrons (ten), one from the hydrogen atom H·, four from the carbon atom ·Ċ·, and five from the nitrogen atom ·N̈·; in accordance with the octet rule. This is an electronic formula; the corresponding structural formula is H—C≡N, and the condensed formula is HCN. (d) Nitric acid, H:Ö:N::Ö:
                                                          :Ö:
A. Formula acceptable; correct total valence electrons (twenty-four), one from the hydrogen atom H·, five from the nitrogen atom ·N̈·, and six each from the three oxygen atoms :Ö·; in accordance with the octet rule. This is an electronic formula; the corresponding structural formula is given in part (e), below; the condensed formulas would be HONO₂ and HNO₃. (e) Nitric acid, H—O—N=O **A.** Formula acceptable; this struc-
        |
        O
tural formula implies the electronic formula given in part (d) above, which was judged to be correct. (f) Lithium fluoride, Li⁺[:F̈:]⁻ **A.** Formula acceptable; correct total valence electrons (eight), one from the lithium atom Li· and seven from the fluorine atom, :F̈·; there is one positive ionic charge and one negative ionic charge; in accordance with the octet rule. This is an electronic formula; the corresponding condensed formula is LiF. (g) Calcium chloride, Ca²⁺[:C̈l:]⁻ **A.** Formula unacceptable. The negative ionic charges do not equal the positive charges. The correct condensed formula is CaCl₂.

Sections 3-16 to 3-18 (Optional)

15. Q. Examine the following formulas and analyze to what extent they are in accordance with the conventions of formula writing.
(a) Ammonia, H—N—H
              |
              H

A. Formula acceptable; this structural formula implies the electronic formula

H:N̈:H
  |
  H

The covalent bonds are in accordance with the combining behavior of hydrogen and nitrogen atoms; correct total valence electrons (eight), five from the nitrogen atom ·N̈· and one each from the three hydrogen atoms H·; in accordance with the octet rule. The condensed formula is NH₃. (b) Beryllium fluoride, :F̈:Be:F̈: **A.** Formula acceptable; covalent bonds in accordance with combining behavior of beryllium and fluorine atoms; correct total valence electrons (sixteen), two from the beryllium atom ·Be·, seven from each of the two fluorine atoms :F̈·; in accordance with the octet rule. Recall that the octet rule prescribes eight electrons as a maximum; fewer are also stable. Thus the four electrons around the beryllium atom, Be, are satisfactory. This is the electronic formula; the corresponding structural formula is F—Be—F, and the condensed formula is BeF₂. (c) Sulfur dioxide, :Ö:S̈:Ö: **A.** Formula unacceptable; total valence electrons (twenty) too many; total should be eighteen, six from the sulfur atom :S̈·, and six each from the two oxygen atoms :Ö·. One of the acceptable resonance structures is :Ö:S̈::Ö:. (d) Boron
              :C̈l:
trichloride, :C̈l:B:C̈l: **A.** Formula acceptable; covalent bonds in accordance with combining behavior of boron and chlorine atoms; correct total valence electrons (twenty-four), three from the boron atom ·B· and seven each from the three chlorine atoms, :C̈l:; in accordance with the octet rule. Recall that the octet rule prescribes eight electrons as a maximum; fewer are also stable. Thus the six electrons around the boron atom, B, are satisfactory. This is an electronic formula; the corresponding structural formula is

              Cl
              |
        Cl—B—Cl

and the condensed formula is BCl₃. (e) Carbon dioxide, :O:::C::Ö: **A.** Formula unacceptable; there are ten valence electrons around the C atom, not in accordance with the octet rule, which holds for these second-period atoms. An acceptable resonance structure is :Ö::C::Ö: or O=C=O (f) Sul-
                                                          O
                                                          ‖
furic acid, H—O—S—O—H **A.** Formula acceptable;
                    ‖
                    O
this structural formula implies the electronic formula

              :Ö:
        H:Ö:S̈:Ö:H
              :Ö:

The covalent bonds are in accordance with the combining behavior of hydrogen, oxygen, and sulfur atoms; correct total valence electrons (thirty-two), six from the sulfur atom $:\ddot{\underset{..}{S}}\cdot$, one each from the two hydrogen atoms H·, and six each from the four oxygen atoms, $:\ddot{\underset{..}{O}}\cdot$. The octet rule is obeyed for the first-period hydrogen atom and the second-period oxygen atom; the sulfur atom is surrounded by twelve valence electrons, but this is acceptable since the sulfur atom is a third-period atom. Condensed formulas would be $HOSO_2OH$, H_2SO_4. (g) Potassium hydroxide, $K^+[O{—}H]^-$ **A.** Formula acceptable; this structural formula implies the electronic formula $K^+[:\ddot{\underset{..}{O}}:H]^-$. The ionic bond is in accordance with the combining behavior of the potassium atom; the covalent bond is in accordance with the combining behavior of the hydrogen and oxygen atoms; correct total valence electrons (eight), one from the potassium atom K·, one from the hydrogen atom H·, six from the oxygen atom $:\ddot{\underset{..}{O}}\cdot$; in accordance with the octet rule; positive and negative ionic charges equal. The condensed formula would be KOH. (h) Phosphoric acid,

$$
\begin{array}{c}
\quad\quad O \\
\quad\quad \| \\
H{—}O{—}P{—}O{—}H \\
\quad\quad | \\
\quad\quad O \\
\quad\quad | \\
\quad\quad H
\end{array}
$$

A. Formula acceptable. This structural formula implies the electronic formula

$$
\begin{array}{c}
:\ddot{O}: \\
\text{..} \\
H:\ddot{\underset{..}{O}}:\overset{..}{\underset{..}{P}}:\ddot{\underset{..}{O}}:H \\
:\ddot{O}: \\
H
\end{array}
$$

PRACTICE EXERCISES

1. To what period and group do the following elements belong? (Consult Table 3-5.)
   (a) aluminum, Al   (b) chlorine, Cl
   (c) neon, Ne   (d) magnesium, Mg
   (e) potassium, K   (f) silicon, Si

2. Consult Tables 3-5 and 3-1 and give names and electronic symbols for (a) the atoms of the elements of the second period, (b) group III in periods 2 and 3, (c) group II in periods 2 through 7, (d) group 0 for periods 1 through 6, (e) group VII for groups 2 through 5.

3. How many electrons and protons are in each of the following? (Consult Tables 2-3, 3-1, and 3-5).
   (a) fluoride ion, F^-
   (b) neon atom, Ne
   (c) magnesium atom, Mg
   (d) magnesium ion, Mg^{2+}
   (e) oxide ion, O^{2-}

The covalent bonds are in accordance with the combining behavior of the hydrogen, oxygen, and phosphorus atoms; correct total valence electrons (thirty-two): five from the phosphorus atom, $\cdot\dot{P}\cdot$, one each from the three hydrogen atoms, H·, six each from the four oxygen atoms, $:\ddot{\underset{..}{O}}\cdot$. The octet rule is obeyed for the first-period hydrogen atoms and the second-period oxygen atoms; the phosphorus atom is surrounded by ten electrons, but this is acceptable since the phosphorus atom is a third period atom. (i) Benzene,

$$
\begin{array}{c}
\quad\quad H \\
\quad\quad | \\
\quad\quad C \\
\quad\quad /\!\!/\quad \backslash \\
H{—}C\quad\quad C{—}H \\
\quad |\quad\quad\quad \| \\
H{—}C\quad\quad C{—}H \\
\quad\quad \backslash\!\!\!/ \\
\quad\quad C \\
\quad\quad | \\
\quad\quad H
\end{array}
$$

A. Formula acceptable; this structural formula implies the electronic formula

$$
\begin{array}{c}
\quad H \\
\quad \ddot{C} \\
H:C \quad\quad C:H \\
H:C \quad\quad C:H \\
\quad C \\
\quad H
\end{array}
$$

The covalent bonds are in accordance with the combining behavior of hydrogen and carbon atoms; correct total valence electrons (thirty): one each from the six hydrogen atoms, H·, and four each from the six carbon atoms, $\cdot\dot{C}\cdot$; octet rule is obeyed.

   (f) silicon atom, Si
   (g) aluminum atom, Al
   (h) aluminum ion, Al^{3+}
   (i) sulfide ion, S^{2-}
   (j) chloride ion, Cl^-
   (k) argon atom, Ar
   (l) potassium atom, K
   (m) potassium ion, K^+
   (n) calcium atom, Ca
   (o) calcium ion, Ca^{2+}

4. By analogy with the examples given in Sec. 3-5, predict a stable species formed from the following atoms in accordance with the octet rule (use Tables 3-1 and 3-5).
   (a) bromine, $:\ddot{\underset{..}{Br}}\cdot$ (b) potassium, K·
   (c) sulfur, $:\ddot{\underset{..}{S}}\cdot$ (d) magnesium, $\cdot Mg\cdot$
   (e) phosphorus, $\cdot\dot{P}\cdot$, in combination with hydrogen atoms, H·
   (f) carbon, $\cdot\dot{C}\cdot$, in combination with hydrogen atoms, H·

5. Identify each of the bonds in the following for-

72

mulas as ionic, single covalent, double covalent, or triple covalent.

(a) H—O—H

(b) Cl—Cl

(c) N≡N

(d) K⁺Cl⁻

(e) H—C=O
     |
     H

(f) Na⁺[O—H]⁻

(g)
```
    H       H
    |       |
H—C—O—C—H
    |       |
    H       H
```

(h) H—C≡N

(i) H—C=O
     |
     O
     |
     H

(j)
$$\left[\begin{array}{c} H—C=O \\ | \\ O \end{array}\right]^- Na^+$$

(i) Allyl chloride, Cl—C—C=C—H
```
          H  H  H
          |  |  |
     Cl—C—C=C—H
          |
          H
```

(j) Methyl cyanide,
```
     H
     |
 H—C—C≡N
     |
     H
```

(k) Methyl formate, H:C::Ö:C:H
with :Ö: above and H's on the right

6. Examine the following formulas and analyze to what extent they are in accordance with the conventions of formula writing:

(a) hydrogen iodide, H—Ï:

(b) potassium iodide, K⁺[:Ï:]⁻

(c) formaldehyde, H—C=Ö
     |
     H

(d) carbon monoxide, :C::Ö:

(e) hydrogen peroxide, H—O—O—H

(f) phosphorus trifluoride, :Ḟ:P::Ḟ:
        :Ḟ:

(g) sodium sulfide, Na⁺[:S̈:]²⁻
         Na⁺

(h) potassium oxide, K⁺[:Ö:]²⁻

(i) magnesium fluoride, Mg²⁺[:Ḟ:]⁻

(j) barium hydroxide, Ba²⁺[O—H]⁻
           [O—H]⁻

7. Examine the following formulas and analyze to what extent they are in accordance with the conventions of formula writing:

(a) Carbon disulfide, S=C=S

(b) Potassium cyanide, K⁺[C≡N]⁻

(c) Carbonic acid, H:Ö:C:Ö:H
         Ö

(d) Methylamine hydrochloride,

$$\left[\begin{array}{c} H\ \ H \\ |\ \ \ | \\ H—C—N—H \\ |\ \ \ | \\ H\ \ H \end{array}\right]^+ [Cl]^-$$

(e) Barium carbonate, Ba²⁺ [:Ö:C:Ö:]²⁻
                  :Ö:

(f) Acetaldehyde,
```
    H  O
    |  ‖
H—C—C—H
    |
    H
```

(g) Lithium nitrate, Li⁺[O—N—O]⁻
                 ‖
                 O

(h) Potassium phosphate,
$$K^+ \left[\begin{array}{c} O \\ ‖ \\ O—P—O \\ ‖ \\ O \end{array}\right]^{3-}$$
K⁺ ... K⁺

Section 3-18 (optional)

8. Using Table 3-11, select atoms which in their principal combining behavior:

(a) do not form covalent bonds

(b) do not form ions

(c) do not form either ions or covalent bonds

(d) form both ions and covalent bonds

(e) form ions with a single positive charge

(f) form ions with a double positive charge

(g) form ions with a triple positive charge

(h) form ions with a single negative charge

(i) form ions with a double negative charge

9. Examine the following formulas and analyze to what extent they are in accordance with the conventions of formula writing, and the principal combining behavior of the atoms.

(a) Formamide,
```
      O
      ‖
 H—C—N—H
      |
      H
```

(b) Sodium hydroxide, Na—O—H

(c) Lithium oxide, LiO

(d) Barium sulfate,
$$Ba^{2+} \left[\begin{array}{c} O \\ ‖ \\ O—S—O \\ ‖ \\ O \end{array}\right]^{2-}$$

(e) Dimethylamine,
```
    H  H  H
    |  |  |
H—C—N—C—H
    |     |
    H     H
```

(f) Magnesium cyanide, Mg⁺[C≡N]⁻

(g) Potassium chloride, K—Cl

(h) Methyl nitrate,
```
    H
    |
H—C—O—N=O
    |
    H     O
```

(i) Aluminum fluoride, AlF₂

(j) Ketene,
```
    H
    |
H—C=C=O
```

SUGGESTIONS FOR FURTHER READING

The following articles are of an introductory nature.

Benfey, Theodor: Geometry and Chemical Bonding, **Chemistry,** May 1967, p. 21.

Bent, Henry A.: The Tetrahedral Atom:Valence in Three Dimensions, **Chemistry,** January 1967, p. 8.

Huntress, W. T., Jr.: The Chemistry of Planetary Atmospheres, **Journal of Chemical Education 53:**204 (1976).

Mellor, D. P.: The Noble Gases and Their Compounds, **Chemistry,** November 1968, p. 8.

Seaborg, Glenn T.: From Mendeleev to Mendelevium—And Beyond, **Chemistry,** January 1970, p. 6.

Stein, Lawrence : Noble Gas Compounds, **Chemistry,** October 1974, p. 15.

Szoketalvi-Nagy, Zoltan: How and Why of Chemical Symbols, **Chemistry,** February 1967, p. 21.

Wallace, H. G.: The Atomic Theory: A Conceptual Model, **Chemistry,** November 1967, p. 8.

Wolfenden, John H.: The Noble Gases and the Periodic Table, **Journal of Chemical Education, 46:**569 (1969).

The following articles are at a more advanced level.

Dye, James L.: Anions of the Alkali Metals, **Scientific American,** July 1977, p. 92.

House, J. E. Jr.: Ionic Bonding in Solids, **Chemistry,** February 1970, p. 18.

Wahl, Arnold C.: Chemistry by Computer, **Scientific American,** April 1970, p. 54.

Wells, A. F.: Topological Approach to Structural Inorganic Chemistry, **Chemistry,** October 1967, p. 22; November 1967, p. 12.

4

WHAT IS THE NATURE OF SOLIDS, LIQUIDS, AND GASES?

4-1 THE EXAMINATION OF MATERIALS

One of the most interesting aspects of chemistry is the relationship between what we see when we look at the various forms of matter and the ideas about the structure of matter chemists use to interpret its behavior. Chemical concepts are constructed at the molecular level and involve dimensions much smaller than the human eye is able to distinguish. Microscopes using ordinary light are able to reveal much useful information about the structure of materials, but their level of viewing falls far short of molecular dimensions. Electron microscopes, on the other hand, reveal structural details which approach the molecular level. What details can be distinguished by the human eye? What further details are revealed by light and electron microscopes? What are the limitations of these instruments, and how close to the molecular level do instrumental methods reach?

As we look about us we have no difficulty in distinguishing the sea from the air above it or the water from the rocky shores. We can easily identify the individual hairs of cats' fur and the tiny threads of a cotton shirt. As a matter of fact, the human eye is able to distinguish bits of matter that have a diameter of only about 0.1 millimeter, which is less than $\frac{1}{250}$ inch. We are able to

FIGURE 4-1
Photomicrograph of crystals of sodium chloride. Examination of salt crystals with a light microscope reveals such details as the irregularities at the edges. This level of viewing, however, falls far short of revealing the actual structure. (*Photomicrograph by Stephen Gunn, courtesy of John Ricci.*)

(*a*)

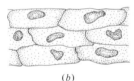

(*b*)

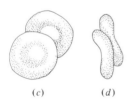

(*c*)   (*d*)

FIGURE 4-2
The shapes of some animal cells as revealed by an optical microscope: (*a*) nerve cell, (*b*) skin cells, and human red blood cell in (*c*) front and (*d*) side view. (Cells are not drawn to the same scale.) Molecular dimensions are much smaller than cellular dimensions; each red blood cell, for example, contains about 280 million molecules of the substance hemoglobin.

see the individual grains of fine sand and the tiny crystals of sugar and table salt, but these are at least 100,000 times wider than even fairly large molecules.

4-2 LIGHT MICROSCOPES

The very best light microscopes are able to distinguish objects with dimensions down to about 0.0002 to 0.0003 millimeter (or about 2000 to 3000 angstrom units†). This level of viewing permits one to distinguish, for example, the irregularities in the edges of crystals of table salt and naphthalene (moth flakes) (Fig. 4-1). Light microscopes have revealed much about individual cells of living organisms. Indeed, it was the development of the light microscope which led to the discovery of the cellular nature of plants and animals.

There is great variety in the shapes and sizes of the cells of living organisms (Fig. 4-2). Skin cells are small and compact, averaging only 0.003 millimeter in length (or 30,000 angstrom units). Nerve cells, however, are unusually long and narrow, sometimes extending more than 1000 millimeters. Human red blood cells are thin round platelets of only about 0.008 millimeter (80,000 angstrom units) in diameter. Although there are

† There are 10 million angstrom units in 1 millimeter and 254 million angstrom units in 1 inch. Units of length, mass, and volume are given in more detail in Supplement 3, Sec. S3-1.

over 200 million of them in each drop, each red blood cell contains about 280 million hemoglobin molecules. Even though the hemoglobin molecule contains about 10,000 atoms, it is only 64 angstrom units across. The dimensions of even such large molecules are considerably smaller than the range of light microscopes.

4-3 ELECTRON MICROSCOPES

The development of the electron microscope greatly extended our ability to examine the world of the very small. This instrument uses beams of high-speed electrons instead of the light waves used in the ordinary microscope. The best electron microscopes are able to distinguish objects with dimensions of about 2 or 3 angstrom units. While this does not bring most individual atoms into focus, it does bring the larger molecules within range. These instruments have made it possible to identify the location of large atoms like thorium and uranium within molecules. They also provide in a dramatically revealing manner much useful information about the structure of matter. Consider, for example, the appearance of human red blood cells in Fig. 4-3.

Figure 4-3
The appearance of human red blood cells when viewed with the scanning electron microscope; magnification ×5000. (*Courtesy of Thomas L. Hayes.*)

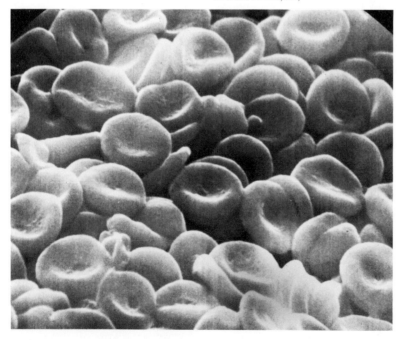

4-4 X-RAY DIFFRACTION ANALYSIS

The detailed internal structure of molecules can be obtained indirectly by a technique known as x-ray diffraction analysis.

X-rays are a form of invisible light rays; when their beams fall on crystalline solids, there is an interaction and the x-rays are scattered in an orderly way. The scattering is recorded on a photographic film and interpreted by means of complicated mathematics (Fig. 4-4). Although the mathematics is formidable, high-speed computers can greatly accelerate the analysis of the data. Molecular structural problems which formerly took months or years are now solvable in a matter of weeks.

One of the common methods of expressing the results of x-ray diffraction analysis is to use electron-density distribution maps, as illustrated by the map of naphthalene in Fig. 4-5. The electron density of a molecule is greatest near the nuclei of the component atoms. The electron-density contours therefore map the terrain of the molecules and identify the position of most atoms.

The use of computers and related automated operations has made x-ray-diffraction analysis a very useful tool for the study of molecular structure. It provides information not only about the sequence of atoms but also about the distances between atoms and the angles involved. X-ray-diffraction analysis has played a big part in important developments in all branches of chemistry. It also provides direct confirmation of the indirect structure-determination procedures in use for more than a century.

FIGURE 4-4
(*a*) Method of obtaining information about molecular structure by means of x-rays. After they are generated in the x-ray tube, the x-rays are made into a narrow stream by the lead screen, fall on the salt crystal, and are scattered to the photographic film. (*b*) Sketch of scattered x-ray photograph.

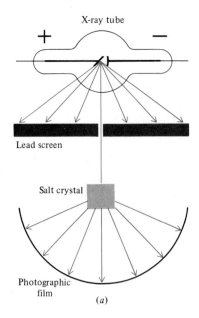

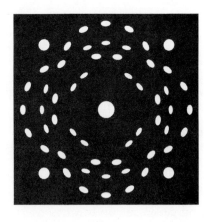

(*a*) (*b*)

FIGURE 4-5

The structure of the naphthalene molecule as obtained from x-ray-diffraction studies. (a) Electron-density map obtained from x-ray analysis [*adapted from S. C. Abraham, J. M. Robertson, and J. G. White, The Crystal and Molecular Structure of Naphthalene, Acta Crystallographica, 2:241 (1949)*]; (b) the corresponding structural formula. Note that in (a) the electron density is greatest near the nuclei of carbon, C, atoms. The electron density near the nuclei of the hydrogen atoms is not obtained in detail.

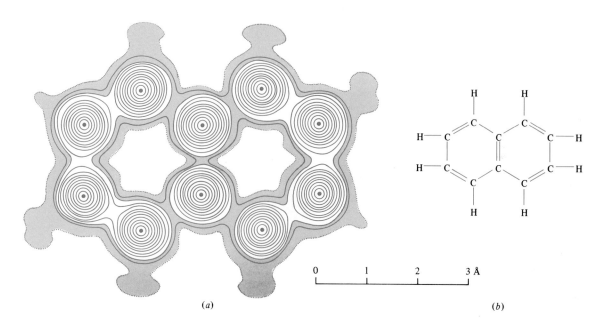

(a)   (b)

4-5 THE NATURE OF SOLIDS, LIQUIDS, AND GASES; THE KINETIC THEORY

When we look at a given specimen of matter without using any instruments, what do we actually see? One of the first things we notice, of course, is whether the sample is a solid, a liquid, or a gas (Fig. 4-6). What determines in which of these three states a given sample of matter exists? How are solids, liquids, and gases constructed?

Solids, e.g., quartz, naphthalene (moth flakes), and ice, are characterized by a definite shape and a definite volume. Liquids, e.g., water and gasoline, on the other hand, have a definite volume but an indefinite shape. But gases, e.g., air, have neither a definite shape nor a definite volume. They fill whatever space is available.

Despite these great differences in behavior, all three states are composed of molecules or ions which are constantly moving about (Fig. 4-7). The spreading of a perfume throughout a room or the pressure of the wind against our face is due to the motion of gas molecules. The slow diffusion of ink in still water suggests the motion of the molecules of liquids (Fig. 4-8). The effects of the more constrained motion of the particles of solids are not so readily observed.

The **kinetic theory** is a group of explanatory concepts which have been elaborated to account for the behavior of gases, liquids, and solids. A fundamental part of this theory postulates that molecules and ions are constantly in motion; the word kinetic is derived from the Greek word for motion. The theory has been formulated with considerable quantitative refinement, but the following summary will be sufficient for our purposes. It

FIGURE 4-6
The three states of matter. A hot poker changes ice into water and steam. *(Photograph by Fritz Goro, Life Magazine. © 1949, Time, Inc.)*

FIGURE 4-7
Schematic representation of the structure of (*a*) solids, (*b*) liquids, and (*c*) gases at the molecular level. The particles of the solid are relatively close together; they vibrate but remain close to fixed positions. The particles of a liquid are also relatively close together, but they are able to move about more freely. The particles of a gas are widely separated, and there is much empty space.

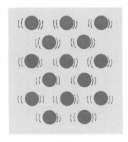
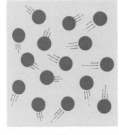
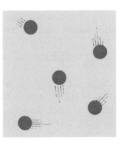

(*a*)   (*b*)   (*c*)

is restricted to some significant qualitative generalizations which have emerged from the kinetic theory and concepts about the structure of matter which are related to it.

1. The kinetic theory proposes that the molecules and ions of solids, liquids, and gases are constantly in motion. Many of the molecules of gaseous carbon dioxide under ordinary conditions are traveling at about 900 miles per hour, or 410 meters per second, faster than the speed of sound in air.

2. The physical state assumed by a substance under ordinary room conditions is determined primarily by the strength of the forces holding the fundamental particles of the substance together. If the forces are relatively strong, the particles are held in relatively fixed positions and the substance is a solid. If the forces are relatively weak, the particles move about very freely and the substance is a gas. If the forces are intermediate, the particles are moderately constrained and the substance assumes the liquid form.

3. The theory maintains that the particles of gases are thinly distributed in a vacuum (Fig. 4-7). The relatively large distances beween the particles account for the low density of gases and the ease with which they ordinarily can be compressed into a small volume. The loosely constrained particles move about very rapidly in all directions (some more than others), colliding very frequently in billiard-ball fashion as they do (in the neighborhood of 1×10^9 collisions per second).† The constant motion accounts for the pressure which gases exert and their extraordinary expansibility.

4. The more cohesively held particles of a liquid are much more compactly arranged than the particles of a gas (Fig. 4-7). Thus, liquids are much denser than gases and much

FIGURE 4-8
Motion of particles in a liquid as revealed by the spontaneous diffusion of colored ink into a beaker of water: (a) a small amount of ink dropped into water; (b) after a lapse of time the ink spreads through the liquid.

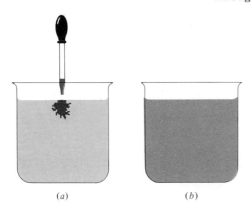

(a)   (b)

† The expression 1×10^9 is an example of exponential notation, explained in Supplement 1.

more difficult to compress. Since their particles are relatively free to move about and readily slip by one another, a sample of a liquid has no rigidity but a well-defined boundary. Close examination reveals that the fastest-moving particles escape from the surface, and this process accounts for the readiness with which most liquids spontaneously evaporate into the gaseous state. This is noticeable, for example, in the case of liquid perfumes, whose molecules leave the liquid state to spread their odor around a room. The continuous motion of the particles of a liquid can be detected by a close examination of oil globules suspended in water. Under a microscope the oil globules exhibit a spontaneous, continuous, haphazard motion, as they are buffeted about by the colliding water molecules (Figs. 4-9 and 4-10). This effect was first observed by the British botanist Robert Brown, who examined pollen grains suspended in water, and is called **brownian motion** in his honor.

5.   The particles of a solid are also constantly in motion, but the strong forces of attraction confine the motion of most of the

FIGURE 4-9
Brownian motion as revealed by a time-exposure photomicrograph of oil globules in water. The multiple images of some of the smaller particles are due to the discontinuous motion caused by molecular bombardment. (*Courtesy of R. W. St. Clair.*)

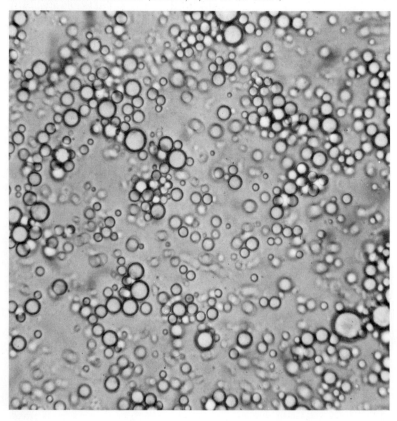

FIGURE 4-10
Schematic representation of causes of brownian motion: (a) collisions occurring equally on all sides of suspended particle result in no particle motion; (b) particle moves in direction of large arrow as a result of unequal collisions on opposite sides.

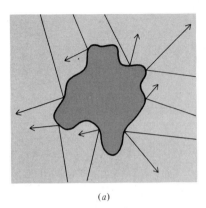

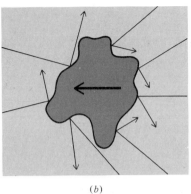

(a)   (b)

FIGURE 4-11
Schematic representation of the relationship between surface tension and the intermolecular attractive forces in liquids. (a) Forces of attraction are exerted on all sides of a particle deep within a liquid. (b) Unequal forces of attraction on a molecule at the surface result in a net pull toward the body of the liquid.

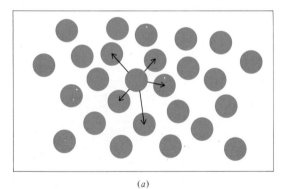

(a)

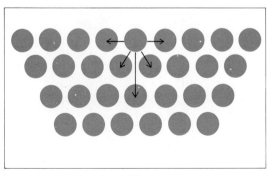

(b)

particles to a vibration around relatively fixed positions. Samples of solids therefore have both sharply defined boundaries and rigid shapes. The motion of the particles is revealed, however, by the evaporation of solids, particularly noticeable in solids like naphthalene (moth flakes) and camphor, both readily detected by their distinctive odors.

4-6 THE SURFACE TENSION OF LIQUIDS

Liquids behave as though they were covered by an invisible elastic membrane. This phenomenon is caused by the unbalanced distribution of the forces acting on the molecules at the surface. Molecules within the body of the liquid are attracted to their neighboring molecules nearly equally in all directions. Molecules at the surface, however, are attracted only to molecules below and beside them in the liquid (see Fig. 4-11). The surface molecules are therefore drawn in a direction toward the interior of the liquid. This inward pull contracts the surface of the liquid and causes it to behave as though it were covered by a thin skin. This property, which is related to the surface tension of a liquid, is particularly noticeable in water, and it permits certain insects to walk on the surface (Fig. 4-12); it is also responsible for the tendency of liquids to minimize their surface area. Small amounts of water, for example, tend to form small beads, and raindrops tend to be spherical.

FIGURE 4-12
Water striders are able to rest on the surface of the water because of the unbalanced distribution of forces acting on the molecules at the surface. (*Courtesy of Syd Radinovsky.*)

FIGURE 4-13
A comparison of the appearance and structure of a crystalline and an amorphous solid. Compare the appearance of (a) quartz crystals (silicon dioxide, SiO_2) with that of noncrystalline quartz glass (b). Schematic representations show (c) the ordered pattern of atoms in quartz and (d) the irregular arrangement in quartz glass. The actual structures are three-dimensional, not flat as represented here. [(a) *Photograph by Hans Oettgen;* (b) *courtesy Corning glass works.*]

(a)

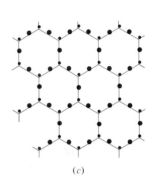

(c)

(b)

Black dots represents oxygen atoms with silicon atoms at intersections

(d)

4-7   CRYSTALLINE SOLIDS

In contrast to the formlessness of liquids, solids have a definite shape, and the outer boundaries of many speciments are characterized by sharp angles and plane surfaces. What causes this crystalline appearance? The primary factor, in addition to the rigid constraints operating between the particles of solids, is that the particles of these solids are often arranged in an orderly, three-dimensional structure, or a lattice, in contrast to the random arrangement of the particles of gases and liquids.

Solids whose particles are in such a fixed pattern are called **crystalline solids** (Fig. 4-13a). The others, whose particles are in a more random arrangement, are called **noncrystalline** or **amorphous solids** (Fig. 4-13b). Crystalline solids are conveniently classified according to the nature of their fundamental particles. Those whose fundamental particles are molecules, appropriately known as **molecular crystals,** include many common substances; sugar is one example, ice and naphthalene are others (Fig. 4-14).

Ionic crystals are crystalline solids composed of a three-dimensional pattern of positive and negative ions (Sec. 3-6). Sodium chloride (table salt), for example, is made up of a lattice of positive sodium ions and negative chloride ions. Each positive ion is surrounded by negative ions, and each negative ion is surrounded by positive ions (Fig. 3-4). In any given sample the number of positive ions equals the number of negative ions, so that the crystal is neutral as a whole. Individual molecules do not occur in an ionic solid.

FIGURE 4-14
Schematic representation of the ordered arrangement of molecules in the crystal structure of naphthalene. The individual molecules are viewed edge on. The molecules themselves are planar. (*From Ralph W. G. Wyckoff, "Crystal Structures," 2d ed., vol. 6, pt. 2, p. 384, John Wiley–Interscience Publishers, 1971, with permission.*)

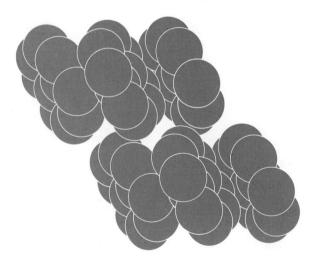

A **covalent-network solid** is another type of crystalline solid. It is composed of an indefinitely extended network of atoms joined by covalent bonds. The network extends throughout a given sample, and there are no discrete molecules (Fig. 4-13c). Whereas solid naphthalene is composed of an arrangement of separated molecules, solid quartz is made up of an arrangement of silicon and oxygen atoms bonded together in a three-dimensional network which extends throughout the crystal. Covalent-network solids have properties which differ significantly from the properties of molecular crystals. The temperatures required to melt them, for example, are much higher. Whereas the molecular crystals of naphthalene can be melted into liquid naphthalene at 81°C, a temperature of more than 1610°C is required to melt the network structure of quartz.

4-8   THE INTERCONVERSION OF SOLIDS, LIQUIDS, AND GASES

Substances which are solids at ordinary temperatures can be converted into liquids by increasing the temperature. The liquids, in turn, can be converted into gases by an additional increase in temperature. Substances which are ordinarily liquids can be changed into gases, usually by more modest temperature increases.

Correspondingly, substances which ordinarily exist as gases can be converted into liquids by lowering the temperature and the liquids, in turn, converted into solids by additional decreases in temperature. Substances which are ordinarily liquids can be solidified, usually by smaller decreases in temperature.

The conversion of solids into liquids is called **melting** or **fusion,** and the temperature at which it occurs is called the **melting point.** The solidification of liquids is known as **freezing,** or **crystallization,** and the temperature at which it occurs is called the **freezing point.** The conversion of a liquid into a gas is called **evaporation** if it takes place at the surface of a liquid and **boiling** if it takes place within the body of the liquid (Fig. 4-21). The temperature required to boil a liquid is called the **boiling point.** The gas obtained by evaporating or boiling a liquid is referred to by the special name **vapor.**

The conversion of a gas into a liquid is called **condensation** or **liquefaction.** The combined two-step process of boiling and condensation is known as **distillation.** Some solids can be

---

Note to Students

Once again the discussion has led to a barrage of new terms. Fortunately, many of them are already familiar to you. But don't attempt to keep them all in mind after the first reading. Return to them after you have consulted the summarizing questions at the end of the chapter and as you work through the practice exercises. By then you will have mastered them.

---

evaporated directly into gases. The combination of this direct conversion with subsequent deposition back to the solid state is called **sublimation.** Ice can evaporate directly into water vapor. Wet clothes hung outdoors in winter, for example, will dry even though the temperature may be so low that the water is in the form of solid ice. The relationships between some of these processes are summarized in Fig. 4-15.

FIGURE 4-15
Relationship between changes in state.

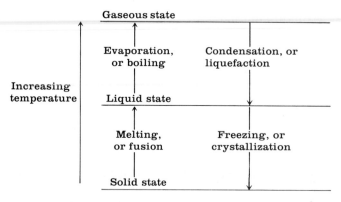

Relationship between changes in state

THE PRINCIPLE OF CLOUD SEEDING

Atmospheric air is capable of dissolving water vapor in appreciable amounts. The exact amount increases with the temperature. The maximum concentration (saturation) at 20°C (a little below room temperature) is about 17 grams per cubic meter. At 0°C the maximum concentration is reduced to less than 5 grams per cubic meter. If a given volume of air saturated with water vapor at a given temperature is cooled, the water vapor in excess of the saturation value at this lower temperature will be deposited as mist, rain, or snow. Water vapor condensed for this reason is common as the dew found on the countryside when chilly nights follow moist humid days.

In order for the condensation actually to take place, however, finely divided particles must be in the air to provide the nuclei on which the drops can form. If the air is perfectly free of any such particles, the excess water vapor over that normally dissolved may remain uncondensed in the air for some time. The excess water vapor is called supercooled moisture.

The principle of cloud seeding for weather modification is to provide finely divided particles to promote the condensation of water vapor in clouds with supercooled moisture. The condensation will then precipitate as rain (or as snow if the temperature is sufficiently low) if the other weather conditions are propitious (Fig. 4-16).

Finely divided silver iodide is commonly used because it is capable of producing between 10^{10} and 10^{14} fine particles per gram. It is usually dispersed by dropping ignited flares containing it from airplanes. Whether rain or snow actually results depends on a combination of conditions, including cloud temperature and how and where the nuclei are dispersed. Any form of weather modification is obviously a very controversial subject because of the difficulty of satisfying everyone who may be affected. Increasing the snowfall may be a welcome source of reservoir replenishment for some, but it may produce avalanches for others.

How are the changes of state interpreted in terms of the structure of matter? According to the kinetic theory, the fundamental particles of solids, liquids, and gases are in spontaneous and constant motion. In solids the molecules or ions are held by strong constraints, and their motion is confined to a vibration around rather fixed positions. In liquids the constraints are less powerful, and the particles are permitted greater freedom to move past each other but at fairly close range. The particles of gases are constrained least of all and are free to move rapidly about in all directions.

In all three states, however, some of the particles at any one temperature are moving faster than others (Fig. 4-17). In solids, for example, some particles are vibrating slowly, and others are vibrating relatively rapidly. The largest fraction of the total

FIGURE 4-16
Growth of a cloud seeded with silver iodide: (*a*) particles of silver iodide are released into a cloud; (*b*) 9 minutes later the cloud has grown; (*c*) the cloud 19 minutes after seeding; (*d*) 38 minutes after seeding the cloud is large enough to produce rainfall. (*Courtesy National Oceanic and Atmospheric Administration.*)

(*a*)

(*c*)

(*b*)

(*d*)

population of particles, however, has vibration rates close to an average value. The speeds of molecules in liquids and gases are similarly distributed over a wide range.

FIGURE 4-17
Plot of distribution of molecular speeds at two temperatures. As the temperature is increased, the average speed increases. The proportion of molecules with higher speeds also increases.

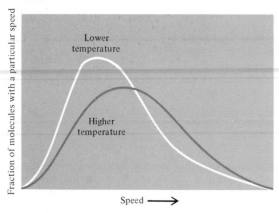

As the temperature of a given sample of matter rises, the average speed of the particles is increased. At the same time the proportion of the whole population at the relatively high speeds is increased (Fig. 4-17). At all temperatures, however, the range of speeds is a broad one. The evaporation of solids involves the escape of the fastest-moving particles from the surface (Fig. 4-18). The escape of these particles from the cohesive forces exerted by their neighbors results in the exertion of **vapor pressure.**

As the temperature of a solid rises, the population of the more rapidly moving particles increases. Eventually, this motion is sufficient to overcome in large measure the cohesive forces between the particles. At this point the particles

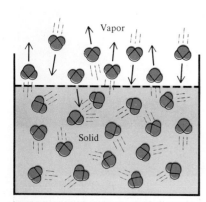

FIGURE 4-18
Schematic representation of the evaporation of solids.

assume the more mobile behavior characteristic of the liquid state, and the solid is observed to melt. The temperature of melting depends on the strength of the constraining forces. The greater the constraining forces, the higher the temperature required for melting. Since the constraining forces are directly related to the structure of the fundamental molecules or ions, melting points are directly related to structure.

The particles of a liquid are freer to move about than the particles of a solid, but they are held by considerably stronger cohesive forces than the particles of a gas. They may be conveniently pictured as moving about in groups or chains, which are free to move past one another. Some are moving slowly and some are moving very rapidly, but most of them have a degree of motion close to an average value. Some particles are moving fast enough to escape the constraining forces of their neighbors when they reach the surface, and this escape accounts for the spontaneous evaporation of liquids and for their vapor pressure (Fig. 4-19).

As the temperature of a liquid is increased, the motion of its particles is increased, resulting in higher populations of the faster-moving particles. The rate of escape of particles from the surface increases, and the vapor pressure increases. The rate of evaporation therefore increases with an increase in temperature, as might be expected.

As the population of the faster-moving particles increases with the temperature, a substantial proportion of particles is eventually able to overcome the cohesive forces between the molecules deep within the body of the liquid. At this point the liquid becomes a gas below the surface, and boiling occurs. This is responsible for the formation of gas bubbles near the bottom of a boiling liquid (Fig. 4-21).

The temperature required for boiling is influenced by two factors, the cohesive forces arising from the structure of the particles and the pressure exerted by the particles of the atmospheric gases above the liquid. The higher the atmospheric pressure, the higher the temperature required for boiling. At sea

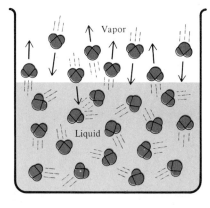

FIGURE 4-19
Schematic representation of the evaporation of liquids.

level, where normal atmospheric pressure is 760 mmHg,† water boils at 100°C. At an elevation of 14,500 feet, where the atmospheric pressure is normally only 445 mmHg, water boils at only 85°C (Fig. 4-22).

The boiling point depends on the structure of the fundamental molecules or ions. If the structure is such that very strong cohesive forces prevail, the boiling point is very high.

† Atmospheric pressures are commonly expressed in terms of the height of a column of mercury which the pressure will support. The normal pressure exerted by the atmosphere at sea level will support a column of mercury 760 millimeters high, and it is described as a pressure of 760 mmHg (read "millimeters of mercury"). (See Supplement 3, Sec. S3-3.)

THE CASE OF THE DIPPING BIRD

An eye-catching item often seen in show windows is a toy bird which with clockwork regularity keeps tilting over to dip its beak in a nearby glass of water unassisted by any mechanical or electrical device. Its operation depends on the evaporation of a highly volatile liquid such as ethyl ether and the cooling effect which accompanies the evaporation of water. When any liquid evaporates, its temperature falls because temperature is directly related to the motion of molecules. In the evaporation process the fastest-moving molecules leave the liquid; those remaining behind are therefore moving more slowly.

The dipping bird contains a bulb at the base and another at the top, connected by a tube (Fig. 4-20). The bottom bulb is filled with a volatile liquid which readily evaporates at room temperature, and the bulb is so arranged that as the liquid evaporates, liquid is forced up the tube to the top bulb. The temperature in the bulb at the top, however, is kept lower by the evaporation of water from a covering on the outside, made wet each time the bird dips. The liquid in the cooler upper bulb has less tendency to evaporate and accumulates. As it does so, its added weight causes the bird to tilt, so that its beak dips into the water. At the same time the tilting also allows the liquid to flow back to the lower bulb and the bird returns to its upright position. And the next cycle begins.

FIGURE 4-20
The four stages in the cycle of the dipping bird: (a) liquid in base; (b) the vapor pressure of the liquid causes it to rise through the tube to the cooler bulb on top; (c) the liquid in the top causes the bird to dip, wetting the beak and allowing the liquid to return to the bottom bulb; (d) as the liquid returns to the bottom bulb, the bird regains its upright position and the cycle repeats.

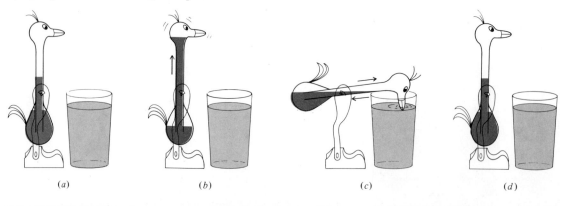

(a)   (b)   (c)   (d)

Sodium chloride, for example, boils at 1413°C. If the cohesive forces are relatively weak, boiling occurs at a low temperature. Ethyl ether, for example, boils at 35°C. If the boiling point of a substance is below ordinary room temperature, the substance will commonly occur as a gas. Oxygen, for example, boils at −183°C.

When gases are cooled, they usually condense into liquids, and when the temperature of liquids is reduced, they usually freeze into solids. What happens during these processes?

The particles of a gas are much freer to move about than the particles of a liquid or a solid. Significant forces of attraction still operate between them, however, and vary according to the structure of the particles themselves. In substances which are gases under ordinary conditions, these forces are extremely weak. As the temperature of a gas is lowered, the speed of its particles is reduced. If the temperature is lowered enough the forces of attraction constrain the gas particles to the motion of the particles of a liquid and the gas liquefies. The temperature at which this occurs depends on the structure of the individual particles, since it is this structure which is responsible for the forces of attraction.

FIGURE 4-21
Boiling water; note that bubbles of vapor form deep within the liquid. (*Photograph by Hans Oettgen.*)

FIGURE 4-22
The boiling point of water at different elevations above sea level. At
the higher elevations the atmospheric pressure decreases and the
boiling point of water is lower.

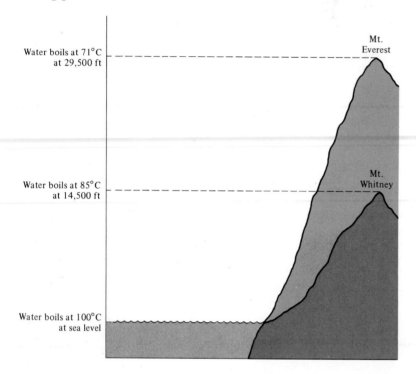

Water boils at 71°C
at 29,500 ft

Mt.
Everest

Water boils at 85°C
at 14,500 ft

Mt.
Whitney

Water boils at 100°C
at sea level

Not surprisingly, the temperature at which a given substance
condenses from a gas to a liquid is the same temperature at
which a substance boils. Both temperatures, of course, are sensitive to the pressure of the atmospheric gases. At normal atmospheric pressure water boils at 100°C, and gaseous water vapor
liquefies at the same temperature.

If the temperature of a liquid substance is lowered, the speed
of its particles is decreased and its tendency to aggregate into
the solid state is increased. If the temperature is lowered sufficiently, the mutual cohesive forces will confine the motion of
the particles to the vibration characteristic of the solid state. The
liquid freezes. The temperature at which solidification takes
place depends, of course, on the structure of the individual particles, since this structure determines the strength of the attractive forces. The structure also determines the shape of the
individual particles. This is often an important factor in solidification, since in cases where a crystalline lattice forms the ease
with which the particles fit together is significant.

The freezing and the melting of a substance take place, of
course, at the same temperature. Water freezes to ice at 0°C, and
ice melts to water at the same temperature. The effect of the atmospheric pressure on the temperature of melting and freezing
is usually very small.

4-10 WHAT STRUCTURAL FACTORS
DETERMINE THE STATE OF A SUBSTANCE?

What factors determine the state of a substance at a given temperature? Why, for example, is water a liquid under ordinary conditions? Why isn't it a solid or a gas? In the discussion of the factors determining the boiling points of liquids and the melt-

FIGURE 4-23
Schematic representations of the structure of (a) liquid water and (b) ice. In ice the water molecules have a regular structure; in liquid water the molecules are somewhat more compactly fitted in a random arrangement.

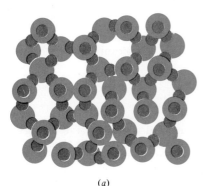

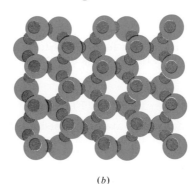

(a)   (b)

WHY DO PONDS FREEZE FROM THE TOP DOWN?

It is fortunate for devotees of ice skating that ponds freeze from the top down. If they froze from the bottom up and had to freeze solid before the surface was safe ice, opportunities for ice skating would be rare. Why do ponds freeze from the top down?

The weight of a given volume of a substance is known as its density; 1 milliliter† of liquid water at 0°C weighs very close to 1.0 gram. The density can therefore be conveniently expressed as 1.0 gram per milliliter at 0°C. (The temperature must be specified since the volume occupied by a given mass of water changes with the temperature.)

The density of ice, however, is less than that of water. At 0°C, for example, the density of ice is only about 0.92 grams per milliliter. At 0°C 1 gram of water occupies 1.00 milliliter whereas 1 gram of ice occupies 1.09 milliliter. This means that as water freezes, it expands and the ice that

forms has a lower density. The ice therefore floats in the water, and if the temperature of a pond remains low and the surface calm, a layer of ice will form over the entire pond. The expansion of water as it freezes also explains why water freezing in pipes can break the pipes. Pipes should be drained of water if there is a danger of freezing.

What features of the structures of ice and liquid water are responsible for ice being less dense than water? The structure of ice is rather open, as the schematic drawing of Fig. 4-23 suggests. Each water molecule forms hydrogen bonds with the water molecules which are its nearest neighbors. This results in more space between the water molecules than if hydrogen bonds were not present. In liquid water some of the hydrogen bonds are broken, allowing many of the water molecules to move closer together and accounting for the increase in density (Fig. 4-23).

† The units of volume are discussed in Supplement 3, Sec. 3-1.

ing points of solids, attention was called to the important role of the cohesive forces operating between the particles of a substance. The stronger these forces, the higher the melting and boiling points. These cohesive forces are electrical and vary in strength over a broad range. The strongest are the forces exerted by the covalent bonds holding the covalent-network solids together. These forces are so strong that the network solids have very high boiling points. The boiling point of quartz, for example, is 2230°C. The electrostatic forces operating between the positive and negative ions of salts are also very strong. As a consequence most ionic compounds are solids with boiling points at relatively high temperatures. Sodium chloride, for example, boils at 1413°C.

The weakest cohesive forces are short-range electric forces, called **van der Waals forces.** They are the only forces found between the particles of many gases. They can be so weak that some gases must be cooled to very low temperatures before they will condense into a liquid. Helium gas, for example, condenses into a liquid only at −269°C.

4-11 DIPOLE FORCES

Molecules are composed of atomic nuclei with positive charges and electrons with negative charges. Moreover, both the nuclei and the electrons are in constant motion. Nevertheless, for some molecules, when the center of positive charge is determined by taking into consideration the magnitude and the geometrical distribution of the charge, it is found to coincide with the center of negative charge determined in an equivalent manner. The term **nonpolar molecule** is used to describe a molecule in which the center of positive charge and the center of negative charge coincide.

Very often, however, the center of positive charge of a molecule is separated from the center of negative charge. One region has an excess of positive charge, and another region has an excess of negative charge. In a **polar molecule** the centers of positive and negative charge are separated from each other.

The distribution of electric charge within a molecule is determined by many structural factors: the number of charged particles, the size of their charges, and the distribution of the charged particles in space. We need not discuss these structural features in detail, but a few simple examples will give some idea of their nature.

Consider the structure of molecular hydrogen, which is represented in an electronic formula as H:H. Each proton has a single positive charge and has an equal attraction for the negative electrons (Fig. 4-24). The equality of the numbers and the sizes of the various charges and the equality of the forces operating between them result in a structure where the center of positive charge coincides with the center of negative charge. The hydrogen molecule is therefore nonpolar. In general, two-atom, or diatomic, molecules in which the atoms are of the same element are nonpolar.

Consider, however, the structure of a molecule of hydrogen chloride, represented in an electronic formula as H:C̈l: (Fig. 4-25). In this molecule the chlorine atom has a much greater attraction for electrons than the hydrogen atom, and as a result

H:H

FIGURE 4-24
The formula and schematic representation of a hydrogen molecule. The center of positive charge overlaps with the center of negative charge, and the molecule is nonpolar.

H:C̈l:

FIGURE 4-25
The formula and schematic representation of the hydrogen chloride molecule. The chlorine atom has a much greater attraction for electrons than the hydrogen atom, and as a result the molecule is polar.

the region of the chlorine atom has an excess of negative charge and the region of the hydrogen atom has an excess of positive charge. The centers of positive and negative charge therefore do not coincide, and the molecule is polar. The polar nature of the molecule can be represented by labeling the region with an excess of negative charge with the symbol δ^- (read "delta minus") and the region with an excess of positive charge with the symbol δ^+. In the structural notation the hydrogen chloride molecule can be represented as shown in the margin.

The structure of a molecule of carbon dioxide is of interest when we consider what determines the overall distribution of electric charge. Let us examine the electronic formula for carbon dioxide and focus on the bonds between the carbon atom and one of the oxygen atoms. These two atoms share four electrons, but the oxygen atom has a stronger attraction for electrons than the carbon atom. As a result the region of the oxygen atom has an excess of negative charge. The region of the carbon atom has an excess of positive charge. This charge distribution can be represented as shown in the margin. Although there are these regions of excess positive and negative charge within the carbon dioxide molecule, the distribution of the regions is such that the center of negative charge for the molecule as a whole coincides with the center of positive charge. (It is important that the three atoms of the molecule lie along a straight line.) The carbon dioxide molecule is therefore nonpolar.

The charge distribution of a molecule of water is of special interest. The structural formula for the molecule is given in Fig. 4-26. The three atoms do not lie in a straight line, and the molecule has a V shape. Since oxygen atoms have a stronger attraction for electrons than hydrogen atoms, the region of the oxygen atom has an excess negative charge and the regions of the hydrogen atoms have an excess positive charge. These regions can be represented in the structural formula shown in the figure. The V shape of the molecule produces a distribution of charge in which the positive and negative centers do not coincide, and the water molecule is polar.

$$\overset{\delta^+ \quad \delta^-}{\text{H—Cl}}$$

Representation of the hydrogen chloride molecule with symbols emphasizing its polar character; the small deltas, δ^+ and δ^-, indicate regions of positive and negative charge

$$:\overset{..}{\text{O}}::\text{C}::\overset{..}{\text{O}}:$$

Electronic formula for carbon dioxide

$$\overset{\delta^- \quad \delta^+ \quad \delta^-}{\text{O}=\text{C}=\text{O}}$$

Representation of the carbon dioxide molecule with symbols emphasizing regions of positive and negative charge; the small deltas, $\delta^-\delta^+$, do not represent a fixed unit of charge; the amount of localized positive charge equals the amount of localized negative charge, even though there are two δ^-'s and only one δ^+; despite the regions of positive and negative charge the molecule is nonpolar

---

Note to Students   Recall from Sec. 3-7 that covalent bonds have a definite orientation in space.

Since a **dipole** is any object in which there is a separation of positive and negative charge, polar molecules are tiny dipoles. Their dipolar nature gives rise to a cohesive force between molecules, due to the attractions between the positive and negative regions, represented schematically in Fig. 4-27. The name **dipole force** is given to the cohesive force operating between polar molecules, resulting from the mutual attraction of regions of excess positive and excess negative charge. The magnitudes of the forces vary widely but in general are close to the range of van der Waals forces. The boiling points of the following polar substances suggest the magnitude of the forces: ozone, O_3, $-112°C$; ethyl ether, $C_4H_{10}O$, $35°C$; and chloroform, $CHCl_3$, $62°C$.

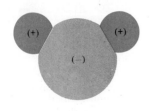

FIGURE 4-26
The structural formula for water and a schematic representation of the molecule. Since the oxygen atom has a stronger attraction for electrons than the hydrogen atoms, the region around the oxygen atom has an excess negative charge and the hydrogen atoms have a slight positive charge. Because the molecule is V-shaped, the center of positive charge does not overlap with the center of negative charge and the molecule is polar.

4-12 HYDROGEN BONDS

Although we cannot discuss in detail the many cohesive forces operating between molecules and ions, the forces called hydrogen bonds deserve special attention. They are the principal forces operating between the molecules of water. **Hydrogen bond** refers to the relatively weak but significant force which arises between hydrogen atoms in certain molecules and nonbonding pairs of electrons in certain neighboring molecules. Consider, for example, several molecules of water rather close together:

$$\text{H—}\ddot{\text{O}}\text{:}\cdots\text{H—}\ddot{\text{O}}\text{:}\cdots\text{H—}\ddot{\text{O}}\text{:}\cdots\text{H—}\ddot{\text{O}}\text{:}\cdots\text{H—}\ddot{\text{O}}\text{:}\cdots\text{H—}\ddot{\text{O}}\text{:}$$

$$\underset{A}{\text{H}}\qquad\underset{B}{\text{H}}\qquad\underset{C}{\text{H}}\qquad\underset{D}{\text{H}}\qquad\underset{E}{\text{H}}\qquad\underset{F}{\text{H}}$$

Representation of several molecules of water, showing how hydrogen bonds form; the strongly electron-attracting oxygen atoms render the hydrogen atoms somewhat positive, causing them to be attracted to the electron pairs on nearby oxygen atoms

One of the hydrogen atoms of molecule B (as shown) is rather close to the nonbonding electrons of the oxygen atom of molecule A, and an attractive force exists between the hydrogen atom of molecule B and the nonbonding pair of electrons on the oxygen atom of molecule A. This attractive force is enhanced by the fact that oxygen atoms have a much stronger attraction

FIGURE 4-27
Schematic representation of the forces operating between the polar molecules of solid hydrogen chloride. The hydrogen atoms are the smaller spheres. The negative regions in one molecule attract the positive regions in adjacent molecules.

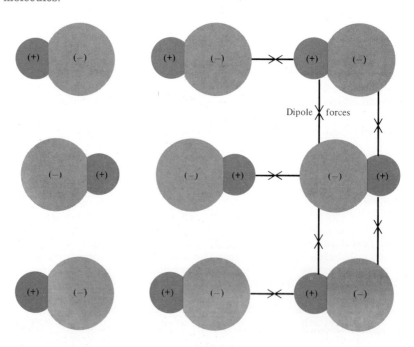

for electrons than hydrogen atoms. Since the oxygen atoms tend to pull the bonding electrons away from the hydrogen atoms, the hydrogen atoms are more positive than ordinarily. This increase in positive charge on the hydrogen atoms strengthens the attractive force between them and the nonbonding electron pairs of oxygen atoms in nearby molecules. This attractive force links the water molecules together.

In general, hydrogen bonds will form if the hydrogen atom in a molecule is bonded to an atom with a strong attraction for electrons, such as oxygen, which has a nonbonding pair of electrons. Most commonly these atoms are those of fluorine, nitrogen, and oxygen. Hydrogen bonds are not as strong as the electrostatic forces operating between ions, but they are often stronger than van der Waals forces. Their existence has a profound effect on the properties of many substances. The cohesive effect of the hydrogen bonds between water molecules, for example, is responsible for water's being a liquid under ordinary conditions rather than a gas.

The molecules of ethyl alcohol are attracted to each other by hydrogen bonds:

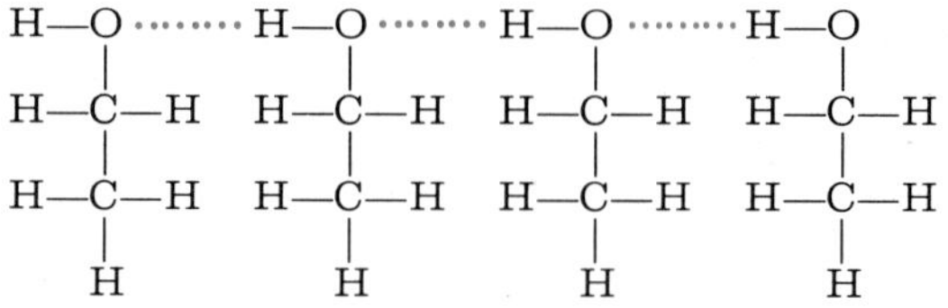

$$
\begin{array}{cccc}
\text{H—O} \cdots\cdots & \text{H—O} \cdots\cdots & \text{H—O} \cdots\cdots & \text{H—O} \\
| & | & | & | \\
\text{H—C—H} & \text{H—C—H} & \text{H—C—H} & \text{H—C—H} \\
| & | & | & | \\
\text{H—C—H} & \text{H—C—H} & \text{H—C—H} & \text{H—C—H} \\
| & | & | & | \\
\text{H} & \text{H} & \text{H} & \text{H}
\end{array}
$$

How hydrogen bonds hold several molecules of ethyl alcohol together

Because of the hydrogen bonds, ethyl alcohol is a liquid under ordinary conditions rather than a gas, and its boiling point is roughly 100°C higher than it would be without the hydrogen bonds.

4-13 METALLIC BONDS

A **metal** is a substance, like copper and silver, which has a metallic luster and the ability to conduct electricity and heat readily. The properties of metals suggest a structure which

---

WHY IS ICE SLIPPERY?

The fact that a given amount of water occupies less volume as a liquid than as solid ice accounts for interesting characteristics of ice. When, for example, we move along on ice skates, the pressure of our movement, magnified by the structure of the blades, compresses the ice. Since water occupies a smaller volume as a liquid than as a solid, it tends to become a liquid under pressure. A thin layer of water forms between the blade and the ice, friction is reduced markedly and the ice becomes slippery.

Moreover, when we walk on snow the pressure of our feet compacts it into a smaller volume. A thin layer of water forms under our feet as we walk, and the sound of our footsteps is muffled. If it is a bitter cold day, however, with the temperature at −23°C (−10°F) or lower, it is too cold for the pressure of our feet to melt the snow, and the snow "squeaks" as we walk.

involves an array of closely packed positive ions in a "sea" of electrons. The positive ions contain the atomic nuclei surrounded by all the electrons except the valence electrons. The valence electrons are very loosely held and are not closely associated with any specific positive ion (Fig. 4-28). The term **metallic bond** is applied to the net forces of attraction operating between the valence electrons and the positive ions of metals. The high strength of metallic bonds is indicated by the boiling points of sodium (892°C), silver (2212°C), and copper (2595°C). Common metals are solids under ordinary conditions, except mercury, which is a liquid.

This picture of the structure of metals explains why they conduct electricity so easily. The electric current consists of a flow of electrons through the metal, and the loosely held valence electrons are free to drift past the positive ions. The electron-sea model also explains why metal wires can be bent and sheets can be shaped without breaking. In bending, layers of the positive ions are forced across each other, but as this happens the cohesive metallic bonds between the ions and the mobile valence electrons remain intact (Fig. 4-29).

4-14 THE STATE OF SOME REPRESENTATIVE SUBSTANCES

Table 4-1 lists the melting points and boiling points of some representative substances to illustrate the relationship between these properties and the cohesive forces operating in the structure of substances. In a given substance more than one type of cohesive force may be operating. Emphasis is placed here on the strongest forces present, which we designate the principal force. As the principal cohesive forces increase from the relatively weak van der Waals forces to the covalent bonding of network solids, the boiling points increase fairly steadily. These sample substances show, however, that van der Waals

FIGURE 4-28
Schematic representation of the structure of a metal. It is essentially an array of positive atomic nuclei in a "sea" of electrons.

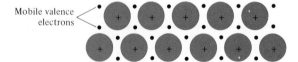

Mobile valence
electrons

FIGURE 4-29
Schematic representation showing how the metallic bonds remain intact as the layers of positive nuclei are shifted. The "sea" of negative valence electrons continues to hold the positive ions together.

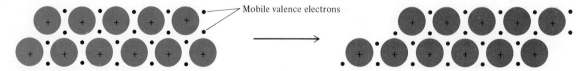

Mobile valence electrons

forces operating in some substances are stronger than hydrogen bonds operating in others. If the data of a broader range of substances were examined, additional examples of the overlapping of the various classes of cohesive forces would be found. Table 4-2 indicates the relationship between the prevailing cohesive force and the state of a substance under ordinary conditions.

Table 4-1
Data of Representative Substances Illustrating Relationship between Melting Points and Boiling Points and the Principal Cohesive Forces Operating in the Structure of the Substances

Substance	Empirical or molecular formula	Principal cohesive force	State at ordinary conditions	Melting point, °C	Boiling point, °C
Hydrogen	H_2	Van der Waals	Gas	−259	−252
Nitrogen	N_2	Van der Waals	Gas	−210	−196
Methane	CH_4	Van der Waals	Gas	−182	−164
Ozone	O_3	Dipole forces	Gas	−193	−112
Ethyl ether	$C_4H_{10}O$	Dipole forces	Liquid	−116	35
Ethyl alcohol	C_2H_6O	Hydrogen bonding	Liquid	−117	79
Chloroform	$CHCl_3$	Dipole forces	Liquid	−64	62
Octane	C_8H_{18}	Van der Waals	Liquid	−57	120
Water	H_2O	Hydrogen bonding	Liquid	0	100
Naphthalene	$C_{10}H_8$	Van der Waals	Solid	81	218
Acetylsalicyclic acid (aspirin)	$C_9H_8O_4$	Hydrogen bonding	Solid	135	
Camphor	$C_{10}H_{16}O$	Hydrogen bonding	Solid	180	204†
Potassium nitrate	KNO_3	Ionic bonding	Solid	334	400‡
Sodium chloride	$NaCl$	Ionic bonding	Solid	801	1413
Silver	Ag	Metallic bonding	Solid	961	2212
Quartz	SiO_2	Covalent bonding§	Solid	1610	2230
Gold	Au	Metallic bonding	Solid	1063	2966
Diamond	C	Covalent bonding§	Solid	>3550	4827

† Sublimes.
‡ With decomposition.
§ Exists as a covalent-network solid.

Table 4-2
Summary of the Relationship between the Principal Cohesive Force Prevailing in a Substance and Its State under Ordinary Conditions

Principal cohesive force	State at ordinary conditions	Example
Van der Waals forces	Gas, liquid, or low-melting solid	Nitrogen, N_2 (gas) Carbon dioxide, CO_2(gas) Iodine, I_2 (solid)
Dipole forces	Gas, liquid, or low-melting solid	Hydrogen chloride, HCl (gas) Chloroform, $CHCl_3$ (liquid)
Hydrogen bonds	Gas, liquid, or low-melting solid	Water, H_2O (liquid) Glucose, $C_6H_{12}O_6$ (solid)
Metallic bonds	Low- or high-melting solid	Copper, Cu (solid) Silver, Ag (solid)
Ionic bonds	High-melting solid	Sodium chloride, NaCl (solid) Barium sulfate, $BaSO_4$ (solid)
Covalent-network bonding	High-melting solid	Quartz, graphite (solids)

"Ammonia! Ammonia!"

Drawing by R. Grossman; © 1962 The New Yorker Magazine, Inc.

KEY WORDS

1. Kinetic theory (Sec. 4-5)
2. Crystalline solid (Sec. 4-7)
3. Amorphous solid (Sec. 4-7)
4. Molecular crystal (Sec. 4-7)
5. Ionic crystal (Sec. 4-7)
6. Covalent network solid (Sec. 4-7)
7. Melting (Sec. 4-8)
8. Melting point (Sec. 4-8)
9. Freezing (Sec. 4-8)
10. Freezing point (Sec. 4-8)
11. Evaporation (Sec. 4-8)
12. Boiling (Sec. 4-8)
13. Boiling point (Sec. 4-8)
14. Vapor (Sec. 4-8)
15. Condensation (Sec. 4-8)
16. Distillation (Sec. 4-8)
17. Sublimation (Sec. 4-8)
18. Vapor pressure (Sec. 4-9)
19. Van der Waals forces (Sec. 4-10)
20. Dipole forces (Sec. 4-11)
21. Nonpolar molecule (Sec. 4-11)
22. Polar molecule (Sec. 4-11)
23. Dipole (Sec. 4-11)
24. Hydrogen bond (Sec. 4-12)
25. Metal (Sec. 4-13)
26. Metallic bond (Sec. 4-13)

SUMMARIZING QUESTIONS FOR SELF-STUDY

Sections 4-1 to 4-4

1. Q. What is the approximate size of the smallest particles the human eye is able to distinguish?

A. Particles of matter which have a diameter of about 0.1 millimeter (less than $\frac{1}{250}$ inch).

2. Q. How does the size of these particles compare with the size of large molecules?

A. They are about 100,000 times wider.

3. Q. What is the approximate size of the smallest objects the very best light microscopes are able to distinguish?

A. Objects with dimensions of about 2000 to 3000 angstrom units (0.0002 to 0.0003 millimeter).

4. Q. What is the approximate size and shape of human red blood cells?

A. Round platelets of about 80,000 angstrom units in diameter.

5. Q. About how many hemoglobin molecules are in each red blood cell?

A. About 280 million.

6. Q. About how many atoms are in a hemoglobin molecule?

A. About 10,000

7. Q. What is the approximate diameter of the smallest objects the very best electron microscopes are able to distinguish?

A. About 2 or 3 angstrom units.

8. Q. Are the best electron microscopes able to distinguish individual molecules?

A. Larger molecules like the hemoglobin molecules.

9. Q. What method of examination of matter can distinguish the individual atoms of molecules and their geometric arrangement?

A. X-ray-diffraction analysis.

10. Q. What is x-ray-diffraction analysis?

A. When x-rays (a form of invisible light rays) fall on crystalline solids, the interaction produces a characteristic scattering of the x-rays. The scattering is recorded on a photographic film and interpreted mathematically.

11. Q. What is an electron-density contour map?

A. It is a map of the electron-density distribution in a molecule which can be derived from x-ray-diffraction data. The areas of highest electron density identify the positions of the atoms.

Section 4-5

12. Q. When we look at a given specimen of matter what do we usually notice first?

A. Whether it is a solid, liquid, or gas.

13. Q. What is the kinetic theory?

A. A theory postulating the constant motion of molecules and ions in all forms of matter which has been formulated to account for the behavior of solids, liquids, and gases.

14. Q. According to the kinetic theory, what is most important in determining whether a given substance is a gas, liquid, or solid at ordinary temperatures?

A. The physical state of a substance at ordinary temperatures is determined primarily by the strength of the forces which hold the fundamental particles of the substance together. If these forces are strong, the substance is a solid. If they are weak, the substance is a gas. If they are of intermediate strength, the substance is a liquid.

15. Q. Why are gases usually very compressible?

A. Because the molecules of a gas are relatively very far apart and separated by a vacuum.

16. Q. Why do gases expand very readily?

A. Because their molecules are moving in all directions, most of them at high speeds.

17. Q. Why are liquids less readily compressed than gases?

A. Because their molecules are much more compactly arranged.

18. Q. Why do solids have more sharply defined boundaries and more rigid shapes than liquids?

A. Because the strong forces of attraction between the particles confine the motion of the particles to a vibration around relatively fixed positions.

19. Q. What is responsible for the evaporation of liquids and solids?

A. The fastest-moving particles escape from the surface.

Section 4-6

20. Q. To what is the surface tension of liquids due?

A. The unbalanced distribution of the forces acting on the molecules at the surface. Molecules within the body of a liquid are attracted to their neighboring molecules nearly equally in all directions. Molecules on the surface, however, are attracted only to molecules below and beside them in the liquid.

Section 4-7

21. Q. What is the difference in fundamental structure between a crystalline solid and a noncrystalline (or amorphous) solid?

A. The particles of a crystalline solid are in a fixed three-dimensional pattern, whereas the particles of a noncrystalline solid are in a more random arrangement.

22. Q. What is the difference in fundamental structure between a molecular and an ionic crystal?

A. The fundamental particles of a molecular crystal are molecules, e.g., the molecules of water, H_2O, in ice crystals, whereas the fundamental particles of an ionic crystal are ions, e.g., the positive sodium ions, Na^+, and negative chloride ions, Cl^-, in the crystals of sodium chloride (table salt).

23. Q. What is the fundamental structure of a covalent-network solid?

A. An extended network of atoms joined together by covalent bonds, e.g., the network of silicon and oxygen atoms that constitute quartz crystals.

Section 4-8

24. Q. Give definitions for (*a*) melting (fusion), (*b*) melting point, (*c*) evaporation, (*d*) boiling, (*e*) boiling point, (*f*) condensation (liquefaction), (*g*) distillation, (*h*) sublimation, (*i*) vapor, (*j*) vapor pressure.

A. (*a*) The conversion of a substance from the solid to the liquid state, (*b*) the temperature at which the melting of a substance occurs, (*c*) the conversion of a liquid or solid to a gas at the surface of the liquid or solid, (*d*) the conversion of a liquid to a gas within the body of the liquid, (*e*) the temperature at which a liquid boils, (*f*) the conversion of a gas into liquid, (*g*) the conversion of a liquid into a gas by boiling and its condensation back into a liquid by cooling (in the process it may be freed of impurities), (*h*) the conversion of a solid into a gas by heating followed by the conversion of it back into the solid state by cooling (in the process it may be freed of impurities), (*i*) the gas obtained from the evaporation or boiling of a liquid, (*j*) the pressure exerted by the escape of the fastest-moving particles from the surface of a liquid or a solid.

25. Q. Why does the rate of evaporation of liquids and solids increase as the temperature rises?

A. Because as the temperature rises the speed of the particles increases and the rate of particles escaping from the surface increases.

26. Q. What two factors influence the temperature at which a liquid boils?

A. The strength of the forces of attraction between the particles and the pressure exerted by the atmospheric gases.

27. Q. What two factors influence the temperature at which a liquid freezes?

A. The strength of the forces of attraction between the particles and the shape of the individual particles.

Sections 4-9 to 4-13

28. Q. What are electrostatic forces?

A. The strong cohesive forces between the positive and negative ions of salts.

29. Q. Name six types of cohesive forces which serve to attract the fundamental particles of substances to each other.

A. Van der Waals forces (the weakest), dipole forces, hydrogen bonds, electrostatic forces, metallic bonds, covalent bonds of covalent network solids (the strongest).

30. Q. What are dipole forces?

A. The cohesive force operating between polar molecules.

31. Q. What are hydrogen bonds, and what structural features must prevail in order for them to be present?

A. The relatively weak but significant force capable of holding molecules together. They operate between certain hydrogen atoms and nonbonding electron pairs in neighboring molecules. The hydrogen atoms must be bonded to atoms, like oxygen, which have a strong attraction for electrons, and the nonbonding electron pairs must be associated with small electron-attracting atoms.

32. Q. What effect does the presence of hydrogen bonds have on the properties of substances?

A. They serve as a cohesive force holding molecules together. They are important to the physical state and behavior of water and ice and are important structural features of many substances of significance in living organisms.

33. Q. What is a metallic bond?

A. The net force of attraction operating between the valence electrons and the positive ions of metals.

PRACTICE EXERCISES

1. Match each definition or other statement with the numbered term above with which it is most closely associated. A numbered item may be used only once.

   1. Condensation
   2. Dipole forces
   3. Sublimation
   4. Kinetic theory
   5. Distillation
   6. Ionic crystal
   7. Polar molecule
   8. Freezing
   9. Covalent-network solid
   10. Van der Waals forces

   (*a*) The constant motion of molecules and ions of solids, liquids, and gases
   (*b*) A solid composed of a three-dimensional pattern of positive and negative ions
   (*c*) The conversion of a gas into a liquid
   (*d*) The weakest cohesive forces operating between the particles of matter
   (*e*) A solid composed of an indefinitely extended network of atoms joined by covalent bonds
   (*f*) The conversion of a liquid into a gas followed by its conversion back into liquid
   (*g*) The cohesive force operating between polar molecules

2. Match each definition or statement with the numbered term above with which it is most closely associated. A numbered item may be used only once.

   1. Dipole forces
   2. Sublimation
   3. Hydrogen bond
   4. Condensation
   5. Molecular crystal
   6. Distillation
   7. Freezing
   8. Polar molecule
   9. Metallic bond

   (*a*) A solid form of matter composed of molecules arranged in an orderly three-dimensional pattern

(b) A molecule in which the center of positive charge and the center of negative charge are separated

(c) The solidification of liquids

(d) The net forces of attraction operating between the valence electrons and the positive ions of metals

(e) Conversion of a solid directly into a gas followed by its conversion directly back into a solid

(f) The relatively weak but significant force which arises between a hydrogen atom bonded to an atom with a strong attraction for electrons and a nonbonding pair of electrons on another atom with a strong attraction for electrons

3. In terms of the qualitative generalizations from the kinetic theory and related concepts account for each of the following:

(a) The low density of gases

(b) The extraordinary expansibility of gases

(c) The exertion of vapor pressure by liquids and solids

(d) The fact that liquids are more difficult to compress than gases

(e) The fact that gases ordinarily condense to liquids when their temperature is lowered

(f) The readiness with which most liquids spontaneously evaporate into the gaseous state

(g) The decrease in volume of a balloon when it is cooled

4. In terms of the qualitative generalizations from the kinetic theory and related concepts, account for each of the following:

(a) The ease with which gases can ordinarily be compressed into a small volume

(b) The fact that liquids are denser than gases

(c) The sharply defined boundaries and rigid shapes of solids

(d) The increase in rate of evaporation of a liquid with temperature

(e) The lack of rigidity of liquids

(f) The fact that liquids ordinarily freeze when their temperature is lowered

5. Explain briefly the difference at the molecular level between the evaporation of a liquid and the boiling of a liquid.

6. How can the pressure exerted by a given sample of oxygen be decreased?

7. Account for the fact that water normally boils at 100°C at sea level but at only 85°C at an elevation of 14,500 feet.

8. How can the average speed of the molecules of a given sample of hydrogen be increased?

9. Explain briefly in terms of the structure of matter why:

(a) Many solids have a crystalline appearance

(b) Not all liquids boil at the same temperature

(c) Metals readily conduct electricity

10. Explain briefly in terms of the structure of matter:

(a) The ease with which metal wires can be bent

(b) Why water is a liquid rather than a gas under ordinary conditions

(c) Why not all solids melt at the same temperature

11. Explain why wet clothes will dry out of doors on a cold day even if the water freezes.

12. Explain why:

(a) The molecules of hydrogen are nonpolar

(b) The molecules of hydrogen chloride are polar

(c) The molecules of carbon dioxide are nonpolar

(d) The molecules of water are polar

13. Which of the substances in each of the following pairs would you speculate has the stronger principal cohesive force operating between its fundamental particles? Why?

(a) Copper or water?

(b) Water or oxygen?

(c) Sodium chloride or nitrogen?

(d) Sodium chloride or ethyl alcohol?

14. In the tables of physical constants of inorganic and organic compounds in either Robert C. Weast (ed.), "Handbook of Chemistry and Physics," The Chemical Rubber Co., or John A. Dean (ed.), "Lange's Handbook of Chemistry," 11th ed., McGraw-Hill, New York, 1973, look up the boiling points of the following pairs of substances and estimate which has the stronger principal cohesive force operating between its fundamental particles:

(a) Ethyl ether, $CH_3CH_2OCH_2CH_3$, or methyl alcohol, CH_3OH

(b) Magnesium, Mg, or carbon in the form of diamond, C

(c) Potassium bromide, KBr, or chloroform, $CHCl_3$

(d) Heptane, C_7H_{16}, or acetic acid, CH_3COOH

15. Explain the difference between a covalent bond and a hydrogen bond.

16. Explain at the molecular level why small amounts of water tend to form beads.

17. Explain why ethyl alcohol, CH_3CH_2OH, has a higher boiling point than butane, $CH_3CH_2CH_2CH_3$.

18. As liquids boil, their individual structural units become widely separated from each other as the compact arrangement of the particles of a liquid is converted into the more open structure of a gas. What is the principal cohesive force overcome when each of the following substances boils?

(a) Sodium chloride, NaCl

(b) Hydrogen chloride, HCl

(c) Water, H_2O

(d) Hydrogen, H_2

(e) Quartz, SiO_2

(f) Silver, Ag

(g) Carbon dioxide, CO_2

SUGGESTIONS FOR FURTHER READING

The following articles are of an introductory nature.

Bragg, Lawrence: The Start of X-Ray Analysis, **Chemistry,** December 1967, p. 8.

Brown, Glenn H.: Liquid Crystals, **Chemistry,** October 1967, p. 10.

Donohue, J.: On Hydrogen Bonds, **Journal of Chemical Education,** 40:598 (1963).

Fletcher, N. H.: Squeak, Skid and Glide: The Unusual Properties of Snow and Ice *in* Robert C. Plumb, Chemical Principles Exemplified, **Journal of Chemical Education, 49:**179 (1972).

House, J. E., Jr.: Weak Intermolecular Interactions, **Chemistry,** April 1972, p. 13.

Kerker, Milton: Brownian Movement and Molecular Reality Prior to 1900, **Journal of Chemical Education, 51:**764 (1974).

Knight, Charles, and Nancy Knight: Snow Crystals, **Scientific American,** January 1973, p. 100.

Runnels, L. K.: Ice, **Scientific American,** December, 1966, p. 118.

Snyder, A. E.: Desalting Water by Freezing, **Scientific American,** December 1962, p. 41.

Webb, Valerie J.: Hydrogen Bond, Special Agent, **Chemistry,** June 1968, p. 8.

The following articles are at a more advanced level.

Alder, B. J., and Thomas E. Wainwright: Molecular Motions, **Scientific American,** October 1959, p. 113.

Bernal, J. D.: The Structure of Liquids, **Scientific American,** August 1960, p. 124.

Bragg Lawrence: X-Ray Crystallography, **Scientific American,** July 1968, p. 58.

Hersh, Reuben, and Richard J. Griego, Brownian Motion and Potential Theory, **Scientific American,** March 1969, p. 66.

Kapecki, Jon A.: An Introduction to X-Ray Structure Determination, **Journal of Chemical Education, 49:**231 (1972).

Mott, Nevill: The Solid State, **Scientific American,** September 1967, p. 80.

5

WHAT IS THE NATURE OF SOLUTIONS?

5-1 INTRODUCTION

Chemists focus much of their effort on the behavior of pure substances, but the materials of nature usually do not occur in pure form. Occasionally we come across pure substances like gold and silver, and in certain regions deposits of essentially pure sodium chloride and sulfur are found underground. Most natural materials, however, are mixtures of more than one substance. Wood, for example, is a mixture of the substances cellulose and lignin. Bone is a mixture of the protein collagen and calcium phosphate.

Mixtures known as solutions have a special significance. Almost all chemical reactions occur in solutions of one kind or another. Human body fluids are solutions which play a key role in the processes of life. The oceans are basically one huge aqueous (water) solution of substances dissolved from land areas. More than half of all the naturally occurring elements are found in the ocean in some form. Although the solution is dilute, the total volume of the oceans is so great (estimated to be about 350 million cubic miles) that the total mass of dissolved substances is enormous. It is estimated that although the concentration of gold ions is only about 0.004 part in 1 billion parts, there is a total of 10 billion tons of gold in all the oceans of the world. (The ocean as a source of commercially important substances is discussed in Sec. 15-6A.)

The term **solution** includes all homogeneous mixtures of two or more substances dispersed in each other at the molecular level. The substances may be solids, liquids, or gases, and so there are solutions of matter of any state in any of the three states, from solutions of gases in solids to solutions of solids in gases. But the word "solution" most commonly suggests a liquid mixture, and we shall be concerned principally with solutions in which gases, liquids, and solids dissolve in liquids to form liquid solutions.

The **solvent** is the material which acts as a dissolving medium. The **solute** is the material dissolved. When gases and solids dissolve in liquids, it is rather easy to identify the liquids as the solvents and the gases and the solids as solutes. When one liquid dissolves in another, the more abundant liquid is arbitrarily called the solvent and the less abundant liquid the solute.

A dilute solution is one in which there is a low proportion of solute. A concentrated solution is one in which the proportion of solute is high. In precise chemical operations it is important to specify the concentration of solutions in quantitative terms, and many systems have been devised for this purpose. A **saturated solution** may be defined for our purposes as a solution in which the maximum amount of solute is dissolved.

Substances differ widely in their solubility in a given solvent. Some, for example, like quartz or granite, will not dissolve in water to any appreciable extent. Others, like table sugar, dissolve in water readily. The quantitative expression of the concentration of solutions is discussed in Supplement 5.

5-2   WHAT ARE THE MECHANISMS OF SOLUTION FORMATION?

For one substance to dissolve in another, the particles of both substances must be able to mingle freely with each other. The particles of each substance must be attracted to those of the other by cohesive forces of about the same strength as the forces which operate between the particles of each substance individually in the pure state. For example, if particles of both substances in the pure state are held together only by very weak forces, it will be easy for the particles of both substances to intermingle. The discussion that follows will illustrate the familiar saying that "like dissolves like."

We are limiting this discussion to solutions in which the solvents are liquids. Substances which are liquids under ordinary conditions are generally those in which the particles are bonded by van der Waals forces, dipole forces, or hydrogen bonds. (With few exceptions, substances whose particles are held by metallic or ionic bonds or in a covalent network are solids under ordinary conditions.) In order to obtain some idea of the various mechanisms by which the close intermingling of particles in solutions is achieved, it will be useful to consider some illustrative examples.

A. Hexane, C_6H_{14}, in Carbon Tetrachloride, CCl_4: A Solution of Weakly Bound Nonpolar Molecules

Carbon tetrachloride is a liquid whose molecules are nonpolar (Sec. 4-11). Its structure is suggested by the formula and schematic sketch of Fig. 5-1a. Since its molecules are nonpolar, they are held together only with relatively weak van der Waals forces (Fig. 5-2a). Hexane is also a liquid whose molecules are nonpolar. Its structure is suggested by the formula and schematic sketch of Fig. 5-1b. As in carbon tetrachloride, the cohesive force between the molecules is the relatively weak van der Waals forces (Fig. 5-2b).

FIGURE 5-1
Formulas and schematic sketches for (a) carbon tetrachloride and (b) hexane molecules.

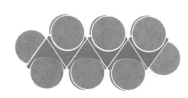

(a)

(b)

FIGURE 5-2
Schematic sketches of (a) molecules of pure carbon tetrachloride, (b) molecules of pure hexane, and (c) a solution of carbon tetrachloride and hexane.

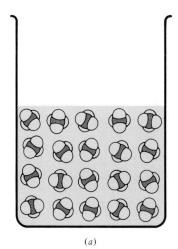
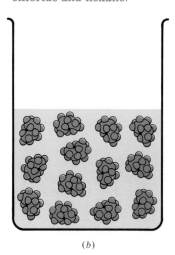
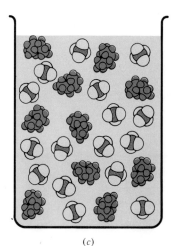

(a)

(b)

(c)

If hexane is stirred into carbon tetrachloride, the molecules of hexane mingle with the carbon tetrachloride molecules as easily as they mingle with other hexane molecules. As a result, the hexane readily dissolves (Fig. 5-2c).

B. Hexane, C_6H_{14}, and Water, H_2O: Not a Solution

The structure of water has already been described as consisting of molecules held together by hydrogen bonds (Sec. 4-12). Since water molecules are polar, dipole forces are also operating. The cohesive forces between the water molecules are therefore of a moderate strength, stronger than the van der Waals forces operating between the molecules of hexane.

If hexane is stirred into water, the moderately strong forces operating between the water molecules "squeeze out" the weakly bound hexane molecules and only a very few penetrate between the water molecules. For practical purposes the hexane does not dissolve but remains in a separate layer (Fig. 5-3). Since the hexane is less dense than the water, it floats on top. In general, nonpolar substances are usually not very soluble in polar solvents. Hexane is representative of the substances found in crude petroleum (Sec. 13-11), and in considering the results of mixing hexane and water we have, in effect, explained why oil and water do not mix.

C. Ethyl Alcohol, CH_3CH_2OH, and Water, H_2O: A Solution of Two Hydrogen-Bonded Substances

The structure of ethyl alcohol has already been described as consisting (like water) of molecules held together by hydrogen bonds (Sec. 4-12). If ethyl alcohol is stirred with water, the ethyl alcohol molecules mingle almost as freely with the water mole-

FIGURE 5-3
Schematic sketches of (a) molecules of pure hexane, (b) molecules of pure water, and (c) hexane and water after mixing.

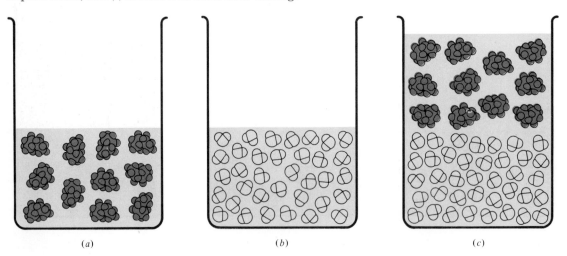

(a) (b) (c)

cules as they do with other molecules of ethyl alcohol, since both kinds of molecules are associated by hydrogen bonding (Fig. 5-4). As a result the ethyl alcohol dissolves very readily.

D. Sodium Chloride, NaCl, and Water, H_2O: A Solution of an Ionic Compound in Water

The structure of sodium chloride has already been described as a three-dimensional lattice of alternating positive sodium ions, Na^+, and negative chloride ions, Cl^- (Sec. 3-6). The electrostatic forces operating between the ions are relatively strong. The ions in the body of the crystal are attracted in all directions by the neighboring ions of opposite charge. The forces acting on the ions at the surface, however, are unbalanced. Because they are polar, the water molecules are attracted to these surface ions. Each sodium ion becomes partially surrounded by the negative ends of water molecules, and each chloride ion becomes partially surrounded by the positive ends of water molecules (Fig. 5-5). The term **ion-dipole force** is used for the cohesive force between an ion and a polar molecule like the water molecule. The combined effect of the ion-dipole forces exceeds the cohesive forces between the ions, and the ions leave the lattice and enter the water (Fig. 5-5b). Since the polarity of the water molecules is of central importance in this process, salts would not be expected to dissolve in nonpolar solvents.

The dissolved ions remain associated with some of the water molecules because of the ion-dipole forces and move among the bulk of the water molecules surrounded by these molecules. **Hydration** is the special attraction between an ion and one or more water molecules by means of ion-dipole forces. All ions in aqueous solution are hydrated.

FIGURE 5-4
Schematic sketches of (a) molecules of pure ethyl alcohol, (b) molecules of pure water, and (c) a solution of ethyl alcohol and water.

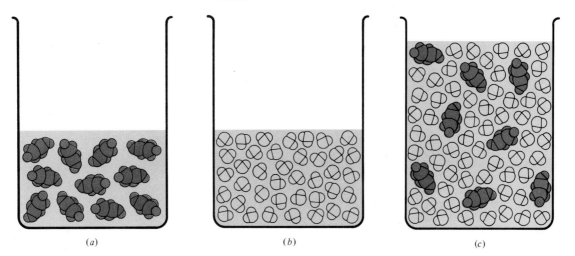

(a) (b) (c)

FIGURE 5-5

Schematic illustration of how a sodium chloride crystal dissolves in water: (a) the crystal mixes with the water; (b) the polar water molecules attract the outer ions from the salt crystal; (c) the ions become hydrated and mingle with the water molecules; and (d) the solution is complete.

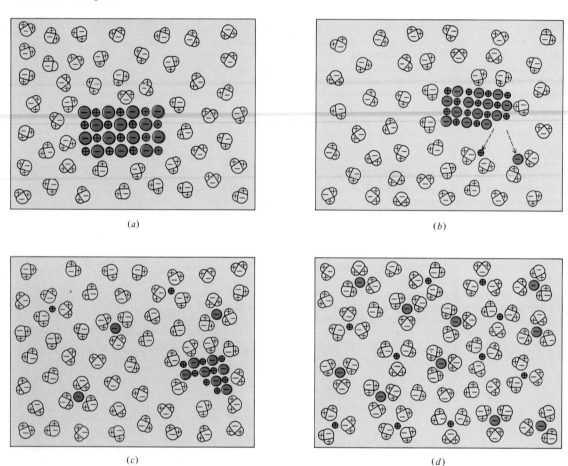

(a)

(b)

(c)

(d)

E. Silver Chloride, AgCl, and Water, H₂O: A Solution of a Salt of Very Limited Solubility; Solubility Equilibrium

The structure of silver chloride resembles the structure of sodium chloride: a three-dimensional lattice of alternating positive silver ions, Ag^+, and negative chloride ions, Cl^-. In silver chloride, however, the electrostatic forces operating between the ions are stronger than the ion-dipole forces which form with the water, and not many ions leave the lattice when the crystals are stirred with water. As a consequence, silver chloride is only very slightly soluble in water; at a temperature slightly below

WHY DOES SALT WATER FREEZE AT A LOWER TEMPERATURE THAN PURE WATER?

It is well known that salt water freezes at a lower temperature than fresh water. Calm water containing 5 percent of dissolved table salt (sodium chloride) will freeze about 3°C below the normal freezing point of pure water. Why is the freezing point depressed this way?

We have mentioned (Sec. 4-9) that the faster moving molecules in solids and liquids escape from the surface. Consider ice crystals at 0°C floating in water at the same temperature. At 0°C the tendency of the ice molecules to escape into the water molecules is the same as the tendency of the water molecules to escape into the ice molecules. Since the two tendencies are equal, there is no overall change in the amount of the crystals or in the amount of water (Fig. 5-6a).

If salt or other soluble substance is dissolved in the water, the particles of the salt (mostly as sodium ions and chloride ions) tend to block the escape of the molecules from the liquid water. (Although the particles of the salt also have an escaping tendency, it is much lower than that of the water molecules and it is not important here.) The escaping tendency of the molecules of the ice crystals remains the same. Since the escaping tendency of the molecules of the ice is now greater than the reduced escaping tendency of the water molecules in the solution, the rate of escape of the ice molecules is higher, and the ice melts (Fig. 5-6b).

The escaping tendency of the molecules of both the ice and the water decreases as the temperature is lowered, but the escaping tendency of the ice molecules decreases more rapidly. Therefore, as the temperature is lowered, a point will be reached where the escaping tendency of the ice molecules again becomes equal to the escaping tendency of the water molecules of the solution (Fig. 5-6c). At this lower temperature the water in the solution will freeze. If the concentration of the salt (sodium chloride) in the solution is about 5 percent, this point is not reached until −3°C. Whether the water will actually freeze depends not only on the concentration of the salt but also on the amount of turbulence.

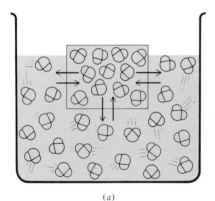

(a)

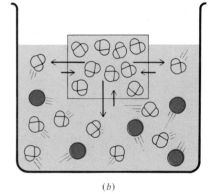

(b)

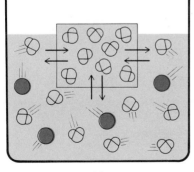

(c)

FIGURE 5-6

Schematic representation of (a) the equal escaping tendencies of water molecules in ice at 0°C and in pure liquid water at 0°C indicated by the arrows of equal length; (b) the excess escaping tendency of water molecules in ice at 0°C compared with that of water molecules in a 5% aqueous solution of NaCl at 0°C indicated by the arrows of unequal length; and (c) the equal escaping tendencies of water molecules in ice at −3°C and in a 5% aqueous solution of NaCl at −3°C indicated by the arrows of equal length.

room temperature, for example, only 8.9×10^{-4} gram will dissolve in 1 liter.

If we were to put on "molecular spectacles" and "examine" a saturated aqueous solution of silver chloride in contact with undissolved silver chloride, we would observe that while the amount of undissolved silver chloride remains constant, the silver and chloride ions are both leaving the lattice to enter the solution and returning to the lattice from the solution. Since these processes take place at the same rate, the amount of silver chloride in the lattice and in the solution remains constant (Fig. 5-7).

The name **chemical equilibrium** is applied to a condition of dynamic balance between two competing processes which take place at equal rates. The overall situation does not change with time. The processes occurring at the molecular level continue but in a balance that causes no change on a large scale. The name **solubility equilibrium** is used for the specific equilibrium condition prevailing when a saturated solution is in contact with undissolved solute. Although the particular example under discussion involves a solute which is an ionic compound, an equivalent situation prevails in saturated solutions of molecular compounds, e.g., table sugar. The solubility equilibrium prevailing in a saturated solution of silver chloride can be represented in symbols as

$$\text{AgCl (undissolved)} \rightleftharpoons \text{AgCl (dissolved)}$$

Notice that the arrows point in both directions indicating that the change occurs in both directions.

FIGURE 5-7
Schematic representation of the solubility equilibrium prevailing in a saturated solution of silver chloride in contact with undissolved silver chloride. The silver and chloride ions are leaving the lattice to enter the solution and returning to the lattice from the solution, both at the same rate.

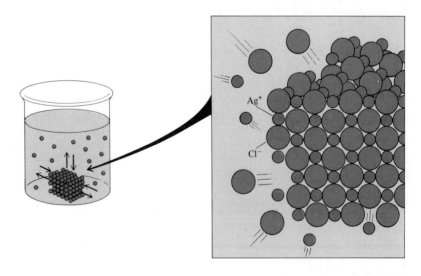

F. Hydrogen Chloride, HCl, and Water, H₂O: Complete Dissociation of a Molecular Substance into Ions

Hydrogen chloride is a gas composed of polar molecules; its structure is suggested by the accompanying formula (Sec. 4-10). When the gas is bubbled into water, the polar water molecules surround both the positive hydrogen atoms and the negative chlorine atoms of the molecules. The attraction for the water molecules breaks the H—Cl covalent bond. Hydrated chloride ions and hydrated hydrogen ions enter the solution (Fig. 5-8). As a consequence the hydrogen chloride readily dissolves in the water. Moreover, as the hydrogen chloride dissolves, it is completely dissociated into ions. The result is an aqueous solution of hydrated hydrogen ions and hydrated chloride ions. This solution is called hydrochloric acid.

Actually almost all the hydrogen ions in an aqueous solution combine with water to form hydronium ions, H_3O^+. The formation of the covalent H—O bond is consistent with the principal combining behavior of hydrogen atoms as discussed in Sec. 3-18. The hydronium ions are in turn closely attracted to, or hydrated with, three water molecules. Although in aqueous solutions hydrogen ions, H^+, are indeed present as hydronium ions, H_3O^+, it is more convenient to refer to them simply as hydrogen ions, with the understanding that actually they are hydronium ions.

$$\overset{\delta^+}{\text{H}}\!\!-\!\!\overset{\delta^-}{\text{Cl}}$$

Representation of a molecule of hydrogen chloride emphasizing its polar nature (Sec. 4-10)

$$\text{H}^+ \;+\; \text{O}\!-\!\text{H} \rightarrow \text{H}\!-\!\overset{+}{\text{O}}\!-\!\text{H}$$
$$\qquad\quad\; | \qquad\qquad\; |$$
$$\qquad\quad \text{H} \qquad\qquad \text{H}$$

Hydrogen   Water   Hydronium
ion   ion

Reaction between a hydrogen ion and a water molecule to form a hydronium ion

G. Acetic Acid, CH₃COOH, and Water, H₂O: Partial Dissociation of a Molecular Substance into Ions; Ionization Equilibrium

Acetic acid is a liquid composed of polar molecules which are bonded together with hydrogen bonds. The hydrogen bonds link the molecules together in pairs:

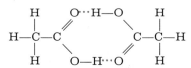

Hydrogen bonding between two molecules of acetic acid in the pure substance

FIGURE 5-8
Schematic representation of a solution of hydrated hydrogen ions, i.e., hydronium ions, and hydrated chloride ions resulting from the dissolving of hydrogen chloride in water.

CH₃COOH

Acetic acid
molecules

⇅

CH₃COO⁻

Acetate
ions

+

H⁺

Hydrogen
ions

**The ionization equilibrium
prevailing in an aqueous
solution of acetic acid**

When the acid is dissolved in water, the molecules become hydrogen-bonded with the water molecules and solution readily takes place. Moreover, a small proportion of the molecules dissociates into ions, which are hydrated in the aqueous solution (Fig. 5-9). If we were able to examine the solution, we would "see" that most of the acetic acid molecules are present as complete molecules hydrogen-bonded to nearby water molecules. Also present are small amounts of hydrated hydrogen ions and hydrated acetate ions. While the amounts of complete acetic acid molecules, hydrated hydrogen ions, and hydrated acetate ions remain constant, the solution is full of activity. The hydrogen and acetate ions are uniting to form acetic acid molecules, and the molecules are dissociating into ions. But since both these processes are taking place at the same rate, the amount of molecules and ions in the solution does not change.

This is another example of chemical equilibrium, discussed in Sec. 5-2E. Equilibrium is a condition of dynamic balance between two competing processes which take place at equal rates. The processes occurring at the molecular level continue but in a balance that causes no change on a large scale. The name **ionization equilibrium** is applied to the specific equilibrium condition prevailing in an aqueous solution of a substance when only some of its molecules are dissociated into ions. The ionization equilibrium prevailing in an aqueous solution of acetic acid is represented in symbols as shown in the margin. Vinegar is a dilute aqueous solution of acetic acid. The next time you toss a salad, you will know that acetic acid molecules, hydrogen ions, and acetate ions are all bouncing around on the lettuce.

5-3   GENERAL SOLUBILITY BEHAVIOR

It is not easy to make generalizations about solubility behavior that are widely applicable, but the summary of Table 5-1 is a useful guide.

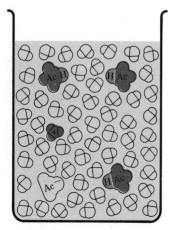

FIGURE 5-9
Schematic representation of the situation prevailing at the molecular level in an aqueous solution of acetic acid. The hydrogen ions are represented by hydronium ions, H₃O⁺, the form in which they are actually found. The acetic molecules are abbreviated HAc and the acetate ions Ac⁻.

Table 5-1
General Solubility Behavior

Description	Example
Nonpolar substances tend to be soluble in nonpolar solvents (free intermingling of weakly bound molecules)	Solubility of hexane in carbon tetrachloride
Nonpolar substances tend not to be soluble in polar solvents (inability of weakly bound molecules to penetrate groups of molecules bound moderately strongly)	Very low solubility of hexane in water; in general, low solubility of oils in water
Salts and substances comprising polar molecules tend to be soluble in polar solvents (free intermingling of polar and hydrogen-bonded molecules; formation of ion-dipole bonds in water)	Solubility of ethyl alcohol and sodium chloride in water (many salts, however, are insoluble)
Salts and substances comprising polar molecules tend to be not very soluble in nonpolar solvents (inability of weakly bound molecules to penetrate groups of molecules bound moderately strongly; lack of formation of ion-dipole bonds with nonpolar solvent molecules)	Low solubility of sodium chloride and water in carbon tetrachloride

WHAT IS ANTIFREEZE?

The expansion of water as it freezes to ice and the hazards of this expansion to pipes containing water have previously been discussed. Most automobile engines are cooled by removing the heat with water or aqueous solutions, which circulate through the engine and a cooling radiator. The aqueous solution of the cooling system must be protected against freezing at low temperatures. The ideal aqueous solution for liquid-cooled automobile engines is one which (1) does not freeze at winter temperatures, (2) does not corrode the cooling system, and (3) does not evaporate significantly at engine temperatures.

In an earlier short essay we discussed the mechanism whereby the freezing point of water is lowered by the presence of dissolved salts. By an equivalent mechanism the freezing point of water is lowered by other dissolved substances. The substance most commonly added to water to form a suitable solution for automobile cooling systems is ethylene glycol:

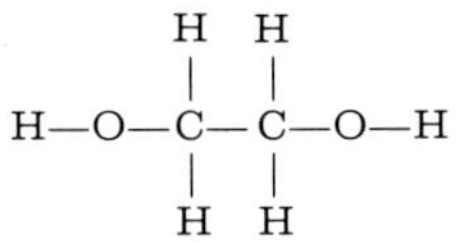

The principal cohesive force between the molecules of pure ethylene glycol is hydrogen bonding (Sec. 4-11), but since each molecule contains two hydrogen atoms bonded directly to oxygen atoms, the opportunities for hydrogen bonding are twice those in ethyl alcohol. The boiling point of ethylene glycol is therefore rather high, 198°C.

The high boiling point of ethylene glycol is an important advantage, greatly reducing its tendency to evaporate at cooling-system temperatures. Ethylene glycol also raises the boiling point of the solution; this retards evaporation of the solution and gives a solution that can be used year round as a cooling-system liquid. The hydrogen bonding in pure ethylene glycol also explains why it dissolves in water so easily (Sec. 5-2C). Solutions can be made concentrated enough to protect cooling systems down to temperatures as low as −40°C. Commercial antifreeze solutions also contain special substances to retard corrosion and plug up small leaks in the cooling system.

```
        H   H
        |   |
H—O—C—C—O—H
        |   |
        H   H
```

Ethylene glycol, the principal ingredient in most "permanent" antifreeze

5-4 HOW CAN SOLUBILITY BEHAVIOR
BE USED TO PURIFY SUBSTANCES?

Optional

The solubility of a substance in a given solvent varies with the temperature. The solubility of many substances in many solvents increases with temperature, in some cases markedly. Consider, for example, the behavior of potassium nitrate in water. At 0°C, 13 grams will dissolve in 100 grams; at 100°C, the solubility increases almost twentyfold to 247 grams in 100 grams. The solubility of some substances, however, decreases with temperature.

Recrystallization is a process of purifying solids which takes advantage of the varying solubilities of substances and their change with temperature. The process is best explained by giving an example. The substance *para*-nitroaniline, $C_6H_6N_2O_2$, is soluble in water to the extent of about 2.2 grams in 100 grams at 100°C. If the temperature of the solution is cooled to 18°C, the solubility shrinks to less than 0.08 grams in 100 grams. Imagine, therefore, an impure sample of *para*-nitroaniline contaminated with some impurities which are practically insoluble in water even at 100°C. It is also contaminated with impurities which are very soluble even at 0°C.

A sample of the impure *para*-nitroaniline is dissolved in a little more water than needed to dissolve it at 100°C. The insoluble impurities remain undissolved and are removed by pouring the solution through a filter at about 100°C. The filtered solution is then cooled in an ice bath to about 0°C. As the temperature of the solution decreases, the solubility of the *para*-nitroaniline decreases and the *para*-nitroaniline, comes out of solution and precipitates. The soluble impurities and a small amount of the *para*-nitroaniline, remain in solution (Fig. 5-10).

The precipitated *para*-nitroaniline, is now removed by filtration and washed free of the soluble impurities clinging to it. The crystals are then dried to remove the excess water. They have now been freed of water-soluble and water-insoluble impurities and are in a purer form than originally. Recrystallization is the most convenient method of purifying solids even though some of the substance is lost in the process (it remains in the solution after cooling). Several successive recrystallizations may be required to obtain the desired purity.

5-5 PURIFICATION BY DISTILLATION

Optional

At a given pressure most pure substances boil at a specific temperature which is characteristic of the substances. Solutions, however, boil over an extended range and at temperatures which depend both on the substances present and on their proportions. The vapor obtained upon the boiling of most solutions contains a higher concentration of the lower-boiling, more volatile constituents than the original mixture. If this is condensed by cooling, the solution obtained will be richer in the lower-boiling constituents than the original solution was. **Distillation** is the process of boiling a substance or mixture of substances and condensing the vapor obtained by cooling it. The product is known as a distillate.

If the original solution contains two substances which differ greatly in volatility, a complete separation can be achieved by

FIGURE 5-10
Steps in the recrystallization of *para*-nitroaniline from water:
(*a*) impure *para*-nitroaniline on a watch glass; (*b*) impure *para*-
nitroaniline stirred into water; (*c*) homogeneous hot solution obtained after removing insoluble material by filtration; (*d*) the
first crystals of *para*-nitroaniline form in the cooled solution;
(*e*) completed crystallization of *para*-nitroaniline. (*Photographs by
Hans Oettgen.*)

(*b*)

(*a*)

(*c*)　　　(*d*)

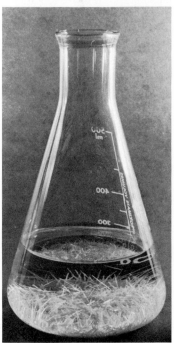

(*e*)

simple distillation. If a solution of sodium chloride in water is distilled, for example, the water boils off and is obtained on condensation in a pure form. The almost nonvolatile sodium chloride remains behind. A sketch of a laboratory apparatus for carrying out such a distillation is given in Fig. 5-11. In many areas where drinking water is scarce, distillation on a large scale is used to obtain fresh water from the sea.

If a solution contains two substances, both of which have appreciable volatility, the distillate will contain more of the more volatile constituent but complete separation is not achieved. If the distillation process is repeated, the second distillate will be even richer in the more volatile substance. A succession of distillations may eventually produce the more volatile substance in pure or nearly pure form. Successive distillations are carried out on a continuous basis in **fractional distillation.** The vapor is passed through a long fractionating column where provision is made for successive condensations and volatilizations. They return the less volatile substance to the distilling flask and permit the more volatile substance to pass out the top to the condenser.

FIGURE 5-11
Laboratory apparatus for a simple distillation.

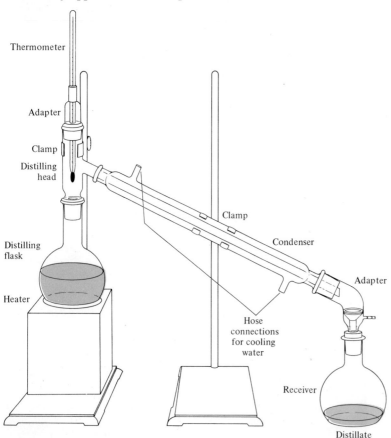

A substance can be freed of both more volatile and less volatile impurities in such a process. The more volatile impurities pass over to the condenser and are removed before the substance passes over. The less volatile impurities collect in the distilling flask. Large-scale fractional distillation is very important in the petroleum industry. Fractionating columns as high as 100 feet are used to separate crude petroleum into several fractions at various boiling-point ranges, such as gasoline (range 30 to 200°C), kerosine (range 175 to 325°C), diesel fuel (about 275°C). Much of the jungle of pipes and columns seen in a petroleum refinery is devoted to distillation. Petroleum refining is discussed further in Sec. 13-11.

"We should be thankful. What if oil and water *did* mix!"

© 1972 American Scientist; reprinted by permission of Sidney Harris.

KEY WORDS

1. Solution (Sec. 5-1)
2. Solvent (Sec. 5-1)
3. Solute (Sec. 5-1)
4. Saturated solution (Sec. 5-1)
5. Ion-dipole force (Sec. 5-2D)
6. Hydration (Sec. 5-2D)
7. Chemical equilibrium (Sec. 5-2E)
8. Solubility equilibrium (Sec. 5-2E)
9. Ionization equilibrium (Sec. 5-2G)

Optional:

10. Recrystallization (Sec. 5-4)
11. Distillation (Sec. 5-5)
12. Fractional distillation (Sec. 5-5)

SUMMARIZING QUESTIONS FOR SELF-STUDY

Section 5-1

1. Q. What is the difference between a substance and a mixture?

A. A substance is a form of matter which is of definite composition and consistent behavior. A mixture contains two or more substances.

2. Q. Give definitions for (a) solution, (b) solvent, (c) solute, and (d) saturated solution.

A. (a) A homogeneous mixture of two or more substances dispersed in each other at the fundamental-particle level; most commonly a liquid mixture of a liquid substance with a solid, a gas, or another liquid. (b) The material which acts as the dissolving medium in a solution; most commonly a liquid. (c) The material which is dissolved in a solution. (d) A solution in which the maximum amount of solute is dissolved.

Sections 5-2 and 5-3

3. Q. What is the basis for the general statement "like dissolves like"?

A. In order for one substance to dissolve in another, the particles of both substances must be attracted to each other by cohesive forces of about the same strength, so that the particles of both substances can mingle freely with each other.

4. Q. Why does water dissolve so many substances?

A. The polar nature of the water molecule favors the solution of polar molecules. The polar nature also makes possible the generation of ion-dipole attractive forces with the ions of a salt. These favor the solution of salts. Water molecules are also able to form hydrogen bonds with many solutes, and these are responsible for water's ability to dissolve many hydrogen-bonded solutes.

5. Q. Why don't petroleum oil and water mix?

A. Petroleum oil is made up of nonpolar molecules (like hexane) held together by van der Waals forces. The moderately strong polar forces and hydrogen bonds operating between the water molecules "squeeze out" the weakly bonded oil molecules, and solution does not occur to any appreciable extent.

6. Q. What are the principal molecular species present in an aqueous solution of sodium chloride?

A. Hydrated sodium ions (commonly called sodium ions), hydrated chloride ions (commonly called chloride ions), and water molecules.

7. Q. What are the principal molecular species present in an aqueous solution of hydrogen chloride?

A. Hydrated hydrogen ions (actually hydronium ions, commonly called hydrogen ions), hydrated chloride ions (commonly called chloride ions), and water molecules.

8. Q. What are the principal molecular species present in an aqueous solution of acetic acid?

A. Acetic acid molecules, water molecules, hydrated hydrogen ions (hydronium ions), and hydrated acetate ions.

9. Q. What is meant by a chemical equilibrium?

A. A condition of dynamic balance between two competing processes which take place at equal rates.

10. Q. What is meant by an ionization equilibrium, and how does it apply to an aqueous solution of acetic acid?

A. An ionization equilibrium is the specific equilibrium condition prevailing in an aqueous solution of a substance only some of whose molecules are dissociated into ions. The ionization equilibrium prevailing in an aqueous solution of acetic acid involves the processes of dissociation and ion combination occurring at equal rates. This condition can be written in symbols as $CH_3COOH \rightleftharpoons CH_3COO^- + H^+$.

11. Q. What is meant by a solubility equilibrium, and how does it apply to a saturated aqueous solution of silver chloride?

A. A solubility equilibrium is the specific equilibrium condition prevailing in a saturated solution in contact with undissolved solute. If the solute is silver chloride, the silver and chloride ions are both leaving the undissolved crystals to enter the solution and returning to the crystals from the solution. These competing processes take place at equal rates. This condition can be represented in symbols: $AgCl$ (undissolved) $\rightleftharpoons AgCl$ (dissolved)

Section 5-4 (Optional)

12. Q. How can recrystallization serve to purify a substance?

A. The impure substance is dissolved at the boiling point of a solvent. The insoluble impurities do not dissolve and are removed by filtration. The solution is cooled, and as the solubility of the solute decreases, the portion no longer soluble precipitates and is removed by filtration. The soluble impurities remain in solution.

Section 5-5 (Optional)

13. Q. How can distillation serve to purify a substance?

A. An impure volatile substance can sometimes be separated from nonvolatile impurities by boiling off the substance and recovering it by condensation. The nonvolatile impurities remain behind in the distillation flask. If the substance to be purified is less volatile than the impurities, the impurities can sometimes be selectively removed by boiling, leaving the substance behind. Successful purifications by distillation often call for an elaborate means of separating substances according to their relative volatility.

PRACTICE EXERCISES

1. Match each definition or other statement with the numbered term above with which it is most closely associated. A numbered item may be used only once.

   1. Solvent
   2. Sublimation
   3. Concentrated solution
   4. Ionization equilibrium
   5. Solution
   6. Saturated solution
   7. Ion-dipole force
   8. Solute
   9. Chemical equilibrium

   (a) A homogeneous mixture of two or more substances dispersed in each other at the molecular level
   (b) The cohesive force between an ion and a polar molecule
   (c) A material which acts as the dissolving medium in a solution
   (d) A condition of dynamic balance between two competing processes which take place at equal rates
   (e) A material which is dissolved in a solution
   (f) A specific equilibrium condition prevailing in an aqueous solution of a substance only some of whose molecules are dissociated into ions.
   (g) A solution in which the maximum amount of solute is dissolved

2. Can a saturated solution be a dilute solution? Explain.

3. Name three types of cohesive forces which can operate between the fundamental particles of a substance which is a liquid under ordinary circumstances.

4. Why are solutions of special importance in chemistry?

5. Account in terms of concepts at the molecular level for the generalization that (a) like dissolves like and (b) oil and water do not mix.

6. From the viewpoint of concepts at the molecular level why wouldn't the ionic solid substance potassium bromide, KBr, be expected to dissolve in the nonpolar liquid substance carbon tetrachloride, CCl_4?

7. Using structural formulas of water, H_2O, and ethyl alcohol, CH_3CH_2OH, represent a portion of a solution of ethyl alcohol in water diagrammatically.

8. In terms of concepts at the molecular level explain the meaning of the statement that all ions in aqueous solution are hydrated.

9. What is the difference between a solubility equilibrium and an ionization equilibrium?

10. Calcium carbonate, $CaCO_3$, is a salt which is only very slightly soluble in water, H_2O. Represent in symbols the chemical equilibrium which prevails in a saturated aqueous solution of calcium carbonate.

11. Hydrogen cyanide, HCN, is a gas which dissolves readily in water and dissociates slightly as it does into hydrogen ions, H^+, and cyanide ions, CN^-. Represent in symbols the chemical equilibrium which prevails in an aqueous solution of hydrogen cyanide.

12. Criticize the following statement: the molecules of hydrogen chloride in an aqueous solution are dispersed homogeneously throughout the solution.

13. Criticize the following statement: the ionic lattice of a crystal of undissolved silver chloride in contact with a saturated aqueous solution of silver chloride remains completely intact.

14. Account briefly for each of the following in terms of concepts at the molecular level:
    (a) Nonpolar substances tend to be soluble in nonpolar solvents.
    (b) Salts and substances composed of polar molecules tend not to be very soluble in nonpolar solvents.
    (c) Salts and substances composed of polar molecules tend to be soluble in polar solvents.
    (d) Nonpolar substances tend not to be soluble in polar solvents.

15. Explain in terms of concepts at the molecular level why sodium chloride, NaCl, dissolves readily in water but silver chloride, AgCl, does not.

Section 5-4 (Optional)

16. When the process of recrystallization is used to purify a substance, how are the impurities removed? What happens to them?

17. Can recrystallization from water be used to purify a substance which is very soluble in water? Explain.

18. Can recrystallization from water be used to purify a substance which is only very slightly soluble in water over a wide range of temperature? Explain.

Section 5-5 (Optional)

19. When the process of distillation is used to purify a substance, how are the impurities removed? What happens to them?

The following questions assume a knowledge of Supplement 5. Express units, show your method clearly, and label the answer.

20. What is the percent by weight of hexane, C_6H_{14}, in a solution with carbon tetrachloride, CCl_4, as the solvent if 33.00 g is dissolved in 86.00 g of solution?

21. How many grams of ethyl alcohol, CH_3CH_2OH, are contained in 76.00 g of solution which is 20.00 percent by weight?

22. How many grams of sodium chloride, NaCl, are contained in 32,000 g of an aqueous solution which has a concentration of 90 ppm (parts per million parts by weight)?

23. If 46.00 g of sodium chloride is contained in 300 mL of an aqueous solution, what is the molarity of the solution?

24. How many grams of acetic acid, CH_3COOH, are contained in 1200 mL of a 1.300 M solution?

6

WHAT IS THE NATURE OF CHEMICAL REACTIONS?

6-1 INTRODUCTION

A current hypothesis concerning the origin of life holds that it emerged spontaneously through the gradual evolution of the complex molecular species of living organisms. Evidence to support this hypothesis is growing. It is proposed that the precursors were substances of very simple structure such as water, H_2O, methane, CH_4, ammonia, NH_3, and hydrogen, H_2, present in the early stages of the earth's formation. Under the influence of energy from the sun, lightning, and the heat of the earth, these substances were converted into larger molecules, which in turn became transformed into the elaborate molecules of living organisms, such as proteins and nucleic acids. When molecules of the necessary complexity and variety had evolved, they became arranged into the organized structures of living cells.

The substances of living organisms and many other substances are very susceptible to the action of oxygen, O_2, and oxygen-containing compounds. Because of such action, paper disintegrates, cracks form in rubber, and dead plants and animals decompose. Yet oxygen is essential to the well-being of mammals, making a critically important contribution to the process whereby energy is obtained from food. Oxygen is both essential for life and a potential hazard.

It has been suggested that the action of oxygen and oxygen-containing compounds damages living cells and is responsible for the process we call aging. Not all the damage is repaired by cellular processes. Aging could involve either an increase in the rate of damage production or a decrease in the rate of repair or both. Provocative supporting evidence for this hypothesis is the fact that the lifespan of mice is noticeably extended if substances capable of counteracting the action of oxygen are included in their daily diet. It is significant that these hypotheses concerning the origin of life and the onset of old age depend on knowledge of chemical changes which matter undergoes.

Chemical changes are, of course, occurring everywhere about us. They generate the energy which propels us in automobiles, jet planes, and rockets (Fig. 6-1); they produce the materials of living organisms; and they capture energy from the sun which is vital to all forms of life (Fig. 6-2). They make it

FIGURE 6-1
Launching of the Atlas-Centaur rocket. The energy to launch the rocket is produced by a chemical reaction. (*Courtesy of NASA.*)

possible to synthesize materials previously unknown, providing tools for combating pain and disease and creating economic wealth. Unfortunately chemical changes produce much of the pollution which is degrading our environment, but they also furnish the means of removing some of the pollutants.

FIGURE 6-2
Green leaves of a bean plant. The process of photosynthesis (see Sec. 11-3), which harnesses the energy of the sun, involves many chemical reactions. (*Photograph by Hans Oettgen.*)

6-2 WHAT ARE CHEMICAL REACTIONS?

A **chemical reaction** is formally defined as a process in which substances are converted into other substances with different sets of properties. **Reactant** is the name applied to a starting substance, and **product** is used to describe a substance formed. Chemical reactions involve gases, liquids, and solids. Gases, for example, react with other gases or with liquids or solids. Solids react with gases, liquids, or other solids. Usually, however, chemical reactions involve liquid solutions of gases, liquids, and solids.

To illustrate a chemical reaction, let us consider a common treatment for one kind of upset stomach. The digestive fluid of the human stomach, called gastric juice, contains many substances in an aqueous solution, one of the more important being hydrochloric acid. Its function is to maintain an environment at the molecular level which permits the digestive processes of the stomach fluid to function properly. The hydrochloric acid is present in the form of the molecular species hydrogen ions, H^+, and chloride ions, Cl^-. These are the same species obtained when the gas hydrogen chloride is dissolved in water (Sec. 5-2F). Under abnormal conditions the concentration of the hydrogen ion may become higher than it should be, causing a stomach upset. To treat the upset some of the acid must be removed. Agents used for this purpose are called gastric antacids. The active ingredient in one of the most effective antacids is calcium carbonate, $CaCO_3$, often administered in the form of chewable tablets.

Calcium carbonate, $CaCO_3$, is a white solid, almost insoluble in water. When it reaches the gastric juice, it converts hydrogen ions into water molecules by a chemical reaction. During the process the carbonate ions, CO_3^{2-}, are converted into molecules of the gas carbon dioxide, CO_2. The reaction which takes place can be summarized as follows:

Hydrogen chloride gas in solution (hydrochloric acid) as H^+ and Cl^- ions + calcium carbonate as a solid containing Ca^{2+} and CO_3^{2-} ions → carbon dioxide gas as CO_2 molecules + water as H_2O molecules + calcium chloride in solution as Ca^{2+} and Cl^- ions

The reaction occurring when the gastric antacid calcium carbonate gets rid of excess hydrochloric acid in the stomach

This reaction can be simulated in the laboratory by stirring calcium carbonate into a dilute solution of hydrochloric acid (Fig. 6-3).

FIGURE 6-3
The chemical reaction between calcium carbonate, $CaCO_3$, and dilute hydrochloric acid, HCl, as carried out in the laboratory: the solid calcium carbonate (a) is added to the hydrochloric acid solution (b), and bubbles of carbon dioxide, CO_2, evolve (c) as the reaction takes place. (*Photographs by Hans Oettgen.*)

(a)   (b)   (c)

In summary, a white solid ($CaCO_3$) and a water-soluble gas in the form of ions (H^+, Cl^-) are converted into a gas (CO_2), a water-soluble solid in the form of ions (Ca^{2+}, Cl^-), and water. A transformation known as a **chemical reaction** takes place. Reactants with characteristic sets of properties are converted into products with different sets of properties.

6-3 WHAT ARE CHEMICAL EQUATIONS?

A **chemical equation** is the representation of a reaction in formulas which specifies each reactant, each product, and the relative number of molecules of each. To illustrate the nature of chemical equations, let us recall our summary of the action of the gastric antacid, calcium carbonate, discussed in Sec. 6-2. Here is the equation for this reaction:

$$2HCl + CaCO_3 \rightarrow CO_2 + H_2O + CaCl_2$$

Hydrochloric acid   Calcium carbonate   Carbon dioxide   Water   Calcium chloride

The reaction between calcium carbonate and hydrochloric acid represented by a chemical equation

The equation specifies succinctly each reactant and each product. All the information about the substances conveyed by formulas is included. Notice, however, that some important details are not specified in the equation. The fact, for example, that the hydrogen chloride and calcium chloride participate as component ions is not explicitly indicated.

The 2 in front of the formula for the hydrochloric acid is of special interest. This becomes part of the equation in the process of making it a balanced equation.

6-4 HOW IS A CHEMICAL EQUATION BALANCED?

To illustrate the process of equation balancing we consider a chemical reaction which is very important in industrial chemistry and in the food supply. As food is grown and harvested, essential plant nutrients are removed from the soil. These important nutrients include substances containing nitrogen. Although some natural processes replenish these nitrogen compounds in part, they usually do not do so in large enough amounts to grow food supplies for the increasing human population (Sec. 11-9). Nitrogen-containing fertilizers are used as soil supplements. Production of these fertilizing substances requires huge amounts of the substance ammonia, NH_3, which does not occur abundantly in nature and must be synthesized. Fortunately, it is readily produced from hydrogen and nitrogen by a chemical reaction represented by the balanced equation on the next page.

This equation tells us that each molecule of nitrogen, N_2, requires three molecules of hydrogen, H_2, to form two molecules of ammonia, NH_3. The number 3 in front of the formula for hydrogen, H_2, applies to the whole formula, and indicates three molecules. The subscript 2 after the H indicates that there are two hydrogen atoms in each molecule of hydrogen. Similarly, the number 2 in front of the formula for ammonia, NH_3, indicates two molecules, and the subscript 3 after the H indicates that there are three atoms of hydrogen in each molecule of ammonia.

$$N_2 \ + \ 3H_2 \ \rightarrow \ 2NH_3$$

Nitrogen   Hydrogen   Ammonia

**Equation for the reaction
between nitrogen and
hydrogen to form ammonia**

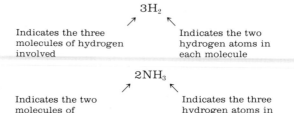

3H₂

Indicates the three   Indicates the two
molecules of hydrogen   hydrogen atoms in
involved   each molecule

2NH₃

Indicates the two   Indicates the three
molecules of   hydrogen atoms in
ammonia involved   a molecule of ammonia

An equation is balanced only if it indicates the same total number of each kind of atom among products as among reactants. The following equation, for example, is not balanced:

$$N_2 + H_2 \rightarrow NH_3$$

**Unbalanced equation for the synthesis of ammonia, NH_3,
from hydrogen, H_2, and nitrogen, N_2**

Two atoms of nitrogen are indicated among the reactants and only one in the product. Two atoms of hydrogen are indicated among the reactants and three in the product. This equation can be brought into balance by writing a 3 in front of the formula for the hydrogen molecule and writing a 2 in front of the formula for ammonia:

$$N_2 + 3H_2 \rightarrow 2NH_3$$

**Balanced equation for the synthesis of ammonia, NH_3,
from hydrogen, H_2, and nitrogen, N_2**

The equation now specifies two atoms of nitrogen in both reactants and product. It also indicates six atoms of hydrogen in the product and in the reactants. In bringing an equation into balance only the number of molecules may be changed. The formula for each substance must remain intact. If the number of atoms within a formula were to be changed, the formula would no longer represent the original substance correctly. (Although no ions are involved in this reaction, we shall see later that for the purposes of equation balancing no distinction is made between atoms and ions.)

A *complete equation* is one in which each product and each reactant is indicated by a formula. Many equations, like the one for the synthesis of ammonia, can be balanced by inspection. Others, however, require special methods for balancing which need not be discussed here.

Many chemical reactions are not understood in sufficient detail to permit writing balanced equations. Complete and balanced equations require the identity of all of the products. For many reactions only one or two of several products have been identified. In such cases the changes are represented by very useful equationlike expressions, which do not represent complete reactions and cannot be brought into balance.

We know, for example, that the vitamin niacin will react with ethyl alcohol in the presence of sulfuric acid to form ethyl nicotinate and some other substances known as by-products. The chemical reaction can be usefully represented in an equationlike manner, but this representation shows only the principal product and does not pretend to be complete or balanced:

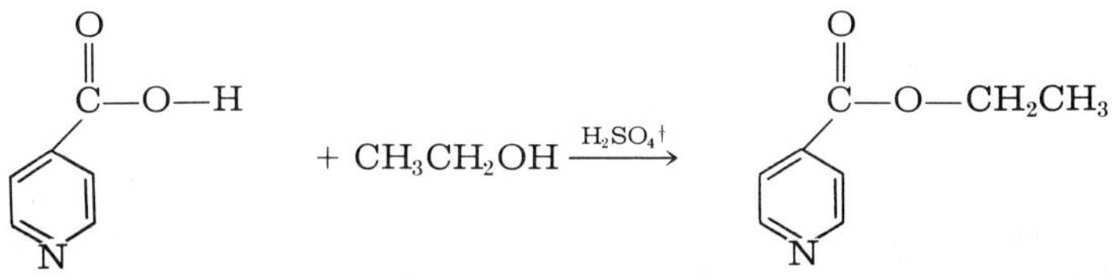

| Niacin | Ethyl alcohol | Ethyl nicotinate |

Equationlike representation of the reaction between niacin and ethyl alcohol; only the principal product is indicated, and no attempt is made to balance the atoms; this kind of representation is commonly used for reactions too complicated to be represented by a balanced equation.

6-5 WHAT IS THE SIGNIFICANCE OF BALANCING AN EQUATION?

The wealth of information about a chemical reaction represented by a chemical equation includes the relative number of molecules of each participating substance. In order for an equation to supply such information it must be balanced. In bringing an equation into balance, we are expressing the reaction in accordance with the law of conservation of mass (Fig. 6-4). It was the eighteenth-century French chemist Antoine Lavoisier (Fig. 6-5) who, in measuring the amounts of substances consumed and produced during chemical changes, first proposed this generalization. The **law of conservation of mass** may be stated as follows: during a chemical reaction the amount of mass remains constant. In the reaction for the synthesis of ammonia, for example, the sum of the amounts of hydrogen and nitrogen consumed is equal to the amount of ammonia produced. While we now know that the law of conservation of mass does not hold strictly, the amount of matter lost or created in most chemical changes is so small that it cannot be detected by the best available scales. For most practical purposes, therefore, the law remains a useful generalization. The significance of the exceptions will be discussed in Sec. 9-7.

In the early nineteenth century, the English chemist John Dalton interpreted the law of conservation of mass by proposing

---

† It is conventional to represent special conditions such as temperature, pressure, and catalysts by words or symbols above and below the arrow. In this case H_2SO_4 is the formula for the sulfuric acid catalyst.

that changes in matter involve only a regrouping of atoms among molecules with none lost or gained; the atoms themselves preserve their identities. In balancing a chemical equation we are following Dalton's insight by making certain that the total number of each kind of atom represented among the reactants is also represented among the products.

FIGURE 6-4
An experiment to demonstrate the law of conservation of mass: (*a*) a candle is burned in a closed flask of oxygen placed on a balance; (*b*) at the end of the experiment most of the candle has burned, consuming oxygen and forming carbon dioxide and water, but the flask and its contents have not changed weight. This indicates that mass was neither created nor destroyed during the chemical reactions of the combustion process. (*Photographs by Fritz Goro, Life Magazine. © 1949, Time, Inc.*)

FIGURE 6-5
Antoine Lavoisier (1743–1794), French lawyer and government administrator, who became interested in an extraordinarily broad range of scientific problems. His emphasis on quantitative relationships, such as his law of the conservation of mass, was of great importance in establishing chemistry as a science. (*Smithsonian Institution.*)

6-6 WHAT AMOUNT OF A SUBSTANCE REACTS, AND HOW MUCH PRODUCT IS OBTAINED?

133

6-6 WHAT AMOUNT OF A
SUBSTANCE REACTS, AND HOW
MUCH PRODUCT IS OBTAINED?

Questions often asked about chemical reactions by chemists, physicians, and chemical engineers relate to the amounts of substances reacting and the quantities of substances formed. To counteract excess hydrochloric acid in the stomach what dosage of antacid should be prescribed? To produce 1000 tons of ammonia, how much nitrogen and hydrogen are needed? To answer such questions one must know enough about the reaction involved to write an equation and one must have information about the weights of atoms and molecules themselves.

A. The Weights of Atoms

The weights of atoms represent the force exerted on their masses by the earth's gravity. Individual atoms have, of course,

---

WHAT IS A SCIENTIFIC LAW?†

The law of the conservation of mass was formulated by Antoine Lavoisier late in the eighteenth century. He had been investigating the weight relationships associated with chemical changes and noticed that in the changes he studied the sum of the weights of the substances undergoing reaction was the same as the sum of the weights of the products. For example, when iron was heated in oxygen, the following change occurred:

$$\text{Iron} + \text{oxygen} \xrightarrow{\text{heat}} \text{iron oxide}$$

Typically, he found that 7.2 grams of iron oxide is formed from 5.3 grams of iron and 1.9 grams of oxygen. He had started with 7.2 grams of matter and had ended with 7.2 grams of matter. During the reaction there was no loss of mass.

As we look back on Lavoisier's work, the observed facts appear to lead directly to the conclusion that during a chemical reaction the amount of mass remains constant. If we could view all the various observations about chemical changes Lavoisier had before him and consider them in the context of what was known about the behavior of matter at the end of the eighteenth century, we would realize that the law of conservation of mass did not emerge spontaneously from the facts. On the contrary, it required considerable intuitive insight for Lavoisier to sort out the important observations and to appreciate which of them were fundamental.

It is significant that Lavoisier formulated the generalization to apply not only to the reactions he had studied, which were few, but to the weight relationships of all possible reactions. The soundness of Lavoisier's assumption was soon confirmed by the experimental observations of other chemists.

More recent observations have indicated, however, that the law is not entirely accurate (Sec. 9-7). While it holds within measurable limits for ordinary chemical changes, it cannot be applied to nuclear changes. The laws of science are frequently found to be in need of revision in the light of new facts, or sometimes abandoned altogether, but this does not invalidate their usefulness. Lavoisier's law of conservation of mass, for example, was one of the key generalizations about the behavior of matter that Dalton used in formulating his theory of the structure of matter. The law has been of immense value in numerous other instances, as it has served to direct attention to important fundamental relationships.

Laws of science are generalizations which express fundamental relationships between factual observations. They are formulated by those who have the background of information and high intelligence to recognize the critical observations. Laws are assumed to apply to all comparable situations beyond those already observed, but since there is always a chance that the future observations may not be in agreement with a law, the provisional nature of a law is never removed. Despite this limitation they are of great assistance in relating and interpreting factual observations.

† Adapted from Charles Compton, "An Introduction to Chemistry," pp. 236–237, Van Nostrand Reinhold Company, New York, 1958.

inconveniently small weights. The weight of a hydrogen atom, for example, is

$$0.0000000000000000000001673 \text{ gram}$$
$$= 1.673 \times 10^{-24} \text{ gram}$$

In order to make the weights of atoms more manageable they are expressed, by international agreement, on a relative scale chosen so that the relative weight of even the smallest atom, hydrogen 1, is greater than 1. This is done by assigning an arbitrary weight to the carbon 12 isotope of 12.0000 units and expressing all other weights relative to this standard. On this scale the weight of a hydrogen 1 atom is 1.0078 units. The absolute and relative weights of some naturally occurring atoms are given in Table 6-1.

Table 6-1
The Absolute and Relative Weights of Some Atoms

Atom	Absolute weight, grams	Relative weight on scale of carbon 12 = 12.0000
Hydrogen 1	1.673×10^{-24}	1.0078
Hydrogen 2	3.345×10^{-24}	2.0141
Hydrogen 3	5.008×10^{-24}	3.0160
Helium 3	5.008×10^{-24}	3.0160
Helium 4	6.647×10^{-24}	4.0026
Carbon 12	1.993×10^{-23}	12.0000
Carbon 13	2.159×10^{-23}	13.0034
Carbon 14	2.325×10^{-23}	14.0032
Uranium 234	3.886×10^{-22}	234.0409
Uranium 235	3.903×10^{-22}	235.0439
Uranium 238	3.953×10^{-22}	238.0508

The atoms of most elements occur in nature as a mixture of isotopes, each with its own characteristic weight. It has been mentioned (Sec. 2-5) that naturally occurring hydrogen atoms are composed of the hydrogen 1 isotope, protium (99.985 percent), the hydrogen 2 isotope, deuterium (0.015 percent), and an extremely small amount of the hydrogen 3 isotope, tritium. Consequently, the atomic weight of a hydrogen atom and of the atoms of other elements is an average value, which reflects the weights of the individual isotopes and their distribution in nature. This value for hydrogen is 1.0079 units. In summary, the **atomic weight** is the weight of an average atom of an element, relative to the weight of a carbon 12 isotope taken as 12.0000 units. Some additional values are given in Table 6-2. A complete list is given in Table S4-1.

Table 6-2
The Relative Weights of
Atoms of a Few Elements
Values averaged over
the natural isotopic
abundancies

Element	Relative weight (carbon 12 = 12.0000)
Hydrogen	1.0079
Helium	4.0026
Carbon	12.011
Oxygen	15.9994
Nickel	58.70
Copper	63.546
Silver	107.868
Uranium	238.029

B. The Weights of Molecules

The relative weight of a molecule of a substance can be obtained by adding the individual relative atomic weights of the atoms in the molecule of the substance. These relative molecular weights are ordinarily referred to simply as molecular weights. The molecular weight of the element hydrogen, H—H, for example, can be obtained by adding together two atomic weights of hydrogen:

135

6-6 WHAT AMOUNT OF A
SUBSTANCE REACTS, AND HOW
MUCH PRODUCT IS OBTAINED?

Atomic weight of hydrogen, H	1.0079
Atomic weight of hydrogen, H	1.0079
Molecular weight of hydrogen, H_2	2.0158

The molecular weight of water, H_2O, is obtained as follows:

Atomic weight of hydrogen, H, 1.0079×2	2.0158
Atomic weight of oxygen, O	15.9994
Molecular weight of water, H_2O	18.0152

One may similarly obtain the molecular weight of ethyl alcohol as follows (molecular formula C_2H_6O):

Atomic weight of carbon, C, 12.011×2	24.022
Atomic weight of hydrogen, H, 1.0079×6	6.0474
Atomic weight of oxygen, O	15.9994
Molecular weight of ethyl alcohol, C_2H_6O	46.069

(The use of significant figures in the expression of the results of numerical calculations is discussed in Supplement 2, Sec. S2.2.)

C. The Weight Relationships in Chemical Reactions

With the use of molecular weights and balanced equations the weight relationships between the reactants and products of a reaction can be determined. Consider, for example, the formation of ammonia from nitrogen and hydrogen. First the relevant balanced equation is written. Then the molecular weights of the reactants and products are determined. Finally the relative weights of the substances involved are found, based on the balanced equation and the molecular weights. (Note that relative weights of the substances involved are in accordance with the law of conservation of mass: the sum of the weights of the reactants is equal to the weight of the product.)

Balanced equation:	$3H_2$	$+$	N_2	$\rightarrow$	$2NH_3$	(1)
Molecular weight of substances:	2.0158		28.0134		17.0304	(2)
Relative weight of substances:	$3 \times 2.0158 = 6.0474$		$1 \times 28.0134 = 28.0134$		$2 \times 17.0304 = 34.0608$	(3)

Steps in determining the relative weights of substances participating in a reaction; note that the sum of the weights of the reactants is equal to the weight of the product

Note to Students

The quantitative aspects of chemistry are very important to the science. They provide the foundation for most of the fundamental concepts and are a significant part of the spirit of chemistry. They are not stressed in the body of this text in order to provide space for topics which are usually of more direct interest to the nonscientist. The weight relationships of chemical reactions are discussed further, however, in Supplement 4.

On the basis of the relative weights of the substances, 6.0474 grams of hydrogen, H_2, is expected to react with 28.0134 grams of nitrogen, N_2, to form 34.0608 grams of ammonia, NH_3, assuming that all the hydrogen and nitrogen undergo reaction.

6-7 WHY IS ENERGY EVOLVED IN SOME REACTIONS BUT ABSORBED IN OTHERS?

Energy is involved in all chemical changes, from the relatively simple combinations of hydrogen and nitrogen to form ammonia to the complex reactions which occur when wood is burned. The consideration of energy and its effects is one of the most

---

WHAT IS A SCIENTIFIC THEORY?†

The English chemist John Dalton proposed his brilliant hypothesis about the structure of matter early in the nineteenth century. In so doing he made use of a considerable background of facts about the behavior of matter and concepts about the structure of matter which had been put forth in preceding years. His thinking was particularly influenced by the proposals of the atomists of ancient Greece and the seventeenth-century concept of the English chemist Robert Boyle and the English mathematician and physicist Isaac Newton that matter is made up of particles. Lavoisier's law of the conservation of mass was a direct stimulation, as well as other laws about the composition and behavior of substances just formulated.

Dalton was most immediately concerned with devising an explanation of the various established facts and laws about matter which would interpret and connect them. In proposing that all substances are composed of atoms and that chemical changes involve only a rearrangement of atoms from one set of substances to another with none lost or gained, he was attempting to account for the law of conservation of mass. But he was also attempting to account for the law by a set of proposals about the structure of matter which would at the same time account for all the other established facts and laws of the behavior of matter.

His proposals were not immediately accepted by the scientific community as a whole. It was not until many years later that the consequences of his hypothesis were tested in a broad enough range of related areas for his tentative proposal to be generally accepted and elevated to a theory. It is ironic that most of the proposals contained in Dalton's original theory were proved wrong as

additional facts about the behavior of matter came to light. Dalton maintained, for example, that the atoms were the ultimate particles of matter and that an atom of a given kind was indivisible and unchangeable. When the electron was discovered, with about $\frac{1}{2000}$ times the mass of the smallest of Dalton's atoms, the concept of atoms as the ultimate particles of matter was destroyed. And when radioactivity was recognized as being the result of the spontaneous decomposition of atoms, the indivisibility and unchangeability of Dalton's atoms became invalid.

But Dalton's theory was of immense use despite these shortcomings, since it served, as all theories in science do, to direct attention to fruitful lines of inquiry and new discoveries. Dalton's proposal, for example, that the atoms of a compound are held together in a group (now called a molecule) immediately touched off very productive inquiries into the nature of the bonds which link the atoms together in these groups. A complete account of the results of Dalton's provocative proposal would fill many pages.

Scientific theories are explanations which interpret and connect numerous relevant facts through broad logical relationships, direct attention to important areas of inquiry, and lead to new discoveries. Human insight and judgment are of paramount importance in the formulation of the initial hypothesis. The hypothesis is elevated to a theory if the consequences deduced from it are proved to be in accordance with observations made to test it. The tentative character of a theory, however, is never completely removed; its continued acceptance remains a matter of probability. A "correct" theory is one which successfully embraces the relevant information of its own time.

† Adapted from "An Introduction to Chemistry," pp. 237–239, Charles Compton, Van Nostrand Reinhold Company, New York, 1958.

important aspects of chemistry, but giving a precise and useful definition of this intangible quality is not easy. We have already described energy as the capacity to do work (Sec. 2-3), e.g., the capacity to move a book from one end of a desk to the other. Setting the book in motion against the resisting force of friction involves work. Work is involved whenever a force is applied through a distance. But work is only one manifestation of energy. Energy can take many forms: heat energy, mechanical energy, electrical energy, light energy, and chemical energy, to name five.

Energy is a characteristic of all forms of matter. Many chemical changes bring about the release of some of this energy, making it available, for example, in propelling rockets, heating buildings, moving of a finger, and blinking an eye. Other reactions absorb energy. Photosynthesis (Sec. 11-3), the chemical process whereby plants convert carbon dioxide and water into sugar and other carbohydrates, involves reactions which absorb energy from the sun.

A. Bond Energies

In discussing energy and chemical changes it is useful to focus on the energy changes involved in breaking interatomic bonds and forming new bonds. In Sec. 2-6 we briefly described the forces which serve to attract atoms to each other in covalent bonds. From an overall point of view the two atoms of a hydrogen molecule, for example, are held together because of their mutual attraction for the pair of electrons they share. We can now add that the arrangement of electrons and nuclei in the hydrogen molecule contains less energy than the arrangement prevailing in two separate hydrogen atoms. The extra energy required by the separate atoms is given off as the bond is formed. It follows that if energy is evolved when a covalent bond forms, energy will be required to break the bond. The **bond energy** is defined as the amount of energy required to break a bond. The bond between the two atoms of a hydrogen molecule is a single covalent bond and involves only one pair of electrons. The same considerations apply, however, to double and triple bonds. Moreover, since double bonds involve more electrons, more energy is required to break a double bond than to break a single bond, and breaking triple bonds requires even more energy.

B. Energy Changes in Chemical Reactions

Case 1 We now turn our attention to the bond breaking and bond formation taking place as a chemical reaction occurs. The equation for the synthesis of ammonia from hydrogen and nitrogen (Sec. 6-4) is repeated in the margin. If we rewrite this equation, using the structural formulas for each molecule involved, we get

$$\text{H—H} + \text{H—H} + \text{H—H} + \text{N≡N} \rightarrow \text{H—N—H} + \text{H—N—H}$$
$$\qquad\qquad\qquad\qquad\qquad\qquad\qquad\quad |\qquad\qquad\quad |$$
$$\qquad\qquad\qquad\qquad\qquad\qquad\qquad\quad \text{H}\qquad\qquad\quad \text{H}$$

Equation for the reaction between hydrogen and nitrogen to form ammonia showing the individual bonds involved

137

6-7 WHY IS ENERGY EVOLVED IN SOME REACTIONS BUT ABSORBED IN OTHERS?

$\text{H·} + \text{H·} \rightarrow \text{H:H} + \text{energy}$

Hydrogen   Hydrogen
atoms   molecule

The arrangement of nuclei and electrons in two separate hydrogen atoms (left) requires more energy than the arrangement in the molecule (right); the extra energy is evolved as the bond is formed

$\text{Energy} + \text{H:H} \rightarrow \text{H·} + \text{H·}$

Hydrogen   Hydrogen
molecule   atoms

The arrangement of nuclei and electrons in the molecule (left) requires less energy than the arrangement in the two separate atoms (right); energy is therefore required to separate the two hydrogen atoms; i.e., energy is required to break the bond

$3\text{H}_2 + \text{N}_2 \rightarrow 2\text{NH}_3$

Hydrogen Nitrogen Ammonia

Equation for the reaction between hydrogen and nitrogen to form ammonia

$3H_2$

Hydrogen

$+$

N_2

Nitrogen

$\downarrow$

$2NH_3$

Ammonia

$+$

heat

Equation for the synthesis of
ammonia from hydrogen
and nitrogen with
representation of its
exothermic nature

This representation reveals that as the reactants are converted into the products, the bonds between the atoms of the reactants are broken and new bonds are formed between the atoms of the product. In other words, we can consider the reaction as involving two processes, (1) bond breaking and (2) bond formation. (These processes do not describe the actual mechanism but are helpful devices for explaining energy changes.) Process (1) involves breaking the following bonds:

$$H:H + H:H + H:H + :N:::N: \rightarrow$$
$$H\cdot + H\cdot + H\cdot + H\cdot + H\cdot + H\cdot + \cdot\ddot{N}\cdot + \cdot\ddot{N}\cdot \qquad (1)$$

Representation of the bond-breaking process

Process (2) involves the formation of the following bonds:

$$H\cdot + H\cdot + H\cdot + H\cdot + H\cdot + H\cdot + \cdot\ddot{N}\cdot + \cdot\ddot{N}\cdot \rightarrow H:\ddot{N}:H + H:\ddot{N}:H \qquad (2)$$
$$\qquad\qquad\qquad\qquad\qquad\qquad\qquad\qquad\qquad H \qquad\quad H$$

Representation of the bond-formation process

Breaking bonds in process (1) absorbs energy, and forming bonds in process (2) evolves energy. In this reaction more energy is evolved in the bond-formation process than is absorbed in the bond-breaking process; so the reaction as a whole evolves energy as heat. **Exothermic reactions** are reactions which evolve heat as they occur. The ammonia synthesis reaction can be written as shown in the margin to emphasize its exothermic nature.

Case 2   Water can be decomposed into hydrogen and oxygen, as shown in the margin. This equation can be written with structural formulas to emphasize the interatomic bonds:

$2H_2O \rightarrow \ 2H_2 \ + \ O_2$

Water   Hydrogen   Oxygen

**Equation for the
decomposition of water
into hydrogen and oxygen**

$$H—O + H—O \rightarrow H—H + H—H + O\!\!=\!\!O$$
$$\quad\ |\qquad\quad |$$
$$\quad\ H\qquad\ \ H$$

**Equation for the decomposition of water into hydrogen and oxygen
showing the individual bonds involved**

This representation reveals that as the water is converted into hydrogen and oxygen, the bonds of the water molecule are broken and the bonds of the hydrogen and oxygen molecules are formed. In this case, the amount of energy required by the bond-breaking processes is greater than the amount of energy evolved in the bond-formation processes, and the reaction as a whole absorbs energy. This energy may be supplied in the form of heat energy, electric energy, or light energy. The term **endothermic reaction** is applied to a reaction which absorbs heat as it takes place. The endothermic nature of the decomposition of water can be represented as

$$\text{Heat} + 2H_2O \rightarrow \ \ 2H_2 \ + \ O_2$$
$$\qquad\qquad \text{Water} \qquad \text{Hydrogen} \quad \text{Oxygen}$$

**Equation for the decomposition of water with representation
of its endothermic nature**

139

6-7 WHY IS ENERGY EVOLVED IN
SOME REACTIONS BUT ABSORBED
IN OTHERS?

Case 3 Water can be formed by the combination of hydrogen and oxygen. This is the reaction which occurs when hydrogen is burned in air, represented as

$$2H_2 \quad + \quad O_2 \quad \rightarrow 2H_2O$$

Hydrogen   Oxygen   Water

**Equation for the formation of water from
hydrogen and oxygen**

This equation can be rewritten with structural formulas to emphasize the interatomic bonds involved:

$$\text{H—H + H—H + O=O} \rightarrow \text{H—O + H—O}$$
$$ \overset{|}{\text{H}} \overset{|}{\text{H}}$$

**Equation for the formation of water from hydrogen and oxygen
showing the individual bonds involved**

Inspection of the equation reveals that as the hydrogen and oxygen are converted into water, the bonds of the hydrogen and oxygen molecules are broken and the bonds of the water molecules are formed. Since the amount of energy required by the bond-breaking processes is less than the amount evolved in bond-formation processes, the reaction as a whole evolves energy. This reaction is readily carried out in such a way that the energy is evolved as heat, and it can be written to emphasize its exothermic nature:

$$2H_2 \quad + \quad O_2 \quad \rightarrow 2H_2O + \text{heat}$$

Hydrogen   Oxygen   Water

**Equation for the formation of water from hydrogen and oxygen with
representation of its exothermic nature**

This is the reaction for hydrogen burning in air and is responsible for the explosive nature of hydrogen, which makes it dangerous to handle. Because of its low density, hydrogen was once used in lighter-than-air zeppelins. The famous German zeppelin "Hindenburg" used 6.7 million cubic feet to keep it in the air. The Hindenburg exploded as it was coming in for a landing at Lakehurst, New Jersey, in May 1937. The airship was destroyed in less than 5 minutes and 35 lives were lost in the fire.

Hydrogen and oxygen are used as the fuels in the last two stages of the Saturn V rocket of the Apollo moon program. (Kerosine and oxygen are used in the initial stage.) Both the hydrogen and oxygen are used in the liquid form, requiring special techniques for handling large amounts of these substances at the low temperatures required ($-253°C$ for liquid hydrogen and $-183°C$ for liquid oxygen).

The reaction for the formation of water is, of course, the reverse of the reaction for the decomposition of water. It is not surprising, then, that since energy is evolved as water is formed from hydrogen and oxygen, energy is absorbed as water is decomposed.

Since the combination of hydrogen and oxygen is highly exothermic, hydrogen has been considered as a possible fuel for the future. The fuel cycle would seem to be a model of simplicity:

$$\text{Energy} + 2H_2O \rightarrow 2H_2 + O_2 \tag{1}$$

$$2H_2 + O_2 \rightarrow 2H_2O + \text{energy} \tag{2}$$

Sequence of reactions illustrating hydrogen's potential as a fuel; hydrogen is generated in reaction (1) and is used in reaction (2) to produce energy

The principal problem in implementing this cycle is that of obtaining a low-cost, abundant supply of hydrogen. To be sure, water can be decomposed by passing an electric current through it with a small amount of sulfuric acid added:

$$\text{Electric energy} + 2H_2O \xrightarrow{\;H_2SO_4\;} 2H_2 + O_2$$

$$\text{Water} \qquad\qquad \text{Hydrogen} \quad \text{Oxygen}$$

Overall reaction for the use of electric energy to decompose water

This process is called **electrolysis,** a term applied to all chemical reactions carried out with the use of electric energy. The production of hydrogen by electrolysis of water, however, is too expensive at present to make the hydrogen obtained a practical source of energy.

Water can also be decomposed by raising it to high temperatures, but the decomposition reaction is too slow to be practical. Light energy also brings about the decomposition of water, but again the process is very slow. Considerable effort is currently being made to find a way of speeding up the decomposition of water by light energy. The perfection of such a process holds the promise of great dividends, for solar energy could then serve to produce an inexpensive and almost endless supply of hydrogen from seawater. The results so far, however, have not been very encouraging.

6-8 THE LAW OF CONSERVATION OF ENERGY

In Sec. 6-7 we mentioned that water could be decomposed by the use of electric energy. Mention was also made of the exothermic nature of the combination of hydrogen and oxygen. If these two reactions are carried out in sequence, electric energy is converted into heat energy through the formation and consumption of hydrogen and oxygen. It is possible to measure the amount of electric energy absorbed and the heat energy produced. When this is done, it is found that if the reactions are carried out carefully, the amount of heat produced is exactly equivalent to the electric energy absorbed:

$$\text{Electric energy} + 2H_2O \rightarrow 2H_2 + O_2 \tag{1}$$

$$2H_2 + O_2 \rightarrow 2H_2O + \text{heat energy} \tag{2}$$

Sequence of reactions which demonstrate the law of conservation of energy; the electric energy used in reaction (1) is converted into the heat energy generated by reaction (2); measurements show that the amount of heat energy generated is exactly equivalent to the amount of electric energy consumed, with no energy lost

This is one way of demonstrating the validity of the **law of conservation of energy:** in energy transformations energy is conserved; it can neither be created nor destroyed but only changed in form. Like the law of conservation of mass, the energy-conservation law is now understood not to hold strictly, but it remains a useful generalization in many practical situations.

6-9 WHAT FACTORS DETERMINE THE SPEED OF CHEMICAL REACTIONS?

One of the most important questions concerning any chemical reaction is: How fast does it go? Some reactions are very fast. The decomposition of a kilogram of TNT (trinitrotoluene), for example, can take place in a split second. The reactions involved in transmitting nerve impulses are very rapid. Other reactions occur very slowly; e.g., the rusting of an iron nail takes years to complete. The elemental carbon of diamond is potentially capable of reacting with the oxygen of the atmosphere, but the rate is so immeasurably slow that for all practical purposes no reaction occurs at all. Several factors control the rate at which a reaction will take place.

A. Structure of Reactants

The structures of the atoms, molecules, and ions that make up reacting substances are important in determining how fast reactions occur. While iron, for example, corrodes in the atmosphere at an appreciable rate, nickel reacts much more slowly under the same conditions. After only a relatively short time, rust appears on the iron, but after the same time the nickel appears unchanged.

B. Concentration of Reactants

The **concentration of a reactant** refers to the relative amount of the reactant in a reaction mixture. If a reaction involves two reactants and both are soluble in the same solvent, the concentration of each reactant can be varied separately. The rate of the reaction ordinarily increases if the concentration of either reactant is raised.

Often, however, the properties of the reactants are such that the concentration of only one of them can be changed. When a piece of wood burns in air, for example, the reactants are substances in the wood and the oxygen of the air (air is about 20 percent oxygen). The concentration of the oxygen available for reaction can be increased by replacing the air with pure oxygen, but the concentration of the substances in the wood is fixed. If a glowing, slow-burning piece of wood in air is transferred to an environment of pure oxygen, the rate of reaction increases and the wood bursts into flame. The burning process requires collision between the oxygen molecules and the wood. As the wood is moved from the air to the pure oxygen, the collision rate increases and the reaction rate increases. This is why the "No Smoking" regulation should be strictly enforced whenever pure

Table 6-3
Summary of
Factors Influencing
the Rate of Reaction

Structure of reactants
Concentration of reactants
Temperature of reactants
Presence of catalysts

oxygen is being used. Hospital patients with heart and lung ailments are frequently placed in tents of pure oxygen to increase the amount of oxygen taken in by each breath. Rigid precautions must be taken to minimize the fire danger. A cigarette burning slowly in air will burst into a vigorous flame in oxygen.

While the concentration of the combustible substances in the wood is fixed, the rate of the reaction of these substances with the oxygen can be increased by grinding the wood into small pieces. Even with the concentration of oxygen in the atmosphere, the rate of collision between the molecules and the combustible substances now increases and the reaction rate increases. The rise in reaction rate as the exposed surface of a solid is increased often introduces an element of danger. Wheat flour, for example, is composed of substances which react with atmospheric oxygen, but under ordinary conditions the reaction rate is slow and unnoticeable. If a room is filled with finely divided flour dust, however, the conditions for an extremely rapid reaction are provided and an explosion can result. Similar conditions prevail when coal, grain, or any other material capable of reacting with oxygen is ground to a fine powder in closed areas. Dust explosions in flour mills, grain elevators, and coal mines have been costly both in human lives and in property (Fig. 6-6).

The increase in the rate of a reaction with increased surface of a reactant can be demonstrated in the laboratory with the

FIGURE 6-6
The destruction caused by a dust explosion in a grain elevator at Westwego, Louisiana in 1977. (*United Press International.*)

reaction between zinc and an aqueous solution of hydrochloric acid:

$$Zn \quad + \quad 2HCl \quad \rightarrow \quad H_2 \quad + \quad ZnCl_2$$

| Zinc, solid insoluble in water | Hydrogen chloride, gas in aqueous solution | Hydrogen gas, insoluble in water, evolves | Zinc chloride, solid, soluble and remains in solution |

The reaction when solid zinc is brought in contact with a solution of hydrochloric acid

The course of this reaction is easy to observe since one of the products is gaseous hydrogen, which is insoluble in the water and evolves. If the zinc is in the form of large lumps, the reaction proceeds at a rate which produces a slow but readily observed evolution of hydrogen. If the zinc is more finely divided, however, the rate of hydrogen evolution speeds up noticeably (Fig. 6-7).

C. Temperature of Reactants

The rate of a reaction usually increases markedly with an increase in temperature. As a general (but not infallible) rule, each increase of 10°C doubles the reaction rate. Cooking food involves chemical reactions. The heat used in cooking accelerates the rate of these reactions. If the cooking is carried out at a high altitude, where the boiling point of water is lower (Sec. 4-9), the slower reaction rates necessitate longer cooking times. The effect of a temperature increase is easy to imagine in terms of collisions between reactants. Higher temperatures speed up the molecules (Sec. 4-9) and increase the probability of effective collision within a given time.

D. Presence of Catalysts

A **catalyst** is a substance in a reaction mixture which changes the reaction rate without being permanently altered in the process. It may be involved in the reaction mechanism, but it is recoverable unchanged from the reaction mixture. Consider the reaction between hydrogen and oxygen to form water. In the absence of a catalyst, this reaction is so slow at room temperature that no appreciable reaction takes place, but if the reaction is carried out in the presence of platinum metal, it proceeds rapidly at the same temperature. Experimental evidence suggests that as the hydrogen molecules come in contact with the surface of the platinum, the attractive force between the platinum and the hydrogen molecules helps decompose the molecules into more reactive individual hydrogen atoms (Fig. 6-8). The hydrogen atoms react with the oxygen molecules more readily, but the platinum remains unchanged.

Another mechanism by which catalysts exert their effect can be summarized in general terms in the following way. Assume that in a given reaction substance A reacts directly but slowly with substance B to form substance AB. The introduction of a catalyst X makes possible a sequence of reactions which occur very rapidly. In the first step of this sequence, the catalyst X

$$2H_2 \quad + \quad O_2 \quad \xrightarrow{Pt} 2H_2O$$

| Hydrogen | Oxygen | Water |

Equation for reaction between hydrogen and oxygen to form water; no appreciable reaction at room temperature, but appreciable rate in presence of platinum metal

$$A + B \xrightarrow{slow} AB$$

$$A + X \xrightarrow[fast]{very} AX$$

$$AX + B \xrightarrow[fast]{very} AB + X$$

FIGURE 6-7
Laboratory demonstration of the effect of surface area on the rate of reaction; the gas evolving is hydrogen, H_2, formed in the reaction between solid zinc, Zn, and hydrochloric acid, HCl ($Zn + 2HCl \rightarrow H_2 + ZnCl_2$): (a) lump zinc, (b) hydrogen evolves slowly as lump zinc reacts with an aqueous solution of hydrochloric acid, (c) finely divided zinc, (d) hydrogen evolves more vigorously as the finely divided zinc is added to some of the same aqueous solution of hydrochloric acid. (*Photographs by Hans Oettgen.*)

(b)

(a)

(d)

(c)

reacts quickly to form the compound AX. The intermediate AX reacts very quickly in turn with substance B to form substance AB and regenerate X. The net result of these steps is that the catalyst X replaces the direct but slow combination of A and B with a sequence of much faster reactions that yields the same end result, product AB. The catalyst X is regenerated and made available for repeated use. A small amount of catalyst X can accelerate the formation of many molecules of AB. In the end the catalyst is obtained unchanged. It is estimated that catalyzed reactions in the chemical industry yield products worth more than $100 billion annually.

FIGURE 6-8
Diagram showing the postulated mechanism of the action of platinum metal as a catalyst. The H—H bond is broken by the attractive force of the platinum atoms. The hydrogen atoms formed are much more reactive. (Colored spheres are hydrogen atoms; white spheres are platinum atoms.)

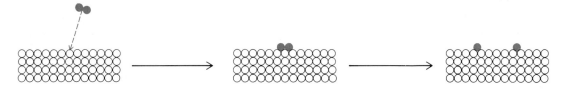

HOW DOES WATER PUT OUT FIRES?

A fire involves rapid chemical processes which release energy in the form of heat and light. The most common examples are the exothermic reactions of the oxygen of the air with such combustibles as paper, wood, and oil. Materials which catch fire in air react with oxygen very slowly at ordinary temperatures in reactions which evolve heat. If a lighted match or other source of flame is brought close to the combustible material, the heat accelerates the rate of the exothermic reaction with oxygen. When the rate becomes fast enough, combustion occurs. At this point the heat from the reaction is being generated fast enough to sustain the process by raising the temperature of adjoining portions of the material. The ignition temperature is the minimum temperature required to bring a material to self-supporting combustion. Materials with low ignition temperatures catch fire very readily.

To extinguish a fire the supply of oxygen must be cut off or reduced, the temperature of the combustible material must be lowered below its ignition temperature, or the combustible material must be removed from the source of heat. Water can accomplish the first two actions and sometimes all three. The evaporation of the water absorbs heat and lowers the temperature of the combustible material. Both the steam formed and the water itself tend to cut off the oxygen supply, and sometimes the force of the stream from a fire hose can actually separate the burning material from the adjacent combustible material.

But water cannot be used on all fires. Since it is an electric conductor, it cannot be used on fires involving live electrical equipment. Such fires require a nonconducting extinguishing agent such as gaseous carbon dioxide to cut off the oxygen supply or a solid powder which generates a gas with the same effect. Fires involving such flammable liquids as gasoline and kerosine are best extinguished by blanketing them with a stabilized foam containing bubbles of carbon dioxide. Carbon dioxide contained in a metal cylinder under pressure is commonly on hand in small fire extinguishers for household use. Many other types of small fire extinguishers are available.

6-10 THE MOLECULAR MECHANISMS OF CHEMICAL REACTIONS

Optional

Chemical equations furnish much useful information about chemical reactions, but they can often be misleading about what happens at the molecular level as a reaction occurs. The equation for the conversion of the gas methane, CH_4, and the gas chlorine, Cl_2, under the influence of light into the gases methyl chloride, CH_3Cl, and hydrogen chloride, HCl, is

$$
\begin{array}{ccccc}
\text{H} & & \text{H} & & \\
| & & | & & \\
\text{H--C--H} + \text{Cl--Cl} \rightarrow & \text{H--C--Cl} & + & \text{H--Cl} \\
| & & | & & \\
\text{H} & & \text{H} & &
\end{array}
$$

Methane   Chlorine   Methyl chloride   Hydrogen chloride

Equation for the reaction of methane and chlorine to produce methyl chloride and hydrogen chloride; this equation is misleading about the actual molecular mechanism (read on)

The **mechanism of a reaction** is the actual sequence of steps which occurs during a reaction. Investigation has shown that the reaction gets under way as chlorine molecules are decomposed into chlorine atoms by light energy:

$$\text{Light energy} + Cl_2 \rightarrow 2Cl\cdot \tag{1}$$

First step of reaction mechanism

In the second step the more forceful of the collisions between the chlorine atoms and the methane molecules produce hydrogen chloride and a molecular species called a methyl radical, $CH_3\cdot$:

$$
\begin{array}{cc}
\text{H} & \text{H} \\
| & | \\
\text{H--C--H} + \cdot\text{Cl} \rightarrow \text{H--C}\cdot + \text{H--Cl} \\
| & | \\
\text{H} & \text{H}
\end{array} \tag{2}
$$

Second step of reaction mechanism

In the third step the more forceful of the collisions between the methyl radicals, $CH_3\cdot$, and the chlorine molecules result in the following reaction:

$$
\begin{array}{cc}
\text{H} & \text{H} \\
| & | \\
\text{H--C}\cdot + \text{Cl--Cl} \rightarrow \text{H--C--Cl} + \text{Cl}\cdot \\
| & | \\
\text{H} & \text{H}
\end{array} \tag{3}
$$

Third step of reaction mechanism

The chlorine atoms formed in the third step can serve to repeat step (2). Steps (2) and (3) follow in succession many times, until one of the following reactions occurs:

$$Cl\cdot \ + Cl\cdot \ \rightarrow Cl-Cl \qquad (4)$$

$$CH_3\cdot + CH_3\cdot \rightarrow CH_3CH_3 \qquad (5)$$

$$CH_3\cdot + Cl\cdot \ \rightarrow CH_3Cl \qquad (6)$$

Processes which consume reactive
molecular species

This mechanism is an example of a **chain-reaction mechanism;** such mechanisms involve a repeating sequence of steps, each generating a reactive molecular species which brings about the succeeding step. Step (1) is the chain-initiating step, steps (2) and (3) are the chain-propagating steps, and steps (4) to (6), which bring the chain to an end, are chain-terminating steps. The principal point of this discussion is to contrast the equation for the reaction with the actual mechanism; i.e., what really happens during a reaction.

Note to Students
Chain-reaction mechanisms occur fairly frequently in chemical reactions, but reactions may occur by many other types of mechanisms.

"Gottlieb, I think I know why we've been receiving so few commissions."

KEY WORDS

1. Chemical reaction (Sec. 6-2)
2. Reactant (Sec. 6-2)
3. Product (Sec. 6-2)
4. Chemical equation (Sec. 6-3)
5. Law of conservation of mass (Sec. 6-5)
6. Atomic weight (Sec. 6-6A)
7. Bond energy (Sec. 6-7A)
8. Exothermic reaction (Sec. 6-7B)
9. Endothermic reaction (Sec. 6-7B)
10. Electrolysis (Sec. 6-7B)
11. Law of conservation of energy (Sec. 6-8)
12. Concentration of a reactant (Sec. 6-9)
13. Catalyst (Sec. 6-9)

Optional:

14. Mechanism of reaction (Sec. 6-10)
15. Chain-reaction mechanism (Sec. 6-10)

SUMMARIZING QUESTIONS FOR SELF-STUDY

Sections 6-1 to 6-4

1. Q. Define (a) chemical reaction, (b) reactant, (c) product, and (d) chemical equation.

A. (a) A process in which substances are converted into other substances with different sets of properties, (b) a starting substance in a chemical reaction, (c) a substance formed in a chemical reaction, (d) the representation of a reaction in formula symbols which specifies each reactant, each product, and the relative number of molecules of each.

2. Q. Balance the following equations:
(a) $H_2 + O_2 \rightarrow H_2O$
(b) $Na + H_2O \rightarrow NaOH + H_2$
   A. (a) $2H_2 + O_2 \rightarrow 2H_2O$
(b) $2Na + 2H_2O \rightarrow 2NaOH + H_2$

Section 6-5

3. Q. What is the law of conservation of mass?

A. During a chemical reaction the amount of mass remains constant.

Section 6-6

4. Q. What is the difference between the absolute weight of a hydrogen 1 atom and the relative weight of a hydrogen 1 atom? What weight is ordinarily used by chemists? Why?

A. The absolute weight is the actual weight, specifically 1.673×10^{-24} gram. The relative weight is the weight on some relative scale. The scale used by chemists is one in which the carbon 12 atom is assigned a value of 12.0000 units. On this scale even the smallest atom, protium (hydrogen 1), has a weight of more than 1 unit. The weights on this scale are more convenient to use than the absolute weights of atoms.

5. Q. Why is the molecular weight of fluorine the same as 2 times the atomic weight of the fluorine 19 isotope?

A. Molecules of the element fluorine are diatomic and the only naturally occurring isotope is fluorine 19.

6. Q. Why is the atomic weight of naturally occurring hydrogen not the same as the atomic weight of the hydrogen 1 (protium) atom?

A. Naturally occurring hydrogen contains three isotopes. The atomic weight of naturally occurring hydrogen is an average value which reflects the relative weight of the three isotopes and their natural proportion. (The atomic weight of naturally occurring hydrogen is 1.0079.) The hydrogen 1 atom is only one of the three isotopes of hydrogen. Its relative atomic weight is not an average value but depends on the weight of only one variety of atom. (The atomic weight of hydrogen 1 is 1.0078. It is very close to the atomic weight of naturally occurring hydrogen since the hydrogen 1 isotope makes up 99.985 percent of natural hydrogen.)

7. Q. Use Table S4-1 to calculate the molecular weight of the element chlorine, Cl_2.

A. Atomic weight of chlorine	35.453
Atomic weight of chlorine	35.453
Molecular weight of chlorine	70.906

8. Q. Use Table S4-1 to calculate the molecular weight of ethyl ether, C_2H_6O.

A. Atomic weight of carbon × 2	24.022
Atomic weight of hydrogen × 6	6.0474
Atomic weight of oxygen	15.9994
Molecular weight of ethyl ether	46.069

Sections 6-7 and 6-8

9. Q. Define (a) bond energy, (b) exothermic reaction, (c) endothermic reaction, and (d) electrolysis.

A. (a) The energy required to break a bond, (b) a reaction which evolves heat as it occurs, (c) a reaction which absorbs heat as it occurs, (d) a chemical reaction carried out with electric energy.

10. Q. What is the principal factor which determines whether a reaction is exothermic or endothermic?

A. A reaction is exothermic if the amount of energy evolved in the bond-formation processes exceeds the amount of energy absorbed in the bond-breaking processes. A reaction is endothermic if the amount of energy absorbed in the bond-breaking processes exceeds the amount of energy evolved in the bond-formation processes.

11. Q. What is the law of conservation of energy?

A. In transformations of energy, energy is conserved; it can neither be created nor destroyed but only changed in form.

Section 6-9

12. Q. What is a catalyst?

A. A substance in a reaction mixture which changes the reaction rate without itself being permanently altered in the process.

13. Q. What four factors influence the rate of a chemical reaction?

A. The structure of reactants, the concentration of reactants, the temperature of reactants, and the presence of catalysts

14. Q. What is the effect of the concentration of a reactant on the rate of a reaction?

A. As the concentration of reactants increases, the reaction rate increases.

15. Q. What is the effect of the temperature of a reaction mixture on the rate of reaction?

A. As the temperature of the reaction mixture rises the reaction rate increases.

Section 6-10 (Optional)

16. Q. What is meant by a reaction mechanism?

A. The actual sequence of steps which occurs during a reaction.

17. Q. What is meant by a chain-reaction mechanism?

A. A reaction involving a repeating sequence of steps, each generating a reactive molecular species which brings about the succeeding step.

18. Q. Give an example of a chain-reaction mechanism.

A. Chain-initiating step:

$$\text{Light energy} + Cl_2 \rightarrow 2Cl\cdot$$

Chain-continuation steps:

$$CH_4 + Cl\cdot \rightarrow CH_3\cdot + HCl$$
$$CH_3\cdot + Cl_2 \rightarrow CH_3Cl + Cl\cdot$$

Chain-ending steps:

$$Cl\cdot + Cl\cdot \rightarrow Cl_2$$
$$CH_3\cdot + CH_3\cdot \rightarrow CH_3CH_3$$
$$CH_3\cdot + Cl\cdot \rightarrow CH_3Cl$$

PRACTICE EXERCISES

1. Match each definition or other statement with the numbered term above with which it is most closely associated. A numbered term may be used only once.

   1. Catalyst
   2. Hydration
   3. Law of conservation of energy
   4. Electrolysis
   5. Concentration of reactant
   6. Law of conservation of mass
   7. Endothermic reaction
   8. Chemical reaction
   9. Decomposition
   10. Atomic weight
   11. Chemical equation
   12. Bond energy
   13. Exothermic reaction

   (a) A process in which substances are converted into other substances with different sets of properties
   (b) The weight of an average atom of an element relative to the weight of a carbon 12 isotope, taken as 12.0000 units
   (c) During a chemical reaction the amount of mass remains constant
   (d) A representation of a reaction in formula symbols which specifies each reactant, each product, and the relative number of molecules of each
   (e) The amount of energy required to break a bond
   (f) A reaction in which heat is evolved as it occurs
   (g) A chemical reaction carried out with the use of electric energy
   (h) In energy transformations energy is conserved; it can be neither created nor destroyed, but only changed in form
   (i) A reaction in which heat is absorbed as it occurs
   (j) A substance in a reaction mixture which changes the reaction rate without itself being permanently altered in the process

2. Balance the following equations:
   (a) $SO_2 + O_2 \rightarrow SO_3$
   (b) $Ca(OH)_2 + HCl \rightarrow CaCl_2 + H_2O$
   (c) $Al + Cl_2 \rightarrow AlCl_3$
   (d) $NO_2 + H_2O \rightarrow HNO_3 + NO$
   (e) $Na + H_2O \rightarrow NaOH + H_2$
   (f) $LiOH + CO_2 \rightarrow Li_2CO_3 + H_2O$
   (g) $NaI + Br_2 \rightarrow NaBr + I_2$
   (h) $H_2S + O_2 \rightarrow SO_2 + H_2O$

3. Using Table S4-2, calculate the molecular weights of:
   (a) ammonia, NH_3
   (b) sulfur dioxide, SO_2
   (c) calcium hydroxide, $Ca(OH)_2$
   (d) lithium carbonate, Li_2CO_3
   (e) methyl alcohol, CH_3OH

4. Using Table S4-2, calculate the molecular weights of:
   (a) barium cyanide, $Ba(CN)_2$
   (b) ethyl ether, $CH_3CH_2OCH_2CH_3$
   (c) aspirin, $C_9H_8O_4$
   (d) nicotine, $C_{10}H_{14}N_2$
   (e) morphine, $C_{17}H_{19}O_3N$

5. Balance the following equations, determine the molecular weights of each of the reactants and products, and check to determine if the sum of the weights of the reactants is equal to the sum of the weights of the products.
   (a) $NaOH + HCl \rightarrow NaCl + H_2O$
   (b) $NO + O_2 \rightarrow NO_2$
   (c) $N_2H_4 \rightarrow H_2O_2 + N_2 + H_2O$

6. The following reaction takes place during the combustion of natural gas, methane, CH_4, and is known to be exothermic:
   $$CH_4 + 2O_2 \rightarrow CO_2 + 2H_2O.$$
   Write the structural formula for each reactant and each product and specify which bonds are broken and which are formed in the course of the reaction. Which processes involve the greater amount of energy, the total of the bond breaking processes or the total of the bond forming processes? How do you know?

7. Hydrazine,

   $$\begin{array}{c} H \quad H \\ | \quad | \\ H-N-N-H \end{array}$$

   is used as a rocket fuel. It reacts with oxygen, O_2, according to exothermic reaction $N_2H_4 + O_2 \rightarrow N_2 + 2H_2O$. Write the structural formula for each reactant and each product and specify which bonds are broken and which are formed in the course of the reaction. Which processes involve the greater amount of energy, the total of the bond-breaking processes or the total of the bond-forming processes? How do you know?

8. It is known that when mercury, Hg, is treated with oxygen, O_2, the following exothermic reaction occurs:

   $$2Hg \quad + \quad O_2 \quad \rightarrow \quad 2HgO$$

   Mercury   Oxygen   Mercury(II) oxide

Would you expect the decomposition of mercury(II) oxide

$$2HgO \quad \rightarrow \quad 2Hg \quad + \quad O_2$$

Mercury(II) oxide   Mercury   Oxygen

to be exothermic or endothermic? Explain.

9. What four factors can influence the rate of a reaction?

10. Describe how each of the following changes will change the rate of a reaction:
    (a) An increase in temperature
    (b) A decrease in concentration of one of the reactants
    (c) The introduction of a catalyst

11. The page of a newspaper burns readily in air, but a tightly rolled newspaper burns almost as slowly as a wooden log. Explain.

12. Why are grain elevators and coal mines particularly prone to explosions?

13. Explain how a small amount of catalyst may exert a profound effect on the rate of a reaction.

14. What is the specific function of platinum when it serves to catalyze the reaction $2H_2 + O_2 \rightarrow 2H_2O$?

Section 6-10 (Optional)

15. What is the difference between the equation for a reaction and the mechanism of a reaction?

16. What is a chain-reaction mechanism?

The following exercises assume a knowledge of Supplement 4. Express units, show your method clearly, and label the answer.

17. Assuming the reaction given in Sec. 6-4, how many grams of ammonia, NH_3, will be formed from 12.096 g of hydrogen, H_2, and an excess of nitrogen, N_2?

18. Assuming the reaction given in Sec. 6-4, how many grams of nitrogen, N_2, are required to form 170.3 g of ammonia, NH_3, if all the hydrogen, H_2, necessary is available?

19. Hydrogen, H_2, and oxygen, O_2, react to form water, H_2O, according to the equation $H_2 + O_2 \rightarrow H_2O$. How many grams of hydrogen are required to form 100.00 g of water if all the necessary oxygen is available?

20. According to the reaction of Question 19, how many grams of hydrogen, H_2, will react with 56.00 g of oxygen, O_2?

SUGGESTIONS FOR FURTHER READING

The following article is of an introductory nature:

Emmons, Howard W.: Fire and Fire Protection, **Scientific American**, July 1974, p. 21.

The following article is at a more advanced level:

Haensel, Vladimir, and Robert L. Burwell, Jr.: Catalysis, **Scientific American**, December 1971, p. 46.

7

HOW ARE CHEMICAL REACTIONS CLASSIFIED AND ORGANIZED?

7-1 INTRODUCTION

It would be enormously useful if we could write a brief set of principles which would predict all the possible reactions between the various forms of matter. Suppose we have isolated and characterized for the first time the substance which is the ingredient of an herb long used as a household remedy for rheumatism. We glance at our short list of principles and predict precisely how the active ingredient will interact with any of the thousands of substances found in a human being. On this basis, the helpful activity could be appraised, the harmful side effects could be anticipated, and a method of use could be devised. Unfortunately, such an all-encompassing list of principles is far beyond the reach of present-day knowledge. Indeed, it is highly doubtful that chemical reactions will ever be summarized so easily.

The number of chemical reactions already characterized runs into the millions; no one has counted them carefully. Moreover since they encompass an astonishing variety of changes, broad classifications are not easily achieved.

Still chemists have been able to organize many reactions so that many useful correlations are possible, and the discovery of more far-reaching principles is being actively pursued. This chapter illustrates the diversity among chemical reactions and describes some simple approaches to classification and organization.

7-2 ION-COMBINATION REACTIONS WHICH FORM PRECIPITATES: THE SYNTHESIS OF LITHIUM CARBONATE, Li_2CO_3, AN IMPORTANT AGENT IN TREATING MANIC DEPRESSION

One of the important uses of chemical reactions is the large-scale conversion of natural raw materials of the earth, sea, and air into substances of greater usefulness. The synthesis of lithium carbonate, Li_2CO_3, is an interesting example. During the last decade this white crystalline salt has been used in the treatment of manic-depressive mental illness, and in many cases it has proved effective. Lithium carbonate does not occur in nature but must be synthesized from lithium-containing substances which do occur naturally.

In the last step of the industrial process for making lithium carbonate, sodium carbonate, Na_2CO_3, is added to an aqueous solution of lithium sulfate, Li_2SO_4, which is prepared by preceding steps from a lithium containing ore. The chemical reaction in the last step can be represented by the equation

$$Li_2SO_4 + Na_2CO_3 \rightarrow Li_2CO_3 + Na_2SO_4$$

| Lithium sulfate | Sodium carbonate | Lithium carbonate | Sodium sulfate |

Equation for the last step of the industrial synthesis of lithium carbonate, Li_2CO_3

Most of the lithium carbonate formed is insoluble and precipitates out. It is isolated by filtration.

Both reactants and both products of this reaction are ionic compounds, and all, except lithium carbonate, are soluble in water. This means that the original mixture of reactants is really a mixture of four ions:

$$Li^+ + SO_4^{2-} \quad + \quad Na^+ + CO_3^{2-}$$

| The ionic species in an aqueous solution of Li_2SO_4 | The ionic species in an aqueous solution of Na_2CO_3 |

Of the two products, sodium sulfate is soluble in water, but lithium carbonate dissolves only to a very slight extent. Let us rewrite the representation of the products recognizing (1) that the soluble sodium sulfate furnishes sodium ions, Na^+, and sulfate ions, SO_4^{2-}, to the solution, but (2) most of the lithium ions, Li^+, and the carbonate ions, CO_3^{2-}, are to be found in the insoluble lithium carbonate. In rewriting we follow the convention of not specifying the ionic character of the lithium carbonate precipitate:

$$Li_2SO_4 + Na_2CO_3 \rightarrow Li_2CO_3 + Na_2SO_4$$

Original representation of reaction

$$2Li^+ + SO_4^{2-} + 2Na^+ + CO_3^{2-} \rightarrow Li_2CO_3 + 2Na^+ + SO_4^{2-}$$

Revised representation emphasizing the ionic species in solution

The revised representation shows that the sodium ions, Na^+, and the sulfate ions, SO_4^{2-}, are in solution both at the beginning of the reaction and at the end and therefore do not really participate in the reaction. The essential reaction can be represented as shown in the margin by indicating only those ions which are involved in the reaction.

In Sec. 5-2 we discussed the condition which prevails in a saturated aqueous solution of the ionic compound silver chloride, AgCl, in contact with undissolved solute. This situation was described as involving a solubility equilibrium:

$$AgCl\ (undissolved) \rightleftharpoons AgCl\ (dissolved)$$

Representation of the solubility equilibrium which prevails in a saturated solution of silver chloride, AgCl (Sec. 5-2)

There is a dynamic balance between the dissolving process and the precipitation process. Since these processes are occurring at equal rates, the amount of silver chloride dissolved and the amount undissolved remain constant.

If we follow the convention of specifying the ions in the solution but not in the precipitate, the equilibrium expression can be rewritten

$$AgCl\ (undissolved) \rightleftharpoons Ag^+ + Cl^-\ (dissolved)$$

Representation of the solubility equilibrium which prevails in a saturated solution emphasizing the ionic species in solution

In discussing the equilibrium in a saturated solution in Sec. 5-2, we thought of the equilibrium as becoming established by dissolving solid silver chloride in water until no more dissolved, i.e., until a saturated solution had been obtained. But in practice the same equilibrium situation would be established if we approached it from the opposite direction; i.e., by bringing together silver ions, Ag^+, and chloride ions, Cl^-, in solution:

$$Ag^+ + Cl^-\ (dissolved) \rightleftharpoons AgCl\ (undissolved)$$

Representation of the solubility equilibrium prevailing in a saturated solution of silver chloride, AgCl, approached from the direction of the soluble ions

How can silver ions and chloride ions be brought together in an aqueous solution? The silver ions can be introduced in the form of a soluble silver salt, e.g., silver nitrate, $AgNO_3$. The chloride ions can be introduced in the form of a soluble chloride salt, e.g., sodium chloride, NaCl.

Let's bring the sources of silver and chloride ions together in an aqueous solution (Fig. 7-1). If we use large amounts of sodium chloride, NaCl, and silver nitrate, $AgNO_3$, concentrations of the silver and chloride ions will be greater than the very limited amount which will remain in solution. The excess will precipitate out as solid silver chloride, AgCl. This change constitutes a chemical reaction:

$$2Li^+ + CO_3^{2-} \rightarrow Li_2CO_3$$

Equation for the last step of the synthesis of lithium carbonate, Li_2CO_3, emphasizing only ions which participate

$$Ag^+ \quad + \quad NO_3^-$$

Silver ion   Nitrate ion

Ionic species present in aqueous solution of silver nitrate, $AgNO_3$

$$Na^+ \quad + \quad Cl^-$$

Sodium ion   Chloride ion

Ionic species present in aqueous solution of sodium chloride, NaCl

$$Ag^+ + NO_3^- \quad + \quad Na^+ + Cl^- \quad \rightarrow \quad AgCl \quad + \quad Na^+ + NO_3^-$$

Ionic species present in aqueous solution of silver nitrate, $AgNO_3$	Ionic species present in aqueous solution of sodium chloride, NaCl	Ionic species in precipitate of solid silver chloride, AgCl	Ionic species which remain in solution

Representation of reaction which occurs when appreciable concentrations of silver ion, Ag^+, and chloride ion, Cl^-, are brought together in aqueous solution

$$Ag^+ + Cl^- \rightarrow AgCl$$

Representation of the essential change taking place as aqueous solutions of silver nitrate, $AgNO_3$, and sodium chloride, NaCl, are mixed

The equation reveals that the sodium ions, Na^+, and nitrate ions, NO_3^-, are in solution both at the start and at the end of the process and therefore do not directly participate. The essential change can be represented as shown in the margin.

By mixing the aqueous solutions of silver nitrate and sodium chloride we have arrived at the same equilibrium situation which prevails when a saturated solution of silver chloride is in contact with undissolved silver chloride:

$$Ag^+ + Cl^- \text{ (dissolved)} \rightleftharpoons AgCl \text{ (undissolved)}$$

Representation of the equilibrium situation which prevails in a solution after addition of silver nitrate, $AgNO_3$, and sodium chloride, NaCl

FIGURE 7-1
The reaction which occurs when an aqueous solution of sodium chloride, NaCl, is added to an aqueous solution of silver nitrate, $AgNO_3$: $NaCl + AgNO_3 \rightarrow AgCl\downarrow + NaNO_3$. (a) Clear aqueous solution of silver nitrate in beaker (left) and clear aqueous solution of sodium chloride in tall cylinder (right). (b) A moment after the first sodium chloride comes in contact with the silver nitrate solution the insoluble silver chloride immediately begins to precipitate out as a finely divided white solid: $Ag^+ + Cl^- \rightarrow AgCl\downarrow$. (c) After the sodium chloride solution has been added. Most of the silver chloride formed has settled to the bottom, but some remains suspended as small particles which impart a slight cloudiness to the solution. A solubility equilibrium prevails between the silver chloride in solution and the silver chloride precipitate. (*Photographs by Hans Oettgen.*)

(a)

(b)

(c)

The lithium carbonate, Li_2CO_3, was formed in a similar manner. By mixing aqueous solutions of lithium sulfate, Li_2SO_4, and sodium carbonate, Na_2CO_3, concentrations of lithium ions, Li^+, and carbonate ions, CO_3^{2-}, were brought together in excess of what will remain in solution. The excess precipitated out as solid lithium carbonate, Li_2CO_3, and the resulting situation involved a chemical equilibrium between the dissolved and undissolved lithium and carbonate ions:

$$2Li^+ + CO_3^{2-} \text{ (dissolved)} \rightleftharpoons Li_2CO_3 \text{ (undissolved)}$$

Representation of the equilibrium situation which prevails in a solution after addition of lithium sulfate, Li_2SO_4, and sodium carbonate, Na_2CO_3

The chemical reactions between aqueous solutions of lithium sulfate and sodium carbonate and aqueous solutions of silver nitrate and sodium chloride illustrate a type of reaction which is common in aqueous solutions. It is called an **ion-combination reaction which forms a precipitate;** ions in excess of their solubility in water combine and form a precipitate. These reactions occur whenever the ions of slightly soluble salts are brought together in aqueous solution, and they are readily predicted from solubility information. The representations of these reactions by molecular equations, ionic equations, and net ionic equations are summarized below.

Molecular equation:

$$Li_2SO_4 + Na_2CO_3 \rightarrow Li_2CO_3 + Na_2SO_4$$

Lithium sulfate   Sodium carbonate   Lithium carbonate (precipitates)   Sodium sulfate

Ionic equation:

$$2Li^+ + SO_4^{2-} + 2Na^+ + CO_3^{2-} \rightarrow Li_2CO_3 + 2Na^+ + SO_4^{2-}$$

Net ionic equation:

$$2Li^+ + CO_3^{2-} \rightarrow Li_2CO_3$$

Lithium ion   Carbonate ion   Lithium carbonate (precipitates)

$$2Li^+ + CO_3^{2-} \rightarrow Li_2CO_3 \downarrow$$
$$2Li^+ + CO_3^{2-} \rightarrow \underline{Li_2CO_3}$$

Net ionic representations of the last step in the synthesis of lithium carbonate, showing two ways of emphasizing the precipitate formation

Sometimes an arrow pointing down or an underline is used to emphasize the precipitate formation. Although the final stage reached among the products involves an equilibrium, the double-arrow representation is not usually used unless the equilibrium is being emphasized.

THE REMARKABLE EFFECT OF LITHIUM ION ON THE HUMAN BRAIN

One of the more striking discoveries in the last 30 years is that very low concentrations of a positive ion of simple structure, the lithium ion, Li^+, have a significant effect on the chemistry of the brain and can be very useful in the treatment of a common form of mental disease. Manic depression is a disturbing mental disorder characterized by alternating moods of exaggerated elation (mania) and severe depression. Oral doses of lithium carbonate, Li_2CO_3, distinctively improve the condition of most people afflicted with this problem. Its administration is extremely simple, but despite considerable investigation its mechanism of action remains obscure.

1949 the Australian physician John F. J. de was investigating the possible presence of oxic nitrogen-containing substances in the urine of mental patients. In an attempt to increase the solubility of the urine components in the guinea pigs being used as test animals, he administered lithium carbonate to the animals. He noticed that the salt made the guinea pigs temporarily lethargic. After some further checking, and with a rare combination of insight and intuition, he gave small doses of lithium carbonate to a number of manic patients. Most of them showed marked improvement. With the further investigation which followed, oral doses of lithium carbonate became established as an effective means of treating manic depression. While the treatment is not always successful, it is effective in about 75 percent of the cases.

It is significant that lithium carbonate in therapeutic doses has almost no effect on people with no manic-depression symptoms. While with some patients it is reported to be effective in preventing both the manic and depression phases, it is most effective in treating the manic phase. For this reason treatment is often combined with an antidepressant agent. While lithium ion occurs naturally in the body, the therapeutic use of lithium carbonate must be monitored carefully since the substance may cause toxic reactions in concentrations greater than the doses normally used. Small concentrations of lithium ion are found in many of the natural mineral waters which have been used beneficially since the days of ancient Rome.

7-3 ANOTHER ION-COMBINATION REACTION FORMING A PRECIPITATE: THE EXTRACTION OF MAGNESIUM FROM THE SEA

Magnesium is a silvery white solid. It is the lightest of all structural metals and is widely used in aerospace projects. It is present in seawater, mostly as magnesium chloride, in the form of magnesium ions, Mg^{2+}, and chloride ions, Cl^-. Its presence in seawater makes it potentially the most available of all the metals. There are almost 6 million tons in each cubic mile of ocean water, and there are an estimated 350 million cubic miles of seawater on the earth.

Although the total amount of magnesium available as magnesium ions is enormous, its concentration in seawater is still a rather low 0.13 percent. That is, the weight of magnesium ions in a kilogram of seawater is only 1.3 grams. In order to extract it from the sea it must be concentrated. This is done by precipitating the magnesium ions in the form of slightly soluble magnesium hydroxide, $Mg(OH)_2$, by the reaction

$$MgCl_2 \quad + \quad Ca(OH)_2 \quad \rightarrow \quad Mg(OH)_2 \downarrow \quad + \quad CaCl_2$$

| Magnesium chloride, in seawater as Mg^{2+} and Cl^- ions | Calcium hydroxide, sufficiently soluble to supply Ca^{2+} and OH^- ions | Magnesium hydroxide, only slightly soluble and precipitates | Calcium chloride, remains soluble as Ca^{2+} and Cl^- ions |

Reaction for extracting magnesium ion, Mg^{2+}, from its dilute solution in seawater

This reaction can be represented by both ionic and net ionic equations:

Ionic equation: $Mg^{2+} + 2Cl^- + Ca^{2+} + 2OH^- \rightarrow Mg(OH)_2 \downarrow + Ca^{2+} + 2Cl^-$

Net ionic equation: $Mg^{2+} + 2OH^- \rightarrow Mg(OH)_2 \downarrow$

Representations of the reaction between magnesium chloride, $MgCl_2$, and calcium hydroxide, $Ca(OH)_2$

The net ionic equation reveals that this reaction is another example of an ion-combination reaction involving formation of a precipitate.

The magnesium hydroxide, $Mg(OH)_2$, is subsequently isolated and converted into elemental magnesium. The extraction of magnesium from the sea is discussed in detail in Sec. 13-7.

7-4 PROTON-TRANSFER REACTIONS: THE SYNTHESIS OF AN IMPORTANT FERTILIZER AMMONIUM NITRATE, NH_4NO_3

Hunger has long been with the human race. Even today at least two-thirds of the people of the world are undernourished. When the problem is viewed in terms of the present rate of population growth, it becomes enormous. Harvesting agricultural products removes nutrients from the soil. The harvesting of 1 ton of wheat, for example, takes away about 44 pounds of nitrogen-containing substances alone. Unless this is replaced, subsequent yields will decline. Natural processes add nitrogen-containing substances to the soil, but if the high crop yields demanded by the world's hunger are to be achieved, synthetic fertilizers must be added. This and other aspects of the relationship between chemistry and the food supply are discussed in Chap. 11.

The most frequently used nitrogen-containing fertilizer is the white solid ammonium nitrate, NH_4NO_3. The key step in its industrial production involves a reaction between the gas ammonia, NH_3, and the liquid nitric acid, HNO_3, both synthetic substances:

$$NH_3 \quad + \quad HNO_3 \quad \rightarrow \quad NH_4NO_3$$

Ammonia, a gas	Nitric acid, in aqueous solution as H^+ and NO_3^- ions	Ammonium nitrate, in aqueous solution as NH_4^+ and NO_3^- ions

Representation of the key reaction in the synthesis of the fertilizer ammonium nitrate, NH_4NO_3

It is illuminating to rewrite this equation to emphasize the identity of the ions in solution in appreciable concentration at the start and at the end of the process:

$$NH_3 + H^+ + NO_3^- \rightarrow NH_4^+ + NO_3^-$$

Ionic species in aqueous solution of nitric acid	Ionic species in aqueous solution of ammonium nitrate

Representation of the key reaction in the synthesis of ammonium nitrate, NH_4NO_3, emphasizing the ionic content of the solutions

A close look at this representation shows that fundamentally the reaction involves only a transfer of a hydrogen ion, H^+, from the nitric acid, HNO_3, to the ammonia molecule, NH_3,

converting it into an ammonium ion, NH_4^+. A hydrogen ion is
another name for a proton (Sec. 2-5). Since a hydrogen atom
contains one proton and one electron, the removal of the electron
to form a hydrogen ion leaves behind only a proton (margin).

Note to Students

As mentioned in Sec. 5-2F, hydrogen ions, H^+, are actually
present in aqueous solutions in the form of hydronium ions,
H_3O^+. We continue to refer to them simply as hydrogen ions,
with the understanding that the actual molecular species is in
fact hydronium ions.

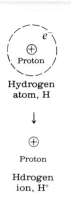

Hydrogen
atom, H

↓

Proton

Hdrogen
ion, H^+

+

e^-

Electron

**Formation of a hydrogen
ion, H^+, demonstrating
that the terms hydrogen
ion and proton are
synonymous**

Since a hydrogen ion and a proton mean the same, this type
of reaction is called a **proton-transfer reaction,** i.e. a reaction
which involves the transfer of one or more protons from one
molecular species to another. Proton-transfer reactions are common, especially in aqueous solutions.

7-5 ACIDS AND BASES

An **acid** can be defined as a substance capable of releasing one
or more protons. In this sense the nitric acid in the ammonium
nitrate synthesis reaction of Sec. 7-4 is behaving as an acid,
as its name suggests. Another acid, hydrochloric acid (an
aqueous solution of hydrogen chloride) behaves equivalently:

$$HCl \quad + \quad NH_3 \quad \rightarrow \quad NH_4Cl$$

Hydrochloric acid in aqueous solution as H^+ and Cl^- ions	Ammonia in aqueous solution	Ammonium chloride in aqueous solution as NH_4^+ and Cl^- ions

**Representation of the proton-transfer reaction between
hydrochloric acid, HCl, and ammonia, NH_3, demonstrating
the acidic properties of hydrochloric acid**

A **base** can be defined as a substance capable of combining
with one or more protons. The ammonia in the ammonium nitrate synthesis reaction of Sec. 7-4 is behaving as a base. Sodium
hydroxide, NaOH, also acts as a base in its reaction with nitric
acid, since it furnishes hydroxide ions, OH^-, which react with
protons, H^+:

$$NaOH \quad + \quad HNO_3 \quad \rightarrow \quad H_2O \quad + \quad NaNO_3$$

Sodium hydroxide in aqueous solution as Na^+ and OH^- ions	Nitric acid in aqueous solution as H^+ and NO_3^- ions	Water in aqueous solution, mostly as H_2O molecules	Sodium nitrate in aqueous solution as Na^+ and NO_3^- ions

**Representation of the reaction between aqueous solutions of
sodium hydroxide and nitric acid**

If the reaction is rewritten to emphasize the ionic content of the solutions involved, the fundamental role of the hydroxide ion is made clearer:

Ionic equation: $Na^+ + OH^- + H^+ + NO_3^- \rightarrow H_2O + Na^+ + NO_3^-$

Net ionic equation: $OH^- + H^+ \rightarrow H_2O$

Representations of the reaction between aqueous sodium hydroxide, NaOH, and nitric acid, HNO₃, emphasizing the ions involved; the proton-combining role of the hydroxide ion is made clear

Some facts about five common acids and five common bases are summarized in Table 7-1.

Table 7-1
Some Interesting Acids and Bases

Name	Condensed formula	Comment
Acids		
Formic acid	$HCOOH$	Liquid; key component of ant sting
Acetic acid	CH_3COOH	Liquid; vinegar is dilute aqueous solution of acetic acid
Benzoic acid	C_6H_5COOH	Solid used as preservative for certain foods
Nitric acid	HNO_3	Liquid; important commercial acid made from ammonia, NH_3, and used to make fertilizers and explosives (Sec. 13-1)
Sulfuric acid	H_2SO_4	Liquid; important commercial acid, used in petroleum refining, in making fertilizers, and so many other substances that its production volume is used as an index of chemical production
Bases		
Sodium hydroxide	$NaOH$	Solid, commonly called lye, used to make soap from fats
Lithium hydroxide	$LiOH$	Solid, used to remove carbon dioxide exhaled by astronauts in the Apollo spacecraft
Magnesium hydroxide	$Mg(OH)_2$	Solid, used in aqueous suspension (milk of magnesia) to neutralize excess stomach acidity and as a laxative
Ammonia	NH_3	Gas, used in dilute aqueous solution as a household cleaner; also used in very large volume to make explosives, fertilizers, and synthetic fibers
Methylamine	CH_3NH_2	Gas, used in making pain-relieving agent meperidine

7-6 PROTON-TRANSFER REACTIONS FORMING VOLATILE SUBSTANCES: THE GENERATION OF THE LETHAL SUBSTANCE HYDROGEN CYANIDE, HCN

Hydrogen cyanide is a highly toxic substance which boils at just about room temperature (26°C), so that it is usually considered a gas. It is quickly fatal to human beings and is used for the control of insects and to rid ships of rats. It has also been used in some states to carry out legal executions in gas chambers.

It is generated by treating sodium cyanide, NaCN, with sulfuric acid, H_2SO_4, as follows:

$$2NaCN \quad + \quad H_2SO_4 \quad \rightarrow \quad 2HCN \quad + \quad Na_2SO_4$$

Sodium cyanide in aqueous solution as Na^+ and CN^- ions	Sulfuric acid in aqueous solution as H^+ and SO_4^{2-} ions	Hydrogen cyanide in aqueous solution as HCN molecules (volatilizes)	Sodium sulfate in aqueous solution as Na^+ and SO_4^{2-} ions

Representation of the proton-transfer reaction for generating hydrogen cyanide, HCN

If the representation is rewritten to emphasize the ionic content of the solutions, its proton-transfer nature becomes clearer:

Ionic equation: $\quad 2Na^+ + 2CN^- + 2H^+ + SO_4^{2-} \rightarrow 2HCN + 2Na^+ + SO_4^{2-}$

Net ionic equation: $\qquad CN^- + H^+ \rightarrow HCN \uparrow$

Representation of the reaction for the generation of hydrogen cyanide, HCN, to emphasize its proton-transfer character; the arrow pointing up emphasizes the volatility of the hydrogen cyanide

7-7 ANOTHER PROTON-TRANSFER REACTION WHICH FORMS A VOLATILE PRODUCT: THE FUNCTION OF CALCIUM CARBONATE, $CaCO_3$, AS A GASTRIC ANTACID

As an introduction to the nature of chemical reactions in Sec. 6-2 we considered the reaction between calcium carbonate, functioning as a gastric antacid, and hydrochloric acid in the stomach. This reaction generates the gas, carbon dioxide, CO_2:

$$2HCl \quad + \quad CaCO_3 \rightarrow CO_2 \quad + \quad H_2O + CaCl_2$$

Hydrochloric acid	Calcium carbonate	Carbon dioxide	Water	Calcium chloride

Equation representing the function of calcium carbonate, $CaCO_3$, as a gastric antacid (from Sec. 6-2)

Actually this reaction proceeds by forming molecular carbonic acid, H_2CO_3, which subsequently decomposes:

$$2HCl + CaCO_3 \rightarrow H_2CO_3 + CaCl_2 \qquad (1)$$

$$H_2CO_3 \rightarrow H_2O + CO_2 \uparrow \qquad (2)$$

The two stages of the reaction between calcium carbonate, $CaCO_3$, and hydrochloric acid, HCl

Step (1) can be written to stress the ionic nature of the solutions:

Ionic equation: $2H^+ + 2Cl^- + CaCO_3 \rightarrow H_2CO_3 + Ca^{2+} + 2Cl^-$

Net ionic equation: $2H^+ + CaCO_3 \rightarrow H_2CO_3 + Ca^{2+}$

Representation of step (1) of the reaction between calcium carbonate, CaCO₃, and hydrochloric acid, HCl, emphasizing the participating ions; following convention, the calcium carbonate is not explicity represented as containing the Ca²⁺ and CO₃²⁻ ions since it is insoluble. The carbonic acid is represented as H₂CO₃ since it is mostly molecular.

A close look at the net ionic equation makes the proton-transfer nature of the reaction clear. The proton from the hydrochloric acid is transferred to the carbonate ion, CO_3^{2-}, of the calcium carbonate to convert it into carbonic acid, H_2CO_3.

7-8 DECOMPOSITION REACTIONS: THE PREPARATION OF THE SURGICAL ANESTHETIC NITROUS OXIDE, N₂O

The inhalation of the gas nitrous oxide, N_2O, brings about a loss of consciousness and a relief from the pain of surgery. It has been widely used as a surgical anesthetic for more than 100 years, and although there have been many recent improvements in the techniques of anesthesia, nitrous oxide still finds important uses, particularly in conjunction with other agents. The anesthetic properties of nitrous oxide were first recognized in the eighteenth century, and this discovery played an important role in the original development of surgical anesthesia. The impact of anesthesia on the healing possibilities of surgical procedures can hardly be overestimated. (Surgical anesthesia is discussed further in Sec. 12-4.)

Nitrous oxide, N_2O, does not occur in nature but can readily be prepared from ammonium nitrate, NH_4NO_3 (Sec. 7-4). If the dry salt is heated to about 200 to 250°C, it decomposes into nitrous oxide, N_2O, and water:

$$NH_4NO_3 \xrightarrow{200-250°C} N_2O + 2H_2O$$

| Ammonium nitrate as dry salt | Nitrous oxide as gas | Water as steam |

Reaction representing the decomposition of ammonium nitrate, NH₄NO₃, to prepare nitrous oxide, N₂O

As the name implies, in a **decomposition reaction** a substance is decomposed into two or more other substances. Decomposition reactions are common in both natural processes and in reactions used by industrial chemistry.

7-9 POLYMERIZATION REACTIONS: THE PREPARATION OF POLYETHYLENE PLASTIC

A **polymer** is a substance whose molecules are made up of many recurring structural units held together by covalent bonds. Polyethylene plastic, for example, is made up of molecules in-

$$-\overset{\displaystyle H}{\underset{\displaystyle H}{C}}-$$

**Structure of fundamental unit
of a polyethylene molecule**

$$\left[-\overset{H}{\underset{H}{C}}-\overset{H}{\underset{H}{C}}-\overset{H}{\underset{H}{C}}-\overset{H}{\underset{H}{C}}-\overset{H}{\underset{H}{C}}-\overset{H}{\underset{H}{C}}-\right]_x$$

**Representation of a
polyethylene molecule;
x = typically 170 to 340**

$$\overset{H}{\underset{H}{C}}=\overset{H}{\underset{H}{C}}$$

Ethylene

volving many units of the structure shown in the margin linked by covalent bonds. The molecules can be represented by a general formula. The number of CH_2 units in a polyethylene molecule varies over a broad range, typically between 1000 and 2000.

Giant polymeric molecules are frequently found in natural substances. They are the materials of animal tissues, blood, and skin. They predominate in the structures of plants and trees. Many polymers have also been put together by chemists, e.g., nylon, Dacron, synthetic rubber, Lucite, and Teflon. They form the world of plastics which has such a dominating influence on our industrial society—often for good but sometimes for bad as far as the environment is concerned. (Polymers are discussed in Chap. 13.)

Polyethylene is made from the gaseous substance ethylene. The production makes use of a **polymerization reaction,** i.e., a reaction in which many molecules of one or a small number of substances of relatively simple structure are linked together to make a very large molecule. The reaction can be represented in general terms in the following equationlike manner:

$$n\ \overset{H}{\underset{H}{C}}=\overset{H}{\underset{H}{C}}\ \xrightarrow[\text{high temp.}]{\text{cat., high press.}}\ \left[-\overset{H}{\underset{H}{C}}-\overset{H}{\underset{H}{C}}-\overset{H}{\underset{H}{C}}-\overset{H}{\underset{H}{C}}-\overset{H}{\underset{H}{C}}-\overset{H}{\underset{H}{C}}-\right]_x$$

n units of
ethylene
molecules,
typically
500 to 1000

Representation of a mixture of
polyethylene molecules of
varying numbers of units;
x = typically 170 to 340

**Representation of the polymerization reaction for the formation
of polyethylene molecules**

Polymerization reactions are fundamental to the production of polymeric substances. Polyethylene is one of the most widely used plastic materials. Because of its inertness it is used for chemical storage bottles. Its electrical insulating properties make it a valuable material for wire insulation. World production amounted to almost 9 billion tons in 1976.

7-10 ELECTRON-TRANSFER REACTIONS: THE PRODUCTION OF IODINE, I_2, A VALUABLE ANTISEPTIC

Iodine is a grayish-black brittle solid which turns solutions reddish-brown. It is an effective antiseptic and has been widely used in solutions for the disinfection of the skin since 1839. It remains a valuable agent.

Iodine does not occur free in nature but must be obtained from a compound containing it, usually from naturally occurring sodium iodide, NaI, by the the action of chlorine, Cl_2:

$$2NaI\ +\ Cl_2\ \rightarrow\ I_2\ +\ 2NaCl$$

Sodium
iodide in
solution as
Na^+ and I^-
ions

Chlorine
gas

Iodine
solid

Sodium
chloride
in solution
as Na^+ and
Cl^- ions

**Representation of the production of iodine, I_2,
from the action of chlorine, Cl_2, on sodium iodide, NaI**

This representation can be rewritten to emphasize the ionic nature of the solutions and the important molecular species:

Ionic
equation: $\qquad 2Na^+ + 2I^- + Cl_2 \rightarrow I_2 + 2Na^+ + 2Cl^-$

Net ionic
equation: $\qquad 2I^- + Cl_2 \rightarrow I_2 + 2Cl^-$

Representation of the reaction for the production of iodine which emphasizes the important molecular species

The net ionic equation reveals that the reaction involves the conversion of two iodide ions, I^-, into an iodine molecule, I_2, and the simultaneous conversion of a chlorine molecule, Cl_2, into two chloride ions, Cl^-. We look at these two processes separately in the margin. A close look at them reveals that all that really happens is that two electrons are transferred from the two iodide ions, I^-, to the chlorine molecule, Cl_2.

Such a reaction is appropriately called an **electron-transfer reaction,** i.e., a reaction which involves the transfer of one or more electrons from one molecular species to another. These reactions are common.

$$2I^- \rightarrow I_2 + 2e^-$$

$$2e^- + Cl_2 \rightarrow 2Cl^-$$

Separation of the net ionic representation into two processes involving electron production and consumption (e^- represents an electron)

7-11 ANOTHER ELECTRON-TRANSFER REACTION: THE PRODUCTION OF CRYSTALS OF SILVER, Ag

This reaction has little practical value, its redeeming feature being the beautiful silver crystals that result (Fig. 7-2). If a strip of elemental copper, Cu, is placed in an aqueous solution of silver nitrate, $AgNO_3$, a reaction occurs in which the silver ions, Ag^+, of the silver nitrate are converted into silver atoms, Ag:

$$Cu \quad + \quad 2AgNO_3 \quad \rightarrow \quad 2Ag \; + \quad Cu(NO_3)_2$$

| As solid copper | Silver nitrate in colorless aqueous solution as Ag^+ and NO_3^- ions | As solid silver | Copper(II) nitrate in aqueous solution as NO_3^- and Cu^{2+} ions |

Representation of the reaction between copper, Cu, and an aqueous solution of silver nitrate, $AgNO_3$

If the reaction is carried out carefully, the silver will appear as brilliantly shiny needles (Fig. 7-2).

When the reaction is written in a net ionic form, its electron-transfer nature becomes clearer:

Ionic
equation: $\quad Cu + 2Ag^+ + 2NO_3^- \rightarrow 2Ag + Cu^{2+} + 2NO_3^-$

Net ionic
equation: $\qquad Cu + 2Ag^+ \rightarrow 2Ag + Cu^{2+}$

Separated
processes: $\qquad Cu \rightarrow Cu^{2+} + 2e^-$

$$2e^- + 2Ag^+ \rightarrow 2Ag$$

Representation of the copper–silver nitrate reaction to emphasize its electron-transfer nature; each copper atom, Cu, donates two electrons to a pair of silver ions, Ag^+

FIGURE 7-2
Steps in the reaction between copper, Cu, and an aqueous solution of silver nitrate, $AgNO_3$: $Cu + 2AgNO_3 \rightarrow 2Ag + Cu(NO_3)_2$. (a) A strip of copper is immersed in aqueous silver nitrate solution; (b) deposit of metallic silver crystals after about 20 minutes; (c) after about 40 minutes. The silver ions, Ag^+, of the solution are converted into silver atoms, Ag, at the surface of the copper. Meanwhile the copper atoms, Cu, are converted into copper ions, Cu^{2+}, which color the solution pale blue. (*Photographs by Hans Oettgen.*)

(a)

(b)

(c)

7-12 ELECTRON-TRANSFER REACTIONS AND THE GENERATION OF AN ELECTRIC CURRENT

Optional

In Sec. 7-11 we discussed the electron-transfer nature of the reaction between copper, Cu, and an aqueous solution of silver nitrate, $AgNO_3$. In this reaction the electrons were transferred directly from the copper atoms to the silver ions. Is it possible to carry out this reaction in such a way that the electrons pass from the copper atoms to the silver ions through a metal wire?

A **galvanic cell** is an apparatus for using an electron-transfer reaction to generate the flow of electrons which constitutes an electric current in a metal conductor. Some of the chemical energy of the reactants is converted into electric energy. A schematic representation of a galvanic cell using the reaction between copper and silver nitrate is given in Fig. 7-3.

The galvanic cell consists of the following parts:

A copper compartment, in which a strip of copper is immersed in an aqueous solution of copper nitrate containing copper ions, Cu^{2+}, and nitrate ions, NO_3^-.

A silver compartment, in which a strip of silver is immersed in an aqueous solution of silver nitrate containing silver ions, Ag^+, and nitrate ions, NO_3^-.

A so-called salt bridge, which connects the two compartments. It contains an aqueous solution of positive and negative ions, e.g., potassium ions, K^+, and nitrate ions, NO_3^-. The salt bridge connects the two compartments in such a way that ions can drift toward or away from the compartments without mixing the solutions in the two compartments.

A copper wire which connects the two metal strips in the compartments and permits the flow of electrons.

FIGURE 7-3
Schematic representation of a galvanic cell using the electron-transfer reaction between copper, Cu, and silver nitrate, $AgNO_3$. The salt bridge permits the necessary drift of ions but prevents the two solutions from mixing, and the copper wire permits the flow of electrons. See the text discussion.

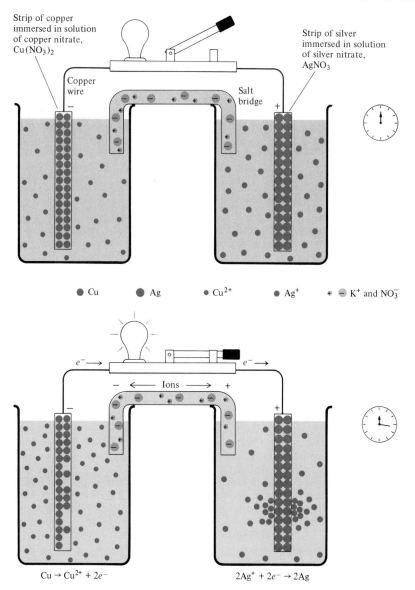

Strip of copper immersed in solution of copper nitrate, $Cu(NO_3)_2$

Copper wire

Salt bridge

Strip of silver immersed in solution of silver nitrate, $AgNO_3$

● Cu　　● Ag　　• Cu^{2+}　　• Ag^+　　• • K^+ and NO_3^-

$e^- \longrightarrow$　　　$e^- \longrightarrow$

$-$ ← Ions → $+$

$Cu \rightarrow Cu^{2+} + 2e^-$　　　　　$2Ag^+ + 2e^- \rightarrow 2Ag$

$$2Ag^+$$

Silver
ions from
aqueous
solution

Cu

Copper atoms
from copper
strip

+

$$\downarrow$$

$$2e^-$$

$$Cu^{2+}$$

Electrons
from
metallic
silver

Copper ions
into aqueous
solution

$$\downarrow$$

+

$$2Ag$$

$$2e^-$$

Silver atoms
deposited
on metallic
silver

Electrons
to connecting
wire

In the copper compartment, copper atoms leave the metallic copper as the spontaneous change occurs. As they do, each gives up two electrons to the connecting copper wire and enters the aqueous solution as a copper ion. At the same time two nitrate ions, NO_3^-, flow from the salt bridge into the copper compartment, accommodating the entrance of the copper ions and maintaining a balance of positive and negative ionic charges in the compartment.

In the silver compartment, meanwhile, two silver ions have picked up one electron each from the silver strip, left the aqueous solution, and become deposited on the silver strip as silver atoms. At the same time, two potassium ions, K^+, flow from the salt bridge to the aqueous solution of the silver compartment, accommodating the departure of the two silver ions, Ag^+, and maintaining a balance of positive ionic charges and negative ionic charges in the compartment.

The electrons given up by the copper atoms in the copper compartment drift through the connecting copper wire and are transferred to the silver ions, Ag^+, through the silver strip. The transfer of electrons from the copper atoms, Cu, to the silver ions, Ag^+, takes place through the connecting wire rather than directly (as it does when the copper wire is immersed directly in an aqueous solution of silver ions). The flow of electrons through the connecting wire constitutes an electric current. Electron-transfer reactions are also used to generate an electric current in dry cell batteries and storage batteries.

7-13 THE PRINCIPLE OF FUEL CELLS

Optional

A galvanic cell, like the one involving the reaction between elemental copper, Cu, and the silver ions, Ag^+, from silver nitrate (Sec. 7-12), makes it possible to obtain electric energy directly from a chemical reaction, and cells of this type have many important practical applications. Unfortunately, however, as electric current is drawn and the electrode reactions occur, the capacity of the cell diminishes and eventually it becomes "dead." The **fuel cell** is a variation of a galvanic cell which can provide electric energy continuously. The special feature of a fuel cell is that provision is made for the continuous addition of reactants and removal of products.

Figure 7-4 is a schematic drawing of a fuel cell which obtains electric energy directly from the reaction between hydrogen and oxygen to produce water (see margin).

$$2H_2 \quad + \quad O_2 \quad \rightarrow 2H_2O$$

Hydrogen   Oxygen   Water

Reaction used to generate electric energy in one type of fuel cell

The hydrogen and oxygen are fed into the cell by diffusing them through porous electrodes of a conducting substance like carbon. The porous electrodes are constructed to permit the gaseous reactants to diffuse into the electrolyte solution† but not to permit the solution to flow back into the gas compartments. The electrolyte is an aqueous solution of potassium hydroxide, KOH, containing potassium ions, K^+, and hydroxide ions, OH^-. As the cell operates, the gaseous hydrogen, H_2, and hydroxide ions, OH^-, are converted into water:

$$2H_2 + 4OH^- \rightarrow 4H_2O + 4e^-$$

† An electrolyte solution is one containing ions.

This reaction provides the electrons which are transported through the conductor to the other electrode. Here oxygen, O_2, and water H_2O, are converted to hydroxide ions, OH^-:

$$4e^- + O_2 + 2H_2O \rightarrow 4OH^-$$

Thus as the fuel cell functions, electrons are transported through the conductor, and charged hydroxide ions are transported through the liquid electrolyte. The overall reaction of the cell involves the combination of hydrogen and oxygen to form water:

Deelectronation reaction: $\quad 2H_2 + 4OH^- \rightarrow 4H_2O + 4e^-$

Electronation reaction: $\quad 4e^- + O_2 + 2H_2O \rightarrow 4OH^-$

Overall reaction: $\qquad\qquad 2H_2 + O_2 \rightarrow 2H_2O$

Provision is made to remove the water as it is formed.

Fuel cells on this principle were chosen for the Apollo space-craft because they are efficient and the technical details of storage and handling of the reactants are familiar to space technologists, who also use hydrogen and oxygen as rocket fuels. Unfortunately, low-cost, high-energy fuel cells, which would find wide application, have not as yet been developed. If perfected, they would provide a very efficient method of obtaining energy from fuels like natural gas.

FIGURE 7-4

Schematic representation of a fuel cell using the reaction between hydrogen, H_2, and oxygen, O_2, which are continuously added through the porous carbon electrodes.

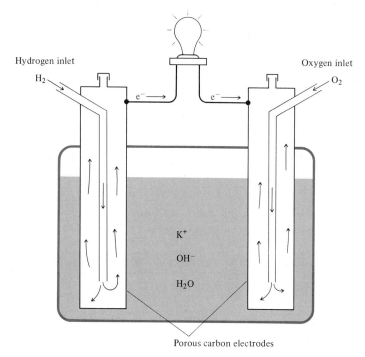

Hydrogen inlet

H_2

Oxygen inlet

O_2

$e^- \longrightarrow$ $\qquad$ $e^- \longrightarrow$

K^+

OH^-

H_2O

Porous carbon electrodes

7-14 ELECTROLYTIC REACTIONS: THE PRODUCTION OF CHLORINE, Cl_2, AN ESSENTIAL DISINFECTANT FOR DRINKING WATER

Optional

Chlorine is a toxic, greenish-yellow gas with a pungent, irritating odor. Added to water in very small concentration (0.2 to 1.0 part per million), it kills disease-causing bacteria. It has been used to sterilize drinking-water supplies since the beginning of the twentieth century. The disinfecting action of chlorine is very important in maintaining adequate sources of safe drinking water in the United States.

Chlorine does not exist in nature but must be prepared from naturally occurring sodium chloride, NaCl, usually by the action of an electric current on a aqueous solution of sodium chloride (electrolysis, briefly discussed in Sec. 6-7). The overall reaction is

$$2NaCl + 2H_2O \xrightarrow[\text{current}]{\text{electric}} 2NaOH + H_2 + Cl_2$$

| Sodium chloride in aqueous solution as Na^- and Cl^- ions | Water | Sodium hydroxide in aqueous solution as Na^+ and OH^- ions | Hydrogen evolves as gas | Chlorine evolves as gas |

Overall reaction for the production of chlorine, Cl_2, by the electrolysis of an aqueous solution of sodium chloride, NaCl

Electrolysis is the process of carrying out a chemical reaction with electric energy. An **electrolytic cell** is the name given to the apparatus for carrying out the electrolysis process. In a galvanic cell (Sec. 7-12) an electron-transfer reaction is used to generate an electric current. In an electrolytic cell an electric current is used to do the reverse: carry out a chemical reaction.

Without going into the details of electrolytic cells we can get an idea of the process from the diagrammatic representation of Fig. 7-5. Two metal electrodes (A and B) are inserted in an aqueous solution of sodium chloride. The principal components of the aqueous solution are sodium ions, Na^+, chloride ions, Cl^-, and water molecules, H_2O. A source of electric current (a flow of electrons) enters the solution through the metal at A. The electrons convert the water molecules into hydrogen molecules, H_2, and hydroxide ions, OH^-. At the same time electrons are being removed at electrode B. These electrons come directly from the chloride ions, Cl^-, which are converted into chlorine molecules. Putting the two processes at the metal electrodes together gives a net ionic equation for the overall reaction, from which the overall molecular equation is readily derived:

$2H_2O + 2e^- \rightarrow H_2 + 2OH^-$

Reaction taking place in electrolysis process at electrode A, where electrons enter solution

$2Cl^- \rightarrow Cl_2 + 2e^-$

Reaction taking place in electrolysis process at electrode B, where electrons leave solution

Reaction at electrode A:	$2H_2O + 2e^- \rightarrow H_2 + 2OH^-$
Reaction at electrode B:	$2Cl^- \rightarrow Cl_2 + 2e^-$
Adding reactions:	$2H_2O + 2Cl^- + 2e^- \rightarrow H_2 + 2OH^- + Cl_2 + 2e^-$
Net ionic equation (eliminating electrons):	$2H_2O + 2Cl^- \rightarrow H_2 + 2OH^- + Cl_2$
Molecular equation (adding the Na^+ ions which do not directly participate):	$2H_2O + 2NaCl \rightarrow H_2 + 2NaOH + Cl_2$

Derivation of the net ionic and molecular equations for the electrolysis process from the reactions which take place at the metal electrodes

Electrolysis processes are widely used to isolate aluminum, magnesium, and other metals from their ores and to prepare chlorine, Cl_2, sodium hydroxide, NaOH, and many other important substances. The investigation of electrolysis has also resulted in landmark discoveries about the nature of electricity and the nature of aqueous solutions.

FIGURE 7-5
Schematic interpretation of the electrolysis of an aqueous solution of sodium chloride, NaCl. Hydrogen gas, H_2, is formed at electrode A and chlorine gas, Cl_2, at electrode B.

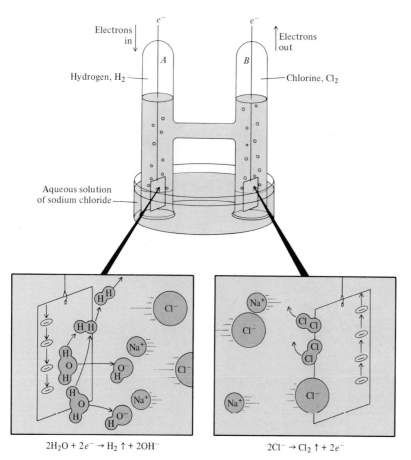

$$2H_2O + 2e^- \rightarrow H_2 \uparrow + 2OH^-$$

$$2Cl^- \rightarrow Cl_2 \uparrow + 2e^-$$

KEY WORDS

1. Ion-combination reaction which forms a precipitate (Sec. 7-2)
2. Proton-transfer reaction (Sec. 7-4)
3. Acid (Sec. 7-5)
4. Base (Sec. 7-5)
5. Decomposition reaction (Sec. 7-8)
6. Polymer (Sec. 7-9)
7. Polymerization reaction (Sec. 7-9)
8. Electron-transfer reaction (Sec. 7-10)

Optional:

9. Galvanic cell (Sec. 7-12)   10. Fuel cell (Sec. 7-13)   11. Electrolytic cell (Sec. 7-14)

"Frankly, I don't see how we can keep it burning through eternity."

© 1975 *American Scientist*; *reprinted by permission of Sidney Harris.*

SUMMARIZING QUESTIONS FOR SELF-STUDY

Sections 7-1 to 7-3

1. Q. Are chemical reactions easily organized and classified? Why?

A. No. Millions are known, and they encompass great variety.

2. Q. What is an ion-combination reaction which forms a precipitate?

A. A reaction which involves a combination of ions in excess of their solubility in a solvent and the resulting formation of a precipitate.

3. Q. Give the molecular equation for the reaction between lithium sulfate, Li_2SO_4, and sodium carbonate, Na_2CO_3, in an aqueous solution.

A. $Li_2SO_4 + Na_2CO_3 \rightarrow Li_2CO_3 + Na_2SO_4$

4. Q. Give the net ionic equation for the reaction in Question 3.

A. $2Li^+ + CO_3^{2-} \rightarrow Li_2CO_3 \downarrow$

5. Q. What is the practical importance of the reaction in Questions 3 and 4?

A. It is the last step in the production of lithium carbonate, Li_2CO_3, used in the treatment of manic depression.

6. Q. Give the molecular equation for the reaction which brings about the concentration of magnesium from the sea.

A. $MgCl_2 + Ca(OH)_2 \rightarrow Mg(OH)_2 \downarrow + CaCl_2$

7. Q. Give the net ionic equation for the reaction in Question 6.

A. $Mg^{2+} + 2OH^- \rightarrow Mg(OH)_2 \downarrow$

8. Q. What is the practical importance of the reaction in Questions 6 and 7?

A. It is the first step in the production of magnesium metal from seawater.

Sections 7-4 to 7-7

9. Q. What is a proton-transfer reaction?

A. A reaction which involves the transfer of one or more protons from one molecular species to another.

10. Q. Give the molecular equation for the reaction between ammonia, NH_3, and nitric acid, HNO_3.

A. $NH_3 + HNO_3 \rightarrow NH_4NO_3$

11. Q. Give the net ionic equation for the reaction in Question 10.

A. $NH_3 + H^+ \rightarrow NH_4^+$

12. Q. What is the practical importance of the reaction in Questions 10 and 11?

A. It is the last step in the production of the most frequently used nitrogen-containing fertilizer, ammonium nitrate, NH_4NO_3.

13. Q. What is an acid?

A. A substance capable of releasing one or more protons.

14. Q. What is a base?

A. A substance capable of combining with one or more protons.

15. Q. Identify the acids and bases: (a) $HCl + NH_3 \rightarrow NH_4Cl$. (b) $NaOH + HNO_3 \rightarrow H_2O + NaNO_3$.

A. (a) Acid HCl; base NH_3; (b) acid HNO_3; base NaOH.

16. Q. Give the molecular equation for the reaction which occurs between calcium carbonate, $CaCO_3$, and hydrochloric acid, HCl.

A. $2HCl + CaCO_3 \rightarrow CO_2 + H_2O + CaCl_2$

17. Q. What is one practical application of the reaction of Question 16?

A. It is the reaction which occurs when the gastric antacid, $CaCO_3$, is used to reduce the amount of hydrochloric acid in the stomach.

Section 7-8

18. Q. What is a decomposition reaction?

A. One in which a substance is decomposed into two or more other substances.

19. Q. Give a molecular equation for the reaction which occurs when ammonium nitrate is heated.

A. $NH_4NO_3 \rightarrow N_2O + 2H_2O$

20. Q. What is the practical significance of the reaction of Question 19?

A. It is the reaction used to prepare the important surgical anesthetic, nitrous oxide, N_2O.

Section 7-9

21. Q. Define (a) polymer and (b) polymerization.

A. (a) A substance whose molecules are made up of many recurring structural units held together by covalent bonds. (b) A reaction in which many molecules of one or a small number of substances of relatively simple structure are linked together to make a very large molecule.

22. Q. Give a representation of the polymerization of ethylene to form the polymer polyethylene.

A.

Sections 7-10 and 7-11

23. Q. What is an electron-transfer reaction?

A. A reaction which involves the transfer of one or more electrons from one molecular species to another.

24. Q. Give a molecular equation for the reaction between sodium iodide, NaI, and chlorine, Cl_2.

A. $2NaI + Cl_2 \rightarrow I_2 + 2NaCl$

25. Q. Give a net ionic equation for the reaction of Question 24.

A. $2I^- + Cl_2 \rightarrow I_2 + 2Cl^-$

26 Q. What is the practical significance of the reaction of Questions 24 and 25?

A. It is the key step in the production of iodine, an effective antiseptic.

Section 7-12 (Optional)

27. Q. What is a galvanic cell?

A. An apparatus for using an electron-transfer reaction to generate the flow of electrons constituting an electric current in a metal conductor.

28. Q. If the electron-transfer reaction

$$Cu + 2AgNO_3 \rightarrow Cu(NO_3)_2 + 2Ag$$

is used in a galvanic cell, what is the source of electrons?

A. $Cu \rightarrow Cu^{2+} + 2e^-$

Section 7-13 (Optional)

29. Q. What is a fuel cell?

A. A variation of a galvanic cell which can

provide electric energy continuously. Provision is made for the continuous addition of reactants and the continuous removal of products.

Section 7-14 (Optional)

30. Q. What is an electrolytic cell and what is electrolysis?

A. An apparatus for carrying out the electrolysis process. The electrolysis process (Sec. 6-7) involves the use of electric energy to carry out a chemical reaction.

31. Q. Give the molecular equation for the electrolysis of an aqueous solution of sodium chloride, $NaCl$.

A. $2H_2O + 2NaCl \rightarrow H_2 + 2NaOH + Cl_2$

32. Q. What is the partial reaction which results in the formation of chlorine, Cl_2?

A. $2Cl^- \rightarrow Cl_2 + 2e^-$

33. Q. What is the practical significance of the reaction of Questions 31 and 32?

A. It is the method of producing elemental chlorine used in disinfecting drinking water supplies.

PRACTICE EXERCISES

1. Match each definition or other statement with the numbered term above with which it is most closely associated. A numbered item may be used only once.

1. Polymerization reaction	2. Acid
3. Electron-transfer reaction	4. Electrolysis
5. Proton-transfer reaction	6. Base
7. Polymer	8. Alcohol
9. Ion-combination reaction	10. Decomposition reaction

(a) A substance capable of combining with a proton

(b) Combining ions in excess of their solubility in water and the resulting formation of a precipitate

(c) A substance capable of releasing one or more protons

(d) A reaction which involves the transfer of one or more protons from one molecular species to another

(e) A reaction in which a substance is decomposed into two or more other substances

(f) A substance whose molecules are made up of many structural units held together by interatomic bonds

(g) A reaction which involves the transfer of one or more electrons from one molecular species to another

(h) A reaction in which many molecules of one or a small number of substances of relatively simple structure are linked together to make a very large molecule

2. Consider the following ionic compounds and their solubility in water:

Calcium bromide, $CaBr_2$, soluble in water
Potassium carbonate, K_2CO_3, soluble in water
Potassium bromide, KBr, soluble in water
Calcium carbonate, $CaCO_3$, insoluble in water

Give the molecular equation and the net ionic equation for any ion-combination reaction which can take place when aqueous solutions of calcium bromide, $CaBr_2$, and potassium carbonate, K_2CO_3, are mixed thoroughly.

3. Consider the following ionic compounds and their solubility in water:

Barium chloride, $BaCl_2$, soluble in water
Sodium sulfate, Na_2SO_4, soluble in water
Barium sulfate, $BaSO_4$, insoluble in water
Sodium chloride, $NaCl$, soluble in water

Give the molecular equation and the net ionic equation for any ion-combination reaction which can take place when aqueous solutions of barium chloride, $BaCl_2$, and sodium sulfate, Na_2SO_4, are mixed thoroughly.

4. Give an example of each of the following. If the example is a substance, give the name and condensed formula. If the example is a reaction, give a balanced equation.
   (a) A polymer
   (b) A polymerization reaction
   (c) A decomposition reaction
   (d) An electron-transfer reaction
   (e) An acid
   (f) A base
   (g) A proton-transfer reaction
   (h) An ion-combination reaction which forms a precipitate

5. Identify the acids and bases in each of the following proton-transfer reactions.
   (a) $KOH + HBr \rightarrow H_2O + KBr$
   (b) $2HCl + Ba(OH)_2 \rightarrow BaCl_2 + H_2O$
   (c) $LiCH_3COO + HNO_3 \rightarrow LiNO_3 + CH_3COOH$†
   (d) $HBr + NH_3 \rightarrow NH_4Br$
   (e) $NaCN + HCl \rightarrow HCN + NaCl$
   (f) $HNO_3 + CH_3NH_2 \rightarrow CH_3NH_3NO_3$‡

6. Distinguish between a decomposition reaction and a polymerization reaction.

7. Distinguish between a proton-transfer reaction and an electron-transfer reaction.

† This can be written $H-\overset{\displaystyle H}{\underset{\displaystyle H}{C}}-\overset{\displaystyle O}{C}-O-H$.

‡ This can be written $(CH_3{}^+NH_3NO_3{}^-)$.

8. Give the net ionic equation for each of the following electron-transfer reactions and identify the molecular species (a) which donates one or more electrons and (b) to which one or more electrons are transferred.

(a) Zn   + 2AgNO₃ →   2Ag   + Zn(NO₃)₂

$$Zn \quad + 2AgNO_3 \rightarrow \quad 2Ag \quad + Zn(NO_3)_2$$

Zinc   Silver nitrate   Silver   Zinc nitrate

(b) Zn   + CuSO₄ →   Cu   + ZnSO₄

$$Zn \quad + CuSO_4 \rightarrow \quad Cu \quad + ZnSO_4$$

Zinc   Copper(II) sulfate   Copper   Zinc sulfate

(c) Pb   + Cu(NO₃)₂ → Pb(NO₃)₂ +   Cu

$$Pb \quad + Cu(NO_3)_2 \rightarrow Pb(NO_3)_2 + \quad Cu$$

Lead   Copper(II) nitrate   Lead nitrate   Copper

(d) Mg   + NiCl₂ →   MgCl₂   +   Ni

$$Mg \quad + NiCl_2 \rightarrow \quad MgCl_2 \quad + \quad Ni$$

Magnesium   Nickel chloride   Magnesium chloride   Nickel

(e) 2Na   + 2HCl →   2NaCl   +   H₂

$$2Na \quad + 2HCl \rightarrow \quad 2NaCl \quad + \quad H_2$$

Sodium   Hydro- chloric acid   Sodium chloride   Hydrogen

Sections 7-12 to 7-14 (Optional)

9. Distinguish between a galvanic cell and an electrolytic cell.

10. In a galvanic cell: (a) By what means are electrons transferred from one cell compartment to the other cell compartment? (b) What is the function of the salt bridge?

11. How does a fuel cell differ from a galvanic cell?

12. For a fuel cell using hydrogen and oxygen as fuels give an equation for (a) the deelectronation reaction, (b) the electronation reaction, and (c) the overall reaction.

13. In the electrolytic reaction

$$2NaCl + 2H_2O \xrightarrow[\text{current}]{\text{electric}} 2NaOH + H_2 + Cl_2$$

(a) What molecular species accepts electrons?
(b) Give an equation for the process involved in part (a).
(c) From what molecular species are electrons removed?
(d) Give an equation for the process involved in part (c).

The following questions assume a knowledge of Supplement 4. Express units, show your method clearly, and label the answer.

14. How many grams of lithium carbonate, Li_2CO_3, can be obtained from 21.988 g of lithium sulfate and an excess of sodium carbonate, assuming the reaction given in Sec. 7.2?

15. How many grams of magnesium hydroxide, $Mg(OH)_2$, can be obtained from 100.00 g of magnesium chloride and an excess of calcium hydroxide, $Ca(OH)_2$, assuming the reaction given in Sec. 7-3?

16. How many grams of ammonia, NH_3, will be required to prepare 32.016 g of ammonium nitrate, NH_4NO_3, assuming an excess of nitric acid, HNO_3, and the reaction given in Sec. 7-4?

17. How many grams of hydrogen cyanide, HCN, will be generated from 200.00 g of sodium cyanide, NaCN, and an excess of sulfuric acid, H_2SO_4, according to the reaction given in Sec. 7-6?

18. How many grams of stomach hydrochloric acid, HCl, will be removed by 1.000 g of the gastric antacid calcium carbonate, $CaCO_3$, assuming the reaction given in Sec. 7-7?

19. How many grams of sodium iodide, NaI, will be required to prepare 1000 g of iodine, I_2, according to the reaction given in Sec. 7-10?

8

WHAT IS ORGANIC CHEMISTRY?

8-1 THE WORLD OF CARBON-CONTAINING COMPOUNDS

Organic chemistry is concerned with the compounds of carbon. What is so distinctive about the compounds of carbon that a separate branch of chemistry is devoted to them? Four points can be made: (1) there is a vast number of carbon compounds; about 2 million have already been characterized; (2) the molecules of numerous carbon compounds are large and of complex structure; many contain thousands of atoms; (3) carbon compounds constitute the major part of the matter of plants and animals (a single-celled organism, for example, contains thousands of carbon-containing compounds); (4) many carbon compounds are of great usefulness, e.g., anesthetics, medicinal drugs, dyes, petroleum products, plastics, textile fibers, and insecticides.

The term organic for the chemistry of carbon compounds dates back to the time when substances were divided into two broad classes, those derived from rocks and minerals (inorganic) and those derived from plants and animals (organic). The substances isolated from plants and animals proved to be compounds of carbon. Even though it has been recognized for years that carbon compounds can be made in the laboratory and do not have to come from living organisms, the term organic remains convenient and useful.

Why are there so many carbon-containing compounds? The answer is found in the extraordinary ability of carbon atoms to form covalent bonds with each other. They can do this to an extent approached by no other kind of atom. Some compounds contain thousands of carbon atoms bonded together in chains, rings, or cross-linked grids.

8-2 THE FUNCTIONAL GROUP PRINCIPLE

Consider some of the questions currently being asked by research scientists: How does penicillin act to cure infections? What is the mechanism of the relief of pain by morphine? What are the molecular mechanisms which control heredity traits? What causes cancer? Any attempt to pursue answers to these questions is immediately confronted with the molecular structure of the substances of living systems. But living organisms are made up of chemical materials of tremendous variety, including complex carbon-containing substances containing thousands of atoms. The proteins are among the most significant. Thousands are present in each living cell, and collectively they account for almost half of the solid material of a person. They serve as the principal materials of skin, blood, nerves, and muscle. They also perform thousands of essential catalytic functions which make it possible for cells to live.

Proteins are fragile molecules of great size, enormous complexity, and awesome diversity. Molecular weights of 10,000 to 100,000 and more are common; even one of the smallest proteins, insulin, contains 777 atoms. For a long time the elaborate structure of proteins appeared to be too complicated to decipher, but not only have the atom-by-atom structures of

THE DISCOVERY OF ETHYL ETHER AS A SURGICAL ANESTHETIC†

Before the use of anesthesia surgical operations were horrifying experiences for the patient and of limited value, since the surgeon had to work in great haste to minimize the agony. Opium, alcoholic beverages, and other pain-relieving agents were employed (with limited success) and even such extreme measures as blows on the head and strangulation. It is ironic that ethyl ether, $CH_3CH_2OCH_2CH_3$, whose use was to change all this in the nineteenth century, had been known for 300 years. Ethyl ether is a low-boiling, highly inflammable, synthetic liquid. Its first synthesis is usually attributed to the traveling apothecary Valerius Cordus around 1540.

In 1798 the English chemist Humphry Davy discovered that the gas nitrous oxide, N_2O, synthesized in 1776 by Joseph Priestley, another English chemist, was able to render a person unconscious and insensitive to pain, but it was not adopted for general use until about 80 years later. In 1818 the English chemist Michael Faraday observed that ethyl ether brought about a similar effect, but physicians considered the substance too dangerous for practical use (Fig. 8-1). Interest in the properties of ethyl ether was kept alive by medical students, who occasionally inhaled the vapor for amusement at "ether parties."

Observation of the effects of ethyl ether under such circumstances presumably led Crawford W. Long, a young doctor in the village of Jefferson, Georgia, to use it as an anesthetic in 1842 in the painless removal of a small tumor from the neck of a volunteering friend. Unfortunately Long did not spread word of his work.

...while, independently of Long, Horace Wells, ...entist in Hartford, Connecticut, used nitrous oxide ("laughing gas") for general anesthesia in his dental practice, acting on a suggestion afforded by a public demonstration of the effect of this gas by the traveling chemist-lecturer Gardner Colton. Wells demonstrated the use of nitrous oxide as a surgical anesthetic at the Massachusetts General Hospital of Boston in 1845, but the gas was not administered properly and the patient regained consciousness during the operation. The use of nitrous oxide in dentistry was later revived, however, and the gas is used today for anesthesia in both dentistry and medicine.

The introduction of anesthesia in dentistry led the Boston dentist and medical student William T. G. Morton, a former dental associate of Wells, to investigate the anesthetic properties of ethyl ether. Morton chose ether at the suggestion of his chemistry teacher Charles T. Jackson, who at one time had accidentally anesthetized himself with the vapor. After trying the anesthetic properties of ether inhalation on cats, chickens, the family dog, and himself, he used it in the successful extraction of a patient's tooth. Later, on the memorable day of October 16, 1846, Morton was permitted to demonstrate the use of ether as a surgical anesthetic at the Massachusetts General Hospital. The success of this demonstration led to the general adoption of ethyl ether as a surgical anesthetic (Fig. 8-2).

A year later, in 1847, James Y. Simpson, a professor of medicine at the University of Edinburgh, introduced chloroform as an anesthetic. This substance had been synthesized 17 years earlier by the German chemist Justus von Liebig. Although chloroform is about 5 times more powerful than ether as an anesthetic, it is also more toxic. But for many years it remained the anesthetic of choice throughout most of Europe.

The introduction of surgical anesthesia, coupled with the antiseptic techniques introduced by Joseph Lister somewhat later, made it possible for surgery to develop into a principal means of healing. In the present century many improved anesthetic agents and techniques have been introduced, greatly extending the usefulness of medical surgery. It is difficult to single out any one person as the discoverer of surgical anesthesia: Long, Wells, Jackson, Morton, and Simpson all share the credit for this far-reaching contribution to humanity.

FIGURE 8-1
A demonstration of the effect of inhaling nitrous oxide, "laughing gas," and ethyl ether, as caricatured in 1802 by James Gillray. (*Courtesy of the National Library of Medicine.*)

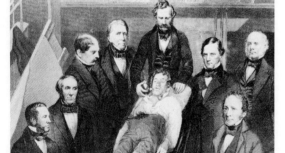

FIGURE 8-2
William T. G. Morton making the first public demonstration of ethyl ether as a general anesthetic at the Massachusetts General Hospital on October 16, 1846. Morton is behind the patient administering the ether. Note the formal attire of the attending physicians and the lack of gowns, masks, and gloves. The germ theory of disease had not yet been established. (Compare with Fig. 12-7). (*Courtesy of the National Library of Medicine.*)

† Adapted from Charles Compton, "An Introduction to Chemistry," pp. 546–547, Van Nostrand Reinhold Company, New York, 1958.

many proteins now been determined in detail, methods of synthesizing these complex substances have been developed.

How do chemists go about attempting to understand the behavior of substances such as proteins from their structure? One useful approach is to zero in on special groups of atoms within the molecule. These **functional groups** are distinctive arrangements of atoms which impart characteristic chemical properties to a substance.

The **functional group principle** assumes that the chemical properties of a substance are determined by the functional groups present in its molecular structure. It is a highly empirical approach and not always applicable, but it provides a useful scheme of classification and often has considerable power. To see how it works, let us consider the reaction which occurs when "substance A" is treated with water at an elevated temperature and with hydrochloric acid as a catalyst.

$$\begin{array}{c}\text{H H}\quad\;\;\text{H H}\\ |\;\;|\qquad|\;\;|\\ \text{H—C—C—C—N—C—C—H}\\ |\;\;|\;\;\|\;\;|\;\;|\;\;|\\ \text{H H O H H H}\end{array}\qquad\qquad\begin{array}{c}\text{CH}_3\text{CH}_2\text{—C—N—CH}_2\text{CH}_3\\ \|\quad|\\ \text{O}\quad\text{H}\end{array}$$

Full structural formula of A   Partially condensed formula of A

The reaction with water can be represented as

$$\begin{array}{c}\text{CH}_3\text{CH}_2\text{—C—N—CH}_2\text{CH}_3 + \text{H—O—H} \rightarrow \text{CH}_3\text{CH}_2\text{—C—O—H} + \text{H—N—CH}_2\text{CH}_3\\ \|\quad|\qquad\qquad\qquad\qquad\qquad\;\;\|\qquad\qquad\qquad|\\ \text{O}\quad\text{H}\qquad\qquad\qquad\qquad\qquad\qquad\text{O}\qquad\qquad\qquad\text{H}\end{array}$$

Close inspection of this equation reveals that the reaction involves the breaking and formation of covalent bonds. To be more specific,

$$\begin{array}{c}\text{CH}_3\text{CH}_2\text{—C}\mid\text{N—CH}_2\text{CH}_3 + \text{H—O}\mid\text{H} \rightarrow \text{CH}_3\text{CH}_2\text{—C}\mid\text{O—H} + \text{H}\mid\text{N—CH}_2\text{CH}_3\\ \|\;\;|\qquad\qquad\qquad\qquad\;\;\;\|\qquad\qquad\qquad|\\ \text{O}\;\;\text{H}\qquad\qquad\qquad\qquad\;\;\;\text{O}\qquad\qquad\qquad\text{H}\end{array}$$

Bond breaks   Bond breaks   Bond forms   Bond forms

Representation of reaction with emphasis on the bonds which are broken and formed; this is not, however, the step-by-step mechanism

For comparison here are the equations for two other reactions carried out under similar conditions:

$$\begin{array}{c}\text{CH}_3\text{C}\mid\text{N—CH}_3 + \text{H—O}\mid\text{H} \rightarrow \text{CH}_3\text{C}\mid\text{O—H} + \text{H}\mid\text{N—CH}_3\\ \|\;\;|\qquad\qquad\qquad\qquad\|\qquad\qquad\qquad|\\ \text{O}\;\;\text{H}\qquad\qquad\qquad\qquad\text{O}\qquad\qquad\qquad\text{H}\end{array}$$

Bond breaks   Bond breaks   Bond forms   Bond forms

$$\begin{array}{c}\text{CH}_3\text{CH—C}\mid\text{N—CH}_3 + \text{H—O}\mid\text{H} \rightarrow \text{CH}_3\text{CH—C}\mid\text{O—H} + \text{H}\mid\text{N—CH}_3\\ |\quad\|\;\;|\qquad\qquad\qquad\qquad\quad|\quad\|\qquad\qquad\qquad\qquad|\\ \text{CH}_3\;\;\text{O}\;\;\text{CH}_3\qquad\qquad\qquad\;\;\text{CH}_3\;\;\text{O}\qquad\qquad\qquad\text{CH}_3\end{array}$$

Bond breaks   Bond breaks   Bond forms   Bond forms

These reactions are strikingly similar; in each case the reaction centers around a rearrangement of the bonding of the following atoms:

$$-\underset{\underset{O}{\|}}{C}-\underset{|}{N}- \ + \ H-O-H \ \rightarrow \ -\underset{\underset{O}{\|}}{C}-O-H \ + \ H-\underset{|}{N}-$$

Bond breaks · · · · Bond breaks · Bond forms · · · · Bond forms

These are the only changes which occur.

Each of the principal reactants of the three reactions contains the same distinctive group of atoms. The *amide group*, shown in the margin is an example of a functional group, and all substances containing such a group belong to a class of compounds called *amides*. All the principal reactants in the preceding reactions are therefore amides.

A convenient method of representing a group of compounds containing the same functional group is to use a *general formula*. The general formula applicable to the amides under consideration is shown in the margin. The two R groups represent all of the structure of the molecule except the functional group. With the help of general formulas the reaction we have been considering can be summarized:

$$R-\underset{\underset{O}{\|}}{C}-\underset{|}{N}-R \ + \ H-O-H \ \rightarrow \ R-\underset{\underset{O}{\|}}{C}-O-H \ + \ H-\underset{|}{N}-R$$

Amide · · · · · · Water · · · · · Carboxylic acid · · · · · Amine

Summary representation of the hydrolysis of amides by means of general formulas

Since water is a reactant, the reaction is called a hydrolysis. Further, since amides are involved, the reaction is referred to more precisely as the hydrolysis of an amide.

The products of this reaction also have distinctive functional groups. The first product (as written) contains the *carboxylic acid group*, and consequently is known as a *carboxylic acid*, with the general formula shown. The second product (as written), contains the amine group and consequently is known as an *amine*, with the general formula shown. The general reaction summarizes the information that the hydrolysis of an amide produces a carboxylic acid and an amine.

8-3 APPLYING THE FUNCTIONAL GROUP PRINCIPLE TO THE STRUCTURE OF PROTEINS

Investigation of protein molecules has shown that while each protein has its own distinctive arrangement of atoms, they all have a structural characteristic in common. A portion of this common structural feature can be represented as

---

Margin notes:

$$-\underset{\underset{O}{\|}}{C}-\underset{|}{N}-$$

Amide group

$$R-\underset{\underset{O}{\|}}{C}-\underset{\underset{H}{|}}{N}-R$$

General formula for an amide

$$-\underset{\underset{O}{\|}}{C}-O-H$$

Carboxylic functional group

$$R-\underset{\underset{O}{\|}}{C}-O-H$$

General formula of a carboxylic acid

$$-\underset{\underset{H}{|}}{N}-H$$

Amine functional group

$$R-\underset{\underset{H}{|}}{N}-H$$

General formula for an amine

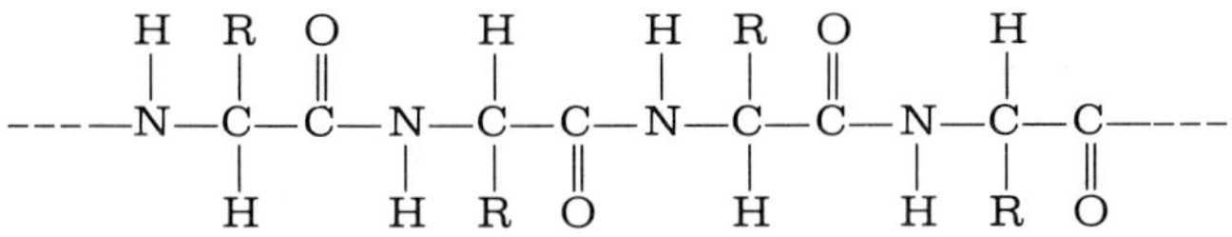

$$\text{----N--C--C--N--C--C--N--C--C--N--C--C----}$$

Representation in general terms of a portion of a protein molecule;
the backbone chain of atoms may extend for hundreds of atoms;
the groups of atoms represented by the R symbols may be as simple as
an individual hydrogen atom or a CH_3 group or much more complex

In order to obtain an idea of the chemical properties suggested by such structures, chemists pay particular attention to
the functional groups present. The distinctive functional
groups in this structure are the amide groups, which appear
at regular intervals along the backbone chain of atoms.

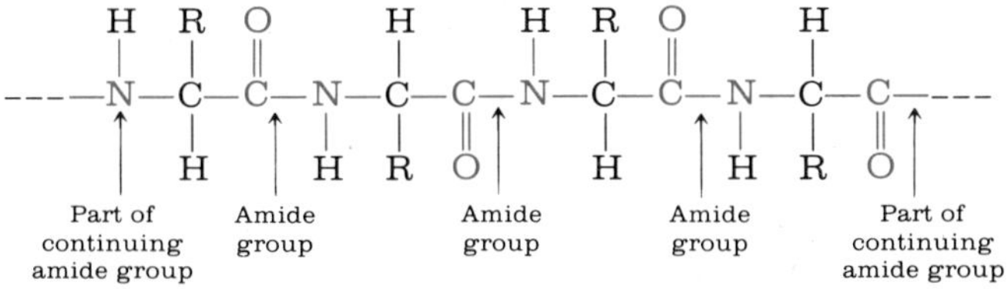

$$\text{----N--C--C--N--C--C--N--C--C--N--C--C----}$$

Part of | Amide | Amide | Amide | Part of
continuing | group | group | group | continuing
amide group | | | | amide group

Representation in general terms of a portion of a protein molecule,
emphasizing the presence of amide groups

Recognition of the presence of the amide groups leads
chemists to understand why protein molecules react readily
with water. Moreover, through application of the functional
group principle, chemists can anticipate what the products of
the reaction will be.

Recall the general equation for the hydrolysis of amides:

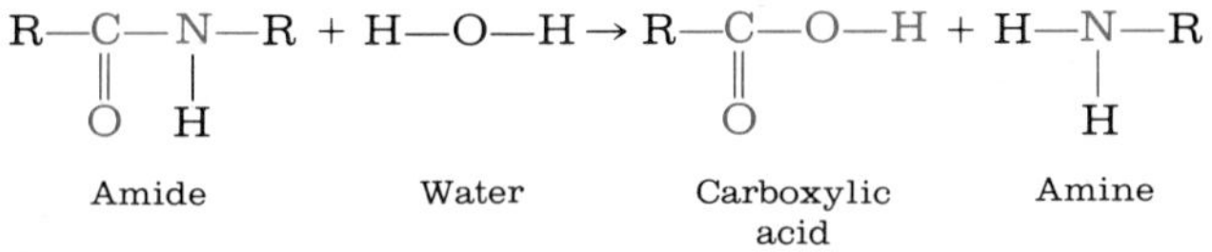

$$\text{R--C--N--R} + \text{H--O--H} \rightarrow \text{R--C--O--H} + \text{H--N--R}$$

Amide | Water | Carboxylic acid | Amine

General representation of hydrolysis of amides

By analogy the products of the hydrolysis of proteins can be
anticipated:

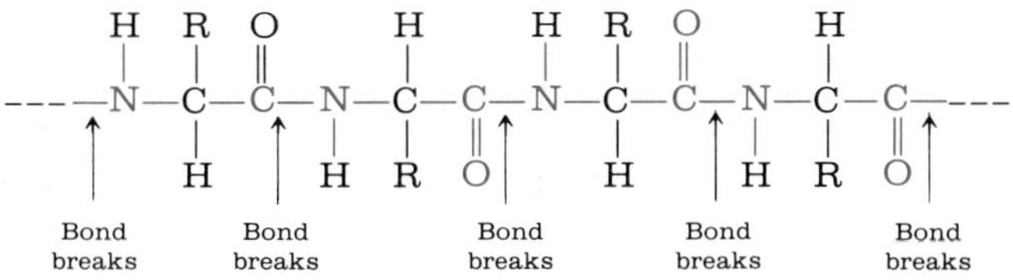

$$\text{----N--C--C--N--C--C--N--C--C--N--C--C----}$$

Bond breaks | Bond breaks | Bond breaks | Bond breaks | Bond breaks

Application of functional group principle to the
hydrolysis of proteins; recognition of the
amide groups makes it possible to know where the
bonds will break in the reaction with water

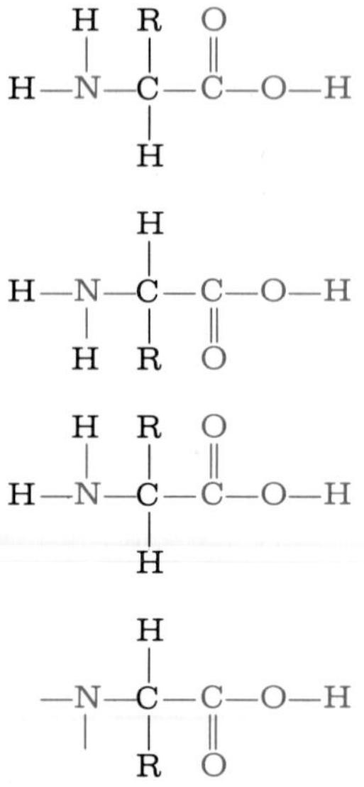

Representation in general form of the products of the hydrolysis of a protein

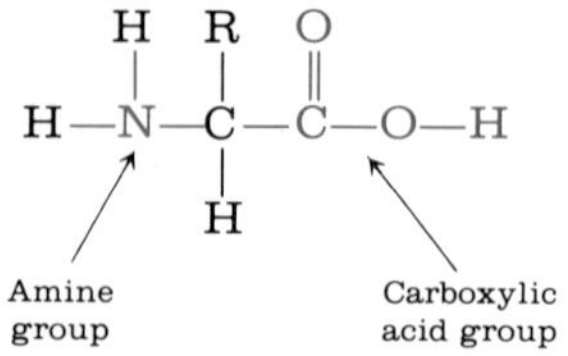

A typical product of the hydrolysis of a protein molecule, known as an amino acid

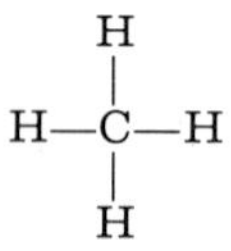

Formula of the simplest hydrocarbon, methane, also known as natural gas, of which it constitutes from 60 to 98 percent

The products of the protein hydrolysis shown in the margin are obtained directly from the formula above with the recognition of the amide groups and a knowledge of the hydrolysis reaction which they undergo. Each product of the hydrolysis contains an amine group and a carboxylic acid group. These products are therefore known as *amino acids*.

The application of the functional group principle involves two principal steps: (1) the recognition of the groups and (2) the derivation of the products of a reaction from a knowledge of the chemical properties associated with each group. This chapter emphasizes the recognition of functional groups and discusses their great usefulness in classifying the compounds of carbon. The discussion of the hydrolysis of proteins in this section demonstrates how functional groups serve as a guide to the important features of a structural formula and how they lead chemists to "where the action is."

8-4 WHAT ARE HYDROCARBONS?

The structure of carbon compounds is used to classify them with the help of functional groups. The name **hydrocarbon** is applied to compounds of carbon which contain only carbon and hydrogen atoms. The simplest hydrocarbon is the gas methane.

Alkanes are hydrocarbons, like methane, which contain only single covalent bonds. A few examples of simple alkanes are given in Table 8-1. In a sense, alkanes contain no distinctive functional group. Sometimes the

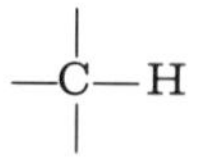

group is considered to be the functional group of an alkane, but it is so common in carbon compounds that it is not useful for classification. Other classes of hydrocarbons besides the alkanes are summarized in Table 8-2.

Hydrocarbons are used for an extraordinary variety of purposes; as fuels they heat buildings, propel automobiles and airplanes, and launch interplanetary rockets; even more significantly they serve as the raw materials for such diverse products as medicinals, dyes, fertilizers, and synthetic rubber. The total influence of hydrocarbons on our economy and life is profound.

The principal sources of hydrocarbons are natural gas, petroleum, and coal. Natural gas is mostly methane, CH_4 (from 60 to 98 percent). There are several different types of coal of varying composition. In general, they are mixtures of many substances, some with molecular weights in the 1000 to 3000 range. The atoms present are chiefly those of carbon, hydrogen, and oxygen, with smaller amounts of sulfur and nitrogen. The

---

Note to Students — **For the sake of brevity the discussion of hydrocarbons has been severely limited despite their importance.**

Table 8-1
Some Representative Alkanes of Simple Structure

Name	Formula	State	
Methane	CH_4	Colorless gas	A prin (natura rial for
Ethane	CH_3CH_3	Colorless gas	Starting duction o pounds
Propane	$CH_3CH_2CH_3$	Colorless gas	Principal component of "bottled gas," used for domestic purposes
Normal octane	$CH_3CH_2CH_2CH_2CH_2CH_2CH_2CH_3$	Colorless liquid	One of many hydrocarbons in gasoline
Iso-octane	$CH_3CCH_2CHCH_3$ with CH_3 above and CH_3 CH_3 below	Colorless liquid	Standard for octane performance rating of gasoline fuels; has octane rating of 100

THE INTRODUCTION OF ANTISEPTICS AND DISINFECTANTS INTO HEALTH CARE

Disease was once thought to be caused by floods, earthquakes, and changes in the weather. Beginning with the investigations and speculations of curious physicians about 400 years ago, the contagious nature of disease gradually came to be recognized. These advances culminated in the germ theory of disease, somewhat past the midpoint of the nineteenth century, by the outstanding work of the French chemist Louis Pasteur, the German physician Robert Koch, and others. Before the bacterial origin of infection was recognized, little attention was paid to cleanliness in surgical operations. The surgeon was often dressed in formal clothes, and there were no gloves, masks, gowns, or hair coverings. While the introduction of anesthesia around the middle of the century made surgery a much more humane means of healing than it had been, the risk of infection from the incision was very great and the loss of life was heavy.

The principles of the germ theory of disease were first applied to surgical procedures by the British surgeon Joseph Lister in 1865. He showed that if surgical instruments and wounds were sterilized with antiseptic phenol, the dreaded infections could be prevented. His success did much to establish the germ theory of disease, then in the process of development.

Centuries before the role of bacteria in infection was recognized chemicals had been used to treat wounds and prevent the spread of contagious diseases. The embalmers of ancient Egypt were aware of many naturally occurring preservatives. The Greeks used wine and vinegar to dress wounds. An alcoholic solution of iodine (tincture of iodine) was introduced by a French surgeon in 1839 and found wide use in the treatment of battle wounds during the Civil War. Once the role of microorganisms in infection was understood, much more effective use was made of antiseptics.

Antiseptics are agents which kill or inhibit microorganisms when applied to the surface of living tissue. They are usually too toxic for internal use. Disinfectants are similar agents used to destroy microorganisms on inanimate objects. One of the most important disinfectants is chlorine, used in treating drinking-water supplies. Its introduction around the turn of the present century had a dramatic impact on the reduction of the incidence of typhoid fever and other waterborne contagious diseases.

Table 8-2
Summary of the Principal Classes of Hydrocarbons

Name and definition	Example			Comment
	Structural formula	Condensed formula	Abbreviation	
Alkanes contain only single bonds	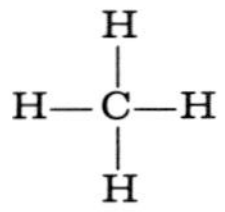 H–C–H (with H above and below) Methane (natural gas)	CH_4		Used as fuel and as raw material for fertilizers
Alkenes contain at least one C=C bond	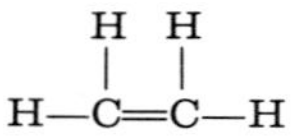 H–C=C–H Ethylene	$CH_2{=}CH_2$		Gas used in making polyethylene
	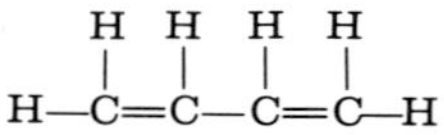 H–C=C–C=C–H 1,3-Butadiene	$CH_2{=}CH{-}CH{=}CH_2$		Gas used in making synthetic rubber
Alkynes contain at least one C≡C bond	H–C≡C–H Acetylene			Gas used to give very hot flame for welding
Alicyclic hydrocarbons contain one or more nonaromatic rings of carbon atoms (see below)	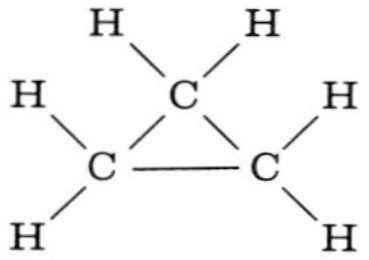 Cyclopropane	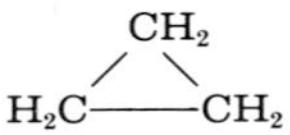 $H_2C{-}CH_2$ with CH_2	 △	Gas used as general anesthetic

182

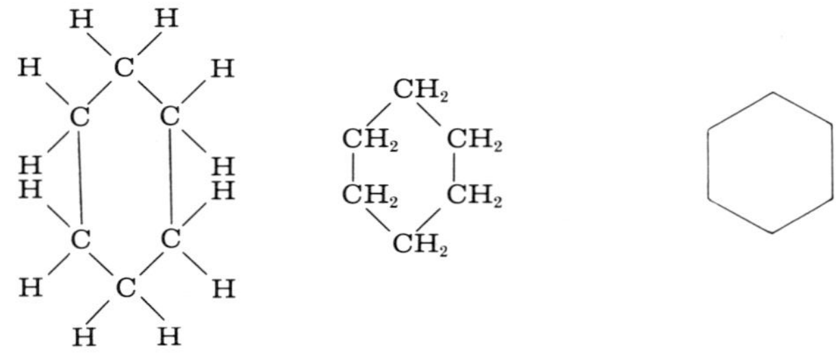

Liquid, used to make nylon

Cyclohexane

Aromatic hydrocarbons contain one or more rings of six carbon atoms linked together by distinctive covalent bond associated with multiple C=C bonds†

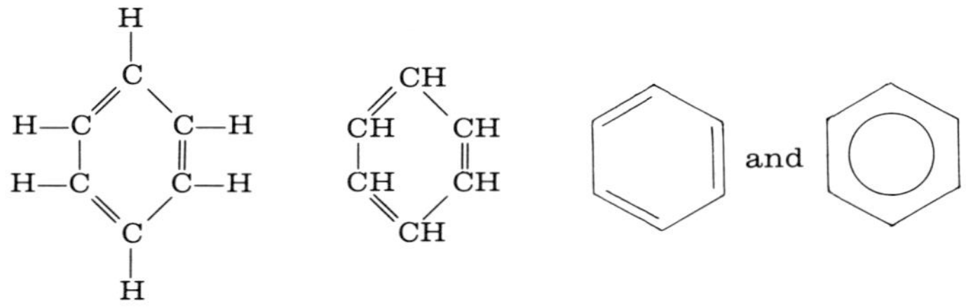

Benzene

and

Liquid, used as raw material for many products, from medicinals to synthetic rubber

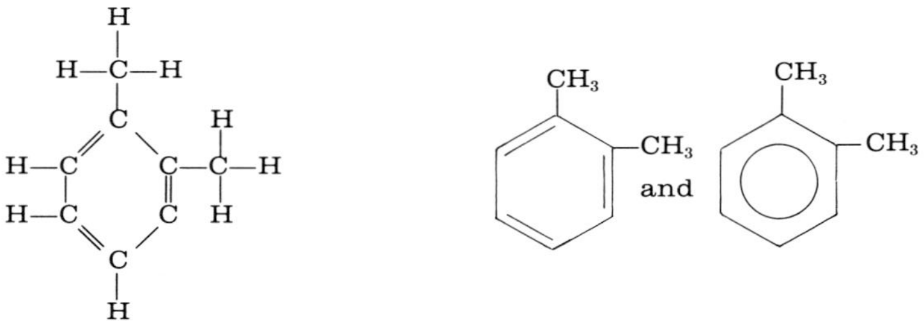

Ortho-Xylene

and

Liquid, used in making house and automobile paints

† This is a simplified definition.

183

$$R-\overset{\overset{\displaystyle O}{\|}}{C}-O-H$$

$$R-\overset{\overset{\displaystyle }{\|}}{\underset{\underset{\displaystyle O}{\|}}{C}}-O-H$$

$$H-O-\overset{\overset{\displaystyle O}{\|}}{C}-R$$

$$R-\overset{\overset{\displaystyle O}{\|}}{\underset{\underset{\displaystyle O--H}{|}}{C}}$$

Some ways of writing the same general formula for a carboxylic acid

$$H-\overset{\overset{\displaystyle H}{|}}{\underset{\underset{\displaystyle H}{|}}{C}}-\overset{\overset{\displaystyle O}{\|}}{C}-O-H$$

$$CH_3-\overset{\overset{\displaystyle O}{\|}}{\underset{\underset{\displaystyle O-H}{|}}{C}}$$

$$H-O-\overset{\overset{\displaystyle O}{\|}}{C}-CH_3$$

$$H-O-\overset{\overset{\displaystyle O}{\|}}{\underset{\underset{\displaystyle H}{|}}{C}}\;\underset{\underset{\displaystyle H}{|}}{\overset{\overset{\displaystyle }{|}}{H-C-H}}$$

Some ways of writing the structural formula for acetic acid; the atoms involved, their sequence, and the types of bonds are the same in each

Note to Students

decomposition of coal at high temperatures supplies an enormous number of substances, many of which are the building blocks of industrial chemicals with important uses.

The composition of crude petroleum also depends on the type and source. Both alkanes and aromatic hydrocarbons are present in varying proportions, and the carbon-atom content of the molecules ranges from 1 to about 35. A typical crude petroleum sample has been estimated to contain at least 400 different compounds. About 150 have actually been isolated from one type of petroleum. (Petroleum refining is discussed in Sec. 13-11.)

Most of our supplies of natural gas, petroleum, and coal are used in the production of energy, but since these fossil mixtures were created in the bowels of the earth, they cannot be replaced. Their use as fuels is in direct competition with their use as sources of hydrocarbon starting materials in the production of many products of substantial significance. The alarming consumption of these hydrocarbons as fuels raises the question: Where will the hydrocarbon raw materials come from in the future?

8-5 HOW ARE FUNCTIONAL GROUPS RECOGNIZED?

More than 200 functional groups have been identified, and with each a class of carbon compounds is associated. To illustrate the functional group approach to classification we shall consider about 10 such groups and the substances related to them.

The distinctive features of some representative functional groups are summarized in Table 8-3. Using the carboxylic acid group as a model, the following list shows the information the table conveys.

Name of class	Functional group	General formula of class	Example
Carboxylic acids	$-\overset{\overset{\displaystyle O}{\|}}{C}-O-H$	$R-\overset{\overset{\displaystyle O}{\|}}{C}-O-H$	Formic acid, HCOOH, a component of ant sting and acetic acid, CH_3COOH, the acid of vinegar

1. Substances containing the functional group $-\overset{\overset{\displaystyle O}{\|}}{C}-O-H$ are called carboxylic acids.

For simplicity the number of functional groups has been severely limited. They have been chosen to include most of the groups which occur in the substances mentioned in later chapters. The discussion here provides only a general idea of the functional group concept and how it is applied to the classification of organic compounds.

2. The general formula for all carboxylic acids is RCOOH. The R symbol may represent (1) something as simple as a hydrogen atom, H, as in formic acid, HCOOH; or (2) any of the numerous groups containing carbon and hydrogen atoms, such as CH_3 in acetic acid, CH_3COOH; or (3) a complex group of atoms so large that it contains other functional groups. In this case the substance will belong to more than one class (see later examples).

3. Condensed formulas are used in the examples to save space.

4. It is important to appreciate that it is the kinds of atoms and the sequence in which they are bonded that distinguish a functional group. How the symbols are written in a specific formula is not important. Thus the general formula for a carboxylic acid can be written in any of the ways shown in the margin of p. 184 (but these do not exhaust the possibilities). The structural formula for acetic acid can be written in any of the ways shown (and many more).

8-6 THE PRINCIPAL
FAMILIES OF CARBON COMPOUNDS

The molecule of each carbon-containing substance contains a basic skeleton of atoms, which furnishes the structural framework. This skeleton is usually made up of carbon atoms, but other kinds of atoms may also be involved. Carbon compounds are classified into families on the basis of this framework of atoms.

Note to Students

The names of substances derived from benzene often include the designations *ortho-*, *meta-*, or *para-*, which indicate the positions of substituent groups on the ring of carbon atoms; e.g.,

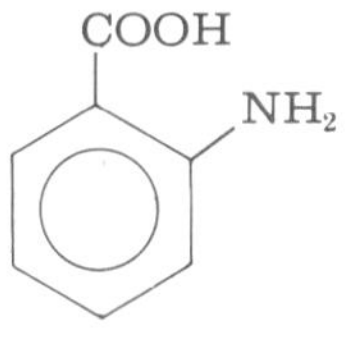

ortho-Aminobenzoic acid (*o*-aminobenzoic acid), substituents on adjacent carbon atoms

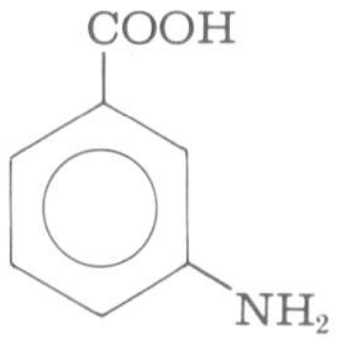

meta-Aminobenzoic acid (*m*-aminobenzoic acid), substituents on carbon atoms separated by one

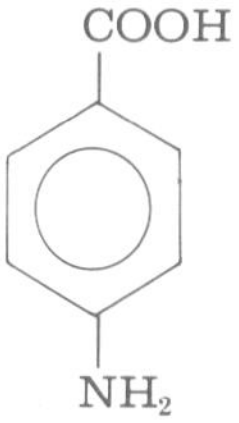

para-Aminobenzoic acid (*p*-aminobenzoic acid), substituents on carbon atoms separated by two

The abbreviations *o*-, *m*-, and *p*, as shown above, are commonly used. In alphabetizing the *ortho-*, *meta-*, and *para-* are disregarded; *p*-aminobenzoic acid, for example, is alphabetized under A. Numbers and Greek and Latin letters (as in *N,N*-dimethylacetamide, Table 8-3) indicate positions in other kinds of organic compounds, and they are also disregarded in alphabetizing.

Table 8-3
Summary of Selected Functional Groups

Name of class	Functional group	General formula of class	Example
Alkane	None	RH	Methane, CH_4, a gas used as fuel and as a raw material for fertilizers; propane, $CH_3CH_2CH_3$, a gas used for domestic purposes as "bottled gas"
Alkene	—C=C—	$\begin{array}{cc} R & R \\ \| & \| \\ R—C=C—R \end{array}$	Ethylene, CH_2=CH_2, a gas used to make polyethylene plastic; 1,3-butadiene, CH_2=CH—CH=CH_2, a gas used to make synthetic rubber
Alkyne	—C≡C—	R—C≡C—R	Acetylene gas, HC≡CH, used to give a very hot flame for welding
Alcohol	—C—O—H	RO	Methyl alcohol, CH_3OH, a poisonous liquid, also called wood alcohol; ethyl alcohol, CH_3CH_2OH, a liquid obtained from carbohydrates by fermentation
Ether	—C—O—	ROR	Ethyl ether, $CH_3CH_2OCH_2CH_3$, a volatile liquid, the first surgical anesthetic
Aldehyde	$\overset{O}{\overset{\|}{—C—H}}$	$R\overset{O}{\overset{\|}{C}}H$	Formaldehyde, HCHO, the principal component of embalming fluid; acetaldehyde, CH_3CHO, found in apples, among other places
Amine	—C—N—	$\overset{H}{\overset{\|}{R—N—H}}$	Methylamine, CH_3NH_2, a gas used to synthesize the pain-relieving agent meperidine
		R—N—R	Dimethylamine, CH_3NHCH_3, a gas used in synthesis of Benadryl, an antihistamine
		R—N—R	Trimethylamine $(CH_3)_3N$, a gas, partially responsible for fishy smell of saltwater salmon

Table 8-3
Summary of Selected Functional Groups (*Continued*)

Name of class	Functional group	General formula of class	Example
Carboxylic acid	$-\overset{\overset{O}{\|\|}}{C}-O-H$	$R-\overset{\overset{O}{\|\|}}{C}-O-H$	Formic acid, HCOOH, a liquid, component of ant sting; acetic acid, CH_3COOH, a liquid, the acid of vinegar
Carboxylic ester	$-\overset{\overset{O}{\|\|}}{C}-O-\overset{\|}{\underset{\|}{C}}-$	$R-\overset{\overset{O}{\|\|}}{C}-O-R$	Ethyl acetate, $CH_3COOCH_2CH_3$, a liquid, used as solvent in spray lacquer
Amide	$-\overset{\overset{O}{\|\|}}{C}-\overset{\|}{N}-$	$R-\overset{\overset{O}{\|\|}}{C}-\underset{\underset{H}{\|}}{N}-H$	Acetamide, a solid, CH_3CONH_2
		$R-\overset{\overset{O}{\|\|}}{C}-\underset{\underset{H}{\|}}{N}-R$	Present in all proteins
		$R-\overset{\overset{O}{\|\|}}{C}-\underset{\underset{R}{\|}}{N}-R$	N,N-Dimethylacetamide, a solid, $CH_3CON(CH_3)_2$, used as solvent

Note to Students	Frequent reference to the information of Table 8-3 will help you classify substances by functional group.

$$CH_3C \ldots \quad \overset{\overset{\displaystyle O}{\|}}{-C} —CH_2CH_2CH_3$$

Ethyl butyrate, a typical aliphatic substance; the fundamental skeleton of atoms has an open structure containing no rings

An **aliphatic substance** has a molecular structure in which the atom skeleton has an open structure, containing no rings. The carboxylic ester ethyl butyrate, used in perfumes for the odor of pineapple, is a typical aliphatic substance.

An **aromatic substance** has a skeleton containing one or more rings of special character. We can consider these rings to contain six carbon atoms linked together by a distinctive covalent bond associated with multiple C=C bonds. The substance benzoic acid, used as a food preservative, is a typical aromatic substance.

An **alicyclic substance** has a skeleton of atoms containing one or more rings of carbon atoms which are not aromatic. Menthol, which occurs in natural peppermint oil, is a typical alicyclic substance.

Two ways of writing the formula for the typical aromatic substance benzoic acid

$$\begin{array}{ccc}
& OH & \\
CH_2\!-\!CH & & \\
CH_3CH & & CHCHCH_3 \\
CH_2\!-\!CH_2 & & CH_3 \\
\end{array}$$

Menthol, from natural peppermint oil, is a typical alicyclic substance

(Heterocyclic sustances, defined in Sec. 10-8, are not considered here.)

8-7 EXAMPLES OF CLASSIFICATION OF SUBSTANCES ACCORDING TO FUNCTIONAL GROUPS

Recognizing the functional groups present in the molecule is the key step in classifying carbon-containing substances. Now we can see how the generalized structures of functional groups presented in Sec. 8-5 are used to identify functional groups in the molecules of actual substances. Several examples are shown in Tables 8-4 and 8-5. In examining these examples the information in Table 8-3 should be used as a reference.

HOW CAN THE AMOUNT OF ALCOHOL IN THE BREATH BE DETERMINED?

In attempts to keep people under the influence of alcohol from driving automobiles, law-enforcement agencies are depending more and more on breath analyzers to determine the alcohol content of a driver's blood. An alcohol content of more than 0.15 percent is typically accepted as evidence of alcoholic influence. This indication is obtained by determining the alcoholic content of the breath because air from deep within the lungs is in equilibrium with the bloodstream. Care must be taken to sample the breath in such a way that the analysis is made on air from deep within the lungs.

One method of breath analysis uses a fast reaction of ethyl alcohol causing a color change in a reaction mixture. The overall reaction takes place with the use of silver nitrate, $AgNO_3$, as a catalyst:

$$3CH_3CH_2OH + 2K_2Cr_2O_7 + 8H_2SO_4 \xrightarrow{AgNO_3}$$

Ethyl alcohol, colorless   Potassium dichromate, yellow in dilute aqueous solution   Sulfuric acid, colorless

$$3CH_3COOH + 2Cr_2(SO_4)_3 + K_2SO_4 + 11H_2O$$

Acetic acid, colorless   Chromium (III) sulfate, green-blue in dilute aqueous solution   Potassium sulfate, colorless   Water, color-less

Overall reaction for determining the amount of ethyl alcohol in a sample of breath

The analytical reagent mixture is pale yellow at the start due to the presence of potassium dichromate, $K_2Cr_2O_7$. As the breath sample is absorbed, the reaction mixture changes color as yellow potassium dichromate, $K_2Cr_2O_7$, is consumed and the green-blue chromium(III) sulfate,

$Cr_2(SO_4)_3$, is formed. The m
sample, the greater the col
the color change is deter
using photoelectric cells,
read off a dial.

NITROGLYCERIN, THE EXPLOSIVE WHICH BECAME A USEFUL DRUG

Nitroglycerin was first synthesized by the Italian chemist Ascanio Sobrero in 1846. It is made by treating the alcohol glycerin (more properly called glycerol) with a mixture of concentrated nitric acid and sulfuric acid. Nitroglycerin, of-

ficially named glyceryl trinitrate, is a nitrate ester of an alcohol (compare with carboxylic esters of carboxylic acids discussed in the body of this chapter).

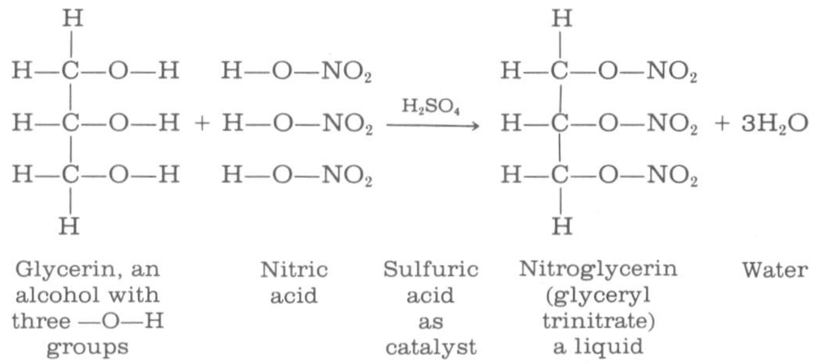

| Glycerin, an alcohol with three —O—H groups | Nitric acid | Sulfuric acid as catalyst | Nitroglycerin (glyceryl trinitrate) a liquid | Water |

Equation for the reaction in Sobero's synthesis of nitroglycerin

To determine the properties of nitroglycerin, Sobrero heated a drop of it in a closed tube. It immediately exploded, cutting his face and hands with a shower of glass splinters. Nitroglycerin quickly became known as a very powerful explosive, easy to make and to use. Unfortunately,

it detonates on slight shock and explodes very unpredictably. All the products are gases, and the sudden formation of a large volume of gaseous products in the very small volume initially occupied by the liquid nitroglycerin generates an explosion wave of tremendous force:

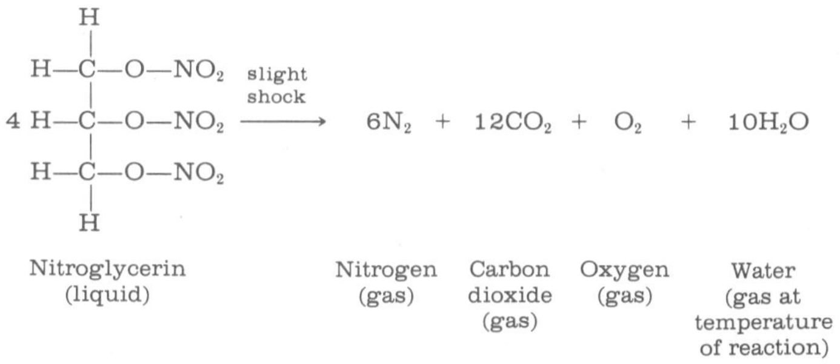

| Nitroglycerin (liquid) | Nitrogen (gas) | Carbon dioxide (gas) | Oxygen (gas) | Water (gas at temperature of reaction) |

Decomposition which occurs spontaneously when nitroglycerin is subjected to slight shock

Despite its hazardous behavior, nitroglycerin was such an effective explosive that it soon found use in mining and construction operations. Its use was plagued with accidents, however, includ-

ing spectacular and costly explosions that blew up ships in the harbors of New York and San Francisco. The Swedish inventor Alfred Nobel discovered around 1867 that if nitroglycerin is

Table 8-4
The Use of General Formulas of Functional Groups
in the Identification of Functional Groups in Substances

Structural formula	Condensed formula	Class and functional group(s) present
Substances with one functional group		

Structural formula	Condensed formula	Class and functional group(s) present
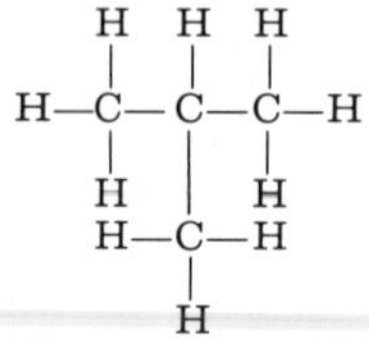 (alkane structure)	CH_3CHCH_3 CH_3	Alkane
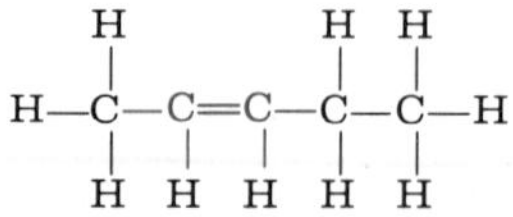 (alkene structure)	$CH_3CH{=}CHCH_2CH_3$	Alkene
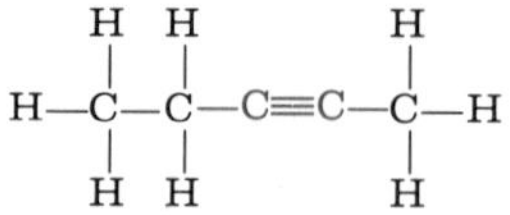 (alkyne structure)	$CH_3CH_2C{\equiv}CCH_3$	Alkyne
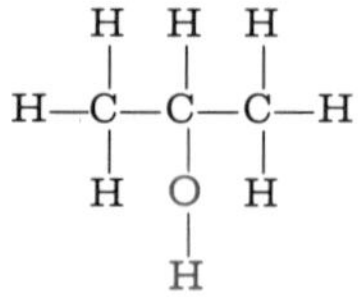 (alcohol structure)	CH_3CHCH_3 OH	Alcohol
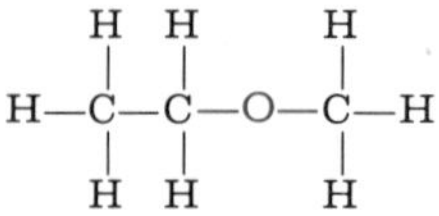 (ether structure)	$CH_3CH_2OCH_3$	Ether
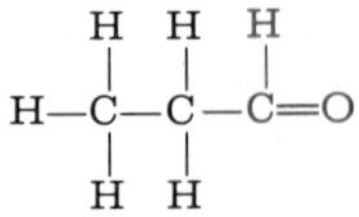 (aldehyde structure)	CH_3CH_2CHO	Aldehyde
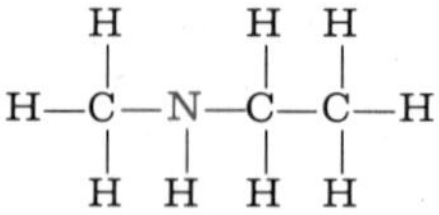 (amine structure)	$CH_3NHCH_2CH_3$	Amine
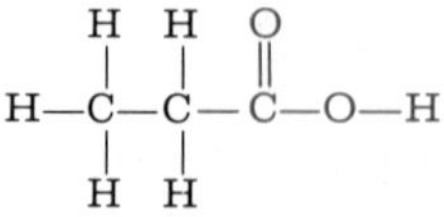 (carboxylic acid structure)	CH_3CH_2COOH	Carboxylic acid
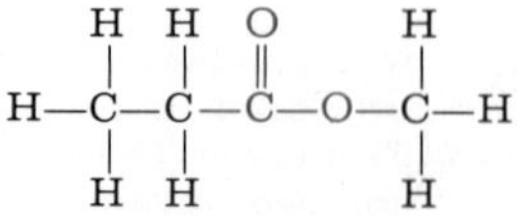 (ester structure)	$CH_3CH_2COOCH_3$	Carboxylic ester

Table 8-4
The Use of General Formulas of Functional Groups
in the Identification of Functional Groups in Substances (*Continued*)

Structural formula	Condensed formula	Class and functional group(s) present

Substances with one functional group (*Continued*)

| 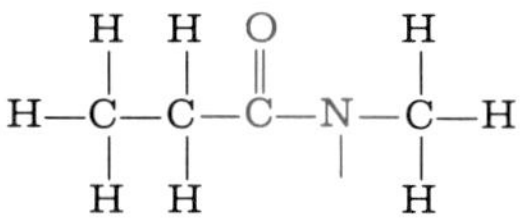 | $CH_3CH_2CONHCH_3$ | Amide |

Substances with more than one of the same type of functional group

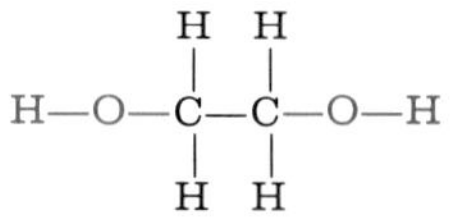	$HOCH_2CH_2OH$	Alcohol
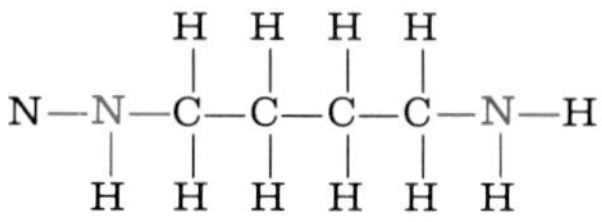	$H_2NCH_2CH_2CH_2CH_2NH_2$	Amine
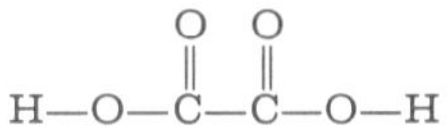	HOOCCOOH	Carboxylic acid

Substances containing more than one type of functional group†

| | H_2NCH_2COOH | An amine and a carboxylic acid |

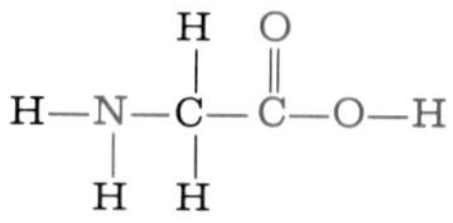

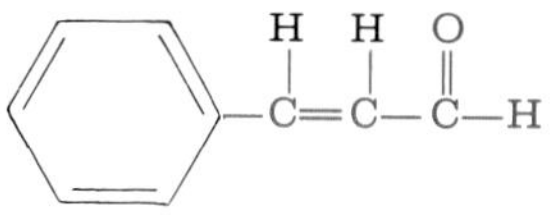

Amine group   Carboxylic acid group

Cinnamaldehyde,‡ principal component of cinnamon oil

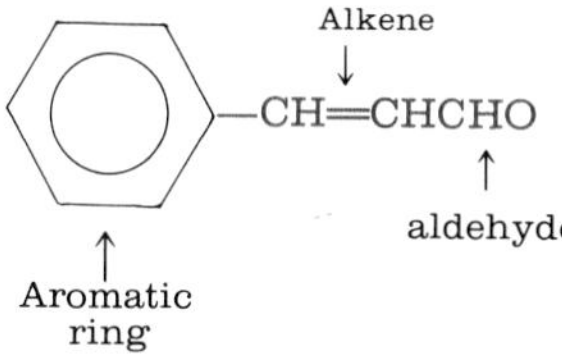

Alkene

—CH=CHCHO   An aromatic alkene aldehyde

Aromatic ring   aldehyde

† Substances whose molecules contain more than one functional group belong to more than one class.

‡ It is customary to refer to aromatic substances specifically as such. Cinnamalde-hyde is usually referred to as an aromatic alkene aldehyde. Aliphatic substances, on the other hand, are not specifically labeled. Although CH_3CH_2CHO is an aliphatic aldehyde, it is usually referred to simply as an aldehyde. Alicyclic substances are sometimes labeled specifically, but more frequently not.

Table 8-5
Identification of Functional Groups
and Ring Structures in 10 Carbon Compounds of Interest

Benzyl acetate, principal component in natural oil of jasmine, used in perfume

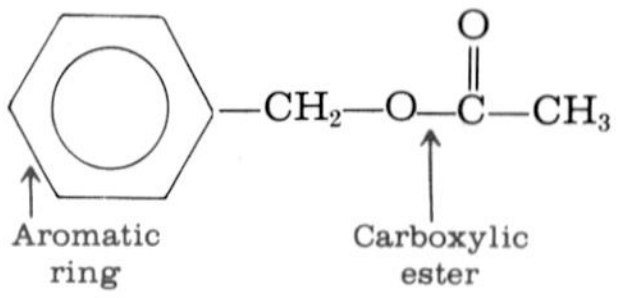

N,N-Diethyl-*m*-toluamide, an insect repellant

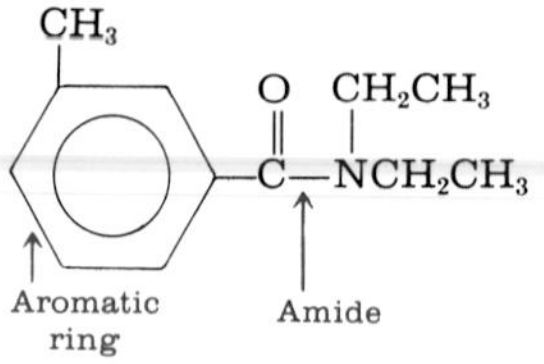

Aspirin (acetylsalicylic acid), a pain-relieving agent

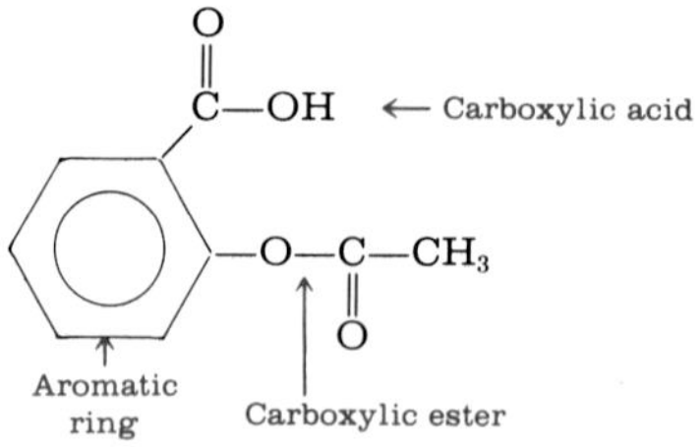

Glucose, a sugar (Sec. 10-2) found in the bloodstream; fed intravenously

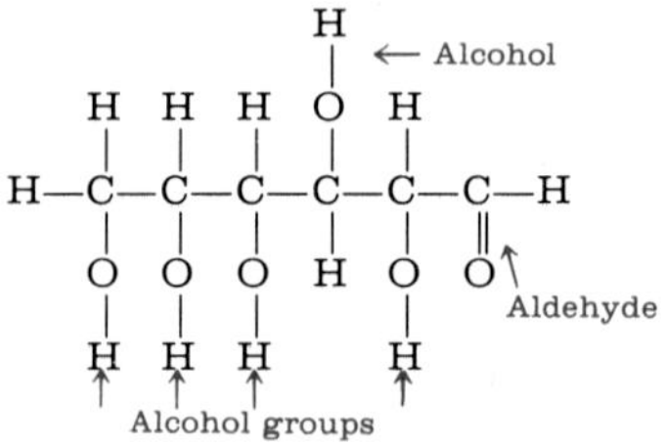

Citric acid, principal acid of citrus fruits

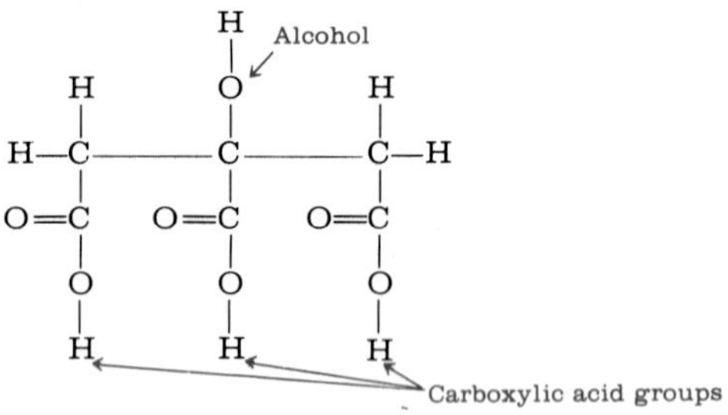

Vitamin A, a dietary essential, important in human growth

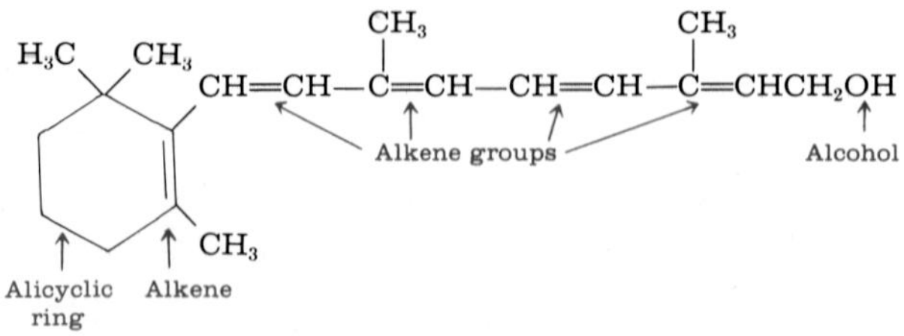

Table 8-5
Identification of Functional Groups
and Ring Structures in 10 Carbon Compounds of Interest (*Continued*)

Amphetamine
(benzedrine), an
addictive stimulant
and widely abused
drug

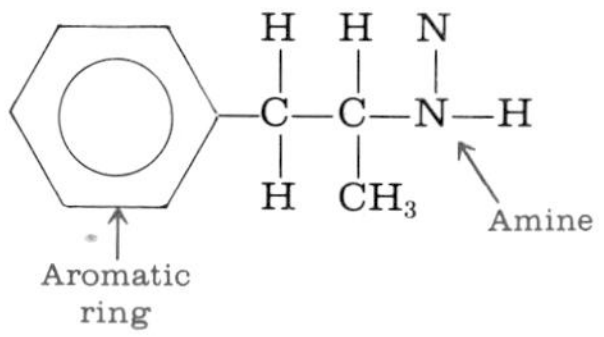

Aromatic ring · C—C—N—H · Amine

Heroin, a strongly
addicting pain-
relieving drug,
widely abused
(Sec. 1-13)

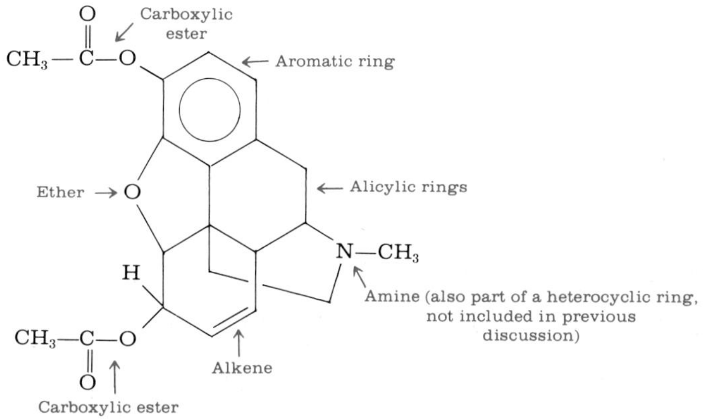

Carboxylic ester · Aromatic ring · Ether → O · Alicyclic rings · N—CH₃ · Amine (also part of a heterocyclic ring, not included in previous discussion) · Alkene · Carboxylic ester

Cholesterol, found
in all body tissues;
principal component
of gallstones

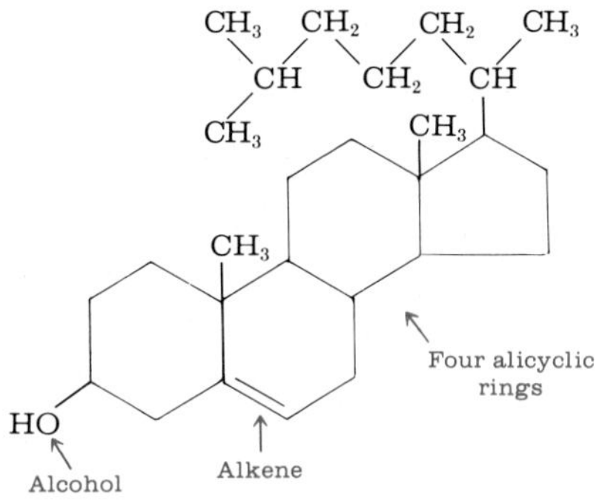

Four alicyclic rings · Alcohol · Alkene

Procaine, readily
converted into procaine
hydrochloride (novocain),
(Sec. 12-4), a local
anesthetic

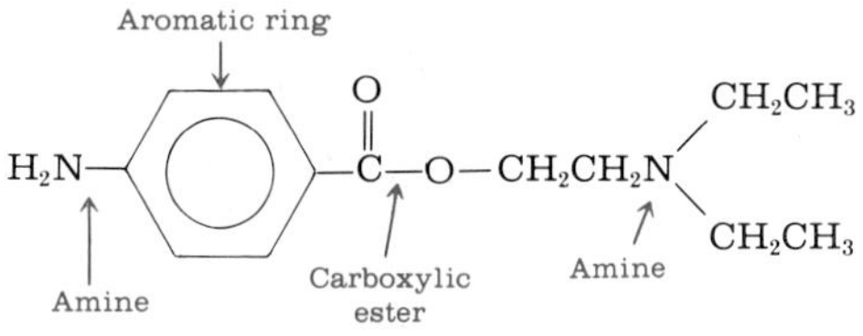

Aromatic ring · Amine · Carboxylic ester · Amine

Note to Students

The identification of the functional groups and ring structures in the examples of this section illustrates how chemists look at molecular structures to understand chemical properties. These examples demonstrate why formulas are the shorthand of chemistry.

absorbed in porous silica, it becomes an easily handled but still very powerful explosive, which he named dynamite. Its manufacture and further development were the principal sources of his great wealth, with which he established the Nobel prizes.

Nitroglycerin was first used as a construction explosive in the United States during the completion of the Hoosac Tunnel in the Berkshire mountains of western Massachusetts. Almost 5 miles long, the tunnel was a very bold engineering venture when it was started in 1851. Using hand drills and black powder (a low-power explosive composed of carbon, sulfur, and potassium nitrate), the workers hacked away at the solid rock, but after 15 years of discouragingly slow progress only 1 mile had been bored in from each end. Nitroglycerin, made on the spot by the daring "Professor" George Mowbray, was put to use beginning in 1866 and brought construction to completion 9 years later. In all, 195 lives were lost, many due to the capricious nature of nitroglycerin. But the tunnel is still in use as a major railway link between Boston and the West (Fig. 8-3).

Sobrero was very much disturbed by the disasters and loss of life which resulted from the use of nitroglycerin. It was he who made the first observations of the physiological effects of nitroglycerin, which eventually led to its widespread use for the relief of the pain in the heart disorder known as angina pectoris. This ailment causes a severely oppressive pain in the chest due to a deficiency of oxygen in areas of the heart muscle. When administered under the tongue, nitroglycerin is able to relieve the pain presumably by increasing the blood flow and reducing the oxygen required by the heart muscle. It remains a preferred treatment for this heart ailment.

FIGURE 8-3
View inside the Hoosac Tunnel, almost 5 miles long, through the Berkshire Mountains of northwestern Massachusetts. Its construction (1851–1874) marked the first use of nitroglycerin for construction in the United States. (*Photograph by Randolph Trabold.*)

KEY WORDS

1. Organic chemistry (Sec. 8-1)
2. Functional group (Sec. 8-2)
3. Functional group principle (Sec. 8-2)
4. Hydrocarbon (Sec. 8-4)
5. Alkane (Sec. 8-4)
6. Alkene (Sec. 8-5)
7. Alkyne (Sec. 8-5)
8. Alcohol (Sec. 8-5)
9. Ether (Sec. 8-5)
10. Aldehyde (Sec. 8-5)
11. Amine (Sec. 8-5)
12. Carboxylic acid (Sec. 8-5)
13. Carboxylic ester (Sec. 8-5)
14. Amide (Sec. 8-2)
15. Aliphatic substance (Sec. 8-6)
16. Aromatic substance (Sec. 8-6)
17. Alicyclic substance (Sec. 8-6)

"It may very well bring about immortality, but it will take forever to test it."

SUMMARIZING QUESTIONS FOR SELF-STUDY

Sections 8-1 and 8-2

1. Q. What is meant by organic chemistry?

A. It is the chemistry of carbon-containing substances.

2. Q. Why is it a separate branch of chemistry?

A. There are an enormous number of substances, many with complex structures, which constitute the major part of plants and animals and form the starting points in the synthesis of many useful compounds.

3. Q. What is a functional group?

A. A distinctive arrangement of atoms which imparts characteristic chemical properties to a substance.

4. Q. What is the functional group principle?

A. The assumption that the chemical properties of a substance are determined by the functional groups in its molecular structure.

5. Q. What is the significance of the functional group principle?

A. It provides a means of classifying the numerous carbon-containing substances and organizing their many reactions.

Sections 8-3 to 8-7

6. Q. Using Table 8-3 as a reference, identify the functional groups present in the following molecules and name the class to which the substance belongs:

(a) $H-\overset{\overset{\displaystyle H}{|}}{\underset{\underset{\displaystyle H}{|}}{C}}-\overset{\overset{\displaystyle H}{|}}{\underset{\underset{\displaystyle H}{|}}{C}}-\overset{\overset{\displaystyle H}{|}}{\underset{\underset{\displaystyle H}{|}}{C}}-H$

(d) $H-\overset{\overset{\displaystyle H}{|}}{\underset{\underset{\displaystyle H}{|}}{C}}-\overset{\overset{\displaystyle H}{|}}{\underset{\underset{\displaystyle H}{|}}{C}}-O-H$

(b) $H-\overset{\overset{\displaystyle H}{|}}{\underset{\underset{\displaystyle H}{|}}{C}}-\overset{}{\underset{\underset{\displaystyle H}{|}}{C}}=\overset{}{\underset{\underset{\displaystyle H}{|}}{C}}-H$

(e) $H-\overset{\overset{\displaystyle H}{|}}{\underset{\underset{\displaystyle H}{|}}{C}}-O-\overset{\overset{\displaystyle H}{|}}{\underset{\underset{\displaystyle H}{|}}{C}}-H$

(c) $H-C\equiv C-\overset{}{\underset{\underset{\displaystyle H}{|}}{C}}-H$

(f) $H-\overset{\overset{\displaystyle H}{|}}{\underset{\underset{\displaystyle H}{|}}{C}}-\overset{\overset{\displaystyle H}{|}}{\underset{\underset{\displaystyle H}{|}}{C}}-\overset{}{\underset{\underset{\displaystyle O}{||}}{C}}-H$

(g)
$$H-\underset{\underset{H-N}{\overset{H}{|}}}{\overset{\overset{H}{|}}{C}}-\underset{\overset{H}{|}}{\overset{\overset{H}{|}}{C}}-\underset{\overset{H}{|}}{\overset{\overset{H}{|}}{C}}-H$$
(with H—N—H at bottom)

(j)
$$H-\underset{\overset{H}{|}}{\overset{\overset{H}{|}}{C}}-\underset{\overset{\overset{O}{\|}}{C}}{}-N\underset{CH_3}{\overset{CH_3}{|}}$$

(m) $H_2NCHCOOH$
$\qquad\quad | $
$\qquad\quad CH_3$

(h)
$$H-\underset{\overset{H}{|}}{\overset{\overset{H}{|}}{C}}-\underset{\overset{H}{|}}{\overset{\overset{H}{|}}{C}}-\overset{\overset{O}{\|}}{C}-O-H$$

(k) [benzene ring]—CH_2CH_3

(n) $H_3C-\overset{\overset{O}{\|}}{C}-\underset{\overset{H}{|}}{N}-\underset{\overset{CH_3}{|}}{\overset{\overset{H}{|}}{C}}-\overset{\overset{O}{\|}}{C}-\underset{\overset{H}{|}}{N}-CH_3$

(i)
$$H-\underset{\overset{H}{|}}{\overset{\overset{H}{|}}{C}}-\overset{\overset{O}{\|}}{C}-O-\underset{\overset{H}{|}}{\overset{\overset{H}{|}}{C}}-H$$

(l) [cyclohexane ring]—CH_3

A. (a) Alkane hydrocarbon (or alkane), (b) alkene hydrocarbon (or alkene), (c) alkyne hydrocarbon (or alkyne), (d) alcohol, (e) ether, (f) aldehyde, (g) amine, (h) carboxylic acid, (i) carboxylic ester, (j) amide, (k) aromatic hydrocarbon, (l) alicyclic hydrocarbon, (m) an amine and a carboxylic acid (commonly called an amino acid), (n) an amide containing two amide groups.

PRACTICE EXERCISES

1. Explain the difference between an aliphatic substance, an aromatic substance, and an alicyclic substance and give two examples of each.

2. Using Table 8-3 as a reference, identify the functional groups in the following molecules and name the class (or classes) to which the substance belongs.

(a) $H_3C-C\equiv C-H$
Methylacetylene

(b)
$$\underset{CH_2OH}{\overset{\overset{H}{|}}{\underset{|}{C}}}$$
H
|
C=O
|
H—C—H
|
H—C—OH
|
H—C—OH
|
CH₂OH

(c)
[benzene ring]—$\overset{\overset{O}{\|}}{C}$—$O$—$\underset{\overset{H}{|}}{\overset{\overset{H}{|}}{C}}$—$\underset{\overset{H}{|}}{\overset{\overset{H}{|}}{C}}$—$H$

Ethyl benzoate

(d)
NH_2
[benzene ring]
$COOH$

p-Aminobenzoic acid, used as a sunscreen agent in suntan oil

(e)
$HN-\overset{\overset{O}{\|}}{C}-CH_3$
[benzene ring]
OCH_2CH_3

Acetophenetidin, used in pain-relieving tablets

(f) $HN\underset{\overset{H}{|}}{CH_2}CHCH_2\overset{\overset{O}{\|}}{C}-O-H$
[benzene ring]

4-Amino-3-phenylbutyric acid, a tranquilizer

(g)
OCH_3
[benzene ring]
$CH=CHCH_3$

Anethole, a flavoring agent

(h)
$\underset{CH_3}{\overset{CH_3}{\diagdown}}C=CHCH_2CH_2C=CH\overset{\overset{H}{|}}{C}=O$
$\qquad\qquad\qquad\qquad\quad | $
$\qquad\qquad\qquad\qquad\quad CH_3$

Citral, a flavoring agent with a lemon odor

(i)

Gladiolic acid, an antifungal antibiotic

(j)

Amoxecaine, a local anesthetic

3. Give structural formulas for one example of each of the following:

(a) An alicyclic hydrocarbon
(b) An alcohol
(c) An alkene
(d) An amide
(e) An aromatic hydrocarbon
(f) An aldehyde
(g) An ether
(h) An aliphatic hydrocarbon
(i) A carboxylic ester
(j) An alkyne
(k) An aromatic carboxylic acid
(l) An amine
(m) An alkane

4. Using Table 8-3 as a reference, identify the functional groups in the following molecules and name the class (or classes) to which each substance belongs. Specify which formulas represent the same substance. (In this exercise the functional group is deliberately written differently from that in Table 8-3, but, of course, the atom sequence of each functional group is the same as in Table 8-3.)

(a) HOCH$_2$CH$_2$
   |
   CH$_3$

(b) H$_2$NCH$_2$
   |
   CH$_2$CH$_3$

(c) CH$_3$CH$_2$OOCCH$_2$CH$_3$

(d) HOOCCH$_2$ ⬡

(e) CH$_3$CH$_2$NCOCH$_3$
   |
   CH$_3$

(f) CH$_3$CHOOCCHCH$_3$
   | |
   CH$_3$ CH$_3$

(g) OHCCH$_2$CH$_2$CH$_3$

(h) CH$_3$CH$_2$NH
   |
   CH$_3$

(i) CH$_3$CH$_2$OCHCH$_3$
   |
   CH$_3$

(j) CH$_3$CH$_2$CH=CCH$_2$CH$_3$
   |
   CH$_2$
   |
   CH$_3$

(k) CH$_3$CH$_2$CH$_2$CHO

(l) CH$_3$CHOCOCHCH$_3$

(m) CH$_3$OCNCH$_3$
   |
   CH$_2$
   |
   CH$_3$

(n) CH$_3$NCH$_2$CH$_3$
   |
   H

(o) CH$_3$CHOCH$_2$CH$_3$
   |
   CH$_3$

5. The "Merck Index," published by Merck and Company, is a useful one-volume encyclopedia of chemicals and drugs, listed alphabetically. Look up the following substances in this encyclopedia, note their uses, and identify the functional groups present in their molecules.

(a) Acetanilide
(b) Ambucetamide
(c) Vitamin B$_{15}$
(d) Citric acid
(e) Aspartame
(f) Phenocoll
(g) p-Anisaldehyde
(h) Phenylethanolamine
(i) Valnoctamide
(j) Butacaine

SUGGESTIONS FOR FURTHER READING

The following articles are of an introductory nature.

Holmes, Linda C., and Frederick J. DiCarlo: Nitroglycerin: The Explosive Drug, **Journal of Chemical Education, 48:**573, (1971).

Treptow, Richard S.: Determination of Alcohol in Breath for Law Enforcement, **Journal of Chemical Education, 51:**651 (1974).

The following articles are at a more advanced level.

Amoore, John E., James W. Johnston, Jr., and Martin Rubin: The Stereochemical Theory of Odor, **Scientific American,** February 1964, p. 42.

Blumer, Max: Polycyclic Aromatic Compounds in Nature, **Scientific American,** March 1976, p. 34.

Breslow, Ronald: The Nature of Aromatic Molecules, **Scientific American,** August 1972, p. 32.

Lambert, Joseph B.: The Shapes of Organic Molecules, **Scientific American,** January 1970, p. 58.

Roberts, John D.: Organic Chemical Reactions, **Scientific American,** November 1957, p. 117.

9

WHAT ARE NUCLEAR REACTIONS?

9-1 WHAT IS RADIOACTIVITY?

Many of the roots of modern chemistry are to be found in the work of the alchemists, whose experiments and speculations centered around the pursuit of an "elixir of life" and a "philosopher's stone." The elixir, once discovered, was supposed to cure all diseases and confer immortality. The stone was supposed to convert metals of low value into gold. The first alchemists are thought to have worked in China over 2000 years ago. Over the centuries they are found among the Egyptians, the Arabs, and later among the Europeans during the twelfth to seventeenth centuries. Although none of them was ever able to find a means of bestowing immortality or of transmuting metals into gold, they made many significant discoveries in the course of their far-reaching experiments. Unfortunately, their writings were often intentionally obscure, and many of the later alchemists were outright frauds.

The influence of the atomic theory during the nineteenth century made the alchemists' dream of transmuting one element into another appear more remote than ever. A cornerstone of the theory held that the atoms remain unaltered during chemical changes. The idea of the indestructibility of atoms caused attempts at transmutation to fall into disrepute.

With the recognition and characterization of radioactivity at the turn of the twentieth century, however, it was soon realized that many of the processes of nature involve changes in which the individuality of atoms is not preserved; atoms of one element lose their identity and become transmuted into atoms of another element. Such changes are called **nuclear reactions,** i.e., changes in which the structures of atomic nuclei are altered. They are also referred to as nuclear transformations or **transmutations.** The discovery of radioactivity is one of the most dramatic examples of how the experienced and imaginative experimenter can capitalize on an unexpected result. When Henri Becquerel was investigating the ability of a uranium salt to emit radiations capable of penetrating opaque paper, he expected it to do so only if it was first exposed to sunlight. A partly cloudy day provided the accidental circumstance which led him to observe that a persistent radiation is emitted from the uranium salt whether it is exposed to light or not. Following through on this discovery, several physicists, led by Marie Curie, Pierre Curie, and Ernest Rutherford, explored the phenomenon. It was discovered that the radiation emanating from the uranium salt was due to the spontaneous decomposition of the nuclei of the uranium atoms. This process is called **radioactivity.**

Radiations from decomposing atomic nuclei are of three types, called alpha (α), beta (β), and gamma (γ) rays (Table 9-1). **Alpha rays** are penetrating streams of high-speed alpha particles. An **alpha particle** is the nucleus of a helium atom containing two neutrons and two protons and having a positive charge of two units. It is represented by He^{2+} or $_2^4He$. **Beta rays** are penetrating streams of high-speed electrons, called **beta particles,** each with a single negative charge and a very small mass. In the symbol for the beta particle, $_{-1}^0e$, the zero mass number indicates the negligibly small mass, and the atomic number of -1 indicates the unit negative charge. **Gamma rays** are a form of highly penetrating, invisible light rays with neither mass nor charge.

Table 9-1
Nature of Radioactivity Rays

Ray emitted	Nature	Symbol
Alpha, α	Helium nuclei	$_2^4He$, He^{2+}
Beta, β	Electrons	$_{-1}^0e$, e^-
Gamma, γ	Light rays	

9-2 HOW ARE NUCLEAR REACTIONS REPRESENTED?

The decomposition of atomic nuclei transmutes an atom of one element into an atom of another element. Consider the first step in the decomposition of uranium:

Uranium atom $\rightarrow$ thorium atom + alpha particle

Mass number 238,   Mass number 234,   Mass number 4,
atomic number 92   atomic number 90   atomic number 2

The first step in the spontaneous radioactive disintegration of uranium 238 atoms

This change can be represented more succinctly by symbols (Sec. 2-5) (see margin).

The nuclei of atoms are composed of protons and neutrons (Sec. 2-4). The subscripts of the symbols represent the atomic numbers of the specific isotopes, which correspond to the number of protons. The superscripts represent the mass numbers of the isotopes, which correspond to the total number of protons and neutrons. In the symbol $^{234}_{90}\text{Th}$, for example, the 90 indicates the atomic number and the number of protons; the 234 indicates the mass number and the total number of protons and neutrons.

A radioactive decomposition does not depend in any way on whether the radioactive substance is chemically combined. Uranium 238 decomposes into thorium 234 whether it is in the form of elemental uranium or present as uranium oxide or uranium fluoride. Correspondingly, the nuclear equation does not specify the state of chemical combination or whether the element is in a neutral or ionic condition.

$$^{238}_{92}\text{U} \rightarrow {}^{234}_{90}\text{Th} + {}^{4}_{2}\text{He}$$

Representation with isotopic symbols of the first step in the radioactive disintegration of uranium 238 atoms

---

Note to Students

Balancing of nuclear equations is discussed in the optional Sec. 9-14.

9-3 WHAT ARE ARTIFICIAL TRANSMUTATIONS?

The first nuclear changes to be investigated were those which occur spontaneously in nature. Early in these investigations (1902) Ernest Rutherford and Frederick Soddy discovered that nuclear changes, i.e., transmutations, can be brought about deliberately by bombarding nuclei with the high-speed alpha particles emitted during a natural radioactive decomposition. The first such transmutation was the conversion of nitrogen 14 into oxygen 17 by means of alpha-particle, $^{4}_{2}\text{He}$, bombardment (margin). A proton, $^{1}_{1}\text{H}$, is also formed (note that the ionized condition of the proton is not specifically indicated in the symbol). The high-speed alpha particles are able to bring about the change as they collide with the nuclei of the nitrogen atoms. An **artificial transmutation** is a nuclear change which does not occur in nature but is deliberately brought about by the bombardment of nuclei with high-speed particles.

The first artificial transmutation was an extraordinary event. For about a century Dalton's idea that atoms are unalterable and untransmutable had been successful in accounting for the observations of chemical changes. Attempts at transmutation were not regarded as respectable science. When Rutherford and Soddy concluded from their experiments that nitrogen had indeed been converted into oxygen, Soddy turned to Rutherford and exclaimed: "Rutherford, this is transmutation!" Rutherford shot back: "For Mike's sake, Soddy, don't call it transmutation; they will have our heads off as alchemists."[†]

$^{4}_{2}\text{He}$

Alpha particle

+

$^{14}_{7}\text{N}$

Nitrogen atom

↓

$^{17}_{8}\text{O}$

Oxygen atom

+

$^{1}_{1}\text{H}$

Proton

The first artificial transmutation; the bombardment of nitrogen atoms with alpha particles converts the nitrogen atoms into oxygen atoms

[†] See Lawrence Baldash, How the "Newer Alchemy" Was Received, **Scientific American,** August 1966, p. 91.

Hundreds of artificial transmutations have now been carried out and characterized. Many kinds of apparatus were developed to accelerate bombarding particles to greater energies than those found in nature and to make them more effective in bringing about nuclear changes. These high-energy accelerators make it possible to generate high-speed neutrons, protons, and deuterons (the nuclei of deuterium atoms), which are not available from the decompositions of natural radioactivity. Neutrons, in particular, were found to be extremely penetrating and effective in bringing about nuclear changes since, being neutral, they are not repelled by the positive charges of the nucleus.

9-4 WHAT IS NUCLEAR FISSION?

The neutron bombardment of uranium atoms attracted considerable interest in the scientific community during the mid 1930s. In the course of such experiments it was discovered by the German chemists and physicists Otto Hahn, Fritz Strassman, Lise Meitner, and Otto Frisch and their associates that when uranium 235 isotopes are bombarded with neutrons, an unusually profound nuclear change takes place. In the nuclear changes discussed so far, the nuclei undergo only relatively small rearrangements, and the products are isotopes of atomic number rather close to the atomic numbers of the bombarded atoms. When uranium 235 atoms are bombarded with neutrons, however, a much more pronounced change occurs, and the nucleus is split into two fragments of substantial size. The specific fragments obtained differ considerably from reaction to reaction, but the most abundant products have mass numbers in the range 89 to 139.

Nuclear fission (or **atomic fission**) is a nuclear change in which an atomic nucleus is split into two nuclei of substantial size. A typical nuclear reaction occurring during the fission of uranium 235 is

$$\underset{\text{Neutron}}{^{1}_{0}n} \quad + \quad \underset{\text{Uranium 235}}{^{235}_{92}U} \quad \rightarrow \quad \underset{\text{Barium 142}}{^{142}_{56}Ba} \quad + \quad \underset{\text{Krypton 91}}{^{91}_{36}Kr} \quad + \quad \underset{\text{Neutrons}}{3\,^{1}_{0}n}$$

Typical nuclear reaction occurring during the fission of uranium 235

The barium 142 and krypton 91 are radioactive and undergo further decomposition before stable nonradioactive isotopes are obtained. Considerable energy is released during the reaction which, as will be shown later, can be harnessed.

An important feature of nuclear fission reactions of this sort is that more than one neutron is produced for each neutron consumed. These product neutrons can initiate subsequent fissions, which in turn produce nucleus-splitting neutrons. A **nuclear chain reaction** is a self-sustaining sequence of nuclear fission reactions. Since each fission takes place in a tiny fraction of a second, the chain reaction (Fig. 9-1) quickly branches out, bringing about the fission of a substantial amount of uranium 235. The fission of a whole kilogram of uranium 235 would require less than a millionth of a second.

FIGURE 9-1
Schematic representation of a chain reaction which uses a moderator to slow neutrons down to velocities more likely to bring fission about. The process shown involves the fission of uranium 235. (*Adapted from G. Tyler Miller, Jr., "Chemistry: A Contemporary Approach," copyright 1976 by Wadsworth Publishing Company.*)

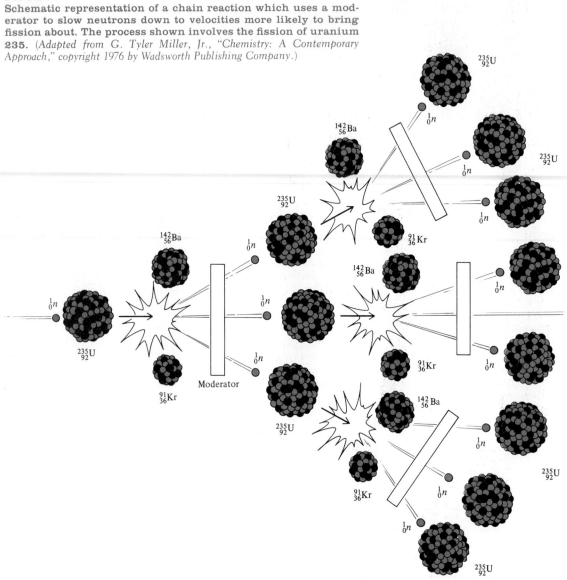

Neutrons, as they are released in nuclear reactions, have a broad range of energies. A **slow neutron** is one with relatively low energy, and a **fast neutron** has a relatively large amount of energy. Both slow and fast neutrons are capable of bringing about the splitting of uranium 235 atoms. The probability of fission is increased, however, if the neutrons are slow. **Moderators** are substances, like graphite, which are capable of slowing down fast neutrons. Using moderators increases the probability that a nuclear chain reaction will keep going. The principal aspects of a moderated nuclear chain reaction are summarized in Figure 9-1.

One of the most interesting and significant transmutations is brought about by the bombardment of uranium 238 nuclei by neutrons. In the first step the unstable isotope uranium 239 is formed:

$$\ _0^1 n \quad + \quad \ _{92}^{238}U \quad \rightarrow \quad \ _{92}^{239}U$$

Neutron   Uranium 238   Uranium 239

The conversion of uranium 238 into uranium 239 by neutron bombardment

The uranium 239 then decomposes into an unstable isotope of atomic number 93, called neptunium, $\ _{93}^{239}Np$, and a beta particle, $\ _{-1}^{0}e$:

$$\ _{92}^{239}U \quad \rightarrow \quad \ _{93}^{239}Np \quad + \quad \ _{-1}^{0}e$$

Uranium 239   Neptunium 239   Beta particle

Spontaneous decomposition of uranium 239 into neptunium 239 and a beta particle

The neptunium 239 isotope in turn decomposes into another unstable isotope, plutonium 239, $\ _{94}^{239}Pu$, and a beta particle, $\ _{-1}^{0}e$:

$$\ _{93}^{239}Np \quad \rightarrow \quad \ _{94}^{239}Pu \quad + \quad \ _{-1}^{0}e$$

Neptunium 239   Plutonium 239   Beta particle

Spontaneous decomposition of neptunium 239 into plutonium 239 and a beta particle

One important feature of these transmutations is that the elements neptunium and plutonium have higher atomic numbers than the most complex naturally occurring atom, uranium. Such elements are called **transuranium elements.** Transuranium elements through atomic number 105 have already been synthesized (Fig. 9-2), and the synthesis of others of even higher atomic number has been predicted.

Plutonium is of very special importance, as subsequent discussion will illustrate. It can be used as a fuel to generate energy (Sec. 9-8A), and it can be used as an explosive of awesome power (Sec. 9-10). It can also play an important role in breeder reactors (Sec. 9-9B).

9-6 WHAT IS NUCLEAR FUSION?

Another important type of nuclear reaction releases enormous amounts of energy as two nuclei of low atomic number combine:

$$\ _1^2 H \quad + \quad \ _1^2 H \quad \rightarrow \quad \ _2^3 He \quad + \quad \ _0^1 n$$

Deuterium   Deuterium   Helium 3   Neutron
nucleus   nucleus   nucleus

$$\ _1^2 H \quad + \quad \ _1^3 H \quad \rightarrow \quad \ _2^4 He \quad + \quad \ _0^1 n$$

Deuterium   Tritium   Helium 4   Neutron
nucleus   nucleus   nucleus

Two representative nuclear fusion reactions

FIGURE 9-2
Workers at the Oak Ridge National Laboratory remove a fuel
element from one of the nation's most powerful research nuclear
reactors. The reactor is used to produce transuranium elements
like curium, Cm (atomic number 96), and californium, Cf (atomic
number 98). The fuel emits a weak bluish-white glow, called
Cerenkov radiation, produced when water is exposed to gamma
rays. (*Courtesy of Union Carbide Corporation.*)

Nuclear reactions like these, in which two nuclei combine to
form a nucleus of higher atomic number, are called **nuclear
fusion** reactions.

For either of the nuclear fusion reactions given above to take
place, the two reacting nuclei must have enough energy to
overcome the repulsive forces arising from the positive charges
they both have. Temperatures of about 100 million degrees
Celsius are required in order for the reacting nuclei to attain
the energy required. Temperatures of this magnitude are
present in the sun and other stars, and it is generally thought
that fusion reactions are the source of the sun's energy. The
precise reactions, however, have not been definitely determined.

9-7 WHAT IS THE NATURE OF NUCLEAR ENERGY?

Until the early part of the present century, investigations of the
behavior of matter suggested that in the course of chemical
changes mass is conserved, as Lavoisier proposed in his law

of the conservation of mass (Sec. 6-5). Consider the reaction of gaseous hydrogen and gaseous oxygen to form liquid water. Insofar as any conventional balance can detect, the weight of water obtained is equal to the sum of the weights of hydrogen and oxygen consumed.

Studies of the conversion of one form of energy into another indicated that energy, like mass, can neither be created nor destroyed. The law of the conservation of energy, first formulated in the middle of the nineteenth century, proposed that if energy is produced in any form, e.g., heat, electricity, or work, an equivalent amount of energy in some other form is consumed (Sec. 6-8). For many decades measurements supported this concept, but further study showed that the idea of the separate conservation of mass and energy is inadequate. A few years after the turn of the twentieth century Albert Einstein announced the concept of the equivalence of mass and energy. Both mass and energy may be regarded as different manifestations of the same fundamental property of matter, called mass-energy; it is this mass-energy which is conserved during transformations.

If a reaction is accompanied by the evolution of energy, an equivalent amount of mass is lost. If energy is absorbed as a reaction takes place, an equivalent amount of mass is gained. The exact equivalence of mass and energy is given by Einstein's mass-energy equation (margin). If we represent the energy by the symbol E, the mass by m, and the speed of light by c, this relationship can be restated as shown. The speed of light in centimeters per second is a very large number (2.998×10^{10} centimeters per second). A very small amount of mass is therefore equivalent to a very large amount of energy. If the mass of 1 gram of matter were completely converted into energy, for example, it would yield enough to drive the average automobile around the earth about 400 times.

The reaction between hydrogen and oxygen to form water gives off energy. More precisely, 3.79 kilocalories of heat is evolved for each gram of water formed. This amount of energy is equivalent to 1.8×10^{-10} gram of matter, and the water actually weighs less by this amount than the combined weights of hydrogen and oxygen consumed. This weight difference is much too small to be detected by any conventional balance, and the measured weights of the reactants and products would support the conservation-of-mass principle. In general, energy changes accompanying nonnuclear chemical reactions are of such a size that the mass equivalents are small enough to neglect. For this reason the conservation of mass remains a useful operating principle.

This is not the case, however, with the energy changes of many nuclear reactions. They are frequently so large that the mass equivalents are significant. Consider the following nuclear fusion reaction, which is accompanied by the evolution of much energy:

$$\ce{_1^2H} + \ce{_1^2H} \rightarrow \ce{_2^3He} + \ce{_0^1}n$$

Deuterium   Deuterium   Helium 3   Neutron
nucleus   nucleus   nucleus

**Nuclear fusion reaction accompanied
by the evolution of much energy**

$$2H_2 + O_2 \rightarrow 2H_2O$$

Hydrogen   Oxygen   Water

Energy
= mass × (speed of light)^2

**Einstein relationship
for the equivalence of
mass and energy**

$$E = mc^2$$

**Einstein relationship
stated in symbols**

If the energy evolved in this reaction were given off entirely as heat, 2.4×10^7 kilocalories would be evolved for each gram of the helium 3 isotope formed. This is about 10 million times more energy than is evolved when oxygen combines with hydrogen to form a gram of water, and the accompanying decrease in mass is appreciable. The evolution of 2.4×10^7 kilocalories of heat corresponds to 0.0011 gram of mass.

The fission of heavy isotopes is also accompanied by the evolution of a large amount of energy. When it was realized that nuclear fission reactions could generate a self-sustaining nuclear chain reaction, the possibility of producing substantial amounts of energy in this way became apparent. The term **nuclear energy** is applied to the energy generated by a nuclear reaction. It is also called **atomic energy.**

9-8 WHAT ARE NUCLEAR REACTORS?

A **nuclear reactor** is an apparatus that carries out a self-sustaining nuclear fission chain reaction in a controlled manner. (It is also referred to as a chain-reaction pile.) The central part of a reactor is called the *core*, and its most important components are as follows.

A. Reactor Fuel

Also called nuclear fuel, this is the substance which undergoes nuclear fission. Naturally occurring uranium contains the uranium 238 isotope (99.27 percent), the uranium 235 isotope (0.72 percent), and a very small amount (0.006 percent) of the uranium 234 isotope. Of the three, uranium 235 is the only one which is readily fissionable, i.e., is a **fissile nuclide.** Two other isotopes, plutonium 239 and uranium 233, are also fissile, but they must be synthesized by nuclear reactions.

Reactor fuel may have many different compositions. It may be (1) a mixture of naturally occurring uranium isotopes which contains 0.72 percent of the fissile uranium 235, (2) a mixture of uranium isotopes in which the content of uranium 235 has been enriched to a proportion above that in the natural mixture, or (3) a mixture containing the synthetic plutonium 239 or uranium 233. When uranium isotopes are used as nuclear fuel, they are generally in the form of pellets of uranium oxide, UO_2, assembled in long thin tubes (Fig. 9-8).

B. Moderator

A moderator serves to slow neutrons down in order to increase the probability of nuclear fissions and the continuation of a nuclear chain reaction (Sec. 9-4). Moderators are necessary with some nuclear fuels, e.g., natural uranium, but not with others. Reactor fuels highly enriched with fissile nuclides (20 to 25 percent) do not require moderators. Materials commonly used as moderators include water of natural isotopic composition (ordinary, or light, water), water with an enriched content of deuterium (heavy water), graphite (a form of elemental carbon), beryllium, and beryllium oxide.

C. Control Elements

It is necessary to be able to absorb neutrons in some way, both to control the rate of a nuclear chain reaction and to interrupt it completely if a shutdown becomes necessary. Compounds of boron and cadmium are commonly used in the control elements, frequently as alloys, and often in the form of rods or plates. The boron 10 isotope and the cadmium 113 isotope absorb neutrons in the reactions shown in the margin.

D. Coolant

Reactors generate heat, which must be removed as it is generated, both to protect the reactor and to extract useful energy. Most commonly, a coolant circulates through the reactor core to maintain a relatively uniform temperature. Such materials as ordinary water, heavy water, liquid sodium, and gaseous helium are used as reactor coolants. The removal of heat is of special significance in the case of reactors designed to generate electric energy. The heat removed by the coolant is used to produce steam (or another hot gas), which, in turn, is used to drive a conventional turbine (Fig. 9-7).

E. Shielding

Since nuclear reactors generate gamma rays and neutrons, provision must be made to contain them within the reactor area. Two shields are commonly used: a few inches of iron and steel to absorb most of the gamma radiation and a layer of concrete several feet thick to absorb gamma rays and neutrons.

F. Radioactive Waste Products

As the reactor operates, the nuclear reactions produce many radioactive isotopes as by-products. In time they accumulate so much that they interfere, and the reactor must be recharged with fresh fuel. The used fuel, with the accumulated radioactive by-products, is processed to recover the fissionable isotopes. The remaining radioactive isotopes are separated as waste products. Since these waste products are highly radioactive they must be stored in a way that will not be dangerous. This involves concentrating them so that they occupy a small volume and removing them to a remote place. The problem of the safe long-term storage of radioactive wastes has not been completely resolved and constitutes one of the most serious problems associated with nuclear reactors. At the moment, about 100 million gallons of waste in liquid form is stored in underground tanks awaiting more adequate disposal. The most promising current proposal is to evaporate the liquids to concentrate the solids into a small volume and store them in abandoned underground salt mines (Fig. 9-3).

$^1_0 n$

Neutron

+

$^{10}_5 B$

Boron 10

↓

$^7_3 Li$

Lithium 7

+

$^4_2 He$

Alpha particle

Reaction of boron 10 responsible for its ability to absorb neutrons and control the chain reaction of a nuclear reactor

$^1_0 n$

Neutron

+

$^{113}_{48} Cd$

Cadmium 113

↓

$^{114}_{48} Cd$

Cadmium 114

Reaction of cadmium 113 responsible for its ability to absorb neutrons and control the chain reaction of a nuclear reactor

9-9 THE USES OF NUCLEAR REACTORS

Nuclear reactors are used to synthesize fissile materials, generate power, synthesize radioactive isotopes for specialized purposes, and carry on research in nuclear science.

FIGURE 9-3
Simplified sketch of a site plan for a deep geological depository
for radioactive wastes in an abandoned dry salt mine. The waste
material would be concentrated to a small volume of solids.
(*Courtesy of Atomic Industrial Forum.*)

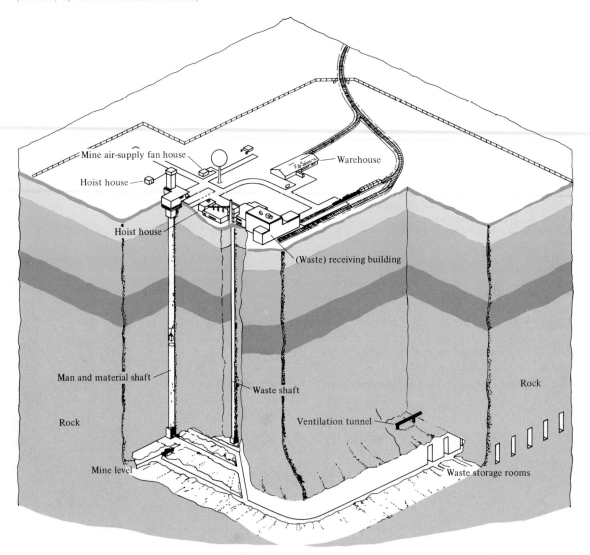

Mine air-supply fan house

Warehouse

Hoist house

Hoist house

(Waste) receiving building

Man and material shaft

Rock

Waste shaft

Ventilation tunnel

Rock

Mine level

Waste storage rooms

A. Synthesis of Radioactive Isotopes

Synthesis of carbon 14 for use as a tracer isotope   Nuclear
reactors are a convenient source of neutrons, which are used
in nuclear reactions to produce radioactive isotopes. Many
different kinds of radioactive isotopes are synthesized in
nuclear reactors. They have been of revolutionary usefulness
in many different ways: in clarifying sequences of chemical
reactions in living organisms, in medical diagnosis and treatment, in chemical analysis, and in many industrial applica-

tions. The production and use of isotopes will doubtless prove to be the most important contribution of nuclear chemistry to humanity.

Carbon 14, for example, can be produced by the bombardment of nitrogen 14 by neutrons (the same reaction which forms the carbon 14 of nature, Sec. 9-16).

Carbon 14 is a radioactive isotope with a half-life of 5730 years (see Sec. 9-15 for a discussion of half-life). It can be incorporated into any carbon-containing compound and used to trace the fate of the compound as it is transformed in biological systems. Since the chemical properties of radioactive isotopes are identical with those of the other isotopes of an element, radioactive isotopes, called *tracers*, can be detected by their radioactivity without disturbing the chemical reaction or the system under study. Carbon dioxide containing carbon 14, for example, has made it possible to learn many important details of photosynthesis.

In photosynthesis green plants take up carbon dioxide and water in the presence of sunlight and convert them into sugar with the evolution of oxygen (Sec. 11-3). One of the questions about the mechanism of this process is: What path do carbon atoms follow as they move from the original carbon dioxide through many intermediate substances to the final product? Using carbon 14 in tracer studies has helped determine this sequence of reactions.

$$^{1}_{0}n$$

Neutron

$$+$$

$$^{14}_{7}N$$

Nitrogen 14

$$\downarrow$$

$$^{14}_{6}C$$

Carbon 14

$$+$$

$$^{1}_{1}H$$

Proton

Reaction which synthesizes carbon 14 in a nuclear reactor

Synthesis of cobalt 60 as a convenient source of gamma radiation Radiation from radioactive isotopes can damage the malignant cells of cancerous growths. Since it is also harmful to normal cells, it must be used with care. But applied judiciously gamma radiation has been effective in treating certain types of cancer. The synthesis of radioactive isotopes in nuclear reactors has made the use of radiation more readily available and less expensive. Cobalt 60, for example, decomposes radioactively with the emission of gamma rays. It is now a source of gamma radiation used for the treatment of certain types of cancer in hospitals throughout the world. Its only source is synthesis in a nuclear reactor by exposing naturally occurring cobalt (100 percent $^{59}_{27}Co$) to neutrons.

$$^{1}_{0}n \quad + \quad ^{59}_{27}Co \quad \rightarrow \quad ^{60}_{27}Co$$

Neutron $\quad$ Natural $\quad$ Cobalt 60
cobalt 59

Reaction for the production of cobalt 60 in nuclear reactors

The gamma rays emitted by radioactive cobalt 60 have also been used effectively in the control of the screwworm fly, an insect causing much damage to livestock. The insects can be sterilized by irradiating their pupae, from which the flies are formed in 8 days. The sterilized male flies are released from airplanes in such large numbers that they greatly outnumber the normal males already present. Most of the normal females therefore mate with sterile males, and the number of fertile larvae produced is sharply decreased. By treating large areas of the southeastern United States in this way, the screwworm fly was eradicated in about a year (1959). Attempts are being made to apply this approach to the control of other insect pests. Although each kind of insect presents distinctive and complex problems, some limited success has been achieved.

B. The Synthesis of Fissile Materials

The production of fissile materials by nuclear reactors is of great significance. Uranium 235 is the only naturally occurring

FIGURE 9-4
Handling plutonium. A chemist examines a "button" of plutonium in a shielded "dry box" protected from radiation by lead glass and lead-impregnated gloves. (*Courtesy of Rockwell International.*)

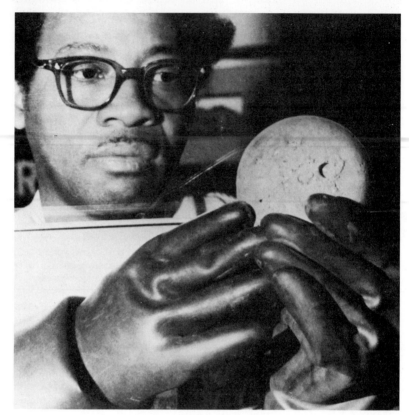

$^{1}_{0}n$

Neutrons

+

$^{238}_{92}U$

Uranium 238

↓

$^{239}_{92}U$

Uranium 239

$^{239}_{92}U$

Uranium 239

↓

$^{239}_{93}Np$

Neptunium 239

+

$^{0}_{-1}e$

Beta particle

$^{239}_{93}Np$

Neptunium 239

↓

$^{239}_{94}Pu$

Plutonium 239

+

$^{0}_{-1}e$

Beta particle

Sequence of steps in the synthesis of plutonium 239 in a nuclear reactor

fissile isotope of practical usefulness. Since there is no known method for regenerating it, the supply of this isotope will eventually be exhausted. Therefore if nuclear reactors are to continue in operation, other fissile isotopes must be synthesized in quantity. Plutonium 239 can be produced in nuclear reactors through the bombardment of the relatively abundant uranium 238 by neutrons. A mixture of uranium 235 and uranium 238 is commonly used as a reactor fuel, and plutonium 239 is produced by the sequence of reactions shown in the margin as the reactor functions to produce energy (Fig. 9-4).

Plutonium 239, however, can be used to produce nuclear bombs (Sec. 9-10). To be sure, it must be separated from the other products formed with it, which requires considerable technical expertise. Nevertheless, the synthesis of plutonium 239 in nuclear reactors represents a political hazard. Not only could it lead to the proliferation of nuclear armaments but it might increase the chance of nuclear bombs becoming available to terrorists. Superrigid control measures are required.

Thorium 232 is not a fissile material but it can be used to produce a fissile isotope, uranium 233, in a nuclear reactor, by the reactions shown in the opposite margin. **Fertile materials** are substances like thorium 232 which are not fissile but can be converted into fissile materials. One of the most remarkable features of nuclear reactors is their ability to synthesize artificial fissile isotopes which can serve as fuels in other reactors.

It is possible to design a **breeder reactor** which produces more nuclear fuel than it consumes. Imagine a reactor fuel composed of plutonium 239 with either uranium 238 or thorium 232 also present. On the average more than three neutrons are produced in each nuclear fission. Only one of these neutrons is required to carry on the subsequent fission of plutonium 239 in the nuclear chain reaction. The extra neutrons are available for the production of plutonium 239 from uranium 238 or uranium 233 from thorium 232. Both nuclear reactions are in the margins of this section. If the loss of neutrons can be controlled, new fissionable isotopes can be produced faster than the plutonium 239 is consumed. Breeder reactors could use both uranium 238 and thorium 232, supplies of which are substantial, and by synthesizing more nuclear fuel than they consume they might in principle provide the means of generating an unlimited supply of energy.

C. The Generation of Electric Energy

As a nuclear chain reaction takes place in a nuclear reactor, considerable heat is evolved. We have already seen that this heat must be removed by circulating a coolant through the reactor. When nuclear reactors are used as a source of energy to generate electricity, the heat removed by the coolant typically is used to produce steam, which in turn is used to generate electricity by driving a conventional turbine.

The exterior view of an atomic power plant is shown in Fig. 9-5, and the principal parts of this power plant are diagrammed in Figs. 9-6 and 9-7. The heat is generated in the reactor core, through which water is circulated under high pressure. The hot water is piped to a heat exchanger, where it generates steam. The reactor core and steam generator are enclosed in a steel vapor sphere. The steam is led through a turbine to drive an electric generator and then condensed back into water using the cold water from a nearby lake.

Figure 9-6 is a side view of the spherical vapor sphere of the Yankee Atomic nuclear power plant at Rowe, Massachusetts, which has been in operation since 1961. It shows the relationship between the 8-ft-high fuel assembly, the 30-ft-high reactor vessel, the steam generators, the shielding, and the vapor sphere.

The fuel assembly in the reactor vessel contains about 23,000 stainless-steel tubes, within which are pellets of uranium oxide, UO_2 (Fig. 9-8). The uranium 235 content has been somewhat enriched. As the reactor operates, the fission products accumulate and about once a year a new charge of fresh enriched uranium oxide fuel must be installed.

$_0^1 n$

Neutron

+

$_{90}^{232}\text{Th}$

Thorium 232

$\downarrow$

$_{90}^{233}\text{Th}$

Thorium 233

$_{90}^{233}\text{Th}$

Thorium 233

$\downarrow$

$_{91}^{233}\text{Pa}$

Protactinium 233

+

$_{-1}^{0} e$

Beta particle

$_{91}^{233}\text{Pa}$

Protactinium 233

$\downarrow$

$_{92}^{233}\text{U}$

Uranium 233

+

$_{-1}^{0} e$

Beta particles

Sequence of steps in the synthesis of uranium 233 in a nuclear reactor

---

Note to Students

The advantages and disadvantages of nuclear fission as a source of energy are discussed in Sec. 14-9 in the chapter devoted to the energy sources of the future.

FIGURE 9-5
General view of the Yankee Atomic plant, Rowe, Massachusetts.
One of the oldest nuclear power plants, it has been in operation
since 1961. (*Courtesy of Yankee Atomic Electric Company.*)

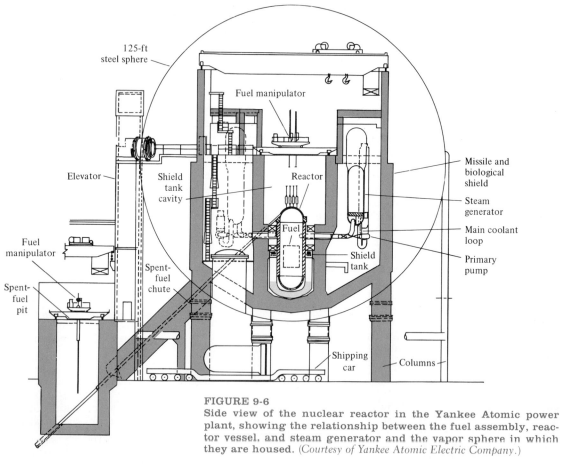

FIGURE 9-6
Side view of the nuclear reactor in the Yankee Atomic power
plant, showing the relationship between the fuel assembly, reactor vessel, and steam generator and the vapor sphere in which
they are housed. (*Courtesy of Yankee Atomic Electric Company.*)

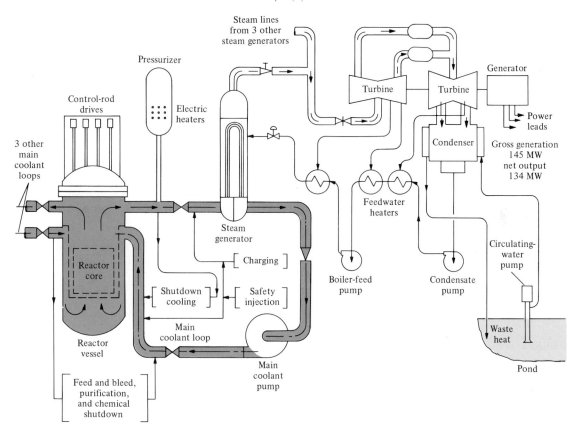

FIGURE 9-7
Diagram of the principal parts of the Yankee Atomic nuclear power plant, Rowe, Massachusetts. *(Courtesy of Yankee Atomic Electric Company.)*

FIGURE 9-8
Inspection of the fuel assembly before installation in the nuclear reactor at the Yankee Atomic plant. The core is made up of 76 square fuel assemblies containing over 23,000 stainless-steel tubes filled with slightly enriched uranium oxide pellets. *(Courtesy of Yankee Atomic Electric Company.)*

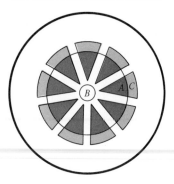

FIGURE 9-9
Schematic diagram of the first type of nuclear fission bomb; A = one of several sections of fissionable material, each smaller than the critical size, B = source of neutrons, C = one of several charges of ordinary explosives. The charges explode simultaneously, forcing the sections of fissionable material, to form an amount considerably larger than the critical size. The nuclear chain reaction is triggered by neutrons from the neutron source.

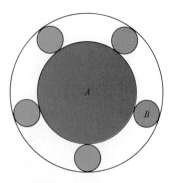

FIGURE 9-10
Schematic diagram of a nuclear fusion bomb. A contains compounds of the hydrogen isotopes deuterium and tritium; B are small fission bombs. When B explode, they generate the very high temperature which triggers the nuclear fusion reaction.

9-10 THE PRINCIPLE OF NUCLEAR FISSION BOMBS

The fission of uranium 235 produces neutrons which may (1) bring about the fission of another uranium 235 nucleus, (2) convert uranium 238 atoms into plutonium 239 in the presence of uranium 238, (3) be absorbed by other particles, or (4) leave the reaction region completely. For the nuclear chain reaction to be sustained, of course, at least one neutron produced from each fission must bring about another fission. If only one neutron is productive in this way, the chain reaction will just barely sustain itself. If, on the average, more than one neutron per fission causes a subsequent fission, the rate of the chain reaction will increase.

The higher the uranium 235 content of a mixture of uranium isotopes, the more readily a nuclear chain reaction will be self-sustaining. The less the fissile isotope is diluted with non-fissile isotopes, the more readily the neutrons will bring about a subsequent fission. But even if the fissile material is pure uranium 235, the surface escape of neutrons from a very small volume can be great enough to prevent the chain reaction from being self-sustaining. As the ratio of the volume to the area of the fissile material is increased, the proportion of neutrons which escape from the fissile material is reduced, and eventually a ratio is reached where the chain reaction will be just barely self-sustaining.

The **critical size** is the size of a sample of fissionable material for which the neutrons lost by surface escape and nonproductive capture are balanced by the neutrons produced in the fission reaction. The critical size of fissionable material depends on the proportion of fissionable isotopes in the material, the nature of the moderator, and the arrangement and shape of the components. The critical size of pure uranium 235 is reported to involve less than a kilogram. Since there are stray neutrons in the atmosphere, it is not possible to prevent a nuclear chain reaction in an amount of fissionable material in excess of the critical size.

If pieces of fissionable material smaller than the critical size are suddenly brought together to form a mass substantially larger man the critical size, it is possible to create the conditions for a very fast accelerating chain reaction which can be triggered by a source of neutrons (Fig. 9-9). An extremely forceful explosion will result, releasing in a fraction of a second a tremendous amount of energy. The high temperature generated converts the materials into gases at very high pressures. These gases expand outward with great force. The combination of very high temperature, very high pressure, and radiation causes frightful destruction. This is the general principle of an early type of fission bomb, or atomic bomb.

9-11 THE PRINCIPLE OF NUCLEAR FUSION (HYDROGEN) BOMBS

Large amounts of energy can be generated by nuclear fusion reactions as well as by nuclear fission reactions (Fig. 9-10). Fusion processes attracted considerable attention early in the development of nuclear energy, but they require temperatures of about 100 million degrees to bring them about; without the means of producing such temperatures, progress was minimal. The development of the fission bomb made it possible to generate such temperatures, and the conditions for carrying out uncontrolled fusion reactions were attained. The result was the fusion bomb, also called the hydrogen or thermonuclear bomb.

In devices of this kind, compounds of the hydrogen isotopes deuterium and tritium (Sec. 2-5) are assembled with the materials of a fission bomb. A fission explosion generates the high temperatures which bring about the fusion reactions. Large amounts of energy are quickly released, and a very powerful explosion results (Fig. 9-11).

9-12 THE PROSPECT OF CONTROLLED NUCLEAR FUSION

Bringing about a fusion process in a controlled fashion has not yet been achieved, but research continues energetically. The possibility of generating energy by controlled nuclear fusion

FIGURE 9-11
This crater formed by the underground explosion of a 100-kiloton hydrogen bomb is 320 feet deep and averages 1280 feet wide. A research team is at work at the bottom of the crater and at the left of the rim. (*Courtesy of Lawrence Radiation Laboratory.*)

holds the promise of replenishing the world's dwindling energy reserves. The principal fuel would be deuterium, found in sea-water. Although the concentration of deuterium in the sea is very small (0.016 percent), the volume of the oceans is large enough (1.5×10^{21} liters) for the total amount of deuterium to be very large (2.6×10^{20} grams). If controlled fusion could be achieved, we would be able to meet the world's energy require-ments for several million years.

9-13 RADIATION HAZARDS

A. The Effect of Radiation on Matter and Living Organisms

Radiation hazards are associated with all radioactive materials, both natural and artificial. Since radiation can neither be seen nor felt, it is an insidious hazard, and severe radiation damage can occur before being noticed.

Nuclear radiation usually refers to the radiation of alpha particles, beta particles, gamma rays, neutrons, and other par-ticles associated with radioactive materials and fission pro-cesses. The harmful effects of these powerful particles and rays are due to their ability to penetrate other forms of matter. Upon penetration they can remove electrons from atoms producing ions and other reactive molecular species. Changes in the struc-ture of substances and the pattern of chemical reactions result.

The interaction of nuclear radiation with living organisms is very complex. By altering the structure of atoms and mole-cules radiation can change the chemical reactions of cells. If enough cells are damaged, the organism cannot survive. The nature and extent of the symptoms in human beings depend on the type of radiation, the length of the exposure, the area of the body exposed, and the depth to which the radiation has pene-trated. One major effect is a marked decrease in the number of white cells in the blood, which causes the body to lose resistance to infection. Overall effects of excessive radiation include nausea, prostration, loss of appetite and weight, diarrhea, fever, and rapid heart action. Long-range effects include the develop-ment of leukemia and shortening of life.

Chemical changes in cells caused by nuclear radiation may alter the structure of the genetic material. A **mutation** is a change in the genetic material of a living organism which can be inherited. Such alterations cause departures from parent types in direct offspring or later generations. Mutations may result even from radiation of very low intensity. This is of special interest because of the radiation from natural radio-active isotopes, which constantly impinges on plants and animals. Radioactive carbon 14 and potassium 40 are widely distributed in living organisms in very small concentrations. The earth also contains such natural radioactive isotopes as radium and thorium and is constantly subjected to the radiation of cosmic rays. Some of the genetic changes which have oc-curred in nature may have resulted from low-level natural nuclear radiation.

B. Nuclear Fallout

The explosion of nuclear bombs in the open atmosphere is another source of radiation hazard. The many radioactive products of these bombs are quickly carried into the upper atmosphere; some soon fall to the earth, causing a dangerous fallout around the blast area, and others may remain in the upper atmosphere for several months before returning to earth.

Among the chief villains in nuclear fallout are the radioactive isotopes strontium 90 and iodine 131. Since strontium 90 behaves chemically like calcium, it becomes concentrated in bone structure. It decomposes with the emission of beta and gamma rays, and has a life of many years. Concentration in the bones of children may cause leukemia and bone cancer. Iodine 131 is a beta- and gamma-ray emitter; although it decomposes in a matter of days, it rapidly finds its way to the iodine-containing material of the thyroid gland, where it can cause irreversible damage.

9-14 HOW ARE NUCLEAR EQUATIONS BALANCED?

The nuclear equation for the first step in the radioactive decomposition of uranium 238 (Sec. 9-2) is

Optional

$$\underset{\text{Uranium 238}}{^{238}_{92}\text{U}} \quad \rightarrow \quad \underset{\text{Thorium 234}}{^{234}_{90}\text{Th}} \quad + \quad \underset{\text{Alpha particle}}{^{4}_{2}\text{He}}$$

**Representation with isotope symbols of the first
step in the radioactive disintegration of
uranium 238 atoms**

The subscripts of the symbols represent the atomic numbers of the specific isotopes and correspond to the number of protons. The superscripts represent the mass numbers of the isotopes and correspond to the total number of protons and neutrons. The term **nucleons** is applied collectively to protons and neutrons, and mass numbers therefore refer to the total number of nucleons in the nucleus of an isotope.

In any nuclear reaction the total number of nucleons remains unchanged. The sum of the mass numbers of the products is equal to the sum of the mass numbers of the reactants. Moreover, the sum of the atomic numbers is the same for both reactants and products. A representation of a nuclear reaction must correspond to this conservation of mass numbers and atomic numbers. The equation for the decomposition of uranium conforms to these principles:

$$
\begin{array}{lccccc}
\text{Mass number:} & 238 & \rightarrow & 234 & + & 4 \\
\text{Atomic number:} & 92 & \rightarrow & 90 & + & 2 \\
 & \text{Uranium} & & \text{Thorium} & & \text{Alpha} \\
 & \text{nucleus} & & \text{nucleus} & & \text{particle}
\end{array}
$$

**The first step in the decomposition of uranium 238 atoms;
as in all nuclear reactions, the sum of the mass numbers
and the atomic numbers of the isotopes is the same
before and after the reaction**

Remember (Sec. 9-2) that a radioactive decomposition does not depend in any way on the state of chemical combination. Moreover, the representation of uranium as $^{238}_{92}U$ does not necessarily mean that the uranium is present as elemental uranium.

The principal product of a radioactive disintegration is called a **daughter element,** and the principal reactant is called the **parent element.** A daughter element formed as the result of the emission of an alpha particle, ^4_2He, has a mass number which is 4 units less than the parent element. The atomic number is 2 units less. The daughter element formed as the result of the emission of a beta particle has an atomic number 1 greater than the parent element, but there is no change in the mass number. One way of accounting for this is to imagine that as the beta particle emerges from the nucleus a proton is created and remains in the nucleus. As a consequence, the daughter element contains one more proton than the parent element. The atomic number is therefore increased by 1 as a beta particle is emitted. The emission of gamma rays is not accompanied by a change in atomic number or mass number. The changes in mass-number units and atomic-number units which accompany radioactive decompositions are summarized in Table 9-1.

These generalizations can be used to determine the missing parts of a representation of a nuclear change. It is known, for example, that radium 226, $^{226}_{88}Ra$, decomposes into radon 222, $^{222}_{86}Rn$. From the mass and atomic numbers we can tell that 4 mass-number units and 2 atomic-number units are needed in the products to bring them into balance. This suggests that the missing product is an alpha particle, ^4_2He. In other words, the complete nuclear equation for the decomposition is as shown.

Thorium 234, $^{234}_{90}Th$, is known to decompose radioactively with the emission of a beta particle, $^0_{-1}e$. Inspection of the mass and atomic numbers indicates that the daughter element is protactinium 234, $^{234}_{91}Pa$, for only in this way is the nuclear equation brought into balance.

$$^1_0n \rightarrow\ ^1_1H\ +\ ^0_{-1}e$$

Neutron   Proton   Beta particle

The hypothetical generation of a proton as a beta particle is ejected from the nucleus; in the symbol for a neutron, ^1_0n, the atomic number is zero since the neutron has no charge

$$^{226}_{88}Ra \rightarrow\ ^{222}_{86}Rn\ +\ ^4_2He$$

Radium 226   Radon 222   Alpha particle

The radioactive decomposition of radium 226

$$^{234}_{90}Th$$

Thorium 234

$\downarrow$

$$^{234}_{91}Pa$$

Protactinium 234

$+$

$$^0_{-1}e$$

Beta particle

The radioactive decomposition of thorium 234

Table 9-1
Summary of the Changes in Mass-Number Units and Atomic-Number Units Accompanying Radioactive Decompositions

Ray emitted	Symbol	Change in mass number as emitted	Change in atomic number as emitted
Alpha, α	^4_2He	Decreases 4 units	Decreases 2 units
Beta, β	$^0_{-1}e$	No change	Increases 1 unit
Gamma, γ		No change	No change

9-15 WHAT IS THE HALF-LIFE OF A RADIOACTIVE ISOTOPE?

Optional   One of the most distinctive characteristics of a radioactive decomposition is the rate at which it takes place. There is no way of predicting the decomposition of an individual nucleus, but the statistical behavior of the many billions of nuclei in any appreciable amount of a radioactive substance is one of the most precisely predictable events in nature.

Experimental investigation of the rate of radioactive disintegrations has revealed that the decomposition rate of an individual isotope depends only on the number of atoms present and is not related to the temperature, pressure, or state of chemical combination of the isotope. If 1.0×10^{10} atoms of a given kind disintegrate at the rate of 1.0×10^2 atoms per minute, 3 times this number will decompose 3 times as fast; that is, 3.0×10^{10} atoms will disintegrate at the rate of 3.0×10^2 atoms per minute.

In 1.000 gram of radioactive iodine 131 there are 4.74×10^{21} nuclei. Measurements show that after 8.07 days one-half of this amount has decomposed and there remains 0.500 gram, or 2.37×10^{21} nuclei. In the next 8.07 days, one-half of the remainder decomposes, leaving 0.250 gram, or 1.18×10^{21} nuclei. The fraction of iodine 131 that decomposes in any 8.07-day period is always the same: one-half of what was present at the beginning of the 8.07-day period. This decomposition behavior is expressed by the plot given in Fig. 9-12. From this plot we see that the time required for complete decomposition of the isotope is infinitely long.

FIGURE 9-12
Graph of the radioactive decay of 1 gram of iodine 131. The fraction of an amount of iodine 131 that decays in an 8.07-day interval (half-life) is always the same: one-half of what was present at the beginning of that interval.

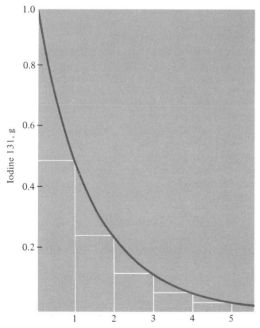

Iodine 131, g

Number of half-lives (each 8.07 days long)

A convenient way of expressing the decomposition rate of a radioactive isotope is provided by the **half-life.** This is the time it takes for one-half of the nuclei in a given sample of a specific radioactive isotope to decompose. It is a characteristic of each radioactive isotope. (For iodine 131 the half-life is 8.07 days.) Although the rate of decomposition depends on the number of nuclei, the time required for one-half of a given number of nuclei to decompose is constant.

The half-life expresses the relative stability of an isotope. The shorter the half-life, the less stable the isotope. Half-lives vary enormously from isotope to isotope, and range from only a fraction of a second to about 5×10^9 years. The half-lives of a few radioactive isotopes are given in Table 9-2.

9-16 THE PRINCIPLE OF RADIOCARBON DATING

Optional

$^{14}_{6}C$

Carbon 14

↓

$^{14}_{7}N$

Nitrogen 14

+

$^{0}_{-1}e$

Beta particle

The radioactive decomposition of carbon 14 into nitrogen 14 with the emission of a beta particle

$^{1}_{0}n$

Neutron

+

$^{14}_{7}N$

Nitrogen 14

↓

$^{14}_{6}C$

Carbon 14

+

$^{1}_{1}H$

Proton

The formation of carbon 14 by the bombardment of nitrogen 14 with neutrons

The carbon atoms found in nature are a mixture of three kinds of isotopes, carbon 12 (98.89 percent), carbon 13 (1.11 percent), and a very small but significant amount of carbon 14, the only isotope which is radioactive. The half-life of carbon 14 is 5730 years; it decays into nitrogen 14 with the emission of a beta particle, $_{-1}^{0}e$. The carbon 14 is formed by the interaction of neutrons from cosmic rays with nitrogen 14, which constitutes 99.6 percent of atmospheric nitrogen. (Cosmic rays are high-speed nuclear particles from outer space.) A balance is established between the formation of carbon 14 and its radioactive decay. The carbon 14 is converted to carbon dioxide, CO_2, and becomes part of the natural pool of CO_2 available to plants for the process of photosynthesis (Sec. 11-3). The carbon 14 thus becomes a part of the numerous carbon compounds of plants and animals, through the cycle of carbon-containing compounds in the living world. As a result, all the carbon compounds of living plants and animals contain the same proportion of carbon 14, and all exhibit radioactivity.

The proportion of carbon 14 in the carbon of plants and animals remains constant as long as the organism participates in the plant-animal carbon cycle. If it is separated from this cycle through death, the replenishing supply of carbon 14 is cut off. The carbon 14 remaining in the dead plant or animal continues to disintegrate. The net result is a continuing decrease in the amount of carbon 14 and a decrease in the radioactivity of the carbon. The longer the interval since death, the lower the amount of carbon 14 and the amount of radioactivity resulting from it. By comparing the radioactivity due to carbon in, say, a leather sandal with the radioactivity due to carbon in living animals, the age of the sandal can be determined.

Radiocarbon dating is the method of determining the age of archeological specimens made from the materials of living organisms by comparing their carbon radioactivity with the carbon radioactivity of current living organisms. Other radioactive isotopes are used in other methods of radiochemical dating.

Radiocarbon dating has made important contributions to history, archeology, and geology. Any remains of once-living material may be used, such as wood, charcoal, leather, textiles, seeds, and charred bones. One of the most interesting deter-

minations was the age (about 1900 years) of the Dead Sea
scrolls, discovered in 1947. Other representative radiocarbon
dates are given in Table 9-3.

221

9-16 THE PRINCIPLE OF
RADIOCARBON DATING

Table 9-2
Half-Lives of Some Radioactive Isotopes

Isotope	Half-life	Isotope	Half-life
Aluminum 24	2.1 seconds	Plutonium 239	2.44×10^4 years
Carbon 11	20.3 minutes	Polonium 214	1.6×10^{-4} second
Carbon 14	5730 years	Potassium 40	1.28×10^9 years
Cobalt 60	5.26 years	Strontium 90	28.1 years
Iodine 131	8.07 days	Uranium 235	7.1×10^8 years
Magnesium 28	21 hours	Uranium 238	4.51×10^9 years
Phosphorus 33	25 days		

Table 9-3
Representative Radiocarbon Dates†

Source	Age, years
Charcoal from the Lascaux cave in France, containing early cave paintings	$15,516 \pm 900$
Bison bones from Lubbock, Texas, associated with prehistoric man	$9,883 \pm 350$
Charcoal from a tree burned during the upheaval that formed Crater Lake, Oregon	$6,453 \pm 250$
Wheat and barley from ancient Egypt	$6,095 \pm 250$
Charcoal associated with early period from Stonehenge, England	$3,798 \pm 275$
Wood from coffin from Ptolemaic period in Egypt	$2,190 \pm 450$
Linen wrappings used for the Book of Isaiah in the Dead Sea scrolls	$1,917 \pm 200$
Ancient Manchurian lotus seeds, still fertile	$1,040 \pm 210$

† From W. F. Libby, p. 70, "Radiocarbon Dating," University of Chicago Press, Chicago, 1952.

KEY WORDS

1. Nuclear reaction (Sec. 9-1)
2. Radioactivity (Sec. 9-1)
3. Alpha ray (Sec. 9-1)
4. Alpha particle (Sec. 9-1)
5. Beta ray (Sec. 9-1)
6. Beta particle (Sec. 9-1)
7. Gamma ray (Sec. 9-1)
8. Artificial transmutation (Sec. 9-3)
9. Nuclear fission (Sec. 9-4)
10. Atomic fission (Sec. 9-4)
11. Nuclear chain reaction (Sec. 9-4)
12. Slow neutron (Sec. 9-4)
13. Fast neutron (Sec. 9-4)
14. Moderator (Sec. 9-4)
15. Transuranium element (Sec. 9-5)
16. Nuclear fusion (Sec. 9-6)
17. Nuclear energy (Sec. 9-7)
18. Atomic energy (Sec. 9-7)
19. Nuclear reactor (Sec. 9-8)
20. Reactor fuel (Sec. 9-8)
21. Fissile nuclide (Sec. 9-8)
22. Fertile material (Sec. 9-9)
23. Breeder reactor (Sec. 9-9)
24. Critical size (Sec. 9-10)
25. Nuclear radiation (Sec. 9-13)
26. Mutation (Sec. 9-13)

Optional:

27. Nucleon (Sec. 9-14)
28. Daughter element (Sec. 9-14)
29. Parent element (Sec. 9-14)
30. Half-life (Sec. 9-15)
31. Radiocarbon dating (Sec. 9-16)

"These days *everything* is higher."

SUMMARIZING QUESTIONS FOR SELF-STUDY

Section 9-1

1. Q. What is a nuclear reaction?

A. A phenomenon in which the structures of atomic nuclei are changed.

2. Q. Why are many nuclear changes also called transmutations?

A. Because they involve the conversion of the atoms of one element into the atoms of a different element.

3. Q. What is meant by radioactivity?

A. The spontaneous decomposition of the nucleus of an atom which is accompanied by the emission of either alpha and gamma rays or beta and gamma rays.

4. Q. What are alpha rays?

A. Streams of high-speed helium nuclei, symbol He^{2+} or ^4_2He.

5. Q. What are beta rays?

A. Streams of high-speed electrons, symbol $_{-1}^0e$.

6. Q. What are gamma rays?

A. Highly penetrating invisible light rays.

Sections 9-2 and 9-3

7. Q. In the symbol for uranium 238, $^{238}_{92}U$, what does the 238 represent?

A. The mass number of the isotope, which is the sum of the number of neutrons and protons.

8. Q. When the symbol $^{238}_{92}U$ is used to represent the uranium 238 isotope, what does it convey about the state of chemical combination of the isotope?

A. It says nothing about the state of chemical combination of the isotope. By convention the symbols used in nuclear equations do not specify the state of chemical combination. The equations can be simplified in this way because nuclear reactions do not depend on the state of chemical combination.

9. Q. What is meant by artificial transmutation?

A. A nuclear change not occurring in nature but deliberately brought about by bombarding nuclei with high-speed particles capable of causing such a change.

Section 9-4

10. Q. What is meant by nuclear fission and atomic fission? Give an example.

A. Both terms refer to the splitting of atomic nucleus into two nuclei of substantial size, for example, $^1_0n + ^{235}_{92}U \rightarrow ^{142}_{56}Ba + ^{91}_{36}Kr + 3^1_0n$.

11. Q. What is a nuclear chain reaction?

A. A self-sustaining sequence of nuclear reactions in which the neutrons formed in one nuclear fission reaction bring about a subsequent neutron-producing nuclear fission reaction.

12. Q. What is the function of a moderator in a nuclear chain reaction?

A. It is a substance, like graphite, that slows down the neutrons produced in a nuclear fission reaction, making them more effective in bringing about subsequent nuclear fission reactions.

Sections 9-5 and 9-6

13. Q. What is a transuranium element? Give the isotopic symbol of an example.

A. An element of atomic number greater than uranium, which has the highest atomic number of the naturally occurring elements; for example plutonium, $^{239}_{94}Pu$.

14. Q. What is nuclear fusion?

A. A nuclear reaction in which two nuclei combine to form a nucleus of higher atomic number.

Section 9-7

15. Q. What is meant by the conservation of mass-energy in a chemical reaction?

A. Both mass and energy are different manifestations of the same fundamental property, mass-energy. It is the mass-energy which remains the same as a chemical reaction takes place rather than either mass or energy independently.

16. Q. Why does the conservation of mass continue to be a useful principle despite its fundamental inaccuracy?

A. Because the mass equivalents of the energy changes during chemical reactions of the ordinary type are so small they cannot be detected by conventional scales.

17. Q. Why is the conservation of mass not a practical operating principle in many nuclear reactions?

A. The large amount of energy involved has a mass equivalent that is significant.

Section 9-8

18. Q. What is nuclear energy (atomic energy)?

A. The energy generated by a nuclear reaction.

19. Q. What is a nuclear reactor?

A. An apparatus for carrying out a self-sustaining nuclear-fission chain reaction in a controlled manner.

20. Q. What is reactor fuel? Give an example.

A. The substance that undergoes fission in a nuclear reactor, e.g., uranium 235 and plutonium 239.

21. Q. What is meant by the control elements of a nuclear reactor? Give an example.

A. The substances which absorb neutrons; they are used to control the rate of the chain reaction of a nuclear reactor or to shut it down completely, e.g., boron and cadmium.

22. Q. What coolants are used in nuclear reactors, and what is their function?

A. Three examples are ordinary water, heavy water, and liquid sodium. They are used to remove the heat produced by the reaction of a nuclear reactor. In a nuclear reactor of a power plant they serve to transfer the heat to the steam-generating equipment.

23. Q. Why must nuclear reactors be shielded?

A. To prevent the escape of harmful radiation, caused principally by their production of neutrons and gamma radiation.

Section 9-9

24. Q. What are three uses of nuclear reactors?

A. Synthesis of fissile materials, synthesis of tracer radioactive isotopes, and generation of nuclear energy.

25. Q. What is one use of the carbon 14 isotope?

A. To trace the fate of carbon-containing compounds as they are transformed in biological systems.

26. Q. What is one use of the cobalt 60 isotope?

A. To serve as a source of gamma radiation in the treatment of cancer.

27. Q. How is cobalt 60 synthesized? Give an equation for the reaction.

A. By bombarding cobalt 59 by neutrons from a nuclear reactor: $^1_0n + ^{59}_{27}Co \rightarrow ^{60}_{27}Co$.

28. Q. What fissile material is synthesized in nuclear reactors which use natural or enriched uranium as a fuel?

A. Plutonium 239 from the bombardment of uranium 238 nuclei with neutrons generated in the fission process.

29. Q. What is the political hazard of the synthesis of this material?

A. Countries which have imported the technological expertise to build nuclear power plants using natural or enriched uranium as a fuel also have a potential source of the fissionable material for an atom bomb if they are able to learn how to separate the plutonium 239 from the other substances formed in a nuclear reactor.

30. Q. What is a breeder reactor?

A. A nuclear reactor which synthesizes more nuclear fuel than it consumes.

31. Q. What two nuclear fuel materials are most likely to be produced in breeder reactors?

A. Plutonium 239 and uranium 233.

Sections 9-10 to 9-12

32. Q. What is the critical size of a fissionable material?

A. The size of a sample of fissionable material for which the neutrons lost by surface escape and nonproductive capture are balanced by the neutrons produced in the fission reaction. Under these conditions the nuclear chain reaction is just barely self-sustaining.

33. Q. What is the principle of the early version of a fission, or atomic, bomb?

A. Pieces of fissionable material smaller than the critical size are suddenly brought together to form a mass of material substantially larger than the critical size. This material is bombarded with neutrons, which trigger a very highly accelerating chain reaction, resulting in an explosion.

34. Q. What is the principle of a fusion, or hydrogen, bomb?

A. Compounds of the hydrogen isotopes deuterium and tritium are assembled with the materials of a fission bomb. A fission explosion generates the high temperatures which bring about the fusion reactions. Large amounts of energy are quickly released, and a very powerful explosion results.

Section 9-13

35. Q. What is nuclear radiation?

A. The radiation of alpha particles, beta particles, gamma rays, neutrons, and other particles associated with radioactive materials and fission processes.

36. Q. To what, fundamentally, are the harmful effects of this radiation due?

A. Its ability to penetrate other forms of matter and to remove electrons from atoms, producing ions and other reactive molecular species.

37. Q. How do these effects change the cells of living organisms?

A. They damage the cells by changing their chemical reactions.

38. Q. Why is strontium 90 a particularly harmful component of the fallout of nuclear bombs?

A. Having a chemistry similar to that of calcium, a component of bone, it becomes incorporated into the bone structure of people, especially of growing children, where its radioactivity can cause leukemia and bone cancer.

39. Q. What are two sources of the natural radiation to which human beings are exposed internally?

A. Radioactive carbon 14 and radioactive potassium 40, widely distributed in living organisms in very small concentrations.

Section 9-14 (Optional)

40. Q. What are nucleons?

A. Protons and neutrons collectively.

41. Q. In balancing the equations which are used to represent nuclear reactions what two structural features should be equal in products and reactants?

A. The sum of the atomic numbers of the products should equal the sum of the atomic numbers of the reactants, and the sum of the mass numbers of the products should equal the sum of the mass numbers of the reactants.

42. Q. What is meant by a daughter element? Give an example.

A. The principal product of a radioactive disintegration, for example, $^{234}_{91}Pa$ in the reaction $^{234}_{90}Th \rightarrow {}^{234}_{91}Pa + {}^{0}_{-1}e$.

43. Q. What is meant by a parent element?

A. The principal reactant in a radioactive disintegration, for example, $^{234}_{90}Th$ in the reaction above.

44. Q. When a beta particle is emitted in a radioactive disintegration, why does the atomic nucleus of the daughter element have an atomic number higher than the atomic nucleus of the parent element?

A. A beta particle is an electron, $_{-1}^{0}e$. One can imagine it as being formed by the decomposition of a neutron: $^{1}_{0}n \rightarrow {}^{1}_{1}H + {}^{0}_{-1}e$. Since the proton, with an atomic number of 1, may be considered to remain behind in the atomic nucleus of the daughter element, the daughter element has an atomic number 1 unit greater than the atomic nucleus of the parent element.

45. Q. Complete the following representations of a nuclear reaction:

(a) $^{1}_{0}n + {}^{27}_{13}Al \rightarrow ? + {}^{1}_{1}H$

(b) Proton + $^{8}_{4}Be \rightarrow ? + {}^{4}_{2}He$

A. (a) $^{1}_{0}n + {}^{27}_{13}Al \rightarrow {}^{27}_{12}Mg + {}^{1}_{1}H$

(b) $^{1}_{1}H + {}^{8}_{4}Be \rightarrow {}^{5}_{3}Li + {}^{4}_{2}He$

Sections 9-15 and 9-16 (Optional)

46. Q. What is the half-life of a radioactive substance?

A. The time it takes for one-half of the nuclei in a given sample to decompose.

47. Q. If a radioactive substance has a short half-life, e.g., a few minutes, what does this indicate about its relative stability?

A. It is relatively unstable.

48. Q. Why is the concentration of the carbon 14 isotope in the naturally occurring carbon isotopes very significant although it is very small?

A. The carbon 14 isotope is the only naturally occurring isotope of carbon which is radioactive.

49. Q. Why does the proportion of the carbon 14 isotopes in the natural mixture of carbon isotopes remain at a constant value even though the carbon 14 atoms, being radioactive, are constantly decomposing?

A. Because the carbon 14 isotopes are constantly being formed as cosmic-ray neutrons bombard nitrogen 14 isotopes: $^{1}_{0}n + {}^{14}_{7}N \rightarrow {}^{14}_{6}C + {}^{1}_{1}H$.

50. Q. Why does the proportion of the carbon 14 isotopes in the mixture of carbon isotopes found in living organisms decline when the organism dies?

A. The chemical reactions which incorporate carbon atoms into the organism stop, but the carbon 14 isotopes continue to decompose.

51. Q. What is the principle of radiocarbon dating?

A. A method of determining the age of archeological specimens made from materials of living organisms by comparing their carbon radioactivity with the carbon radioactivity of current living organisms.

PRACTICE EXERCISES

1. Match each definition or other statement with the numbered term above with which it is most closely associated. Each numbered term may be used only once.

   1. Nuclear radiation
   2. Fissile nuclide
   3. Nuclear reaction
   4. Nuclear chain reaction
   5. Transuranium element
   6. Radioactivity
   7. Artificial transmutation
   8. Beta rays
   9. Alpha rays
   10. Fertile material

   (a) A nuclear change which does not occur in nature but is deliberately brought about by the bombardment of nuclei with high-speed particles

   (b) Penetrating streams of high-speed alpha particles

   (c) An element of atomic number greater than uranium

   (d) An isotope which is readily fissionable

   (e) A phenomenon in which the structures of atomic nuclei are changed

   (f) The radiation of alpha particles, beta particles, gamma rays, neutrons, and other particles associated with radioactive materials

   (g) The spontaneous decomposition of atomic nuclei

   (h) A self-sustaining sequence of nuclear reactions

2. Match each definition or other statement with the numbered term above with which it is most closely associated. Each numbered term may be used only once.

   1. Nuclear fusion
   2. Nuclear energy
   3. Alpha rays
   4. Critical size
   5. Nuclear fission
   6. Gamma rays
   7. Moderator
   8. Beta rays
   9. Nuclear reactor
   10. Artificial transmutation

   (a) A nuclear reaction in which an atomic nucleus is split into two nuclei of substantial size

   (b) Penetrating streams of high-speed electrons

   (c) A substance capable of slowing down fast neutrons

225

(d) An apparatus for carrying out a self-sustaining nuclear-fission chain reaction in a controlled manner

(e) A nuclear reaction in which two nuclei combine to form a nucleus of higher atomic number

(f) Energy generated by a nuclear reaction

(g) The size of a sample of fissionable material for which the neutrons lost by surface escape and nonproductive capture are balanced by the neutrons produced in the fission reaction

3. How did Henri Becquerel discover the phenomenon of radioactivity?

4. Give the number of protons and the number of neutrons in each of the nuclei represented by the following:

(a) ^3_1H (b) ^4_2He

(c) $^{238}_{92}U$ (d) $^{239}_{94}Pu$

(e) $^{234}_{90}Th$

5. Give the isotopic symbol for (a) a beta particle and (b) a helium 4 ion.

6. For the following isotopic symbols:

$^{234}_{90}Th$ $^{235}_{92}U$ $^{239}_{94}Pu$ $^{233}_{92}U$ $^{239}_{93}Np$ $^{238}_{92}U$

(a) Which isotopes have the same atomic number?

(b) Which isotopes have same number of neutrons?

(c) Which isotopes have same number of protons?

(d) Which are transuranium elements?

7. How, briefly, was nuclear fission discovered?

8. Give a nuclear reaction representing a nuclear fission. Explain how it may lead to a nuclear chain reaction.

9. What, briefly, is the source of nuclear energy?

10. What is the basic principle of a nuclear reactor?

11. Give the isotopic symbols of three isotopes which can serve as nuclear fuel.

12. In nuclear reactors what is the function of the (a) moderator, (b) control elements, (c) coolant, and (d) shielding. Give two examples of materials which fill the function.

13. Describe three uses of nuclear reactors and give nuclear equations for a specific process involved in two of the uses.

14. What is the most distinctive characteristic of a breeder reactor?

15. What is the special problem associated with the waste products of nuclear reactors? What is one proposed solution?

16. What, briefly, is the principle of an atom bomb? Why does it not explode during construction?

17. What is the principle of a hydrogen bomb? What is the role of the nuclear fission process in the bomb?

18. What is nuclear radiation and

(a) To what fundamental property is its harmful effects on living organisms due?

(b) What are the overall effects of excessive nuclear radiation on human beings?

(c) What are two natural sources of nuclear radiation?

19. What is meant by nuclear fallout? What two isotopes are among the chief harmful agents? Describe their effects briefly.

Section 9-14 (Optional)

20. Give the meaning of (a) nucleon, (b) daughter element, and (c) parent element.

21. Complete the following nuclear equations:

(a) $^7_3Li + ^2_1H \rightarrow ? + ^1_1H$

(b) $? \rightarrow ^{241}_{95}Am + ^{0}_{-1}e$

(c) $? \rightarrow ^{225}_{88}Ra + ^4_2He$

(d) $^4_2He + ^9_4Be \rightarrow ? + ^1_0n$

(e) $^1_1H + ^{14}_7N \rightarrow ? + ^4_2He$

(f) $^{233}_{92}U \rightarrow ? + ^4_2He$

Section 9-15 (Optional)

22. What is the half-life of a radioactive isotope?

23. The half-life of plutonium 241 is 13.2 years at 100°C. Will the half-life be longer or shorter at −50°C? Explain.

Section 9-16 (Optional)

24. (a) Give the nuclear equation for the formation of carbon 14 from nitrogen 14.

(b) Give the nuclear equation for the radioactive decomposition of carbon 14.

(c) Why does the concentration of the carbon 14 isotope in the carbon atoms of living plants and animals decline after the plant or animal dies?

(d) Why does the concentration of the carbon 14 isotope in the carbon atoms of living plants and animals remain constant as long as the plant or animal is alive?

(e) What is the principle of radiocarbon dating?

SUGGESTIONS FOR FURTHER READING

The following articles are of an introductory nature.

Bebbington, William P.: The Reprocessing of Nuclear Fuels, **Scientific American**, December 1976, p. 30.

Choppin, Gregory R.: Nuclear Fission, **Chemistry**, July-August 1967, p. 25.

Clark, Herbert M.: The Origin of Nuclear Science, **Chemistry**, July-August 1967, p. 8.

Cohen, Bernard L.: The Disposal of Radioactive

Wastes from Fission Reactors, **Scientific American,** June 1977, p. 21.

Epstein, William: The Proliferation of Nuclear Weapons, **Scientific American,** April 1975, p. 18.

Gough, William C., and Bernard J. Eastland: The Prospects of Fusion Power, **Scientific American,** February 1971, p. 50.

Hudis, J.: Nuclear Reactions, **Chemistry,** July-August 1967, p. 20.

Hyde, Earl K.: Nuclear Models, **Chemistry,** July-August 1967, p. 12. Clues to the nature of nuclear forces.

Johnsen, Russell H.: Radiation Chemistry, **Chemistry,** July-August 1967, p. 31.

Leachman, R. B.: Nuclear Fission, **Scientific American,** August 1965, p. 49.

McIntyre, Hugh C.: Natural-Uranium Heavy-Water Reactors, **Scientific American,** October 1975, p. 17.

Parker, E. N.: The Sun, **Scientific American,** September 1975, p. 43.

Ross, Leonard: How "Atoms for Peace" Became Bombs for Sale, **The New York Times Magazine,** December 5, 1976, p. 39. The effort of the United States to assist other nations in developing nuclear power.

Seaborg, Glenn T., and Justin L. Bloom: Fast Breeder Reactors, **Scientific American,** November 1970, p. 13.

Severo, Richard: Too Hot to Handle, **The New York Times Magazine,** April 10, 1977, p. 15. Recycling nuclear fuel.

Ullrich, R. L., J. M. Holland, and J. B. Storer: Cancer-causing Radiation, **Chemistry,** April 1977, p. 6.

Vendryes, Georges A.: Superphenix: A Full-Scale Breeder Reactor, **Scientific American,** March 1977, p. 26.

York, Herbert F.: The Debate over the Hydrogen Bomb, **Scientific American,** October 1975, p. 106.

The following articles are at a more advanced level.

Barschall, H. H.: The Production and Use of Neutrons for Cancer Treatment, **American Scientist,** 64: 668 (1976).

Bethe, Hans A.: What Holds the Nucleus Together? **Scientific American,** September 1953, p. 58.

Cowan, George A.: A Natural Fission Reactor, **Scientific American,** July 1976, p. 36. A rich deposit of uranium ore operated as a nuclear reactor in West Africa 2 billion years ago.

Emmett, John L., John Nuckells, and Lowell Wood: Fusion Power by Laser Implosion, **Scientific American,** June 1974, p. 24.

Lubin, Moshe J., and Arthur P. Fraas: Fusion by Laser, **Scientific American,** June 1971, p. 21.

Ralph, Elizabeth K., and Henry N. Michael: Twenty-five Years of Radiocarbon Dating, **American Scientist,** 62:553 (1974).

Seaborg, Glenn T., and Harold R. Fritsch: The Synthetic Elements: III, **Scientific American,** April 1963, p. 68. Synthesis of the transuranium elements.

Seaborg, Glenn T., and Justin L. Bloom: The Synthetic Elements: IV, **Scientific American,** April 1969, p. 56. Synthesis of transuranium elements.

Zare, Richard N.: Laser Separation of Isotopes, **Scientific American,** February 1977, p. 86.

10

WHAT ARE THE SUBSTANCES OF LIVING ORGANISMS?

10-1 INTRODUCTION

Chemists have always been intensely curious about the nature of plants and animals, but the day is still to come when chemistry will be able to explain completely the processes of life in terms of the reactions occurring between the substances of living organisms. The task is a formidable one since even the simplest one-celled bacterium is made up of thousands of substances of tremendous variety. Some are ions of simple composition. Many are large, complex molecules containing thousands of atoms. Most are involved in intricate sequences of chemical changes which have not yet been completely characterized. Much present research is devoted to unraveling molecular structures and tracking down chemical interactions. **Biochemistry** is the study of the chemistry of plants and animals. Despite the enormous problems, discoveries are following in quick succession.

About 60 percent of the adult human body is water. What else is present? This chapter surveys the principal kinds of substances found in human beings and other living organisms. We begin with the carbohydrates.

10-2 WHAT ARE CARBOHYDRATES?

A. Glucose

Carbohydrates are among the most important classes of substances in plants and animals. Their principal function in human beings is to provide energy for the body. A simple definition of the term carbohydrate is not easy to formulate (we shall get to one in Sec. 10-2G). As a beginning we consider the structure of the most important carbohydrate, the white crystalline solid glucose. Its structure, fortunately, is relatively simple as carbohydrate structures go. Glucose is one of the carbohydrates known as sugars, and it is widely distributed in plant juices. A small concentration is normal in human blood, about 0.1 percent. Aqueous solutions of glucose are fed directly into the bloodstream in intravenous feeding.

The first features we notice in the glucose formula are (1) the skeletal chain of six carbon atoms running down the middle, (2) the aldehyde group at the top, and (3) the five alcohol groups, one on each carbon atom below the aldehyde group. (Four of these have been placed on the right side of the carbon atom chain, and one on the left. Although this is of some significance, we may ignore it.) Since alcohol groups are also called hydroxyl groups, glucose can correctly be described as a polyhydroxyaldehyde.

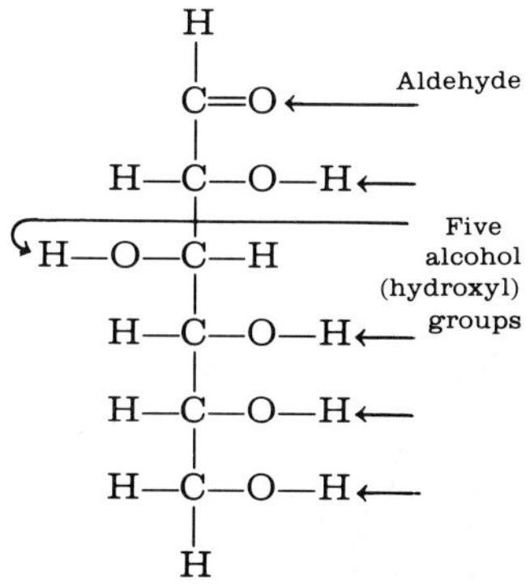

The structure shown, called the *open-chain* structure, however, is only one of three possible structures of glucose. In the other two the chain of carbon atoms forms an alpha (α) or beta (β) ring which includes one of the oxygen atoms. In forming the ring the

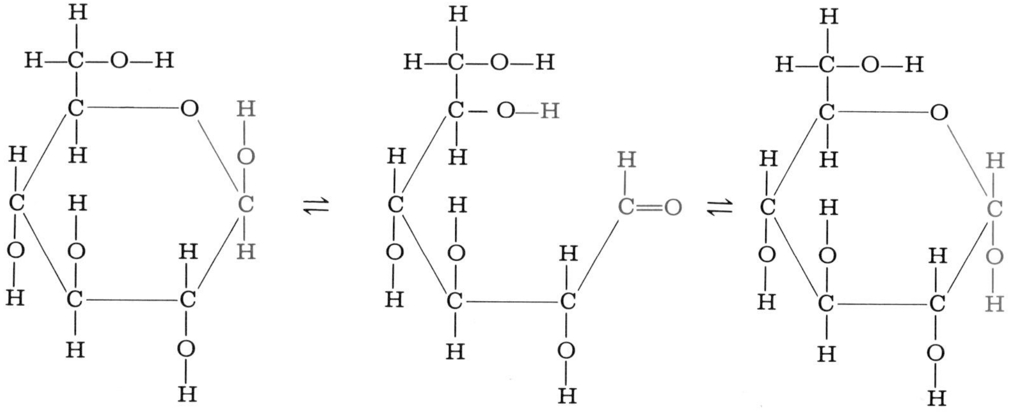

Functional groups in the "open-chain" structure of glucose, emphasizing its polyhydroxyaldehyde nature

| β-ring form (about 64%) | Open-chain form arranged to show relationship to other structures (less than 0.5%) | α-ring form (about 36%) |

The three forms of glucose present in an aqueous solution are in equilibrium with each other; in the ring forms the aldehyde group has been combined with an alcohol group to form a ring of five carbon atoms and one oxygen atom; the aldehyde group has been modified; the β-ring form differs from the α-ring form only in the spatial arrangement of the hydroxyl group and the hydrogen atom on the carbon atom at the right (color).

aldehyde group of the open-chain structure is converted into a modified aldehyde group. The molecules of crystalline glucose contain one of the ring structures. An aqueous solution contains all three forms in an equilibrium mixture.

B. Fructose

As a second step toward a definition of a carbohydrate we consider the structure of another sugar, the white crystalline solid fructose, widely distributed in nature (honey is a mixture of fructose and glucose). Fructose, incidentally, is considered the sweetest of all the sugars. Like the open-chain form of glucose, the formula has a skeletal chain of six carbon atoms which is easily recognized. It is also fairly easy to spot the five alcohol, or hydroxyl, groups. The molecule of the open-chain form of fructose also contains a functional group which is new to us—the ketone group.

Since fructose contains five hydroxyl (alcohol) groups and a ketone group, it is known as a polyhydroxyketone. Like glucose, fructose also exists in ring forms, of which we consider only one. The ring contains four carbon atoms and an oxygen atom, and in its formation the ketone group is converted into a modified ketone group.

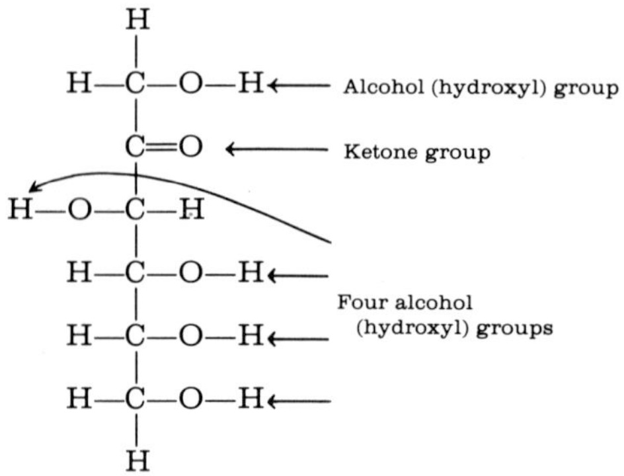

Functional groups in the
open-chain structure of
fructose, emphasizing its
polyhydroxyketone nature.

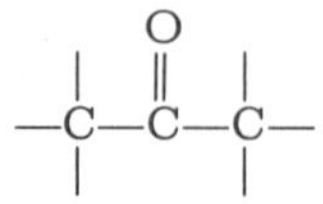

The ketone functional group

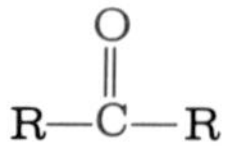

General formula of a ketone;
both R groups must contain
at least one carbon atom
attached directly to the
carbon atom of the ketone group

**Representations of a ketone
functional group**

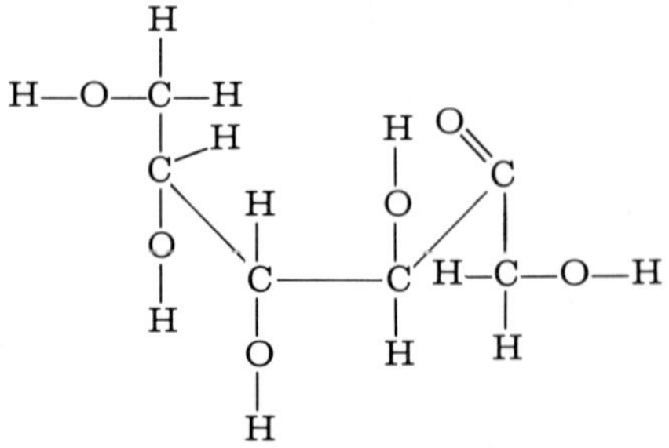

Open-chain form written
to show its relationship to
ring structure

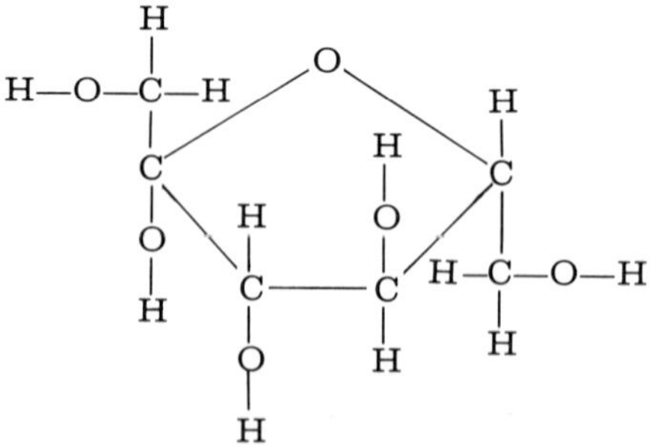

One of the ring forms; the
ketone group has been modified

Structural formulas representing two of the forms of fructose

C. Maltose

As a third step toward understanding carbohydrates we consider the structure of the sugar maltose, which like glucose and fructose is a white crystalline solid. Maltose is not as widely distributed in nature as glucose and fructose, but it is present, for example, in sprouting grain seeds. It can also be obtained from the hydrolysis of starch.

It is informative to consider the structure of maltose in relation to the structure of glucose:

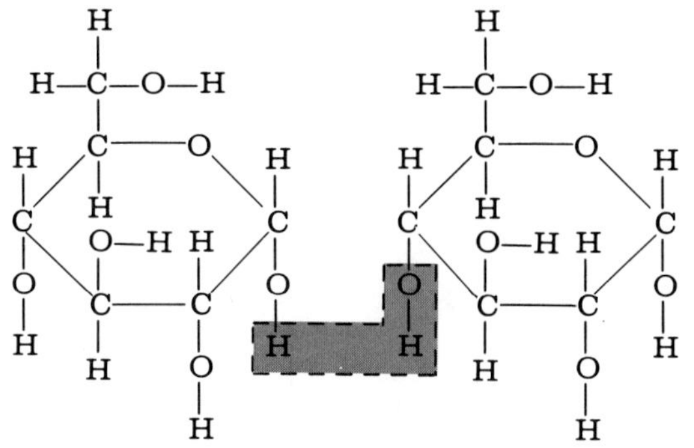

Structures of two molecules of the
α-ring structure of glucose

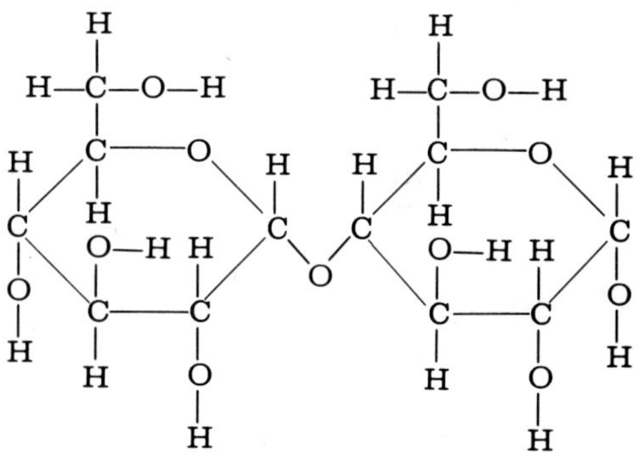

Structure of a molecule of maltose

Comparison of the structure of a molecule of maltose with the structures of two molecules of glucose; the structure of maltose can be obtained by removing the atoms of a molecule of water from two molecules of glucose

Since the atoms of a molecule of water are removed as a molecule of maltose is formed from two molecules of glucose, it is not surprising that the action of water can reverse the process and serve to form two molecules of glucose from a molecule of maltose. This reaction is another example of a hydrolysis reaction (Sec. 8-2).

D. Sucrose

We now examine the structure of another sugar, sucrose. This is the substance of the common sweetening agent table sugar. Sucrose is found in many plants, notably sugar cane and sugar beets. Most of our supplies of table sugar come from these two plants. Sugar cane is about 15 to 20 percent by weight sucrose. Sugar beets average from 10 to 20 percent sucrose.

It is helpful to relate the structure of sucrose to the structures of glucose and fructose:

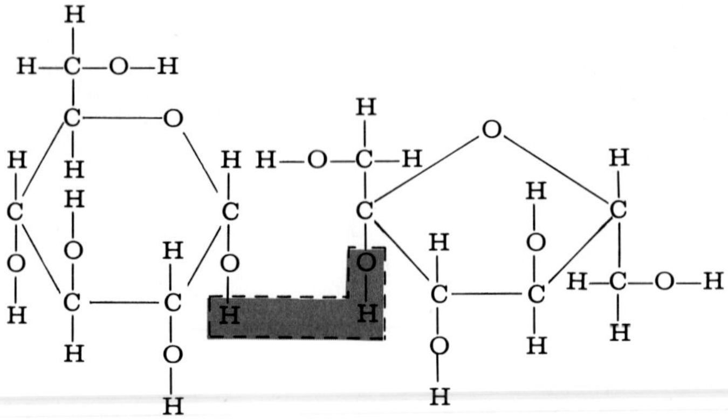

Glucose   Fructose

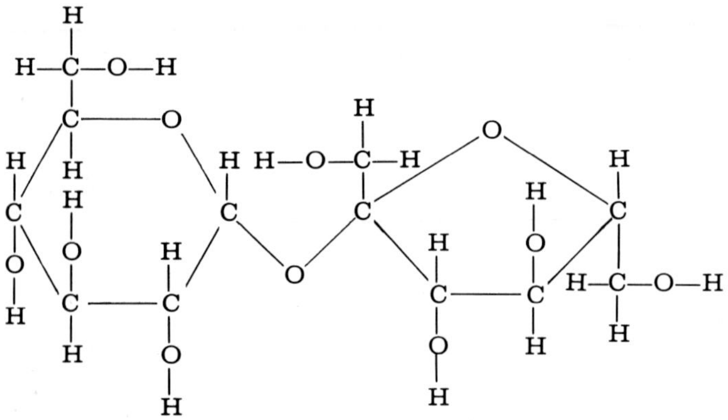

Sucrose

**Comparison of the structure of a molecule of sucrose with that of a
molecule of glucose and fructose; the structure of sucrose can be
obtained by removing the atoms of a molecule of water from one
molecule of glucose and one molecule of fructose**

Like maltose, a molecule of sucrose readily undergoes hydrolysis to form a molecule of glucose and a molecule of fructose.
Such reactions are important in the digestion of sugars (Sec.
11-5).

E. Starch and Glycogen

The structure of starch is of special importance in understanding carbohydrates. Starch is a tasteless white solid only partially soluble in water. It is found widely distributed in the
seeds, roots, and fibers of plants. Seeds contain up to 70 percent
starch and roots up to 30 percent. Starch is the most significant

source of carbohydrates in the human diet. Supplies for commercial uses are obtained principally from corn, potatoes, wheat, and rice. They are used in the production of sweetening agents, adhesives, stiffening agents, and explosives.

The composition of starch depends on its source. In general, it contains two different groups of molecules. One group, collectively called amylose, is soluble in water and makes up about 20 percent. The other group, collectively called amylopectin, does not dissolve in water and makes up about 80 percent. All starch molecules are polymers containing large molecules with the α-ring form of glucose as their repeating structural unit. (A polymer was defined in Sec. 7-9 as a substance whose molecules are made up of many recurring structural units held together by covalent bonds.)

The structure of the amylose (water-soluble) molecules is simpler to describe than that of the amylopectin (water-insoluble) molecules. Amylose molecules are polymers of glucose containing between 1000 and 4000 glucose units in a molecule. The glucose units are in the α-ring form and are joined together as they are in maltose:

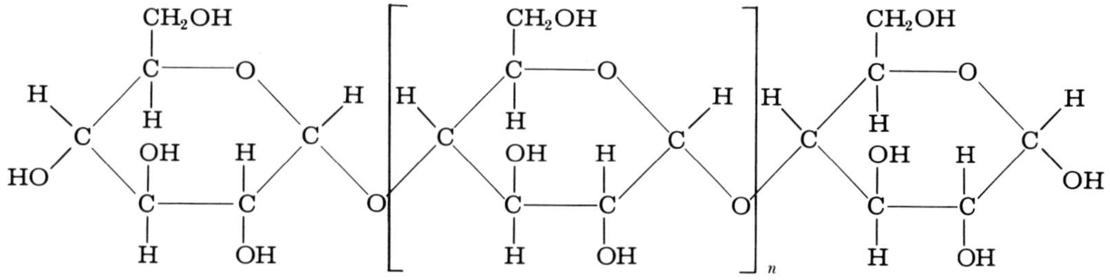

Representation of the structure of amylose starch molecules; they are polymers of the α-ring form of glucose; the value of n varies from about 1000 to 4000

The useful schematic representation of part of an amylose starch molecule below emphasizes the straight-chain arrangement of the glucose units:

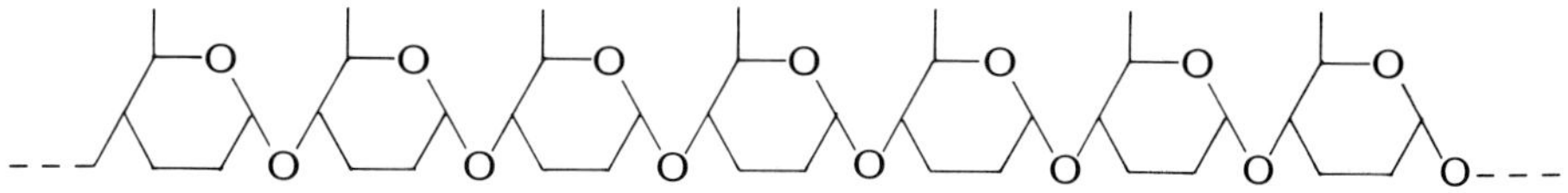

Schematic representation of part of an amylose starch molecule, emphasizing the straight-chain arrangement of α-glucose units

The molecules of amylopectin (water-insoluble) are also polymers of α-glucose units, but they contain many more glucose units than amylose. Evidence suggests a number in the range of

6000 to 40,000. Moreover, the glucose units are linked in a branched-chain manner, represented schematically as follows:

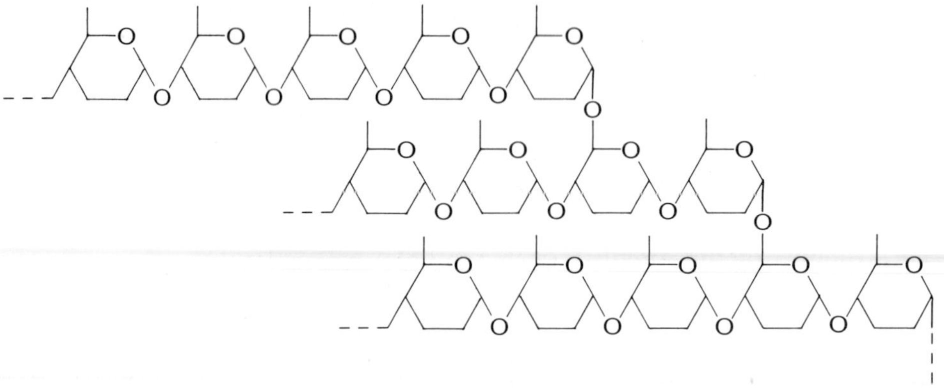

Schematic representation of a portion of an amylopectin starch molecule emphasizing the branched-chain nature of the linkages; between 6000 and 40,000 glucose units are found in individual molecules

Glycogen is the name given to the group of starch molecules found in animals. They are formed in the animal from α-glucose units and serve as a reserve supply of carbohydrates for the animals' chemical processes. Their structure apparently resembles the structure of amylopectin molecules except that there is more chain branching in the overall arrangement.

Amylose starch, amylopectin starch, and glycogen are all readily susceptible to hydrolysis. Many different products can be obtained, depending on conditions, but if all the links between the fundamental structural units are broken, all three classes are converted into a single substance, glucose. Hydrolysis reactions are significant in the digestion of carbohydrate foodstuffs (Sec. 11-5).

F. Cellulose

Cellulose is a principal component of plants. Most woods contain over 50 percent, and cotton fiber is 98 percent cellulose. Wood pulp, the principal raw material of paper production, is prepared by removing the binding materials from wood and consists almost entirely of cellulose.

Cellulose, like starch, is a polymer of a large number of glucose units. Evidence suggests that there are at least 1500 units in a molecule, and the number may go to 5000 and beyond. The glucose units of cellulose, however, have the β-ring form, and not the α-ring form of the glucose units in starch. This means that while the atom sequence is the same, the arrangement in space is different, as suggested in the formula

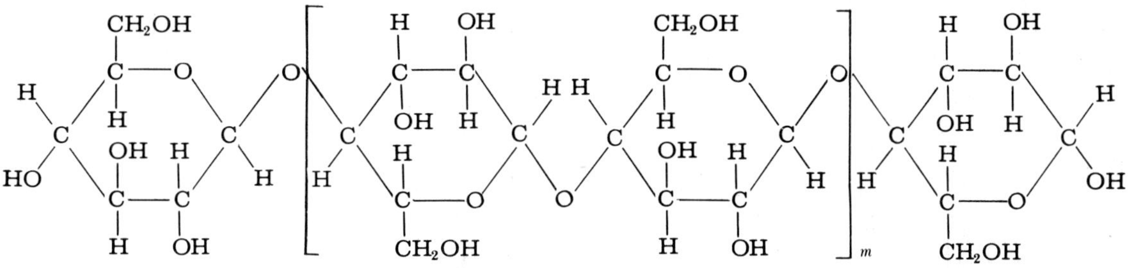

Representation of the structure of cellulose molecules; they are polymers of the β-ring form of glucose; the value of *m* varies over the approximate range of 700 to 3000

Cellulose, like starch, can be hydrolyzed into its glucose units. Human beings and flesh-eating animals cannot use cellulose as a source of glucose because their digestive fluids lack the necessary enzymes to catalyze the hydrolysis reactions.

G. What, Then, Is a Carbohydrate?

Now that we have considered the structures of some important and representative carbohydrates, what definition of the term carbohydrate will embrace them all? In considering the structure of glucose we learned that from a functional group viewpoint the open-chain form is a polyhydroxyaldehyde. The ring forms contain modified aldehyde groups and hydroxyl groups. The open-chain form of fructose proved to be a polyhydroxyketone, and in the ring form we considered it contains a modified ketone group and hydroxyl groups. Maltose contains two glucose units, and sucrose contains a glucose and a fructose unit. Starch, glycogen, and cellulose are polymers containing hundreds of glucose units. **Carbohydrates** can therefore be defined for our purpose as substances with polyhydroxy molecules containing aldehyde groups, modified aldehyde groups, ketone groups, or modified ketone groups or molecules containing a sequence of such structural units from two to thousands. This definition is not only cumbersome, it is not completely adequate. Taken together with the preceding discussion of carbohydrate structures, however, it should nevertheless give a useful idea of the nature of this important but complex class of substances.

A **monosaccharide** is a carbohydrate, like glucose and fructose, which contains one structural unit. A **disaccharide** is a carbohydrate, like maltose and sucrose, which contains two

---

Note to Students   The definition of a carbohydrate, as promised, but you can probably see why it was postponed.

---

structural units. A **polysaccharide** is a carbohydrate, like starch and cellulose, which contains many structural units. A **sugar** is a carbohydrate of simple structure, like glucose, fructose, maltose, and sucrose (Table 10-1). Table sugar itself is sucrose. Information about representative carbohydrates is summarized in Table 10-2.

10-3 WHAT IS THE NATURE OF FATS?

Fats make up about 18 percent of the total weight of the average adult. [The remainder is protein (about 16 percent), minerals (about 5 percent) and water (about 60 percent).] The amount of fat is variable, however, and in obese people fats may account for as much as 70 percent of the total weight. They are the most important means of storing a potential source of energy in animals. They are found in foods of both plant and animal origin, e.g., butter, margarine, fatty meats, nuts, and olives. Fats are also synthesized in the human body. They are important commercially, e.g., as the starting substances in the production of soap (Sec. 13-8) and in the preparation of some detergents (Sec. 13-9).

The structural formula of a typical fat will demonstrate the structural characteristics of this class of substances:

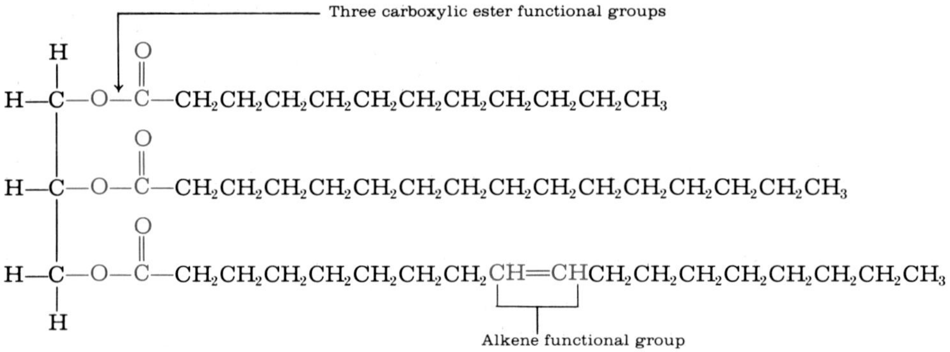

Structural formula of a typical fat with the functional groups identified

Table 10-1
Classes of Carbohydrates

Name of class	Structural characteristic	Examples
Monosaccharide	One structural unit	Glucose, fructose
Disaccharide	Two structural units	Maltose, sucrose
Polysaccharide	Polymers up to thousands of structural units	Starch, glycogen, cellulose
Sugar	Simple structure; usually only one, two, or three structural units	Glucose, fructose, maltose, sucrose

Table 10-2
Some Representative Carbohydrates

Class and name	Molecular formula	Comment
Monosaccharides:		
Glucose	$C_6H_{12}O_6$	Also called dextrose; in plant juices and honey; about 75% as sweet as sucrose (table sugar); only sugar present in human blood (0.1%); aqueous solution fed intravenously to the ill; obtained from complete hydrolysis of starch and cellulose
Fructose	$C_6H_{12}O_6$	Only naturally occurring polyhydroxy ketone; found in fruits and honey; sweetest of all sugars
Disaccharides:		
Maltose	$C_{12}H_{22}O_{11}$	Not widespread in nature but found in germinating grain; principal product of the enzyme-catalyzed hydrolysis of starch; hydrolysis yields glucose
Sucrose	$C_{12}H_{22}O_{11}$	Also called cane sugar; common table sugar, widespread in plants; chiefly used as food sweetener; world production exceeds 60 million tons; hydrolysis yields glucose and fructose
Polysaccharides:		
Starch	$(C_6H_{10}O_5)_n$	Found in most plants; most important source of carbohydrates in diet; over 3 billion pounds isolated per year in United States from corn; used in commercial production of adhesives, stiffening agents, and explosives; converted by hydrolysis into corn syrup for use as sweetening agent
Glycogen	$(C_6H_{10}O_5)_n$	Also called animal starch; reserve carbohydrate of animals; synthesized in body from glucose and stored in liver and muscles
Cellulose	$(C_6H_{10}O_5)_n$	Found in all plants; principal structural component of plant cell walls; most abundant of all carbohydrates; used in production of paper and as a starting material for making plastics (cellulose acetate) and explosives (gun cotton); cotton fiber almost pure cellulose

Since such formulas occupy so much space, an abbreviated notation has been devised. Its nature will be apparent if we use it to rewrite the formula for the fat molecule under consideration. The structure of this fat is representative of all fats.

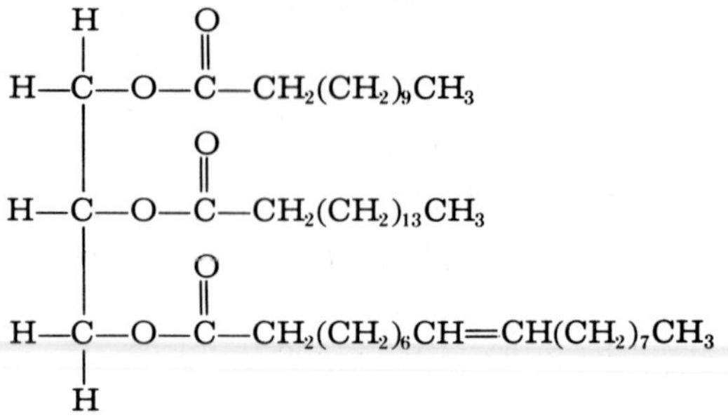

$$
\begin{array}{l}
\text{H}\quad\;\;\text{O} \\
|\quad\;\;\; \| \\
\text{H---C---O---C---CH}_2(\text{CH}_2)_9\text{CH}_3 \\
| \\
\quad\;\;\;\text{O} \\
\quad\;\;\;\| \\
\text{H---C---O---C---CH}_2(\text{CH}_2)_{13}\text{CH}_3 \\
| \\
\quad\;\;\;\text{O} \\
\quad\;\;\;\| \\
\text{H---C---O---C---CH}_2(\text{CH}_2)_6\text{CH}{=}\text{CH}(\text{CH}_2)_7\text{CH}_3 \\
| \\
\text{H}
\end{array}
$$

Structural formula for the same fat as above using an abbreviated notation to save space [(CH$_2$)$_9$ stands for nine connected CH$_2$ groups, etc.]

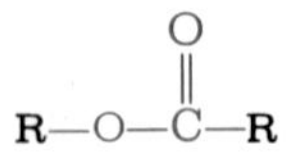

$$
\begin{array}{c}
\text{O} \\
\| \\
\text{R---O---C---R}
\end{array}
$$

General formula for a carboxylic ester

The formula shows that they are carboxylic esters of an alcohol with three hydroxyl groups known as glycerol. Let us recall the general formula for a carboxylic ester (Sec. 8-5). One of the reactions which carboxylic esters undergo is a reaction with water, known appropriately as a hydrolysis reaction. (Compare the hydrolysis of amides, Sec. 8-2.) This reaction can be represented in general terms as

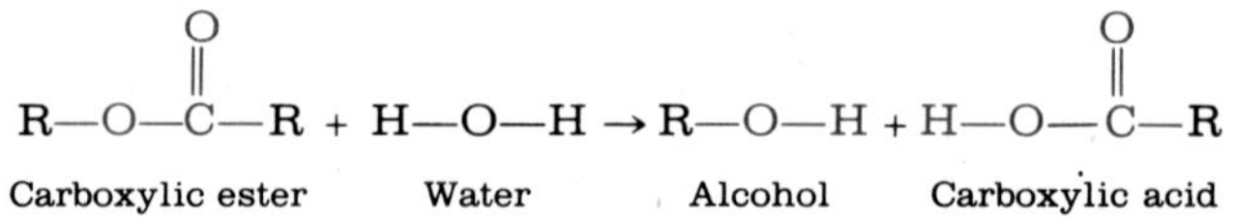

$$
\begin{array}{ccccc}
\text{O} & & & & \text{O} \\
\| & & & & \| \\
\text{R---O---C---R} & + \;\text{H---O---H} \rightarrow & \text{R---O---H} & + & \text{H---O---C---R}
\end{array}
$$

Carboxylic ester Water Alcohol Carboxylic acid

Representation in general terms of the hydrolysis of a carboxylic ester; the products are an alcohol and a carboxylic acid

---

Note to Students

It would probably be a good idea at this point to review at least part of Sec. 8-5 on the recognition of functional groups.

This general representation reveals that the hydrolysis of a carboxylic ester yields an alcohol and a carboxylic acid. Let us examine the representation more closely, paying particular attention to which bonds break and which bonds form during the reaction:

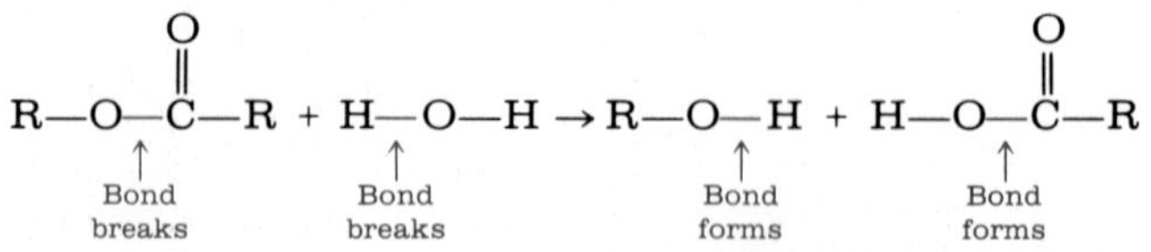

$$
\begin{array}{ccccc}
\text{O} & & & & \text{O} \\
\| & & & & \| \\
\text{R---O---C---R} & + \;\text{H---O---H} \rightarrow & \text{R---O---H} & + & \text{H---O---C---R}
\end{array}
$$

Bond breaks Bond breaks Bond forms Bond forms

Representation in general terms of the hydrolysis of a carboxylic ester, with emphasis on breaking and forming bonds

For the hydrolysis of a specific carboxylic ester:

$$CH_3CH_2O\overset{O}{\overset{\|}{-C}}-CH_3 + H-O-H \rightarrow CH_3CH_2O-H + H-O-\overset{O}{\overset{\|}{C}}-CH_3$$

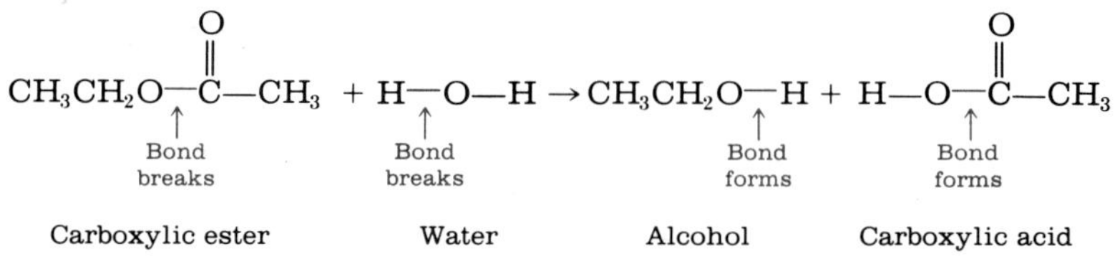

Bond breaks	Bond breaks	Bond forms	Bond forms
Carboxylic ester	Water	Alcohol	Carboxylic acid

Hydrolysis of a specific carboxylic ester

We are now prepared to examine the results of the hydrolysis of the typical fat molecule we have chosen to consider, using the abbreviated notation to save space:

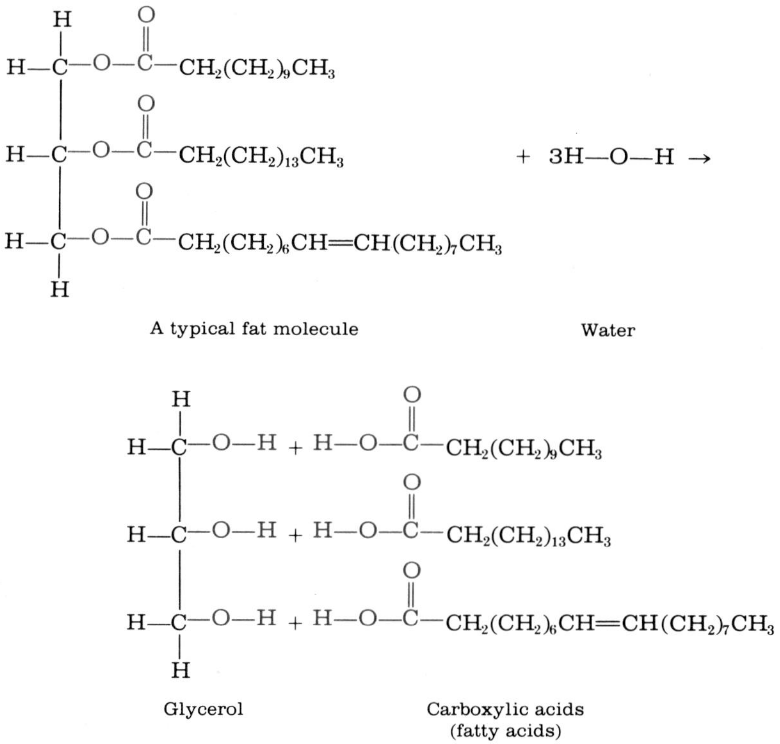

$$H-\overset{\overset{\displaystyle H}{|}}{\underset{|}{C}}-O-\overset{O}{\overset{\|}{C}}-CH_2(CH_2)_9CH_3$$

$$H-\overset{|}{\underset{|}{C}}-O-\overset{O}{\overset{\|}{C}}-CH_2(CH_2)_{13}CH_3 \qquad\qquad + \; 3H-O-H \;\rightarrow$$

$$H-\overset{|}{\underset{\underset{\displaystyle H}{|}}{C}}-O-\overset{O}{\overset{\|}{C}}-CH_2(CH_2)_6CH{=}CH(CH_2)_7CH_3$$

A typical fat molecule   Water

$$H-\overset{\overset{\displaystyle H}{|}}{\underset{|}{C}}-O-H + H-O-\overset{O}{\overset{\|}{C}}-CH_2(CH_2)_9CH_3$$

$$H-\overset{|}{\underset{|}{C}}-O-H + H-O-\overset{O}{\overset{\|}{C}}-CH_2(CH_2)_{13}CH_3$$

$$H-\overset{|}{\underset{\underset{\displaystyle H}{|}}{C}}-O-H + H-O-\overset{O}{\overset{\|}{C}}-CH_2(CH_2)_6CH{=}CH(CH_2)_7CH_3$$

Glycerol   Carboxylic acids
(fatty acids)

The hydrolysis of a typical fat molecule; the alcohol molecule produced is a substance with three hydroxyl groups, called glycerol; the molecules of three different carboxylic acids are also produced

This reaction leads to the general statement that the hydrolysis of fat molecules yields a molecule of glycerol and three carboxylic acid molecules. The carboxylic acids obtained are given the special name fatty acids and have some distinctive characteristics: (1) they contain a long chain of carbon atoms, usually with at least 12 carbon atoms; (2) they almost always contain an even number of carbon atoms; and (3) they may contain several alkene groups or none.

We are now in a position to define what we mean by a fat. A **fat** is a carboxylic ester of the alcohol glycerol and three long-chain carboxylic acids. The **carboxylic acid residue** is that part of a carboxylic acid molecule incorporated into a carboxylic ester. The carboxylic acid residues may be the same in a specific molecule, but usually they are different. The **alcohol residue,** or **glyceryl residue,** is that part of the glycerol molecule which is incorporated into the carboxylic ester.

If the fat is a liquid at room temperature (about 20°C), it is customary to refer to it as an oil. Fats obtained from plants are usually liquids at room temperature, whereas fats from animal sources are usually solids. Thus it is common to speak of vegetable oils and animal fats.

Although all fats have the same general structure as the typical fat molecule we have considered, there are many different fats. Since the alcohol residue incorporated into the fat is always derived from glycerol, the differences in the structures of fats must be found in the carboxylic acid residues (or fatty acid residues) involved. Table 10-3 lists fatty acids whose residues appear most frequently in fats of animal and vegetable origin.

A **saturated fatty acid** is one with no alkene groups. An **unsaturated fatty acid** is one which contains one or more alkene groups. An **unsaturated fat** or **polyunsaturated fat** is one which contains substantial amounts of unsaturated fatty acid residues, and a **saturated fat** is one which contains fatty acid residues which are all or almost all saturated. These terms have recently become familiar because of the possible connections between saturated fats in the diet and diseases of the arteries (Chap. 11).

The melting points of fatty acids in Table 10-3 illustrate a significant generalization: the presence of alkene groups lowers the melting points of fatty acids. Stearic acid (18 carbon atoms, no alkene groups) has a melting point of 70°C. Linoleic acid (18 carbon atoms, two alkene groups) has a melting point of 75°C lower, at −5°C. Fats with a high proportion of saturated fatty acid residues tend to have melting points above room temperature and to be solids under ordinary conditions. Fats with a high proportion of unsaturated fatty acid residues tend to have melting points below room temperature and to be liquid oils under ordinary conditions. Butter, with about 70 percent saturated residues and 30 percent unsaturated residues, is a solid. Corn oil, with about 20 percent saturated residues and about 80 percent unsaturated residues, is a liquid. The fatty acid residue composition of some common fats is given in Table 10-4.

10-4 WHAT ARE PROTEINS?

Proteins are among the most important substances found in living systems. The name is appropriately derived from the Greek word *proteios* meaning "of first importance." Collectively proteins make up 40 percent of the solid material of the human body. Brief mention has already been made (Sec. 8-2) of their function as the structural substances of skin and muscle and

Table 10-3
Fatty Acids Whose Residues Occur Most Frequently in Fats

Name	Number of carbon atoms	Abbreviated formula	Melting point, °C	Fat containing substantial amount of residue
Lauric acid	12	$CH_3(CH_2)_{10}COOH$	44	Coconut oil
Myristic acid	14	$CH_3(CH_2)_{12}COOH$	58	Butter fat
Palmitic acid	16	$CH_3(CH_2)_{14}COOH$	63	Palm oil
Stearic acid	18	$CH_3(CH_2)_{16}COOH$	70	Beef tallow
Oleic acid	18	$CH_3(CH_2)_7CH=CH(CH_2)_7COOH$	13	Olive oil
Linoleic acid	18	$CH_3(CH_2)_4CH=CHCH_2CH=CH(CH_2)_7COOH$	−5	Corn oil

Table 10-4
Fatty Acid Residue Composition of Representative Samples of Some Common Fats; Percentages by Weight

Fat	Lauric (12 C atoms, no alkene groups)	Myristic (14 C atoms, no alkene groups)	Palmitic (16 C atoms, no alkene groups)	Stearic (18 C atoms, no alkene groups)	Oleic (18 C atoms, 1 alkene group)	Linoleic (18 C atoms, 2 alkene groups)	Other
Coconut oil	48	18	9	2	6	3	14
Palm oil		1	48	4	38	9	
Olive oil		1	13	4	28	54	
Cottonseed oil		1	29	4	24	40	2
Butter fat	4	12	29	11	25	2	17
Beef tallow		3	25	24	42	2	4
Human fat	1	3	25	8	46	10	7

of their special role as catalysts for the many reactions of
living cells. Besides serving as structural materials, they are
involved in the transport of essential substances within the
living organism, in the regulation of essential reactions, and
in the protection of the organism against harm.

$$-\overset{\overset{\displaystyle O}{\|}}{C}-\overset{\displaystyle |}{N}-$$

Amide or peptide group

A. The Component Amino Acid Residues

The molecules of proteins are characterized by their large size
and complexity. Structures with thousands of atoms are common. Despite the structural intricacies they all have a characteristic arrangement of atoms in common, a backbone of carbon
atoms linked by amide groups (Sec. 8-2). When it occurs in
proteins, the amide group is frequently called a peptide group.
A portion of the backbone structure can be represented as
follows:

$$
--- -\overset{\overset{\displaystyle H}{|}}{N}-\overset{\overset{\displaystyle R}{|}}{\underset{\underset{\displaystyle H}{|}}{C}}-\overset{\overset{\displaystyle O}{\|}}{C}-\overset{\overset{\displaystyle H}{|}}{\underset{\underset{\displaystyle H}{|}}{N}}-\overset{}{\underset{\underset{\displaystyle R}{|}}{C}}-\overset{\overset{\displaystyle H}{}}{\underset{\underset{\displaystyle O}{}}{C}}-\cdots
$$

General representation of a portion of the backbone chain of
atoms in proteins, which may extend for hundreds of atoms

The R symbols in this general formula represent hydrogen
atoms or groups of atoms called side chains. Each protein has
its own characteristic sequence of side chains, which impart a
specific chemical identity.

All amide groups are very susceptible to hydrolysis. In
general, when an amide group is hydrolyzed, the products are
a carboxylic acid and an amine (Sec. 8-2):

$$
R-\underset{\underset{\displaystyle O}{\|}}{C}-\underset{\underset{\displaystyle H}{|}}{N}-R + H-O-H \rightarrow R-\underset{\underset{\displaystyle O}{\|}}{C}-O-H + H-\underset{\underset{\displaystyle H}{|}}{N}-R
$$

Amide   Water   Carboxylic   Amine
                              acid

General representation of the hydrolysis of an amide (Sec. 8-2)

By analogy the products of the hydrolysis of proteins can be
anticipated:

$$
--- -\overset{\overset{\displaystyle H}{|}}{N}-\overset{\overset{\displaystyle R}{|}}{\underset{\underset{\displaystyle H}{|}}{C}}-\overset{\overset{\displaystyle O}{\|}}{C}-\overset{\overset{\displaystyle H}{|}}{\underset{\underset{\displaystyle H}{|}}{N}}-\overset{}{\underset{\underset{\displaystyle R}{|}}{C}}-\overset{\overset{\displaystyle H}{}}{\underset{\underset{\displaystyle O}{}}{C}}-\cdots
$$

Bond   Bond   Bond   Bond   Bond
breaks   breaks   breaks   breaks   breaks

Application of the general representation of the hydrolysis
of an amide to the hydrolysis of the amide groups in proteins

The products of the hydrolysis of this particular structural fragment are substances which can be represented in general terms as shown in the margin. As we saw in Sec. 8-3, each product of the hydrolysis is an amino acid, i.e., the substance contains an amine group and a carboxylic acid group. Since the amino groups are attached to what we call the alpha (α) carbon atom in the nomenclature of carboxylic acids, these acids are also known as α-amino acids. (The structure of amino acids is discussed further in the optional Sec. 10-10.)

The **amino acid residue** is that part of an amino acid molecule found in a protein molecule (margin). The residues of somewhat more than 20 different amino acids occur commonly in proteins. Still others are found in individual proteins. The names and structures of 24 important amino acids are given in Supplement 7. Table 10-5 gives the structures and other information about six representative amino acids.

Table 10-5
Representative Amino Acids

Amino acid	Comment
Glycine	One of most common building blocks of proteins, constituting for example, about one-third of the amino acid residues in collagen, a structural protein of skin, tendons and, cartilage; one of the principal "ancestor" substances from which compounds of living cells have evolved
Alanine	One of the most common building blocks of proteins, constituting for example, more than 40 percent of the amino acid residues in fibroin, the protein of natural silk; one of the "ancestor" substances from which compounds of living cells have evolved
Phenylalanine	Residues occur, among other proteins, in pepsin, the enzyme protein involved in the digestion of proteins, and in myosin, a protein of muscle tissue; one of the seven amino acids which must be in the diet for normal health
Lysine	One of the most common building blocks of proteins, occurring, for example, in myosin, a protein of muscle tissue; one of the seven amino acids which must be in the diet for normal health; another "ancestor" amino acid; note that it contains two amino groups

(Continued on next page)

General representation of the products of the hydrolysis of a portion of a protein molecule

A typical product of the hydrolysis of a protein molecule; since it contains both an amine group and a carboxylic acid group, it is known appropriately as an amino acid

Amino acid molecule

Amino acid residue

Comparison of an amino acid molecule and an amino acid residue

243

Table 10-5
Representative Amino Acids (*Continued*)

Amino acid	Comment
$$\begin{array}{ccccc} & O & H & H & O \\ & \parallel & \mid & \mid & \parallel \\ H-O-&C-&C-&C-&C-O-H \\ & & \mid & \mid & \\ & & H & N-H \\ & & & \mid & \\ & & & H & \end{array}$$ Aspartic acid	One of the most commonly occuring building blocks of proteins; found, for example, in keratin, the protein of hair and nails; another "ancestor" substance; note that it contains two carboxylic acid groups
$$\begin{array}{cccc} H & H & O \\ \mid & \mid & \parallel \\ H-C-&C-&C-O-H \\ \mid & \mid & \\ & N-H \\ S & \mid & \\ \mid & H & \\ S & H & O \\ \mid & \mid & \parallel \\ H-C-&C-&C-O-H \\ \mid & \mid & \\ H & N-H \\ & \mid & \\ & H & \end{array}$$ Cystine	One of the residues in, for example, the protein of human hair and the hormone insulin; unique in containing two amino groups and two carboxylic acid groups, permitting it to play a special role in overall shape of protein molecules (Sec. 10-9); largely responsible for whether hair is straight or curly

A protein molecule, then, contains a long chain of amino acid residues, typically between 50 and 1000 (or more) linked together in a specific sequence. Not all known amino acid residues are present in every protein molecule, and some proteins contain a high proportion of a given residue. For example,

WHAT IS LIFE?

What is meant by a living organism in contrast to an aggregation of molecules which is not living? It is easy enough to distinguish a living dog from the couch on which he is lying or a tree from a wooden chair. But a definition of life which distinguishes clearly between the living and the nonliving and applies to all cases is not easily formulated. This is especially true if very small forms of life are taken into consideration.

We can take a step toward a definition of life by considering the principal characteristics of those plants and animals which can be readily recognized as living. One such characteristic is reproduction, an ability which all living organisms have in common. Another is the capacity for growth, for development in size and substance. Growth leads directly to the capacity for metabolism, i.e., the ability to break down nutrients and from them synthesize the substances and generate the energy required by growth. Other properties associated with most forms of life, if not all, are an ability to move, a responsiveness to stimulation, and a capacity for adapting to changes in the environment.

In describing these characteristics we have been moving toward a definition of life in terms of the processes associated with living organisms. The minimum requirements of a living system are probably the capacity for self-replication and the ability to carry out the reactions required by metabolism. As evidences of life have been investigated and evaluated, scientists have discovered that the transition from aggregations of molecules without life to aggregations with the characteristics of life is a gradual and continuous one. But it is also keenly appreciated that even the smallest and simplest living organisms contain an extraordinary variety of complex substances in large numbers. One cell of a certain kind of bacteria, for example, is estimated to contain about 3000 different proteins, about 1000 different nucleic acids (Sec. 10-6), and at least 1000 other substances, including fats and carbohydrates. Moreover, these substances are organized into a dynamic system which is able to regulate and reproduce itself. Thus living organisms are not only aggregations of molecules, they are highly organized systems of complex structures involved in intricate sequences of chemical reactions which enable them to carry out the processes of life.

almost one-third of the amino acid residues in collagen, the fribrous structural protein, are glycine residues. The amino acid sequence in a protein molecule is of critical importance to its chemical behavior. Even though a protein may contain about 500 residues, a change in just one residue can bring about a marked change in properties.

Although the number of different amino acids isolated from proteins is only about 25, they can be assembled in an enormous number of combinations. Imagine, for example, a series of proteins with 50 amino acid residues in the polypeptide chain. If there are 20 different possibilities for each residue, the number of possible substances, each with a different amino acid sequence, becomes 20^{50}. The number of possible protein molecules is therefore almost infinitely large. A few representative proteins are listed in Table 10-6.

Table 10-6
Some Representative Proteins

Name	Number of amino acid residues†	Comment
Collagen	~3000	A principal structural protein, found in skin, cartilage, and bone; forms fibers with high tensile strength; most abundant protein in mammals (about 25% of total protein content); when boiled in water, it is converted into gelatin
Myosin	~5000	One of the major proteins of muscle fiber; coagulates on death to a hard material, causing rigor mortis
Ovalbumin	~300	One of the proteins stored in mammals for use when needed; found also in hens' eggs; when an egg is cooked, heat changes the structure and the protein coagulates (denaturation, Sec. 10-9)
Insulin	51‡	A hormone, secreted by the pancreas, which regulates reactions involved in use of glucose in mammals; deficiency causes diabetes; used to control this ailment
Serum albumin	~600	Protein found in human blood; serves to transport fatty acids from fatty tissues to other body areas
Hemoglobin	574‡	Protein which imparts red color to blood of vertebrates; serves to transport oxygen from the lungs to tissues
Hemocyanin	~3000–70,000	Protein found in blood of squid, octopuses, lobsters, and other invertebrates; serves to transport oxygen
Thrombin	~300	Protective protein, necessary for blood clotting
Pepsin	~30	Enzyme, found in stomach gastric juice, which catalyzes hydrolysis of peptide linkages during digestion of proteins
Ptyalin (α-amylase)	~850	Enzyme, found in saliva, which catalyzes hydrolysis of carbohydrates at start of digestion of foods
Luciferase	~800	Enzyme which catalyzes a reaction using oxygen to produce the luminescent light of fireflies
Phosphorylase	~3000	Enzyme which catalyzes breakdown of certain carbohydrates

† Some of the proteins contain a nonpolypeptide group. The approximate number of amino acid residues is given to suggest the size and complexity of the structure (the symbol ~ is read "approximately").
‡ Known exactly.

B. Polypeptides

Peptide bond is a special name given to the amide bonds of protein molecules. **Polypeptide structure** refers to the poly-amino acid residue structure of a protein. A **dipeptide** is a molecule containing two amino acid residues. Correspond-ingly, a tripeptide contains three; a tetrapeptide contains four, etc. Since **polypeptide** applies to molecules containing several amino acid residues, proteins contain polypeptide structures. By convention polypeptides with fewer than 40 amino acid residues are referred to as polypeptides, and those with more than about 40 residues are referred to as proteins, but this classification is not adhered to rigidly. Many relatively simple polypeptides, with as few as about eight amino acid residues, have significant physiological activity.

C. Prosthetic Groups

Some proteins contain a structural unit which is not a poly-peptide structure, called a **prosthetic group.** Proteins with such a group are called **conjugated proteins** and those without such a group are called **simple proteins.**

Prosthetic groups may be simple ions such as the zinc ion, Zn^{2+}, or they may be structurally complex units, such as the molecule heme, which forms the prosthetic group of hemo-globin, the protein which carries oxygen from the lungs to the tissues.

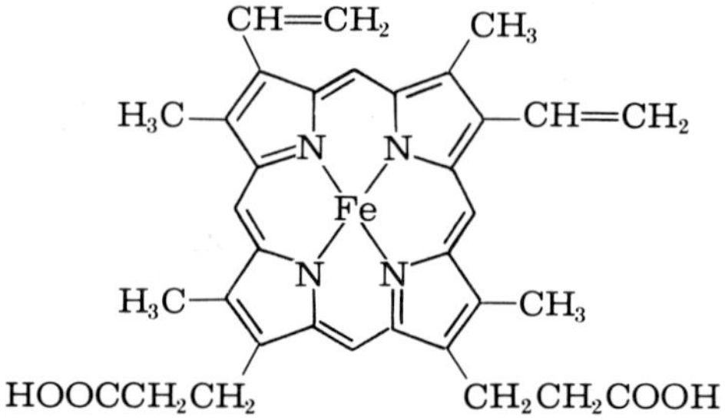

The structure of heme, the prosthetic group of the protein hemoglobin, which carries oxygen in the bloodstream

D. A Definition of the Term Protein

Having described what is meant by a polypeptide structure, a prosthetic group, and something of the significance of pro-teins, we are now in a position to consider a brief definition of the term protein. **Proteins** are substances with large molecules which contain a long chain of amino acid residues joined by peptide bonds (amide groups) and sometimes, in addition, a prosthetic (nonpolypeptide) structural unit. They perform many critically important functions in living cells, e.g., serving as structural materials, transporting essential substances, and catalyzing essential reactions.

Most of the reactions of living cells would take place very slowly, if at all, in the absence of catalysts to speed their rates. **Enzymes** are substances which catalyze the reactions of living systems. All enzymes are proteins, and the molecules of most of them are large enough to contain at least 100 amino acid residues. Their amazing catalytic power makes it possible for the numerous reactions essential to living organisms to take place at a practical rate under the rather mild temperatures and concentrations prevailing in plants and animals. Enzymes make it possible to transform hundreds of thousands of molecules in a minute, increasing the reaction rate at least 1 million times. Thousands of different enzymes are required by the chemical processes of the human body (Fig. 10-1).

One of the most remarkable characteristics of enzymes is that they are specific for the reactions they catalyze. A given enzyme will catalyze only reactions which are closely related, and many will catalyze only one reaction of a given compound. Pepsin, one of the enzymes catalyzing reactions involved in the digestion of proteins, catalyzes the hydrolysis of the amide linkages between certain amino acid residues only, even though all the amide groups are susceptible to hydrolysis. Ordinary catalysts of amide hydrolysis, such as hydrogen ion, do not discriminate between amide groups and catalyze the hydrolysis of them all.

Enzymes, like all catalysts, act only to change the rate at which reactions take place. But because the reactions which enzymes accelerate are the ones which will predominate, enzymes control the transformations which constitute the chemistry of life. Since enzymes catalyze the chemical reactions of living organisms which produce proteins, nucleic acids, fats and carbohydrates forming the organisms, speculations about the origin of enzymes play a central role in hypotheses about the origin of life.

An enzyme, like any catalyst, functions by providing a reaction mechanism which takes place faster than the mechanism

FIGURE 10-1
Crystals of the enzyme phosphorylase as seen through a light microscope. [From J. Biol. Chem., **151**:23 (1943), courtesy of Carl F. Cori.]

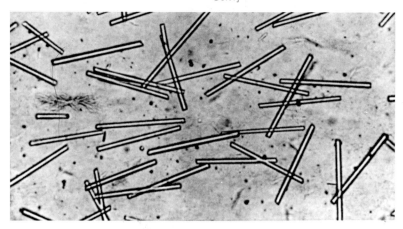

which would prevail in its absence. A **substrate** is a reactant whose participation in a specific reaction is catalyzed by an enzyme. The key step in the catalytic action of an enzyme involves its ability to form a temporary association with the substrate, and the weak association between enzyme and substrate is called an **enzyme-substrate complex.** Its formation is an intermediate step in the mechanism of a reaction catalyzed by an enzyme.

HOW DID LIFE BEGIN?

Investigations suggest that the earth was formed about 5 billion years ago as clouds of gases and dust condensed to form the planets of the solar system. The most abundant atoms on the surface of the very primitive earth were probably those of hydrogen, nitrogen, oxygen, and carbon, but the identity of the molecules into which they were incorporated is uncertain. It is frequently assumed, however, that the atmosphere of the primitive earth contained substantial amounts of nitrogen, N_2, hydrogen, H_2, ammonia, NH_3, methane, CH_4, and water, H_2O. Elemental oxygen was absent and did not appear until the process of photosynthesis developed.

It has long been speculated that out of this mixture of a few simple molecules the complex substances of living organisms somehow evolved. The question is: how? About 50 years ago the biochemists A. O. Oparin of the Soviet Union and J. B. S. Haldane of England suggested that the energy provided by lightning flashes or the ultraviolet rays of the sun brought about the conversion of the simple molecules of the early atmosphere into those of amino acids and sugars. These in turn interacted to form the more complicated molecules of living organisms. In 1953 the American chemist Stanley Miller continuously circulated a mixture of hydrogen, H_2, methane, CH_4, ammonia, NH_3, and water, H_2O, in a closed flask at 80°C for more than a week, through an electric sparking between electrodes in an imitation of lightning (Fig. 10-2), thus simulating the conditions of atmospheric composition and lightning thought to prevail in the early stages of the earth's existence. Analysis of the contents of the flask after the experiment identified the gaseous products carbon monoxide, CO, carbon dioxide, CO_2, and nitrogen, N_2, plus significant amounts of carboxylic acids and amino acids, including four of those whose residues are commonly found in proteins. Subsequent experiments by many investigators, using different mixtures of simple molecules and many different forms of energy likely to have been available in primitive times, have resulted in the formation of several hundred different carbon-containing substances, including simple sugars, all the amino acids commonly found in proteins, and the heterocyclic bases of nucleic acids (Sec. 10-6).

These provocative demonstrations of the probable origins of the building blocks of the substances of living organisms have led to further experimentation and some striking hypotheses about the emergence of the first living organisms.

Fossilized bacteria have been discovered in places which date them with some certainty back to about 3 billion years ago, and it is thought that living organisms may actually have come into existence several hundred million years earlier. One line of speculation suggests that the evolution of living systems took place in a series of developments, beginning with the formation of primordial molecules of simple sugars, amino acids, and heterocyclic bases. Such substances concentrated in the water of lakes or of the oceans, and out of this "primal soup" the proteins, polysaccharides, fats, and nucleic acid-like substances evolved from a sequence of chemical combinations. Gradually the basic molecules of a living system were formed, and from them developed the first organized molecular aggregations which could carry out the processes of life.

Of course, there are many questions to be answered. How, for example, did molecular systems emerge which were capable of supervising their self-replication? The provocative work of the American biochemist Sydney Fox and his colleagues may have provided at least part of an answer. They have found that when mixtures of dry amino acids are heated above 100°C, proteinlike polypeptides, called proteinoids, are formed. Moreover, when hot concentrated aqueous solutions of these proteinoids and certain salts are cooled, cell-like structures, called proteinoid microspheres, form spontaneously. If aqueous suspensions of these microspheres are allowed to stand, buds form on them after a week or two (Fig. 10-3). If such suspensions are heated, the buds are released from the microspheres. When the free buds are transferred to a saturated proteinoid solution, a second-generation microsphere is formed (Fig. 10-3c). In other words, the proteinoid microspheres spontaneously proliferate by a budding process. This work may very well provide answers to such questions as how the first cells came into existence and how the reproduction of cells began.

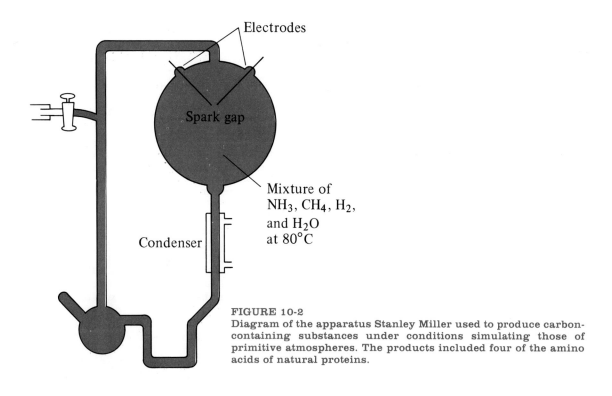

Electrodes

Spark gap

Mixture of
NH_3, CH_4, H_2,
and H_2O
at 80°C

Condenser

FIGURE 10-2
Diagram of the apparatus Stanley Miller used to produce carbon-containing substances under conditions simulating those of primitive atmospheres. The products included four of the amino acids of natural proteins.

FIGURE 10-3
The reproduction of cell-like proteinoid microspheres: (*a*) buds form spontaneously on parent microspheres standing in the solution in which they were formed; (*b*) buds are released as the solution is warmed; (*c*) a second-generation microsphere is formed when a bud is transferred to a saturated proteinoid solution; (*d*) bud forms on second-generation microsphere. (*Photomicrographs courtesy of Sydney Fox.*)

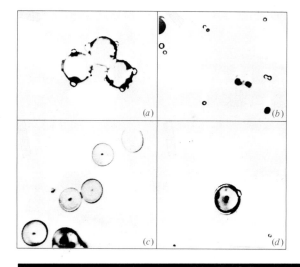

In overall terms the role of the formation of an enzyme-substrate complex can be summarized as follows:

Extremely slow
(no enzyme
present): Substrate → products

Extremely fast
(enzyme present): Substrate + enzyme ⇌ enzyme-substrate complex

Extremely fast: Enzyme-substrate complex → products + enzyme

**General representation of the role of enzyme-substrate formation
in the catalytic action of enzymes; the temporary association
between enzyme and substrate facilitates the reaction, enormously
increasing its rate, and the enzyme is regenerated**

Since the enzyme is regenerated for repeated use, a small amount of enzyme may have a great effect on the reaction rate.

The number of amino acid residues in an enzyme is very large, but only a few of them are involved in the critically important enzyme-substrate complex. The **active site** is the specific region on the surface of an enzyme to which the substrate is bonded in an enzyme-substrate complex. The three-dimensionality of the active-site region is important to its function. The region is usually located at a crack or a fissure on the surface of the enzyme. The atom sequence of the acid residues at the active site is an important factor in the formation of the complex although the attractive forces operating between the substrate and the active site are relatively weak.

The formation of an enzyme-substrate complex is represented by a useful but highly simplified schematic diagram in Fig. 10-4. In this drawing the substrate is given a specific shape which fits the shape of the active site on the enzyme. The two come together in a sort of lock-and-key fashion. Actually, however, the active site may change shape somewhat to accommodate the substrate, and the shape of the substrate may be strained to fit the active site. Still, the formation of the enzyme-substrate complex requires a reasonably close three-dimensional fit between the two. The spatial features of the substances of living systems, the bond lengths and the bond angles, are

FIGURE 10-4
Diagrammatic representation of the role of the enzyme-substrate complex in the catalytic action of an enzyme in a decomposition reaction. The geometries of the substrate and the active site are depicted as coming together in a lock-and-key fashion. Actually the shapes of both substrate and active site may change somewhat as the complex is formed.

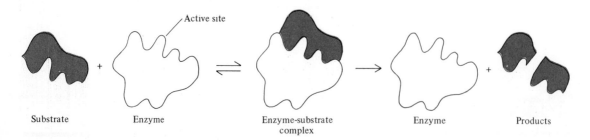

Substrate   Enzyme   Enzyme-substrate   Enzyme   Products
                           complex

therefore important in determining the mechanism of the reactions which occur.

The close fit required in forming a complex is the main reason why the action of enzymes is so specific. An enzyme will form a complex with a substrate only if the structure of the substrate complements the active site of the enzyme. Therefore each enzyme catalyzes only the reaction of one specific substance or a group of substances with close structural relationships. Yet the normal chemistry of a single living cell involves 1000 or more different reactions, almost all catalyzed by enzymes. The normal functioning of each cell therefore requires the action of many different enzymes.

10-6 WHAT ARE NUCLEIC ACIDS?

The sequence of amino acid residues in a protein molecule is of decisive importance to the properties exhibited by the protein. Perhaps no more dramatic illustration of this relationship can be found than in the properties of an abnormal variety of the protein hemoglobin. Normal hemoglobin is the protein of red blood cells which serves to transport oxygen from the lungs to the tissues. The molecules of normal human hemoglobin contain 574 amino acid residues. If only 2 of the 574 residues are of the wrong kind, the hemoglobin behaves abnormally even though all of the others are the correct residues in the correct sequence. The abnormal hemoglobin molecules tend to aggregate, causing the blood cells to become rigid and distorting their shape. Many assume a cresent or sickle shape instead of the round shape of normal red blood cells. The rigid cells block the smaller blood vessels, restricting the flow of blood and depriving the tissues of oxygen. This serious ailment, called sickle-cell anemia, is caused when only 2 of the 574 amino acid residues in hemoglobin are of the wrong identity.

SICKLE-CELL ANEMIA, A DISEASE CAUSED BY ABNORMAL PROTEIN MOLECULES

An estimated 2 million people in the world suffer from sickle-cell anemia, a disease causing recurring severe pain, fever, dizziness, skin ulcers, and malfunction of the liver. It strikes in childhood and shortens the life-span to about 30 or 40 years. The disease is inherited. If a gene for the disease (a segment of a DNA molecule) is inherited from only one parent, the child does not have the disease but carries the sickle-cell trait. To have sickle-cell anemia, a child must inherit genes for the disease from both parents.

The disease is especially prevalent in Africa. Occurrence of the sickle-cell gene among populations there correlates with the incidence of malaria. People with the sickle-cell trait are protected against some forms of malaria and have a greater chance of survival in malarial regions. The sickle-cell trait is found predominantly among black populations.

The disease has been widespread in malarial regions for centuries, but it was first characterized medically in 1910 by a Chicago physician, James B. Herrick. Its cause was tracked down to abnormal hemoglobin molecules in 1949, by a group of chemists at the California Institute of Technology under the leadership of Linus Pauling. The abnormality of the hemoglobin molecules was subsequently narrowed down to the substitution of only 2 of the 574 amino acid residues in normal hemoglobin by wrong residues. This work was done by Vernon Ingram, then at Cambridge University. For the first time the cause of a disease was understood in terms of molecular structure.

The presence of hemoglobin in the blood permits the blood to transport the large quantities of oxygen required by the human body from the lungs to the tissues. In this process a hemoglobin molecule combines with an oxygen molecule. The hemoglobin-oxygen combination dissociates when the oxygen is released to the tissues. The hemoglobin molecules are conjugated proteins containing the prosthetic group heme (Sec. 10-4C) and a polypeptide structure of amino acid residues. The molecules are contained in the red blood cells, with about 280 million molecules in each cell.

As mentioned briefly in Sec. 10-6, the abnormal hemoglobin molecules tend to aggregate, causing the red blood cells to become rigid and giving them distorted shapes. This happens when the hemoglobin is not carrying oxygen and is in its deoxygenated form. The rigid and distorted cells block the smaller blood vessels. This blockage creates a local region of low oxygen concentration, resulting in more deoxygenated forms of the abnormal hemoglobin and more aggregation. The clogging of the blood vessels by the rigid, distorted red cells causes the pain of the disease and produces the anemia. Because some of the blood cells are distorted into shapes of a crescent or sickle shape, the disease was called sickle-cell anemia (Fig. 10-5).

FIGURE 10-5

(*a*) Oxygenated red blood cells from a patient with sickle-cell anemia enlarged 1600 diameters. In the presence of oxygen most of the cells have a normal shape; only the cell at the lower right is abnormally shaped. (*b*) Deoxygenated cells from a sickle-cell patient have a variety of unusual shapes. The upper cell at the far right has the classic sickle shape for which the disease was named. (*c*) Cyanate-treated red blood cells from a sickle-cell patient after removal of oxygen. Most of the cells have retained the normal shape. (*Photographs by James Jamieson, courtesy of Anthony Cerami, Rockefeller University.*)

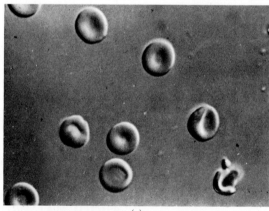

(*a*)

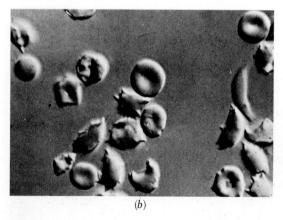

(*b*)

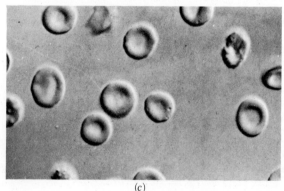

(*c*)

Sodium cyanate is being investigated as a treatment for this disease. It has been found that given orally to patients, the cyanate ion, NCO⁻ or [N=C=O]⁻, increases the affinity of the abnormal cells for oxygen. Since sickling of the cells occurs only when the abnormal hemoglobin molecules are dissociated from oxygen, the treat... cyanate ion reduces sickling. Other carbo... taining antisickling agents have also show... promise. It is hoped that these developments will lead to an effective treatment for sickle-cell anemia.

The properties of proteins are of vital importance to the processes of living systems, especially if the proteins are enzymes. The proteins of a living organism are synthesized by the organism from structural units obtained in the digestion of foodstuffs. How does the organism manage to fabricate these large and complex molecules in the precise way necessary? How are the correct amino acid residues assembled in the required order?

The answers to these questions are to be found in the role played by a group of substances in living systems known as nucleic acids, which guide the synthesis of proteins in living cells. They determine the number, the identity, and the sequence of amino acid residues in each individual protein. In so doing they implement the transmission of hereditary traits. **Genes,** the basic units of heredity, are actually segments of nucleic acid molecules. Current investigations of how nucleic acids preserve the information of heredity (Sec. 10-8F) and control the synthesis of proteins (Sec. 10-8G) are among the most exciting fields of research today.

Nucleic acids are polymers composed of very large molecules containing many structural units (which may number in the millions) with carbohydrate- and nitrogen-containing cyclic subunits. The units are linked together by phosphate ester groups (Sec. 10-8A; compare with the amide linkages in proteins). The identity and the sequence of their structural units contain information for guiding the synthesis of proteins. There are two general classes of nucleic acids of related structure, deoxyribonucleic acids (DNA) and ribonucleic acids (RNA). Their structure and function are discussed in Sec. 10-8.

10-7 THE COMPOSITION OF THE HUMAN BODY

The chemical composition of the human body varies, of course, from person to person, and depends, among other factors, on the sex of the person and what they eat. Table 10-7 lists the approximate composition of the average adult. About 50 percent of the protein material is in the form of muscle, about 20 percent is in the skeleton, about 9 percent is skin, and about 5 percent is blood. Most of the fats are found as deposits of adipose (fatty) tissue. The carbohydrate material is mainly glycogen (Sec. 10-2E). Most of the mineral content is found in the bone structure.

Table 10-8 lists the composition, as percent by weight, of the average adult human body in terms of the atoms present in the various substances. In view of the high concentrations of proteins, fats, and water, it is not surprising that more than 95 percent of the body is composed of oxygen, carbon, hydrogen, and nitrogen atoms. Substantial concentrations of calcium and

Table 10-7
Chemical Composition of the Average Adult Human Body

Class of substance	Approximate percent by weight
Proteins	16
Fats	18
Carbohydrates	<1
Minerals	5
Water	60

Table 10-8
Composition† of the Average Adult Human Body in Percent by Weight of Atoms

Atom	Percent by weight
Oxygen	65
Carbon	18
Hydrogen	10
Nitrogen	3.0
Calcium	2.0
Phosphorus	1.2
Potassium	0.35
Sulfur	0.25
Sodium	0.15
Chlorine	0.15
Magnesium	0.05
Iron	0.004

† Plus traces of iodine, fluorine, silicon, manganese, cobalt, copper, zinc, boron, aluminum, vanadium, and molybdenum.

phosphorus are required by the bone structure. The fluid in cells contains an appreciable concentration of potassium ions, K^+. The fluid outside cells contains a significant concentration of sodium ions, Na^+.

Table 10-9 compares, on an atom percent basis, the composition of the human body with the composition of the universe, the earth's crust, and seawater. The preponderance of hydrogen atoms in the body is made clearer on this basis than on a weight percent basis since they have such a small weight. This comparison reveals that carbon and phosphorus atoms, which are very important in the human body, are rather scarce elsewhere.

Table 10-9
Comparison of the Atom Composition of the Universe, the Earth's Crust, Seawater, and the Average Adult Human Body
Percent of total number of atoms present†

Universe		Earth's crust		Seawater		Human body	
Hydrogen	91	Oxygen	47	Hydrogen	66	Hydrogen	63
Helium	9.1	Silicon	28	Oxygen	33	Oxygen	25.5
Oxygen	0.057	Aluminum	7.9	Chlorine	0.33	Carbon	9.5
Nitrogen	0.042	Iron	4.5	Sodium	0.28	Nitrogen	1.4
Carbon	0.021	Calcium	3.5	Magnesium	0.033	Calcium	0.31
Silicon	0.003	Sodium	2.5	Sulfur	0.017	Phosphorus	0.22
Neon	0.003	Potassium	2.5	Calcium	0.006	Potassium	0.06
Magnesium	0.002	Magnesium	2.2	Potassium	0.006	Sulfur	0.05
Iron	0.002	Titanium	0.46	Carbon	0.0014	Chlorine	0.03
Sulfur	0.001	Hydrogen	0.22	Bromine	0.0005	Sodium	0.03
		Carbon	0.19			Magnesium	0.01
All others	< 0.01	All others	< 0.1	All others	< 0.1	All others	< 0.01

† Adapted from E. Frieden, The Chemical Elements of Life. Copyright 1972 by *Scientific American, Inc.* All rights reserved.

Note to Students   The summarizing questions for self-study at the end of the chapter are of particular help in distinguishing the points of greatest importance in a far-ranging chapter like this.

THE MECHANISM OF CARBON MONOXIDE POISONING

Carbon monoxide, CO, an odorless gas, is a lethal poison. Exposure to air containing only 240 parts per million for an hour is very dangerous. The gas is found, among other places, in automobile exhaust, and is formed when coal and fuel oil are burned. The carbon monoxide produced by the burning of construction materials is a major cause of death in fires. Carbon monoxide can kill without any flames reaching the body, and it can bring on unconsciousness, preventing escape from flames.

Carbon monoxide is poisonous because it in-terferes with the transport of oxygen in blood by combining with the hemoglobin molecules of red blood cells. Carbon monoxide has a greater affinity for hemoglobin than oxygen has, and it combines at the same part of the molecule where oxygen combines. A hemoglobin molecule combined with carbon monoxide cannot combine with oxygen. When too many hemoglobin molecules are combined with carbon monoxide, not enough oxygen reaches the brain to maintain normal functioning. Loss of consciousness and death result.

10-8 THE STRUCTURE AND FUNCTION OF NUCLEIC ACIDS

The nucleic acids are found in cells as nucleoproteins, so named because they were once thought to occur only in the nuclei of plant and animal cells. The nucleoproteins are essentially salt-like combinations of proteins and nucleic acids. The molecules of nucleic acids, like the molecules of proteins, are large and complex polymers, but the structure of the individual units differs considerably from the amino acid residue units of proteins.

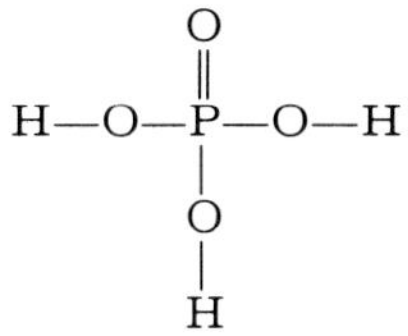

Optional

A. Phosphate Esters

One of the principal structural features of nucleic acids is derived from the structure of phosphoric acid. Like a carboxylic acid, phosphoric acid will form esters with alcohols. The products are called **phosphate esters:**

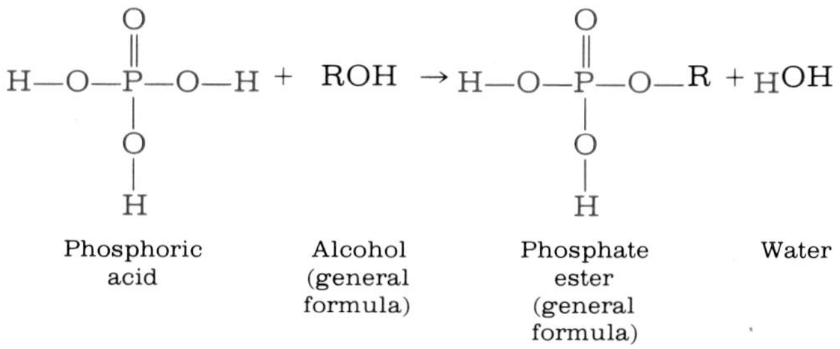

$$H—O—\overset{\overset{\displaystyle O}{\|}}{\underset{\underset{\displaystyle H}{|}}{P}}—O—H$$

Structural formula for
phosphoric acid

Phosphoric acid	Alcohol (general formula)	Phosphate ester (general formula)	Water

The formation of a phosphate ester from phosphoric acid and an alcohol

In corresponding reactions two or all three of the —O—H groups of the phosphoric acid molecule will react to form esters:

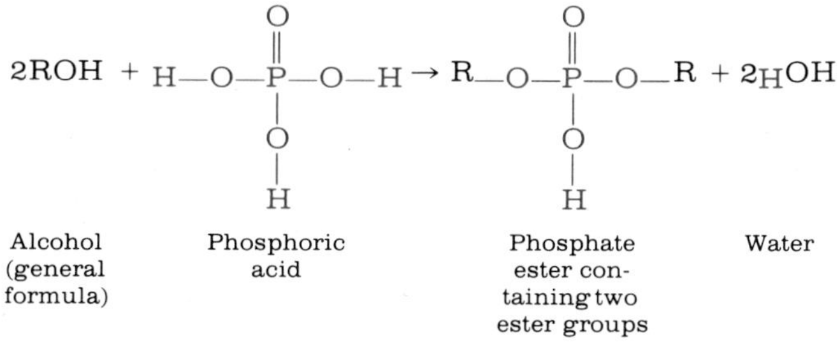

Alcohol (general formula)	Phosphoric acid	Phosphate ester containing two ester groups	Water

Formation of a phosphate ester containing two ester groups from one molecule of phosphoric acid and two molecules of an alcohol

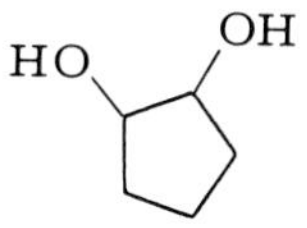

The phosphate esters found in nucleic acids are derivatives of small sugar molecules which contain several hydroxyl, —O—H, groups. The sugar molecules can be abbreviated as shown in the margin. The phosphate esters of nucleic acids involve two of the —O—H groups in each phosphoric acid molecule and two of the —O—H groups in the sugar molecules. The phosphate esters are incorporated into the following structure:

Simplified representation of a sugar molecule, derivatives of which are found in nucleic acids; only two of the hydroxyl, —O—H, groups are shown

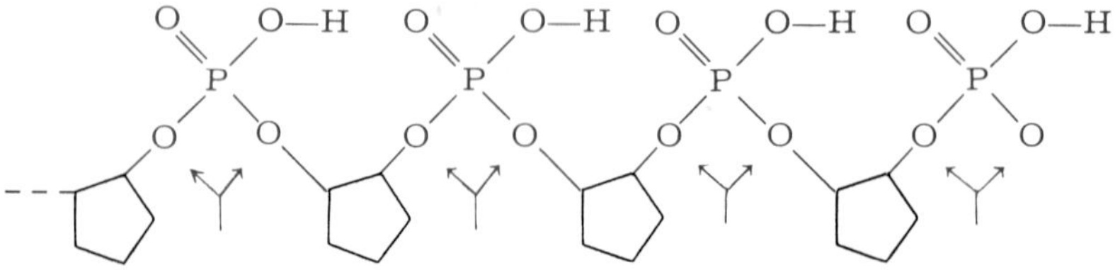

Simplified representation of the phosphate ester linkages in nucleic acids; the arrows point to the phosphate ester groups; each group was formed from an —O—H group from the sugar and an —O—H group from the phosphoric acid

Under the conditions prevailing in living cells the phosphate ester groups actually are present as ions formed by the departure of hydrogen ions, one hydrogen ion from each structural unit. These ionic structures may be represented as follows:

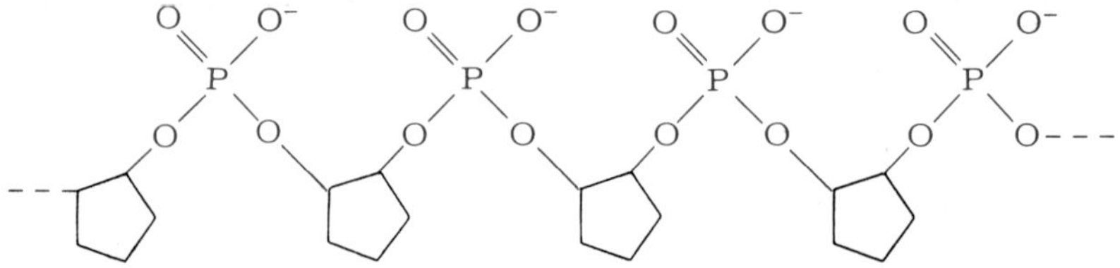

Simplified representation of the phosphate ester linkages in nucleic acids; under the conditions prevailing in living cells ionic charges are formed by the departure of hydrogen ions, as shown

Full structural formula

Abbreviated formula

Formulas for pyrimidine; derivatives of this substance furnish important structural units in nucleic acids

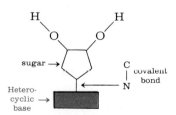

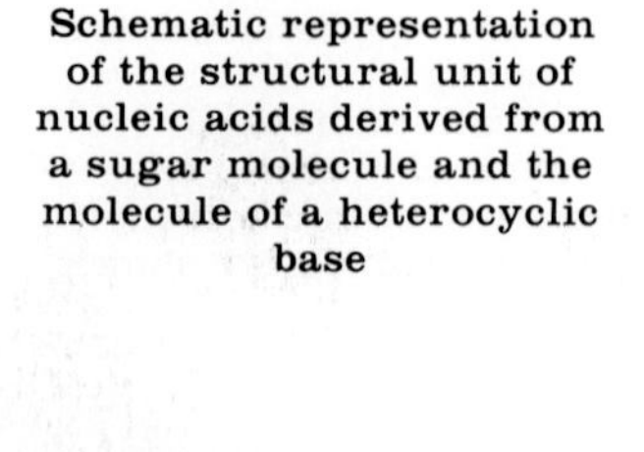

Schematic representation of the structural unit of nucleic acids derived from a sugar molecule and the molecule of a heterocyclic base

B. Heterocyclic Bases

Nucleic acids also contain units derived from substances containing ring structures involving both nitrogen and carbon atoms. The substances in turn are derivatives of a substance known as pyrimidine. Pyrimidine and its derivatives found in nucleic acids are bases in the same sense that ammonia, NH_3, is a base: their molecules are capable of reacting with protons (Sec. 7-5). They are called heterocyclic bases because their structures contain one or more heterocyclic rings. A **heterocyclic ring** is a structural ring unit involving carbon atoms and other atoms. A **heterocyclic compound** is any substance whose molecules contain heterocyclic rings. A **heterocyclic base** is a basic compound, like pyrimidine, which contains a heterocyclic ring.

Nucleic acids contain structures derived from heterocyclic bases. The heterocyclic base groups are linked directly to the sugar groups by covalent bonds between carbon atoms in the sugar molecules and nitrogen atoms in the heterocyclic rings of the bases. We can represent the link in schematic fashion, as shown in the margin.

C. The Fundamental Polyester Backbone of Nucleic Acids

The fundamental structure of a nucleic acid can be represented schematically as

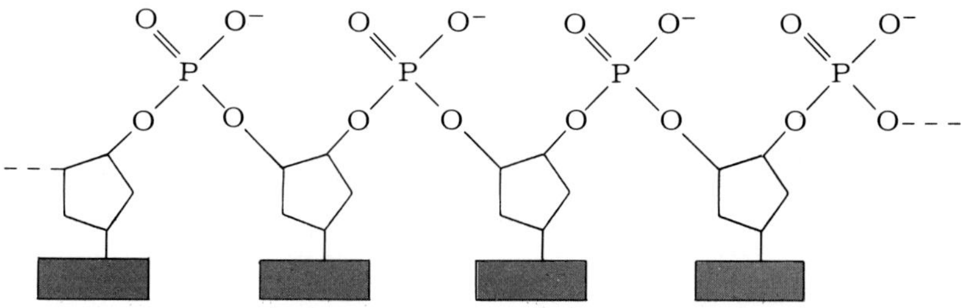

Schematic representation of the fundamental structure of a portion
of a nucleic acid; this polyester chain is the backbone of the structure;
each nucleic acid has its own characteristic sequence of structural units

The fundamental structure of nucleic acids contains a chain of
subunits linked by phosphate ester groups. This polyester
chain is the backbone of nucleic acids just as a polyamide chain
serves as the backbone of the structures of proteins. Attached to
the polyester backbone are various groups. Each nucleic acid
has its own characteristic sequence of groups.

D. The Structure of RNA

There are two general classes of nucleic acids, ribonucleic acids
(RNA) and deoxyribonucleic acids (DNA). The sugar unit in
RNA is derived from the one-ring sugar, ribose. The sugar unit
in DNA is derived from another one-ring sugar, deoxyribose.
(The prefix *deoxy-* indicates the absence of one of the —O—H
groups.) All RNA is derived from four different heterocyclic
bases; their names and abbreviated symbols are given in Table
10-10. A portion of the structure of an RNA molecule can be
represented schematically as shown below. The heterocyclic
bases are identified by their symbols. The number of polyester
units in a given ribonucleic acid depends on the particular acid
but tends to range from 70 to 4000.

Table 10-10
Heterocyclic Bases in RNA

Name	Symbol
Adenine	A
Cytosine	C
Guanine	G
Uracil	U

Schematic representation of a portion of the structure of RNA;
the different shapes of the symbols for the bases suggest their different
character; A = adenine, C = cytosine, G = guanine, U = uracil

E. The Structure of DNA

The sugar unit in DNA is derived from deoxyribose, as mentioned in Sec. 10-8D. Like RNA, DNA is derived from only four
different heterocyclic bases. Three of these are the same as those
in the RNA, and the fourth is thymine (which replaces uracil).
Their names and abbreviated symbols are given in Table 10-11.
  A portion of the structure of DNA is represented schematically by Figs. 10-6 and 10-7 (see also Fig. 10-8). Each acid contains two polyester chains held together by hydrogen bonds
between the heterocyclic bases. The base adenine (A) always

Table 10-11
Heterocyclic Bases in DNA

Name	Symbol
Adenine	A
Cytosine	C
Guanine	G
Thymine	T

pairs with the base thymine (T), and the base cytosine (C) always pairs with the base guanine (G). Figure 10-6 emphasizes the double-stranded nature of the structure. The hydrogen bonds are represented by dotted lines. Figure 10-7 emphasizes the spiral nature of the double-stranded structure. The term **double helix** is applied to the double-stranded spiral structure of DNA. The number of ester units in the strands depends on the particular type of DNA, ranging from about 5000 to more than 3 million. The molecules of DNA are therefore very long.

In summary, nucleic acids can be described as a large and important group of substances whose molecules contain long-chain polyester structures to which are attached characteristic sequences of heterocyclic bases. They contain information which serves to guide the synthesis of proteins, and they possess the means of preserving that information.

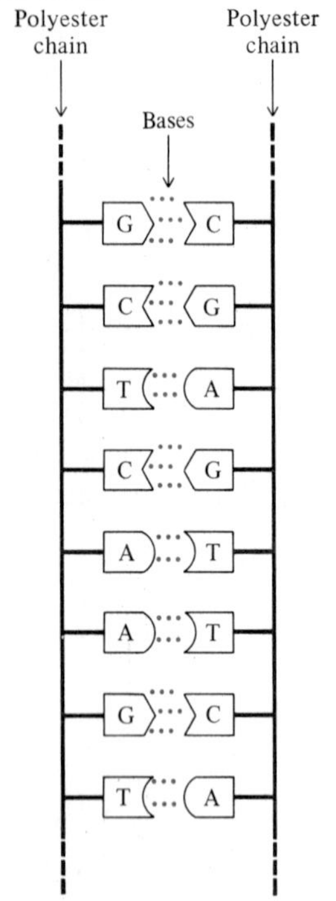

F. How Do Nucleic Acids Preserve the Information for Synthesizing Proteins?

As living cells divide to form daughter cells, the information of heredity contained in the DNA structures must be passed on intact to the new cells. This is accomplished by the ability of the DNA molecules to duplicate themselves; i.e., they can supervise the synthesis of DNA molecules with exactly the same structure as themselves in the following steps: (1) The two strands "unzip," leaving the sequence of base units intact. (2) The sequence of base units of each strand guides the sequence of base units as another strand is assembled. The sequence of base units in the new strand will complement the sequence in the old strand. (3) The net result is two double-stranded DNA molecules with exactly the same sequence of bases as the parent DNA molecule (Fig. 10-9). The structure of each DNA molecule is therefore preserved, to be passed on to the next generation. In the overall process, called replication, enzymes play an important role in catalyzing each step.

G. How Do Nucleic Acids Guide the Synthesis of Proteins?

The synthesis of proteins is guided specifically by RNA molecules; they obtain their information from DNA molecules, which store it. The information is in the form of sequences of the base units. The RNA molecules are synthesized by a series of steps which resemble those in the duplication of the DNA molecules (each step is catalyzed by enzymes): (1) The two strands of the parent DNA molecule "unzip," leaving the sequence of base units intact. (2) The sequence of base units in each strand guides the assembly of base units in a molecule of RNA. (3) The assembled RNA molecules separate from the single strands of the DNA molecules. (4) The net result is a new RNA molecule with a sequence of bases determined by a DNA molecule. The protein-synthesizing guidelines contained in the structure of the DNA molecule are passed on to the RNA molecule as sequences of bases.

There are about 20 different amino acid residues in proteins, but there are only 4 different residues in a given nucleic acid.

FIGURE 10-6
Schematic representation of a portion of the structure of DNA, emphasizing the double-stranded nature of the structure. The dotted lines represent hydrogen bonds between the heterocyclic bases. The number of units varies from about 5000 to more than 3 million; A = adenine, C = cytosine, G = guanine, T = thymine.

FIGURE 10-7
Schematic representations of a portion of the structure of DNA, emphasizing the spiral arrangement of the two polyester chains.
(From M. J. Sienko and R. A. Plane, "Chemistry," 5th ed., p. 553. Copyright 1975 by McGraw-Hill, Inc.; used by permission of McGraw-Hill Book Company.)

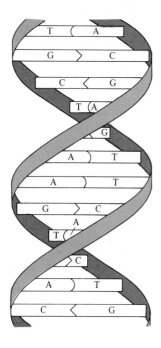

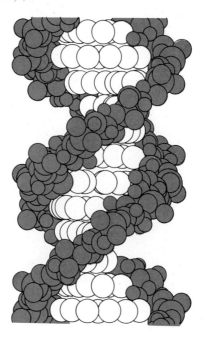

FIGURE 10-8
Electron micrograph of a closed loop form of a DNA molecule. The enlargement is about 100,000 diameters. *(Courtesy of Arthur Kornberg.)*

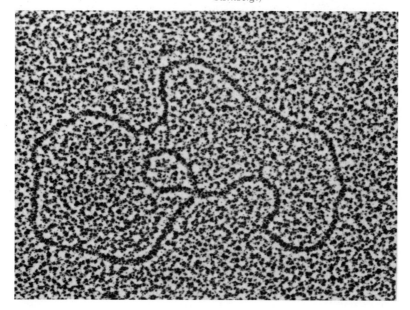

FIGURE 10-9
DNA is replicated by "unzipping" the two strands and assembling a new complementary strand for each. The result is two
double-stranded DNA molecules with **exactly the same sequence
of bases as the parent DNA molecule.** (*From James D. Watson, "Molecular Biology of the Gene," 2d ed., W. A. Benjamin, Inc. Copyright 1970 by
J. D. Watson.*)

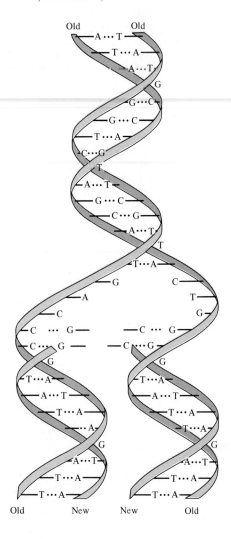

The order in which the amino acid residues are assembled
during the synthesis of a protein depends on the sequence of
base units in the RNA chain. Each sequence of three base units
is a code for a specific amino acid residue; i.e., for each amino
acid residue there is a specific code sequence of three RNA base
units. Some residues have more than one code sequence. The
code for alanine, for example, is the sequence guanine (G)–
cytosine (C)–uracil (U), or GCU. The code for glycine is guanine

(G)–guanine (G)–uracil (U), or GGU. Table 10-12 gives the base-unit code for a few amino acid residues. The base-unit sequence GCU brings up an alanine residue (Table 10-12). The base-unit sequence GGU brings up a glycine residue. In this way the structure of RNA molecules determines the identity and sequence of amino acid residues in protein molecules. (Every step is catalyzed by enzymes.) The **genetic code** is the relationship between the base-unit sequence in RNA molecules (or their DNA parents) and the sequence of amino acid residues in proteins. The genetic code is the same for all organisms.

In guiding the synthesis of all-important proteins, the nucleic acids control the processes of life. Ultimately the vital influence of the nucleic acids resides in the sizes, the shapes, and the structures of their molecules.

Table 10-12
The Base-Unit Sequence
Code for Five Amino Acids

Amino acid	Code sequence†
Alanine	GCU
Aspartic acid	GAU
Glycine	GGU
Lysine	AAA
Phenylanine	UUU

† Only one is given here; some amino acids have more than one code sequence.

10-9 THE OVERALL STRUCTURE OF PROTEIN MOLECULES

It is useful to consider the structure of proteins in terms of levels of organization. The **primary level** refers to the fundamental atom-by-atom sequence in the polypeptide chains and any prosthetic groups present. The **secondary level** considers how chains of amino acid residues are arranged in three dimensions to form coils, sheets, or more compact arrangements. The **tertiary and quaternary levels** consider the ordering of protein molecules into larger aggregates, e.g., the assembly of several coils into ropelike structures (Fig. 10-11).

The actual shape of a protein molecule involves such structural features as hydrogen bonds (Sec. 4-12) and disulfide cross-links. Hydrogen bonds arise most frequently between the hydrogen atom of one amide group and the oxygen atom of another amide group. The second amide group may be in the same or another polypeptide chain. The basic structural situation of the hydrogen bond can be represented as shown in the margin. Figures 10-10 and 10-11 illustrate how hydrogen bonds serve to bond polypeptide chains.

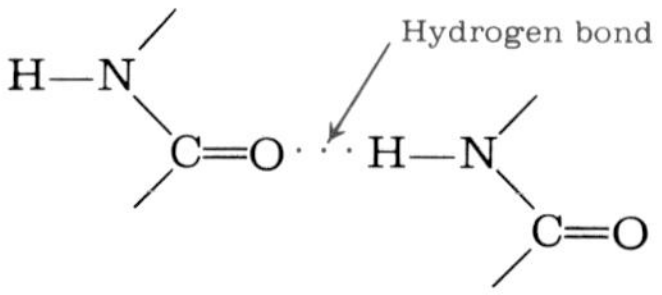

The presence of disulfide bonds between polypeptide chains or two sections of the same chain is another important factor determining the overall shape of a protein. Cystine is the only amino acid with two amino groups and two carboxylic acid groups. These unique features make it possible for this acid to serve a special role: one cystine residue can become part of

Optional

Representation of a hydrogen bond which can form between two amide groups; the amide groups may be in the same chain or in different chains

FIGURE 10-10
Schematic representation of how hydrogen bonds hold two polypeptide chains together (or two parts of the same chain). The
hydrogen bonds are represented by dots. In an actual protein
structure the crowding due to the side chains (R groups) causes
the polypeptide chain to be somewhat distorted from this hypothetical representation.

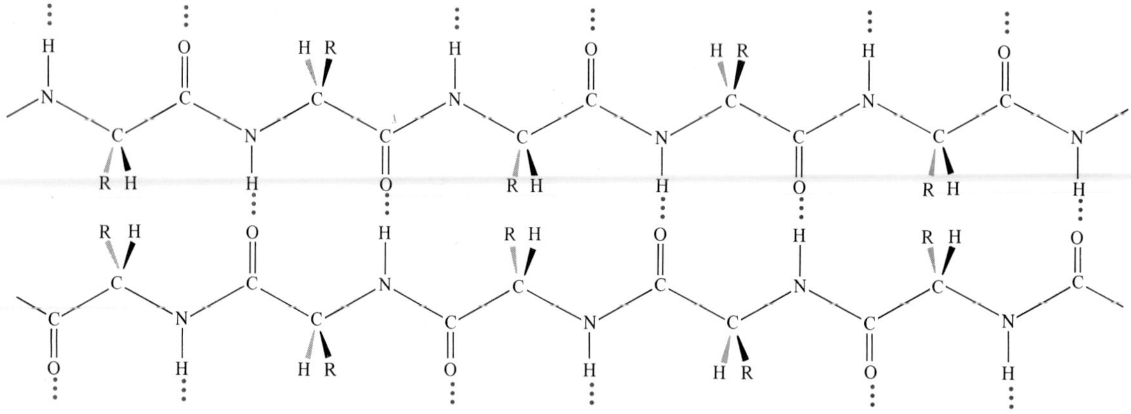

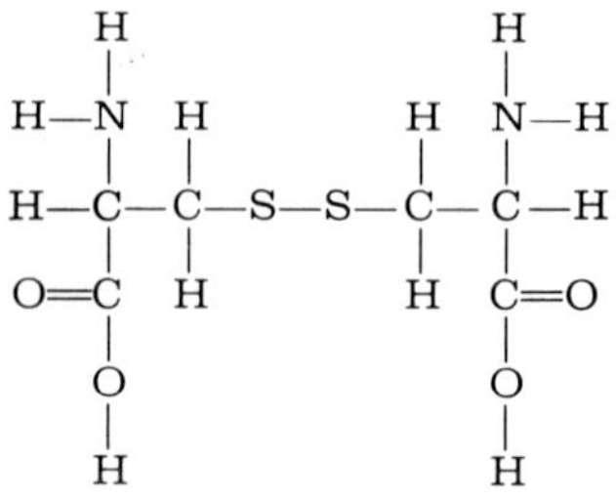

The structure of the amino
acid cystine with its unique
combination of two amino
groups and two carboxylic
acid groups

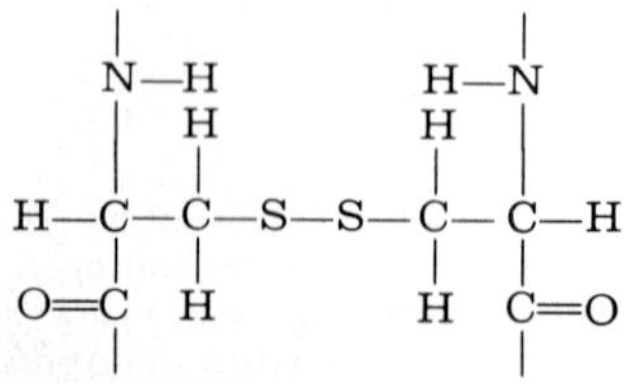

Disulfide linkage between
two different parts of the
same polypeptide chain, or
two different chains,
provided by the single
amino acid residue of
cystine

two different polypeptide chains, or part of different regions
of the same chain. In so doing, it provides a link involving a
disulfide, S—S, group. The representation of a protein fragment,
shown in the margin focuses on the important structural relationship. How hydrogen bonds and the disulfide linkages
determine the shape of a polypeptide chain is represented in
Fig. 10-12.

Chains of different polypeptides may be joined together solely
by hydrogen bonds, as in fibroin, the protein of silk. The chain
of a single polypeptide may be connected at several points by
hydrogen bonds to form a spiral, as in keratin, the protein of
hair and wool (Fig. 10-11). How polypeptide spirals can be organized into larger units is illustrated in the structure of a hair
fiber (Fig. 10-11c).

Denaturation of a protein is an alteration in its structure
brought about by heat, chemical changes, or other changes in
the protein's environment. The amino acid residue sequence of
the protein is believed to remain intact, only forces which hold
the acid residue sequence into a definite shape being affected.
Denaturation of a protein leads, among other things, to coagulation and a change in solubility. When an egg is cooked, the
heat denatures proteins and causes coagulation.

The physiological activity of proteins is determined as much
by their overall shape as by their fundamental atomic sequence,
e.g., the shape of the active site of an enzyme in forming an
enzyme-substrate complex as the enzyme exerts its catalytic
effect (Sec. 10-5). The overall shape is, of course, determined by
the atom-by-atom sequences, the number and spacing of the
side chains of the amino acid residues, and such linkages as
hydrogen bonds and the disulfide, S—S, groups of cystine
residues.

The structure of the enzyme ribonuclease illustrates the complexity possible in protein molecules. The overall shape is suggested by Fig. 10-13. In this drawing each amino acid residue is represented by a numbered circle, and the disulfide linkages are represented by double lines between residues 40 and 95, 26 and 84, 65 and 72, and 58 and 110. There are a total of 124

FIGURE 10-11
The structure of keratin, the protein of hair. (a) How the hydrogen bonds hold the polypeptide chain in the formation of a coil, called an alpha helix; (b) schematic sketch of the main atom chain; (c) how coils are incorporated into increasingly complex structures; (d) diagram of a hair fiber. [(a) and (b) from M. J. Sienko and R. A. Plane, "Chemistry," 5th ed. p. 541. Copyright 1975 McGraw-Hill, Inc.; used by permission of McGraw-Hill Book Co.; (c) and (d) from R. E. Dickerson and I. Geis, "The Structure and Action of Proteins," pp. 37 and 39, Harper & Row, 1969.]

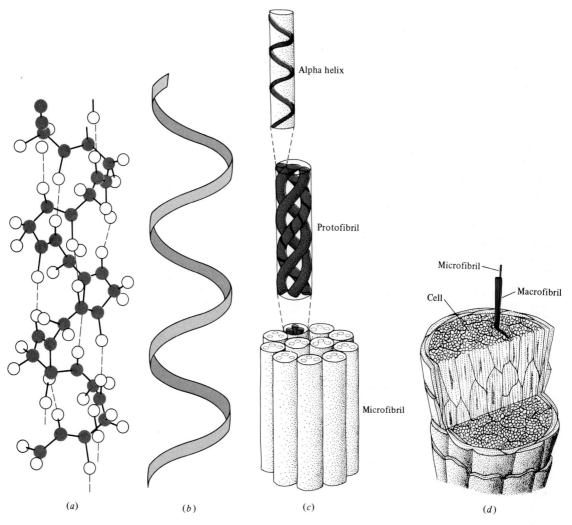

(a)   (b)   (c)   (d)

amino acid residues and 1876 atoms (587 of carbon, 909 of
hydrogen, 197 of oxygen, 171 of nitrogen, and 12 of sulfur).
An atom-by-atom representation would, of course, be even more
complex.

FIGURE 10-12
Schematic representation of the cross-linking of a portion of a
polypeptide chain by hydrogen bonding and the disulfide portion
of a cystine residue. The hydrogen bonds are represented by dots.
Such cross-links are a major factor in determining the shape of a
protein.

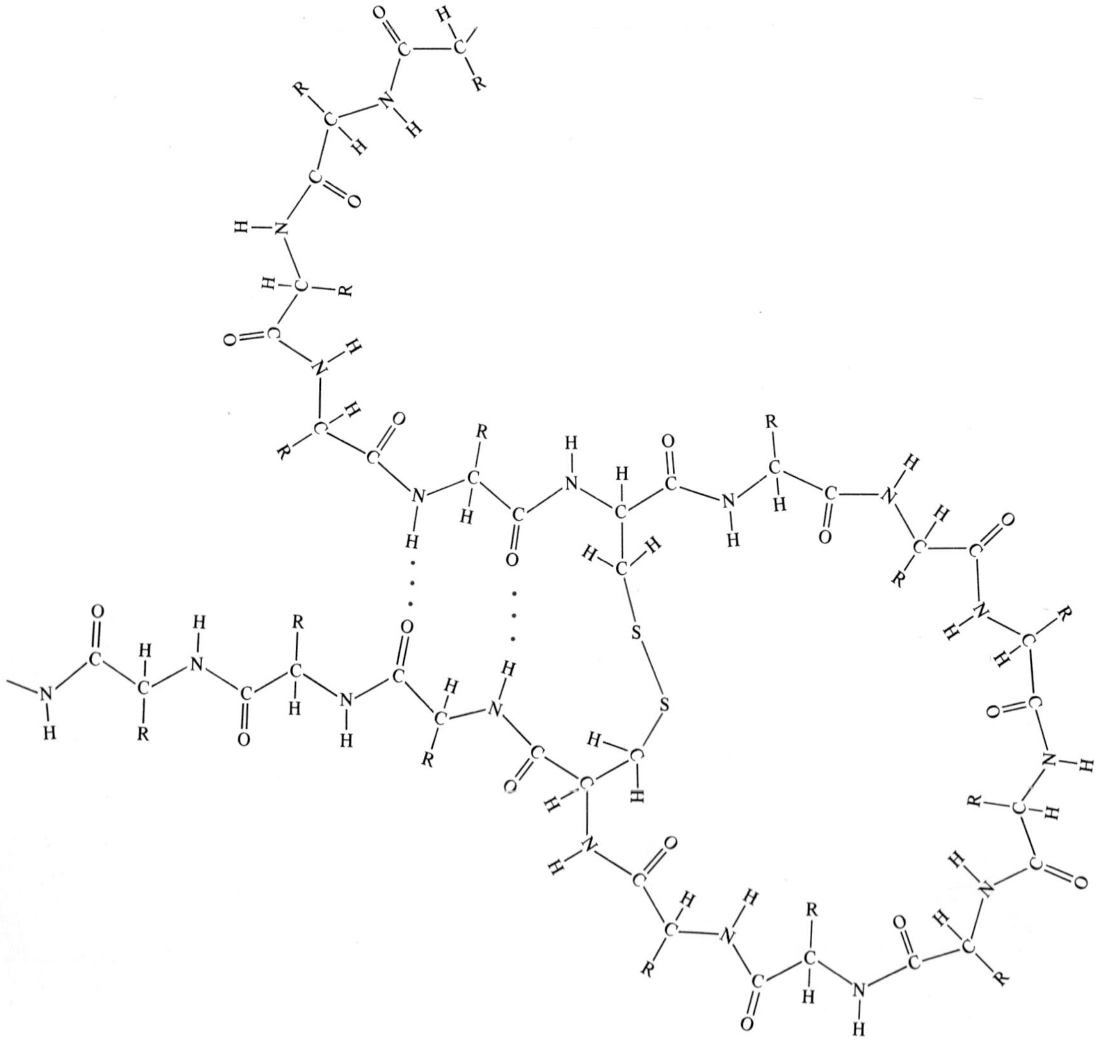

Although we have been representing amino acids as though they contained an amino group and a carboxylic acid group (below, form B), their melting points, solubility behavior, and chemical properties indicate that they are actually ionic compounds with the structure of form A. These two forms differ in the location of a hydrogen ion. The hydrogen ion of the carboxylic acid group in form B is actually on the nitrogen atom of form A. A structure like A is known as a **dipolar ion,** i.e., an ion which contains both positive and negative ionic charges.

Optional

$$
\begin{array}{ccc}
& \text{H} \quad \text{H} & \\
& | \quad\quad | & \\
\text{H}-\text{N}^+-\text{C}-\text{C}-\text{O}^- \\
& | \quad\quad | \quad\quad \| & \\
& \text{H} \quad \text{R} \quad \text{O} &
\end{array}
\qquad
\begin{array}{ccc}
& \text{H} & \\
& | & \\
\text{H}-\text{N}-\text{C}-\text{C}-\text{O}-\text{H} \\
& | \quad\quad | \quad\quad \| & \\
& \text{H} \quad \text{R} \quad \text{O} &
\end{array}
$$

Form A   Form B

**Comparison of the actual dipolar structure (form A)
of an amino acid with the conventional
representation (form B)**

FIGURE 10-13
The overall shape of the enzyme protein ribonuclease, suggested by the sequence and spatial arrangement of the amino acid residues. Each residue is represented by a numbered circle (124 in all). The disulfide linkages are represented by double lines. (*From R. E. Dickerson and I. Geis, "The Structure and Action of Proteins," p. 80, Harper & Row, 1969.*)

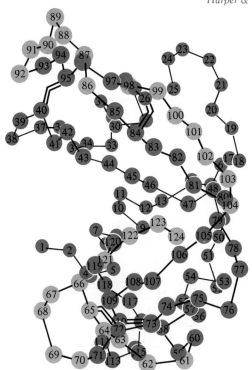

© 1974 *American Scientist; reprinted by permission of H. Martin.*

KEY WORDS

1. Biochemistry (Sec. 10-1)
2. Ketone (Sec. 10-2B)
3. Carbohydrate (Sec. 10-2G)
4. Monosaccharide (Sec. 10-2G)
5. Disaccharide (Sec. 10-2G)
6. Polysaccharide (Sec. 10-2G)
7. Sugar (Sec. 10-2G)
8. Fat (Sec. 10-3)
9. Carboxylic acid residue (Sec. 10-3)
10. Alcohol residue (Sec. 10-3)
11. Saturated fatty acid (Sec. 10-3)
12. Unsaturated fatty acid (Sec. 10-3)
13. Saturated fat (Sec. 10-3)
14. Unsaturated fat (Sec. 10-3)
15. Polyunsaturated fat (Sec. 10-3)
16. Amino acid (Sec. 10-4A)
17. Amino acid residue (Sec. 10-4A)
18. Peptide bond (Sec. 10-4B)
19. Polypeptide structure (Sec. 10-4B)
20. Dipeptide (Sec. 10-4B)
21. Polypeptide (Sec. 10-4B)
22. Prosthetic group (Sec. 10-4C)
23. Conjugated protein (Sec. 10-4C)
24. Simple protein (Sec. 10-4C)
25. Protein (Sec. 10-4D)
26. Enzyme (Sec. 10-5)

27. Substrate (Sec. 10-5)
28. Enzyme-substrate complex (Sec. 10-5)
29. Active site of an enzyme (Sec. 10-5)
30. Nucleic acid (Sec. 10-6, optional Sec. 10-8E)
31. Gene (Sec. 10-6)

Optional

32. Phosphate ester (Sec. 10-8A)
33. Heterocyclic ring (Sec. 10-8B)
34. Heterocyclic compound (Sec. 10-8B)
35. Heterocyclic base (Sec. 10-8B)
36. Double helix (Sec. 10-8E)
37. Genetic code (Sec. 10-8G)
38. Primary level of protein structure (Sec. 10-9)
39. Secondary level of protein structure (Sec. 10-9)
40. Denaturation of a protein (Sec. 10-9)
41. Dipolar ion (Sec. 10-10)

SUMMARIZING QUESTIONS FOR SELF-STUDY

Sections 10-1 and 10-2

1. **Q.** What is meant by the term biochemistry?
   A. The study of the chemistry of plants and animals.

2. **Q.** From what viewpoint can glucose be described as a polyhydroxyaldehyde?
   A. One of the three forms of glucose, the open-chain structure, contains an aldehyde group and five alcohol (hydroxyl) groups.

3. **Q.** How many forms of glucose are present in an aqueous solution of the substance?
   A. Three, in equilibrium with each other.

4. **Q.** How many atoms are in the ring of the ring forms of glucose, and what kinds of atoms are they?
   A. Five carbon atoms and one oxygen atom, six atoms in all.

5. **Q.** Give the structural formula of the α-ring form of glucose.
   A.

6. **Q.** What is a ketone?
   A. A carbon-containing substance which contains a ketone group.

7. **Q.** Give the general formula for a ketone.
   A.

$$R-\overset{\overset{\displaystyle O}{\|}}{C}-R$$

8. **Q.** From what viewpoint can fructose be described as a polyhydroxyketone?
   A. One of the forms of fructose, the open-chain form, contains a ketone group and five alcohol (hydroxyl) groups.

9. **Q.** How many atoms are there in the ring of the ring form of fructose included in the text discussion, and what kinds of atoms are they?
   A. Four carbon atoms and one oxygen atom, five atoms in all.

10. **Q.** Give the structural formula of the ring form of fructose included in the text discussion.
    A.

11. **Q.** What structural units are contained in a molecule of maltose, and how many are there?
    A. Two glucose units (minus the atoms of a molecule of water).

12. **Q.** What is the product of the hydrolysis of maltose?
    A. Glucose.

13. **Q.** What structural units are contained in a molecule of sucrose, and how many are there?
    A. One glucose unit and one fructose unit (minus the atoms of a molecule of water).

14. **Q.** What are the products of the hydrolysis of sucrose?
    A. Glucose and fructose.

15. **Q.** What structural units are contained in a molecule of the starch obtained from plants, and how many are there?
    A. Starch is composed of glucose units. One

form of plant starch (amylose) contains between 1000 and 4000 units. The other form (amylopectin) contains somewhere between 6000 and 40,000.

16. Q. What is glycogen, and what structural units does it contain?

A. Glycogen is the starch formed in animals as a reserve supply of carbohydrates. It contains glucose units.

17. Q. What is the product of the complete hydrolysis of plant starch and glycogen?

A. Glucose.

18. Q. What structural units are contained in a molecule of cellulose, and how many are there?

A. Glucose units of uncertain number, estimated at between 1500 and 5000 and beyond.

19. Q. What is the product of the complete hydrolysis of cellulose?

A. Glucose.

20. Q. What is the difference between cellulose and starch?

A. Both are made up of glucose units, but the forms of the glucose rings are different.

21. Q. What is a carbohydrate?

A. A substance with polyhydroxy molecules containing aldehyde groups, modified aldehyde groups, ketone groups, or modified ketone groups or molecules containing a sequence of such structural units from two to thousands.

22. Q. What is the principal function of carbohydrates in human beings?

A. To provide energy for the body.

23. Q. What is a polysaccharide?

A. A carbohydrate which contains many structural units.

24. Q. What is a sugar?

A. A carbohydrate of simple structure. For the most part sugar molecules contain only one, two, or three structural units.

25. Q. What is table sugar?

A. Sucrose.

26. Q. What are the sources of table sugar?

A. Sugar cane and sugar beets.

Section 10-3

27. Q. What is a fat?

A. A carboxylic ester of the alcohol, glycerol, and three long-chain carboxylic acids which contain one or more alkene groups.

28. Q. What is the principal function of fats in animals?

A. They are the most important means of storing a potential source of energy.

29. Q. What is a vegetable oil?

A. It is a liquid fat obtained from a plant.

30. Q. What is a fatty acid?

A. It is a carboxylic acid obtained from the hydrolysis of a fat.

31. Q. What is a saturated fat?

A. A fat which contains fatty acid residues with few, if any, alkene groups.

32. Q. What is a polyunsaturated fat?

A. It is another name for an unsaturated fat, i.e., a fat whose fatty acid residues contain substantial numbers of alkene groups.

Section 10-4

33. Q. What is an amino acid?

A. A compound containing an amine group and a carboxylic acid group.

34. Q. What is an amino acid residue?

A. The part of an amino acid molecule that is actually found in a protein molecule.

35. Q. What is a peptide bond?

A. A special name given to the amide bonds of proteins which link the amino acid residues.

36. Q. What is a polypeptide?

A. A molecule containing several amino acid residues linked by amide groups.

37. Q. What are the hydrolysis products of polypeptides?

A. Amino acids.

38. Q. What is a prosthetic group?

A. The nonpolypeptide structural unit of a conjugated protein.

39. Q. What is a conjugated protein?

A. A protein which contains a prosthetic group in addition to a polypeptide chain.

40. Q. What is the name of the prosthetic group in the protein hemoglobin?

A. Heme.

41. Q. What is a simple protein?

A. A protein which contains a polypeptide structure only, with no prosthetic group.

42. Q. What is a protein?

A. A substance made up of large molecules which contain a long chain of amino acid residues joined by peptide bonds (amide bonds) and sometimes also a prosthetic (nonpolypeptide) structural unit.

43. Q. What are some of the functions of proteins in living cells?

A. They serve as structural materials, transport essential substances, and catalyze essential reactions.

Section 10-5

44. Q. What is the function of an enzyme?

A. To catalyze the reactions of living systems.

45. Q. To what class of substances do enzymes belong?

A. Proteins.

46. Q. What is a substrate?

A. Any substance whose participation in a specific reaction is catalyzed by an enzyme.

47. Q. What is an enzyme-substrate complex?

A. The weak, temporary association between an enzyme and a substrate.

48. Q. What is the role of an enzyme-substrate complex?

A. Its formation provides a reaction mechanism which acts more rapidly than the mechanism which would take place in its absence.

49. Q. What is the active site of an enzyme?

A. The specific region on the surface of an enzyme to which the substrate is bonded in an enzyme-substrate complex.

50. Q. What feature of the mechanism of the action of enzymes accounts for their ability to be specific in the reactions which they catalyze?

A. The close fit between enzyme and substrate required in forming the enzyme-substrate complex. Each enzyme therefore catalyzes only the reaction of one specific substance or a group of substances with close structural relationships.

Sections 10-6 and 10-7

51. Q. What are nucleic acids?

A. Polymers composed of very large molecules containing many structural units (up to a few million), with carbohydrate and nitrogen-containing subunits, linked together by phosphate ester groups.

52. Q. What is the function of nucleic acids?

A. The identity and the sequence of their structural units contain information for guiding the synthesis of proteins in living organisms.

53. Q. What does the abbreviation DNA represent?

A. A class of nucleic acids, the deoxyribonucleic acids.

54. Q. What is a gene?

A. A segment of a nucleic acid molecule (more precisely a DNA molecule).

55. Q. What does the abbreviation RNA represent?

A. A class of nucleic acids, the ribonucleic acids.

56. Q. What is the approximate percent by weight of water in the average adult human body?

A. 60 percent.

57. Q. What are the other principal classes of compounds in the adult human body?

A. Fats, proteins, minerals, and a small amount of carbohydrates.

58. Q. What are the four kinds of atoms which predominate in the adult human body?

A. Hydrogen, oxygen, carbon, and nitrogen.

59. Q. Is the predominance of these kinds of atoms consistent with the classes of substances in the human body?

A. It is consistent with the large amounts of proteins, fats, and water present.

60. Q. Which of the predominant kinds of atoms in the human body are rather scarce in the earth's crust?

A. Carbon, nitrogen, and hydrogen.

Section 10-8 (Optional)

61. Q. What is a phosphate ester?

A. The product of a reaction between phosphoric acid and an alcohol which contains one or more units of the distinctive group

$$
\begin{array}{c}
\overset{\displaystyle O}{\overset{\|}{-P}}-O-\overset{\displaystyle |}{\underset{\displaystyle |}{C}}- \\
\end{array}
$$

62. Q. Give the general formula for a phosphate ester.

A.
$$
\text{H}-\text{O}-\overset{\overset{\displaystyle O}{\|}}{\underset{\underset{\displaystyle H}{|}}{\underset{|}{\text{P}}}}-\text{O}-\text{R} \quad \text{or} \quad \text{R}-\text{O}-\overset{\overset{\displaystyle O}{\|}}{\underset{\underset{\displaystyle H}{|}}{\underset{|}{\text{P}}}}-\text{O}-\text{R}
$$

$$
\text{or} \quad \text{R}-\text{O}-\overset{\overset{\displaystyle O}{\|}}{\underset{\underset{\displaystyle R}{|}}{\underset{|}{\text{P}}}}-\text{O}-\text{R}
$$

63. Q. What is a heterocyclic ring?

A. A ring of atoms in the structure of a molecule which involves both carbon atoms and atoms other than carbon.

64. Q. What is a heterocyclic base?

A. A basic bompound which contains a heterocyclic ring.

65. Q. Give the name and abbreviated formula of the heterocyclic base whose derivatives are important components of nucleic acids.

A. Pyrimidine,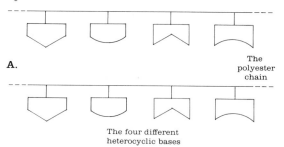

66. Q. In the following highly simplified representation of the fundamental structure of a portion of a nucleic acid, identify in general terms the symbols used.

A.

The polyester chain

The four different heterocyclic bases

67. Q. What are the principal differences between the structures of RNA and DNA?

A. RNA molecules contain a single strand of structural units. DNA molecules contain a double strand. DNA molecules tend to contain many more structural units (estimated at 5000 to 3 million or more) than RNA molecules do (estimated at 70 to about 4000). Each class contains only four different heterocyclic bases. Three of the bases are the same in both, but one is different.

68. Q. What is the principal bonding force holding the two strands of DNA molecules together?

A. Hydrogen bonds (between the heterocyclic bases).

69. Q. What is a double helix in regard to nucleic acid structure?

A. It refers to the spiral and double-stranded nature of DNA molecules.

70. Q. What is a nucleic acid?

A. A member of a group of substances whose molecules contain long-chain polyester structures to which characteristic sequences of heterocyclic bases are attached.

71. Q. What are the principal functions of nucleic acids?

A. They contain information in the form of heterocyclic base sequences which serve to guide the synthesis of proteins, and they possess the means of preserving that information.

72. Q. How do the nucleic acids preserve the information for guiding the synthesis of proteins?

A. The ability of the DNA molecules to unzip and duplicate themselves makes it possible for identical molecules with the same heterocyclic base sequence to be passed on to successive generations of cells.

73. Q. How do the nucleic acids guide the synthesis of proteins?

A. RNA molecules are synthesized in a cell, their sequence of heterocyclic bases being determined by DNA molecules, which guide the synthesis. The sequence of heterocyclic bases in the RNA molecules in turn determines the sequence of amino acid residues in the proteins, the synthesis of which the RNA molecules guide. Each sequence of three base units in the RNA molecules is a code for a specific amino acid residue. In determining the sequence of amino acid residues the RNA molecules determine the structure of a protein.

74. Q. What is meant by the genetic code?

A. It is the relationship between the base-unit sequence in RNA molecules (or their DNA parents)

and the sequence of amino acid residues in proteins.

Sections 10-9 and 10-10 (Optional)

75. Q. What is meant by the primary level of protein structure?

A. The fundamental atom-by-atom sequence in the polypeptide chain and any prosthetic groups which may be present.

76. Q. What is meant by the secondary level of protein structure?

A. How the chains of amino acid residues are arranged in three dimensions to form coils, sheets, or other arrangements.

77. Q. What are two of the types of linkages which can bond two different polypeptide chains or two regions of the same chain together?

A. Hydrogen bonds (between amide groups) and disulfide linkages (introduced by the presence of residues of the amino acid cystine).

78. Q. Name three of the principal factors which determine the shape of protein molecules.

A. (1) The atom-by-atom sequences, (2) the number and spacing of the amino acid residues, and (3) linkages like hydrogen bonds and disulfide links.

79. Q. What is one example of the role the overall shape of a protein plays in determining its properties?

A. The significance of the shape of the active site of an enzyme in the formation of an enzyme-substrate complex as the enzyme exerts its catalytic effect.

80. Q. What is meant by the denaturation of a protein?

A. The change in the structure of a protein brought about, for example, by heat; the structural features which hold the protein into a definite shape are changed.

81. Q. What is a dipolar ion?

A. An ion which contains both positive and negative ionic charges.

82. Q. What is meant by the dipolar nature of amino acids?

A. Under many conditions, e.g., in the pure solid form, amino acids have a dipolar structure, compared with the conventional representation:

$$H-\overset{\overset{\displaystyle H}{|}}{\underset{\underset{\displaystyle H}{|}}{N^+}}-\overset{\overset{\displaystyle H}{|}}{\underset{\underset{\displaystyle R}{|}}{C}}-\overset{}{\underset{\underset{\displaystyle O}{||}}{C}}-O^-$$

PRACTICE EXERCISES

1. Match each definition or other statement with the numbered term above with which it is most closely associated. Each numbered term may be used only once.

1. Saturated fat
2. Amino acid residue
3. Peptide group
4. Oil
5. Monosaccharide
6. Polyunsaturated fat
7. Enzyme
8. Sugar
9. Polysaccharide
10. Biochemistry
11. Polypeptide
12. Amino acid
13. Disaccharide
14. Prosthetic group
15. Carbohydrates
16. Nucleic acid
17. Fat

(*a*) A carbohydrate of simple structure

(b) A substance which contains an amine group and a carboxylic acid group

(c) The study of the chemistry of plants and animals

(d) Substances with polyhydroxy molecules containing aldehyde groups, modified aldehyde groups, ketone groups, or modified ketone groups or molecules containing a sequence of such structural units from two to thousands

(e) A carboxylic ester of the alcohol glycerol and three long-chain carboxylic acids which may contain one or more alkene groups

(f) A carbohydrate which contains two structural units

(g) That part of an amino acid molecule actually found in the structure of a protein molecule

(h) A fat which is liquid at room temperature

(i) A carbohydrate which contains one structural unit

(j) Special name given to the amide groups of protein molecules

(k) A fat which contains fatty acid residues with no alkene groups

(l) A carbohydrate containing many structural units

(m) A fat with fatty acid residues containing alkene groups

(n) A molecule containing many amino acid residues

2. Match each definition or other statement with the numbered term above with which it is most closely associated. Each numbered term may be used only once.

1.	Enzyme-substrate complex	2.	Active site
3.	Gene	4.	Conjugated protein
5.	Polypeptide	6.	Protein
7.	Substrate	8.	Tripeptide
9.	Polysaccharide	10.	Prosthetic group
11.	Peptide bond	12.	Nucleic acid
13.	Simple protein	14.	Enzyme

(a) A reactant whose participation in a specific reaction is catalyzed by an enzyme

(b) A polymer composed of very large molecules containing many structural units, with carbohydrate and nitrogen-containing cyclic subunits, linked together by phosphate ester groups

(c) The nonpolypeptide unit of a conjugated protein

(d) A molecule containing three amino acid residues

(e) The specific region on the surface of an enzyme to which the substrate becomes bonded

(f) A segment of a nucleic acid molecule

(g) A protein containing a prosthetic group

(h) A protein without a prosthetic group

(i) A substance which catalyzes a reaction of living systems

(j) A substance whose large molecules contain a long chain of amino acid residues joined by peptide bonds and sometimes a nonpolypeptide structural unit as well

(k) A weak association between an enzyme and a substrate

3. Name the functional groups in the open-chain form of glucose.

4. Name the functional groups in the open-chain form of fructose.

5. "An aqueous solution contains three forms of glucose in an equilibrium mixture." Explain.

6. Describe in words the structural relationship between glucose and maltose.

7. Describe in words the structural relationship between glucose, fructose, and sucrose.

8. What is the difference between glucose and starch? What do they have in common?

9. What is the difference between starch and cellulose? What do they have in common?

10. What is the difference between an amino acid and an amino acid residue? Give the structural formula of an example of each.

11. In structural-formula notation illustrate the relationship between an amino acid residue and a protein.

12. In structural-formula notation illustrate the relationship between a fatty acid residue and a fat.

13. In structural-formula notation illustrate the relationship between a saturated fat and an unsaturated fat.

14. What is obtained when (a) cellulose, (b) glycogen, and (c) maltose are hydrolyzed?

15. (a) What classes of substances are obtained when a fat is hydrolyzed?
    (b) Give an equation for the hydrolysis of a typical fat.

16. (a) What class of substance is obtained when the polypeptide portion of a protein is hydrolyzed?
    (b) Give the names and formulas for two examples of the answer to part (a).

17. Using words rather than formulas, account for the mechanism of the action of an enzyme in increasing the rate of a reaction. Make clear why a small amount of enzyme may have a great affect.

18. What is meant by the statement "Enzymes are specific in the reactions which they catalyze"? Give an example of a nonspecific catalyst.

19. On the basis of its mechanism of action, what is the principal reason an enzyme is so specific in the reaction it catalyzes?

20. What are the four most abundant atoms in the human body? In what principal classes of substances are these atoms found?

Section 10-8 (Optional)

21. Give structural formulas for two different general types of phosphate esters.

22. How is the structure of a nucleic acid related to the structure of phosphate esters?

23. Explain briefly the meaning of (*a*) a heterocyclic ring, (*b*) a heterocyclic compound, and (*c*) a heterocyclic base.

24. How is the structure of a nucleic acid related to the structure of heterocyclic bases?

25. In what way are the structures of ribonucleic acids (RNA) and deoxyribonucleic acids (DNA) similar? How do they differ? What role is played by hydrogen bonds in the structure of deoxyribonucleic acid (DNA)? (Use the drastically simplified representation of structure described in Sec. 10-8C in formulating answers.)

26. What is meant by the double helix as applied to nucleic acids?

27. List the three principal steps of the mechanism whereby nucleic acids preserve the information for guiding the synthesis of proteins.

28. What is the genetic code? What is its relationship to the mechanism whereby nucleic acids guide the synthesis of proteins?

Sections 10-9 and 10-10 (Optional)

29. Give an example of how hydrogen bonds serve to determine the three-dimensional shape of protein molecules.

30. Give an example of how disulfide linkages serve to determine the three-dimensional shape of protein molecules.

31. Distinguish between the primary and secondary level of protein structure.

32. What is meant by the denaturation of a protein? What general structural changes take place?

33. What is a dipolar ion?

34. Write a structural formula for the dipolar structure of the amino acid alanine

$$H-\underset{\underset{H}{|}}{\overset{\overset{H}{|}}{N}}-\underset{\underset{CH_3}{|}}{\overset{}{C}}-\overset{\overset{O}{\|}}{C}-O-H$$

35. What is the relationship of the dipolar structure of an amino acid to the actual structure of an amino acid?

SUGGESTIONS FOR FURTHER READING

The following articles are of an introductory nature.

Bolin, Bert: The Carbon Cycle, **Scientific American**, September 1970, p. 124.

Cerami, Anthony, and Charles M. Peterson: Cyanate and Sickle-Cell Disease, **Scientific American**, April 1975, p. 44.

Cloud, Preston, and Aharon Gibor: The Oxygen Cycle, **Scientific American**, September 1970, p. 110.

Cohen, Stanley N.: The Manipulation of Genes, **Scientific American**, July 1975, p. 24.

Davies, David R.: X-Ray Diffraction and Nucleic Acids, **Chemistry**, February 1976, p. 8.

Delwiche, C. C.: The Nitrogen Cycle, **Scientific American**, September 1970, p. 136.

Doty, Paul: Proteins, **Scientific American**, September 1957, p. 173.

Finley, K. Thomas: Landenburg and the Cup of Hemlock, **Chemistry**, January 1968, p. 10.

Frieden, Earl: The Chemical Elements of Life, **Scientific American**, July 1972, p. 52.

Grobstein, Clifford: The Recombinant-DNA Debate, **Scientific American**, July 1977, p. 22.

Keller, Eugenia: The Origin of Life, **Chemistry**, December 1968, p. 6; January 1969, p. 12; April 1969, p. 8.

Kendrew, John C.: The Three-Dimensional Structure of a Protein Molecule, **Scientific American**, December 1961, p. 96.

Penman, H. L.: The Water Cycle, **Scientific American**, September 1970, p. 98.

Phillips, David C.: The Three-dimensional Structure of an Enzyme Molecule, **Scientific American**, November 1966, p. 78.

Riesenberg, Laura B.: Recombinant DNA: The Containment Debate, **Chemistry**, December 1977, p. 13.

Stein, William H., and Stanford Moore: The Chemical Structure of Proteins, **Scientific American**, February 1961, p. 81.

Thompson, E. O. P.: The Insulin Molecule, **Scientific American**, May 1955, p. 36.

Wald, George: The Origin of Life, **Scientific American**, August 1954, p. 44.

The following articles are at a more advanced level.

Crick, F. H. C.: The Structure of the Hereditary Material, **Scientific American**, October 1954, p. 54.

Crick, F. H. C.: The Genetic Code, **Scientific American**, October 1962, p. 66.

Crick, F. H. C.: The Genetic Code: III, **Scientific American**, October 1965, p. 55.

Fox, Sidney W., Kaoru Harada, Gottfried Krampitz, and George Mueller: Chemical Origins of Cells, **Chemical and Engineering News**, June 22, 1970, p. 80.

Holley, Robert W.: The Nucleotide Sequence of a Nucleic Acid, **Scientific American**, February 1966, p. 30.

Kornberg, Arthur: The Synthesis of DNA, **Scientific American**, October 1968, p. 64.

Merrifield, R. B.: The Automatic Synthesis of Proteins, **Scientific American**, March 1968, p. 56.

Nirenberg, Marshall W.: The Genetic Code: II, **Scientific American**, March 1963, p. 80.

Perutz, M. F.: The Hemoglobin Molecule, **Scientific American**, November 1964, p. 64.

Rich, Alexander, and Sung Hou Kim: The Three Dimensional Structure of Transfer RNA, **Scientific American**, January 1978, p. 52.

11

WHAT IS THE RELATIONSHIP BETWEEN CHEMISTRY AND FOOD?

11-1 INTRODUCTION

The balance between the food supply and the population of the world affects not only people who live where food is perennially scarce but everyone. In countries where food is usually abundant, like the United States (Fig. 11-1), shortages bring the penalties of higher prices, which may mean undernourishment for the poorest people. In regions where limited food supplies are the rule rather than the exception, food shortages mean outright starvation. The production and distribution of food is complicated not only by the weather, the fertility of the soil, food spoilage, and the ravages of insects and plant diseases but by economics and politics as well.

There are differences of opinion about the long-term food prospects for the world's population. Those who are inclined to be optimistic point to the almost constant increase in the food supply. They note that present production increases are keeping ahead of population increases, although they concede that reserves are limited and distribution methods in many areas are very poor, making substantial segments of the population vulnerable to stretches of bad weather. The pessimists emphasize the large proportion of the population currently undernourished (probably 50 percent), the rapid increase in the world's population, and the frequency of severe food shortages in certain regions. Whatever the long-term prospects, there is no doubt that food supplies are where they are today largely because the application of scientific knowledge and technology has led to improvements in plant varieties, improvements in irrigation and farming practices, and the increased use of fertilizers and insecticides. Further increases in the food supply to meet the rising population will probably require more intensive application of all these. Limitations to continuing increases in food production, however, are becoming apparent. The environmental costs of insecticides, weed killers, and possibly nitrogen fertilizers may prove forbidding; some of the new high-yielding plant varieties may prove to be excessively susceptible to diseases and pests; and the production of nitrogen fertilizers depends on the shrinking supplies of natural gas and petroleum. In short, even with the judicious use of all the technology we can muster, it is not at all clear that hunger will cease to be a cause of human misery.

This chapter surveys the relationship between chemistry and food supplies, principally fertilizers, insecticides, and food additives. We shall begin by considering the substances found in food and what happens to them in the human body. In closing we shall return to the long-term problems of the world's food supply.

FIGURE 11-1
Harvesting wheat in the state of Washington. (*Courtesy of the Department of Agriculture, photograph by Doug Wilson.*)

11-2 THE PRINCIPAL SUBSTANCES
IN FOODS: CARBOHYDRATES, FATS, AND PROTEINS†

275

11-2 THE PRINCIPAL
SUBSTANCES IN FOODS:
CARBOHYDRATES, FATS, AND
PROTEINS

A. Carbohydrates

The diet of Americans normally averages about 43 percent carbohydrates, 44 percent fats, and 13 percent proteins. **Carbohydrates** are substances containing only carbon, hydrogen, and oxygen. They include a wide range of compounds, from the sugars, which are commonly used as sweetening agents, to substances of complex structure, e.g., starch and cellulose. Since the molecular formulas of most carbohydrates can be reduced to the general formula $C_x(H_2O)_y$, they were once considered to contain the units of water in a weakly combined form. For this reason they were named "hydrates of carbon" since the term hydrate is applied to such substances. The name remains although it has long been established that they are not actually carbon hydrates.

For our immediate purposes the most important thing to know about the structure of carbohydrates is that they are made up of residues of the sugars of relatively simple structure, most commonly glucose, molecular formula $C_6H_{12}O_6$. The structure of glucose is most conveniently represented by one of its ring structures (margin).

A **sugar** is a carbohydrate of simple structure, e.g., glucose, sucrose (table sugar), and fructose. A **monosaccharide** is a carbohydrate, like glucose, which contains only one structural unit. A **disaccharide** is a carbohydrate made up of two structural units, e.g., sucrose (table sugar) and maltose, molecular formulas $C_{12}H_{22}O_{11}$ or $(C_6H_{11}O_5)$—O—$(C_6H_{11}O_5)$. A **polysaccharide** is a carbohydrate containing many structural units. The polysaccharides starch and cellulose are made up of units of glucose residues of molecular formula $C_6H_{10}O_5$. The molecular formulas for starch molecules can be written $(C_6H_{10}O_5)_n$, and the structure of a portion of a starch molecule can be represented in simplified fashion as follows:

Structural formula for
glucose in one of its
ring forms

Representation of a portion of a starch molecule; the number of
glucose residues in starch molecules varies from range of
1000 to 40,000

The action of water breaks the units of polysaccharides up into their component sugar units, in a reaction appropriately called hydrolysis. The hydrolysis of a starch molecule, for ex-

$$(C_6H_{10}O_5)_n \xrightarrow{\text{hydrolysis}} n\,C_6H_{12}O_6$$

Starch   Glucose

Representation in very
general terms of the
hydrolysis of starch
molecules into glucose
molecules

† A more detailed discussion is given in Chap. 10, Secs. 10-2 to 10-5. If you have
studied Secs. 10-2 to 10-5, you may proceed directly to Sec. 11-3.

H
|
H—C—O—H
|
H—C—O—H
|
H—C—O—H
|
H

Glycerol

O
‖
H—O—C—R

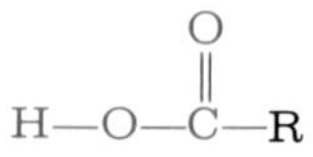

General formula for a carboxylic acid

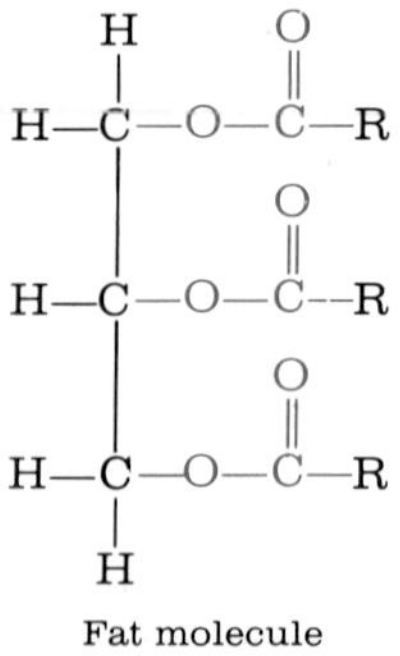

H   O
| ‖
H—C—O—C—R
|
| O
| ‖
H—C—O—C--R
|
| O
| ‖
H—C—O—C—R
|
H

Fat molecule

General formula for a fat molecule. The R groups may be the same or different and usually contain 12 to 18 carbon atoms

ample, produces the molecules of just one product, glucose. This reaction, as we shall see, is a fundamental part of the digestion of carbohydrates.

B. Fats

For the most part **fats** are carboxylic esters (Sec. 10-3) of carboxylic acids containing from 12 to 18 carbon atoms and the alcohol glycerol, which contains three hydroxyl groups. If we use the general formula of a carboxylic acid (Sec. 8-5) the general formula for a fat molecule can be written as shown.

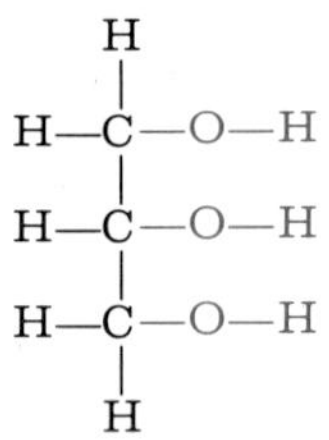

For our immediate purposes the most important chemical property of fats is that they undergo hydrolysis (reaction with water). The focus of the reaction is the carboxylic ester group, which reacts with water to give a carboxylic acid and an alcohol:

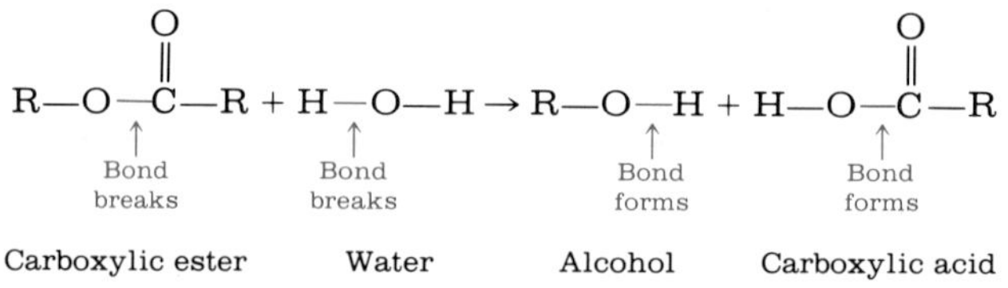

$$R—O—\overset{\overset{O}{\|}}{C}—R + H—O—H \rightarrow R—O—H + H—O—\overset{\overset{O}{\|}}{C}—R$$

Bond breaks   Bond breaks   Bond forms   Bond forms

Carboxylic ester   Water   Alcohol   Carboxylic acid

General representation of the hydrolysis of a carboxylic ester to form an alcohol and a carboxylic acid

When fats undergo hydrolysis, the reaction may occur in stages (see bottom of page).

The hydrolysis of fats is the fundamental reaction occurring during the digestion of the fats in foodstuffs. The complete hydrolysis of a fat molecule yields one molecule of glycerol and three molecules of carboxylic acids. A **fatty acid** is a carboxylic acid obtained from the hydrolysis of fats. The formulas and names of some of the most common fatty acids were given in

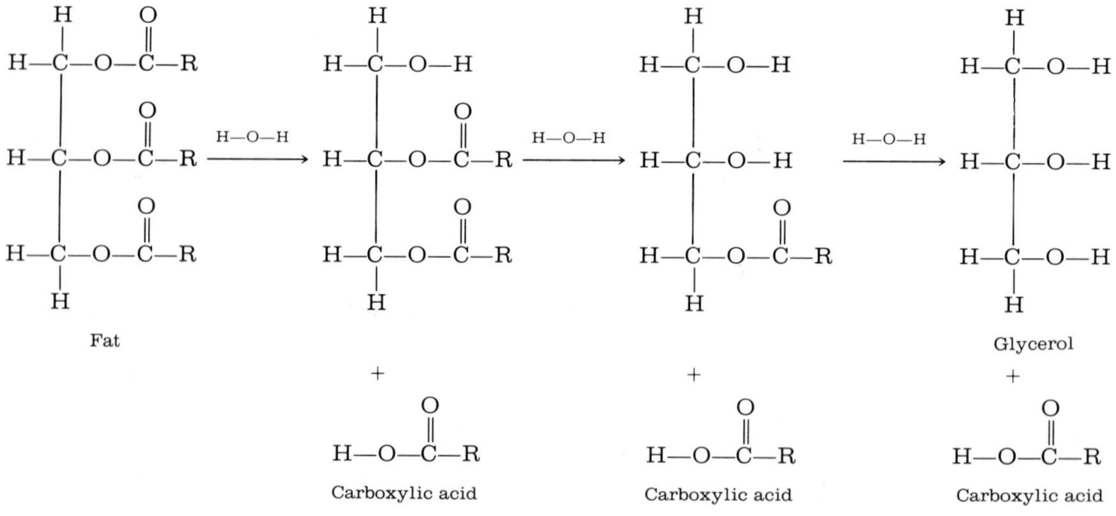

The stepwise hydrolysis of a fat molecule

Table 10-3, and fatty acids obtained from the hydrolysis of some representative fats were summarized in Table 10-4.

C. Proteins

Proteins are large polymeric molecules made up of structural units known as amino acid residues. A portion of a protein molecule may be represented in general terms as

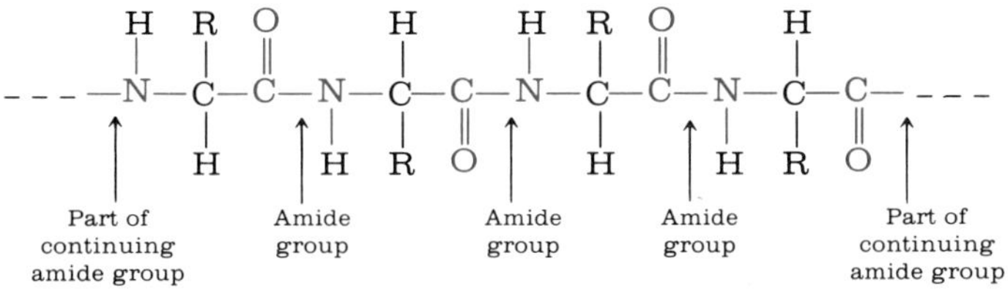

General representation of a portion of the backbone chain of atoms which may extend for hundreds of atoms

The structure includes a backbone of atoms linked by the covalent bonds of amide groups, which may extend for hundreds of atoms. The R symbols represent hydrogen atoms or groups of atoms called side chains.

The amide groups, like all amide groups, are very susceptible to hydrolysis. When, in general, an amide group is hydrolyzed, the products are a carboxylic acid and an amine (Sec. 8-2):

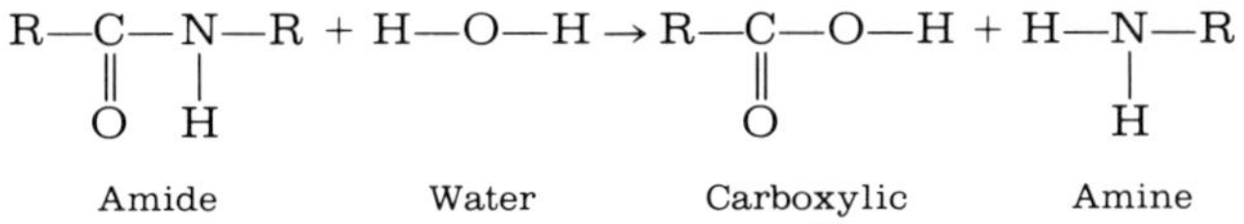

General representation of the hydrolysis of an amide (Sec. 8-3)

By analogy the products of the hydrolysis of proteins can be anticipated:

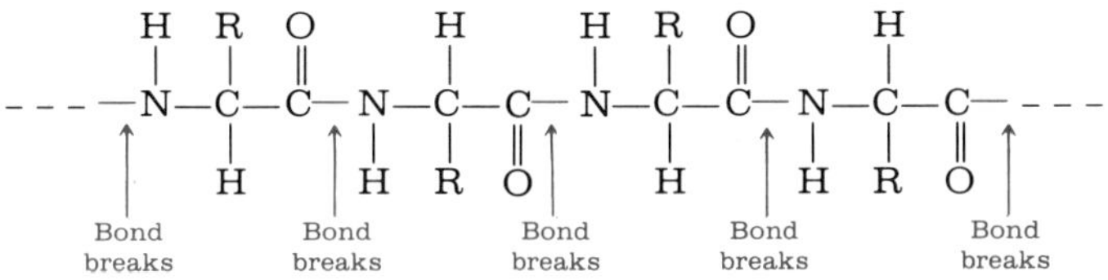

Application of the general representation of the hydrolysis of an amide to the hydrolysis of the amide groups in proteins

IS THERE A RELATIONSHIP BETWEEN THE TYPE OF FATS IN THE DIET AND HEART ATTACKS?

The carboxylic acid, or fatty acid, portions of fat molecules may contain one or more alkene groups. If the fatty acid portions contain relatively few alkene groups, the fat is called a saturated fat. If there are many alkene groups, the fat is called an unsaturated or a polyunsaturated fat. Statistical evidence suggests that a diet relatively high in saturated fats is one of the many factors increasing the probability of heart attacks.

Atherosclerosis, the most common form of

of the arteries, is characterized by the inner linings of blood vessels. ...sits contain fatty materials and choles- ...11-2). Cholesterol is a normal com- ...blood and all body cells. The normal ...ions continuously produce significant amounts, and it is also found in many foods, especially animal fats. The amount of cholesterol in the blood may increase if the cholesterol content of the diet is excessive. There is also evidence to suggest that a high dietary intake of unsaturated fats tends to lower blood cholesterol levels. Although disturbances of the blood vessels and heart are known to occur even when the blood level of cholesterol is low and such disturbances do not necessarily occur when the blood level of cholesterol is high, overall statistical evidence suggests that a high blood level of cholesterol is one of the risk factors increasing the probability of heart disease caused by disturbances of the arteries. (The other risk factors include excessive weight, high blood pressure, emotional stress, family history, lack of physical exercise, and excessive smoking.) Moreover, on the basis of the statistical evidence, a high blood level of choles-

terol appears in turn to be related to the level of saturated fats and cholesterol in the diet, but no specific cause and effect relationship has yet been established between the incidence of heart attacks and the dietary levels of saturated fats and cholesterol.

The structure of cholesterol, a normal component of the blood and all body cells; it is also found in the deposits on the inner lining of blood vessels which cause atherosclerosis; its structure is related to that of vitamin D and certain sex hormones

FIGURE 11-2
Stages in the gradual development of deposits on blood-vessel linings, leading to atherosclerosis: (*a*) normal artery cross section, (*b*) deposits of cholesterol and fatty materials forming on inner lining of artery, (*c*) deposits harden, (*d*) the narrowed channel of the artery is blocked by a blood clot, depriving the heart of blood and resulting in a heart attack. (*Courtesy of the American Heart Association.*)

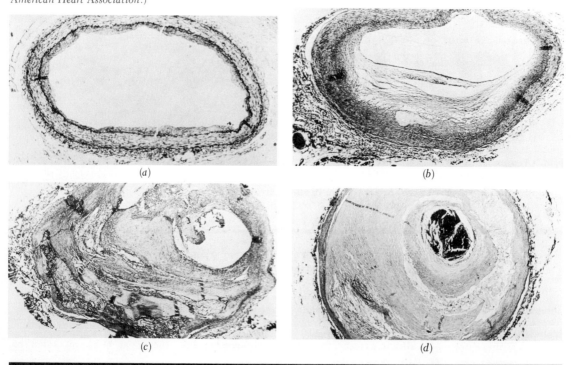

(*a*)　　　　(*b*)

(*c*)　　　　(*d*)

The hydrolyis of protein molecules is the fundamental reaction occuring during the digestion of the proteins in foodstuffs. The products of this hydrolysis are substances with the general formulas shown in the margin. Each product of the hydrolyis contains an amine group and a carboxylic acid group. The term **amino acid** is appropriately applied to these products. The **amino acid residue** is part of the amino acid molecule found in the protein molecule (margin). The residues of more than 20 different amino acids occur commonly in proteins. Still others are found in individual proteins. The names and structures of common amino acids are given in Supplement 7. The structures of six representative amino acids were given in Table 10-5.

Peptide bond is the special name given to the amide bonds of protein molecules. The term **polypeptide structure** is given to the poly-amino acid residue structure of a protein. Some proteins contain a structural unit which is not a polypeptide structure; it is called a **prosthetic group. Conjugated proteins** are those proteins with a prosthetic group, and **simple proteins** are those without. Hemoglobin, the protein of red blood cells which transports oxygen in the blood, is a conjugated protein containing the prosthetic group heme (Sec. 10-4C).

The name **enzyme** is given to the innumerable proteins which serve as the highly important catalysts for the reactions of living systems (Sec. 10-5). Proteins also serve as structural materials, e.g., skin and muscle, transporters of essential substances in the body fluids, e.g., hemoglobin, regulators of essential reactions, and protectors of the organism against harm. The general nature of some representative proteins was given in Table 10-6.

Representation in general
form of the products of
the hydrolysis of the portion
of a protein molecule
shown on page 277

An amino acid, a typical
product of the hydrolysis of
a protein molecule

11-3 WHAT IS PHOTOSYNTHESIS?

Photosynthesis is a chemical process on which all life depends. Its vital function is to trap solar energy and convert it into chemical energy. More specifically, **photosynthesis** is the process by which light energy from the sun is converted into chemical energy as glucose is synthesized from carbon dioxide and water. It takes place in microscopic algae and bacteria of lakes and oceans and in green plants on land. The step-by-step details of the process reveal that it is highly complex, involving almost 100 reactions. A useful general idea of the photosynthesis process can be obtained by considering the overall reaction which occurs in green plants:

$$6CO_2 \;+\; 6H_2O \;+ \text{ solar energy} \rightarrow\; C_6H_{12}O_6 \;+\; 6O_2$$

| Carbon dioxide from atmosphere | Water from environment | Glucose, a building block of carbohydrates | Oxygen to atmosphere |

The overall reaction of the photosynthesis process of
green plants (the actual process involves almost 100 steps);
solar energy is initially absorbed mostly by chlorophyll, a pigment
of green plant cells; the solar energy absorbed is converted
into the chemical energy of the glucose molecules

During this overall reaction carbon dioxide and water from the environment are converted into a sugar, glucose, and oxygen, which is released to the atmosphere. Each year the green plants

Amino acid molecule

Amino acid residue

Comparison of the structure
of an amino acid molecule
with an amino acid residue;
it is the residues which are
found in protein molecules

of the earth bring about the combination of about 550 billion tons of carbon dioxide and 225 billion tons of water to form 375 billion tons of glucose and 400 billion tons of oxygen. Photosynthesis is the only significant way of replenishing the oxygen in the atmosphere.

The major sunlight-absorbing substances of the plants are two chlorophyll molecules of closely similar structure, named chlorophyll a and chlorophyll b, which convert the light energy into chemically useful energy. Their structures are given in Fig. 11-3. The structures of the chlorophylls are closely related

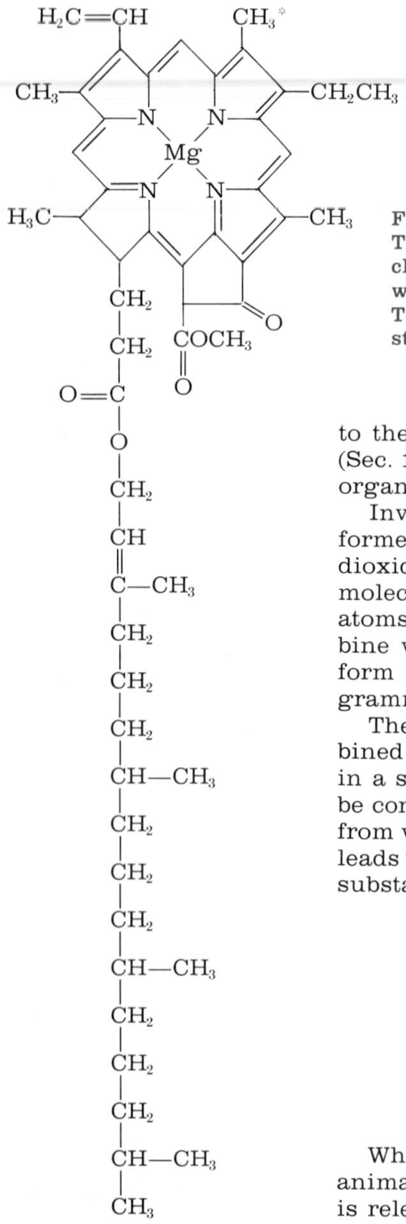

FIGURE 11-3
The structural formula for chlorophyll a. The structure of chlorophyll b differs only at the position marked by the asterisk, where the —CH_3 group is replaced by an aldehyde group, —CHO. Together the chlorophylls are the major light-absorbing substances in green plants.

to the structure of heme, the prosthetic group of hemoglobin (Sec. 10-4). The chlorophylls and heme are synthesized in living organisms by similar pathways.

Investigations have shown that the oxygen produced is formed from the oxygen atoms of the water (not the carbon dioxide). In effect, the light energy of the sun splits the water molecules into hydrogen atoms and oxygen atoms. The oxygen atoms become oxygen molecules, O_2. The hydrogen atoms combine with the carbon and oxygen atoms of carbon dioxide to form glucose, $C_6H_{12}O_6$. These processes are represented diagrammatically in Fig. 11-4.

The glucose molecules formed by photosynthesis can be combined to form the much larger molecules of starch or combined in a somewhat different way to form cellulose. They may also be converted into other simple structures to give a mix of units from which other carbohydrates are assembled. Photosynthesis leads therefore to the synthesis of a whole range of carbohydrate substances:

$$C_6H_{12}O_6 \quad \rightarrow \quad C_{12}H_{22}O_{11} \quad \rightarrow \quad (C_6H_{10}O_5)_n$$

Glucose, a monosaccharide with one structural unit	Maltose, a representative disaccharide	Aproximate formula of starch; n varies from 1000 to 40,000

Molecular formulas of two carbohydrate substances formed from glucose in plants

When the glucose or other carbohydrates are taken into animal bodies as food, the energy absorbed in photosynthesis is released as the carbohydrate material undergoes a complex series of reactions. For glucose this energy-releasing process can be represented by the overall reaction

$$C_6H_{12}O_6 \;+\; 6O_2 \;\rightarrow\; 6CO_2 \;+\; 6H_2O \;+ \text{energy}$$

| Glucose, as a representative carbohydrate | Oxygen, breathed in by animal | Carbon dioxide, exhaled by animal | Water, released in animal body | Available for body processes |

Overall reaction of the processes which make the chemical energy stored as carbohydrate available for body processes

In the photosynthesis of plants solar energy is absorbed while carbon dioxide and water are converted into glucose and then other carbohydrates. We obtain the energy we need by eating the plants (or the flesh of animals which ate the plants) and subjecting the carbohydrate material to a series of reactions which releases the energy. From the standpoint of human survival, the energy-absorbing process of photosynthesis is one of the most important chemical processes occurring about us.

11-4 WHAT IS MEANT BY NITROGEN FIXATION?

Substances containing nitrogen are necessary for the synthesis of proteins, and proteins are essential for all plants and animals. Although we go through life surrounded by a pool of elemental nitrogen (it makes up more than 75 percent of the atmospheric gases), we are unable to use it despite our vital need for nitrogen-containing compounds. As far as natural processes are concerned, we and most other living organisms are mainly dependent on certain types of bacteria and blue-green algae to convert the elemental nitrogen of the atmosphere into nitrogen-containing substances.

Nitrogen fixation is any process converting elemental nitrogen into a nitrogen-containing compound. The energy supplied by lightning (or the intense heat of an internal combustion engine) can bring about the direct combination of nitrogen with oxygen in the atmosphere to form oxides of nitrogen (Sec. 15-2D). A much greater contribution is made, however, by certain nitrogen-fixing bacteria and blue-green algae, which are able to convert the elemental nitrogen of the atmosphere into ammonia, NH_3. We are so used to thinking of bacteria as being harmful, it may come as a surprise to find that some types are helpful. Just how they manage the chemical transformation is not completely understood, but it is known that enzymes are of central importance.

Microorganisms that carry out nitrogen fixation are found in a wide variety of places: the seas, on rocks, in the roots of plants, in decaying wood, in the gut of insects. Some of them live by their own devices; others live in a **symbiotic relationship** with other living organisms, i.e., in a way that is advantageous to both. The outstanding examples are the bacteria which invade the roots of soybean, alfalfa, clover, and other legumes (Fig. 11-5). The bacteria live within nodules on the roots. Through photosynthesis the plants provide substances which supply energy to the microorganisms. The microorganisms, on their part, supply the plants with nitrogen-containing compounds.

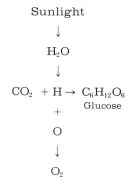

FIGURE 11-4
Representation of the overall process of photosynthesis, emphasizing that the sunlight serves to split the water molecules into hydrogen atoms, H, and oxygen atoms, O.

The nitrogen-fixation reaction caused by the action of lightning on the nitrogen and oxygen in the atmosphere

FIGURE 11-5
Sketch of soybean plant indicating position of the root nodules in which nitrogen-fixing bacteria live in a symbiotic relationship with the plant.

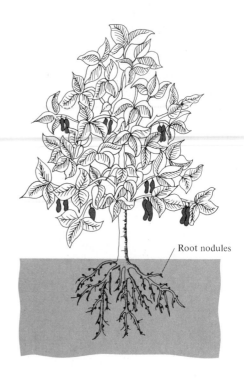

Root nodules

N₂

Nitrogen

+

3H₂

Hydrogen

↓

2NH₃

Ammonia

The overall reaction of the nitrogen-fixing Haber process

The ammonia produced is converted by the plants into proteins. Some of this plant protein is either ultimately returned to the air as nitrogen (with the aid of other bacteria) or finds its way back to the soil as fertilizing substances. The remainder is consumed by animals and humans and by way of many transformations is also ultimately returned either to the air as elemental nitrogen or to the soil as fertilizing substances.

The nitrogen-fixing microorganisms initiate the production of nitrogen-containing substances which are important at several stages of the cyclic processes of nature. Their function as plant fertilizers is one of the most important. In order to increase food supplies farmers have tried to increase soil fertility. For centuries this was limited to adding natural nitrogen-containing substances to the soil in the form of compost and manure, but for about the last 70 years large-scale industrial production of ammonia directly from nitrogen and hydrogen has made nitrogen-containing fertilizers much more plentiful. This has been made possible by the development of the Haber process (Sec. 13-1A). Each year about 55 million tons of nitrogen are converted into ammonia by this process, but even this great amount is less than one-third of the estimated total production of the microorganisms.

11-5   THE PROCESSES OF DIGESTION

Digestion is the series of changes whereby the complex substances of foods are converted into substances of simpler structure which can be absorbed by the body. Digestion takes place in the organs constituting the digestive tract (Fig. 11-6). In humans this includes principally the mouth, the esophagus, the stomach, and the intestines. Absorption of the products of digestion takes place mainly in the small intestine. Molecules of food substances are too large to pass through the membranes of the digestive tract. They must be broken up into smaller molecules before they can be absorbed. For the most part the reactions of digestion involve the hydrolysis of carbohydrates, fats, and proteins (Table 11-1).

FIGURE 11-6
The digestive tract consists principally of the mouth, esophagus, stomach, and intestines.

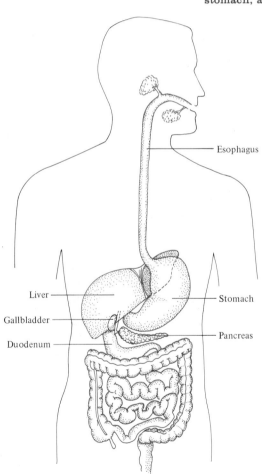

Esophagus

Liver

Stomach

Gallbladder

Pancreas

Duodenum

Food	Weight percent				Kilocalories per 100 grams
	Water	Proteins	Fats	Carbohydrates	
Whole milk	87	3.7	3.7	4.9	66
Cheddar cheese	37	25	32	3.6	411
Whole eggs	74	12	12	Trace	160
Butter	16	0.9	81	0.9	717
Margarine	16	0.9	81	0.9	721
Broiled hamburger	54	25	20	0	288
Broiled chicken	71	24	3.5	0	135
Roast lamb	54	26	19	0	276
Canned tuna fish	61	28	8.2	0	200
Shelled almonds	5	18	54	20	599
Peanut butter	2	25	50	19	594
Cooked lima beans	71	7.6	0.6	20	112
Canned corn	81	2.0	0.8	16	66
Cooked peas	82	5.6	0.6	12	72
Baked potatoes	75	3.0	Trace	21	90
Cooked white rice	73	2.0	Trace	24	110
Raw apples	85	Trace	Trace	12	47
Raw bananas	76	0.6	Trace	15	57
Raw pineapple	85	0.7	Trace	14	54
White bread	36	8.6	3.3	50	270
Whole wheat bread	36	9.0	2.6	49	242
Cola	90	0	0	10	393
Prepared gelatin dessert	84	1.7	0	14	58
Cheese pizza	45	9.3	8.0	36	411

A. Carbohydrates

$$(C_6H_{10}O_5)_n \xrightarrow{\text{hydrolysis}} nC_6H_{12}O_6$$

Starch　　　　　　　Glucose

The hydrolysis of starch to glucose, which takes place during digestion

We ingest carbohydrates mostly in the form of sugars and starch. The molecules of starch can be represented as $(C_6H_{10}O_5)_n$ (Sec. 11-2A), and the digestive process converts them into glucose molecules by hydrolysis (Sec. 11-2A). The sugars of food are hydrolyzed at the same time into three monosaccharides, glucose, fructose, and galactose. After the action of the liver, however, they are all converted into glucose, the only sugar in the bloodstream.

Digestion begins in the mouth. (It may start earlier, when food is marinated and when it is cooked.) The teeth break up the food particles into small units, and the enzyme ptyalin in the saliva catalyzes reactions which start the hydrolysis of the starch and sugars. Most hydrolysis of starch, however, takes place in the small intestine, catalyzed by the enzymes amylopsin and maltase.

B. Fats

For the most part the digestion of fats is delayed until the food particles reach the upper portion of the small intestine. Here the digestive process is facilitated by bile from the gallbladder, which converts the large water-insoluble units of fats into smaller units more susceptible to hydrolysis. Enzymes from the pancreas catalyze the hydrolysis of the ester groups of the fat molecules (Sec. 11-2B). The products of hydrolysis are not entirely understood, but probably the fats do not have to be completely hydrolyzed in order to be absorbed by the intestine wall.

C. Proteins

The digestion of proteins involves hydrolysis of their peptide bonds (amide bonds) (Sec. 11-2C). It begins in the stomach, where the enzymes of the gastric juice (principally pepsin) catalyze partial hydrolysis of the proteins. The process is completed in the small intestine, aided by the enzymes trypsin and chymotrypsin. The end products are individual amino acids, which can be absorbed by the intestinal wall.

To summarize, the principal products of human digestion are glucose from carbohydrates; glycerol, fatty acids, and partially hydrolyzed esters from fats; and amino acids from proteins (Table 11-2). These products are absorbed into the body fluids and become the starting substances for the synthesis of the innumerable compounds needed for various bodily functions.

Table 11-2
The Principal Products
of Human Digestion

Class of substances in food material	Principal products of digestion
Carbohydrates	Glucose
Fats	Glycerol, fatty acids†
Proteins	Amino acids

† After the hydrolysis of fats has been completed following digestion.

11-6 WHAT HAPPENS TO THE PRODUCTS OF DIGESTION?

Digestion converts the carbohydrates into glucose, the fats into glycerol, fatty acids, and partially hydrolyzed fats, and the proteins into amino acids. These are the basic starting substances for the numerous reactions essential to our well-being. **Metabolism** refers to all the chemical processes which food substances undergo. We are concerned here with the chemical changes which the products of digestion undergo during the synthesis of the substances needed for our bodily functions and the generation of the needed energy. Many of these reactions are understood rather well, but uncertainties are numerous.

A. Metabolism of the Digestion Product of Carbohydrates

Glucose, a key substance in the metabolism of all animals, performs many different roles. It is directly involved in the generation of energy. Mention was made in Sec. 11-3 of the overall reaction between glucose and oxygen, which produces considerable energy:

$$C_6H_{12}O_6 + 6O_2 \rightarrow 6CO_2 + 6H_2O + energy$$

| Glucose | Oxygen, inhaled by lungs | Carbon dioxide, exhaled by lungs | Water, released in body | Available for bodily processes |

Overall reaction of the oxidation of glucose, which provides energy for bodily processes

This overall reaction parallels the oxidation of such common fuels as gasoline, coal, and wood when they are burned to gen-

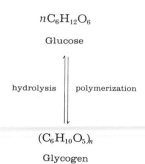

$$n\,C_6H_{12}O_6$$

Glucose

hydrolysis | polymerization

$$(C_6H_{10}O_5)_n$$

Glycogen
(*n* ranges from
about 10,000 to
about 60,000)

**Glucose is stored in the liver
and muscle tissue in the
form of a polymer, glycogen;
the glucose is obtained
when needed by hydrolysis**

erate heat. The oxidation of glucose, however, takes place in a series of individual steps which release the energy gradually and mostly in a better form than heat for use in the body's reactions.

Glucose is also the starting point in the body for the synthesis of other carbohydrates, as well as fats and proteins. Some of it is converted into a polymer of many glucose units, called glycogen, or animal starch. Most glycogen is stored in the muscles and liver, serving as a reserve supply of glucose, which is readily obtained from it by hydrolysis.

B. The Metabolism of the Digestion Products of Fats

The hydrolysis of fats into glycerol and fatty acids is only partially carried out during digestion, but it is completed as the digestion products are acted upon after absorption. Fats serve as the richest energy source in the body. This energy is obtained in a stepwise fashion as the glycerol and fatty acids undergo reaction with oxygen in a series of reactions.

Many of the fatty acids are reassembled into fats, which are stored as a reserve supply under the skin in many areas of the body. These depots of reserve fat are called **adipose tissue.** Some of the reassembled fats protect vital organs like the kidneys against injury and insulate the body against loss of heat energy. Fatty acids also serve as a starting point for the synthesis of substances with important special functions in the body, e.g., components of nerve and brain tissue. Although people can survive on an almost fat-free diet, obtaining their energy requirements from carbohydrates and proteins, certain fatty acids cannot be synthesized in the human body and must be obtained from fats. The term **essential fatty acids** is applied to those fatty acids which the human body must obtain from food.

Our bodies can convert carbohydrates into fats but lack the necessary enzymes to convert fats into carbohydrates. When someone eats more carbohydrates and fats than are needed for immediate bodily functions, much of the excess will probably be stored as fat. If the excess is appreciable, the person gains too much weight. The insulating effect of the excess fat causes overweight people to lose heat less readily, increasing their discomfort in hot weather. Excess fat also develops stresses in the body which shorten the life-span, possibly through increasing the risk of coronary heart disease. To reduce the amount of stored fat, the person must eat fewer carbohydrates and fats, forcing the body to use stored fat for energy.

C. The Metabolism of the Digestion Products of Proteins

The body uses the amino acids obtained from digestion to synthesize the innumerable protein substances it needs for structural materials, enzymes, and other special purposes. The body can also consume amino acids to generate energy or use them as starting materials in the synthesis of glycogen and fats.

The liver is the principal organ where amino acids obtained from digestion are converted into the amino acids needed for proteins. Certain **essential amino acids,** which the body cannot synthesize at all or fast enough, must be obtained directly from the proteins in food. Examples are listed in Table 11-3. Many of these essential amino acids can be obtained from both plant and animal proteins, but plant proteins tend to be deficient in two, lysine and methionine. If the diet is deficient in any of the essential amino acids, the body does not have all of the amino acids required to synthesize the needed proteins and a wide range of disorders may result.

Significant metabolic functions of glucose, fatty acids, glycerol, and amino acids are summarized in Table 11-4. Note the considerable overlapping in the functions of these substances. All of them, for example, can be drawn upon as a source of energy. Glucose can be converted into both fats and proteins, and amino acids can be converted into fats and carbohydrates.

11-7 VITAMINS

It was discovered at the beginning of this century that for good health the diet must contain small quantities of specific substances called vitamins, in addition to carbohydrates, fats, and proteins. This became apparent during the investigation of certain diseases which were eventually traced to the lack of certain vitamins in the diet. The term **deficiency disease** is applied to ailments which are caused not by poisons or invading parasites but by the absence of an essential dietary substance.

Scurvy, the first deficiency disease to be recognized, occurred among sailors on long voyages. The symptoms were general weakness, loss of weight, bleeding gums, and loosening teeth. It was learned early in the eighteenth century that the juice of citrus fruits is effective in preventing this disease. The daily

Table 11-3
Essential Amino Acids
in Human Adults†

Arginine
Isoleucine
Leucine
Lysine
Methionine
Phenylalanine
Threonine
Tryptophan
Valine

† The names and formulas of all amino acids are given in Table S7-1. A group of six were discussed in Table 10-5.

Table 11-4
Significant General Aspects of Human Metabolism

Fundamental substances and sources	Functions
Glucose, from reactions of carbohydrates in digestion and some subsequent processes; polymerized into glycogen for storage as reserve supply	Source of energy; starting substance for synthesis of fats, proteins, and substances needed for special purposes
Fatty acids and glycerol, from reactions of fats in digestion and some subsequent processes; converted into body fats for storage as reserve supply	Source of energy; starting substances for synthesis of (1) body fats for insulation and protection of organs; and (2) for special substances, especially for nerve and brain tissue
Amino acids, from reactions of proteins during digestion	Starting substances for synthesis of (1) structural proteins, (2) enzymes and other substances needed for special purposes, and (3) fats and carbohydrates; source of energy

ration of lemon or lime juice which in 1804 became mandatory for all members of the British navy, is responsible for the nickname "limey" given British sailors. Scurvy is now known to be caused by a deficiency of vitamin C. This substance is widely distributed in foods, but since it is sometimes destroyed during storage and cooking, one of the best sources is fresh citrus fruits.

Beriberi, another deficiency disease, was among the first to be corrected by changes in diet. It is accompanied by deterioration of the nervous system and emaciation. In the late nineteenth century the Japanese navy checked the high incidence of this disease among its sailors by replacing a portion of the diet staple, polished rice, with wheat and barley. About the same time a Dutch army surgeon, Christiaan Eijkman, showed that chickens with beriberi, caused by an experimental diet of polished rice, were cured if the polishings removed from the rice were added to their diet. In 1911 the Polish biochemist Casimir Funk isolated a nutritionally active concentrate from rice polishings. Because investigations showed that it contained a substance with an amine group, he coined the word "vitamine." Even though subsequent investigations have showed that most vitamins do not even contain nitrogen, the name, without the e, has stuck.

Within the next three decades additional essential dietary factors came to be recognized, and eventually the substances responsible for the activities were isolated and characterized. The essential factors were originally designated by letters of the alphabet. The dietary factor essential for the prevention of an eye disease was called vitamin A. The factor found in rice polishings was called vitamin B. Later investigations showed that some of these factors contained more than one substance, and subscripts in the letter designations began to be used. Vitamin B became a group, or complex, of vitamins designated by names like vitamin B_1 and vitamin B_2.

The name **vitamin** is now applied to a group of about 20 carbon-containing substances known to be necessary in the diet for normal health in addition to the usual carbohydrates, fats, and proteins. Some of them have names which include the word vitamin, such as vitamin A and vitamin B_6, but others have special names, such as niacin and folic acid. Still others are known by two names; vitamin C, for example, is also called ascorbic acid.

Present evidence suggests that most vitamins are essential for body chemistry because they are involved in vital enzyme systems. This probably accounts for their ability to be effective in very small amounts. Thiamine (vitamin B_1), for example, is structurally related to a substance which is part of an enzyme important in the metabolism of carbohydrates.

The isolation and characterization of the vitamins represent a series of triumphs for chemists; since many of them are extremely complex, unraveling their structures has required much chemical ingenuity. The structures of vitamins are widely varied. Vitamin C, isolated from the dietary factor which prevents scurvy (Fig. 11-7), has one of the simplest structures and is closely related to the structure of some sugars. Vitamin B_1 (thiamine), isolated from the dietary factor which prevents beriberi, has a more complex structure. It contains the amine

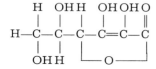

Vitamin C (ascorbic acid); this structure is related to that of glucose (Sec. 11-2A) and contains only carbon, hydrogen, and oxygen atoms

group which led to the name vitamin. Molecular formulas, dietary sources, and deficiency consequences for a number of vitamins are given in Table 11-5.

Structure determinations of the molecules of most vitamins have led to a method of synthesis suitable for large-scale production although a few are still easier to obtain from natural sources. Large-scale production has made many vitamins much more available and less expensive than they would otherwise be.

Most normal diets in regions of ample food supplies provide all the necessary vitamins in the amounts required by the body, but a vitamin deficiency may arise in several circumstances. The scarcity of food in many areas of the world leads to diets inadequate in vitamin content. Chronic alcoholism can lead to a low vitamin intake, and sometimes extreme dieting to lose weight can have the same result. In such cases or where there are abnormalities in food absorption or body requirements adding vitamins to the diet is appropriate. In a few instances large doses of vitamins are helpful in treating some disorders related to vitamin deficiency, but excess vitamins are to be avoided. Some of them, especially those insoluble in water, can cause toxic reactions. Large doses of vitamin D, for example, are reported to cause vomiting, diarrhea, and fatigue.

Vitamin B_1 (thiamine); note the presence of the amine group which led to the name vitamin

FIGURE 11-7
Crystals of vitamin C as seen through a light microscope. (*Courtesy of Merck and Co., Inc.*)

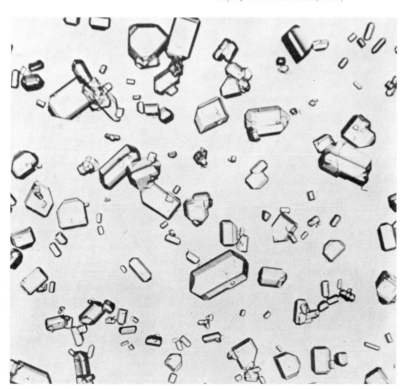

Table 11-5
Molecular Formulas, Dietary Sources, and Deficiency Consequences for Several Vitamins

Vitamin	Molecular formula	Some deficiency consequences	Some dietary sources
Vitamin A	$C_{20}H_{30}O$	Low resistance to infection, night blindness, blindness	Liver, kidneys, egg yolk, butter, fortified margarine
Vitamin B₁ (thiamine)	$C_{12}H_{17}ON_4SCl$	Impairment of nervous system, poor appetite, mental depression, beriberi	Meats, poultry, beans, peas, bread, milk
Vitamin B₂ (riboflavin)	$C_{17}H_{20}O_6N_4$	Skin lesions, retarded growth, burning sensitive eyes	Liver, meats, milk, eggs, vegetables
Niacin	$C_6H_5O_2N$	Skin lesions, diarrhea, pellagra	Meat, poultry, fish, vegetables, rice
Vitamin B₆ (pyridoxine)	$C_8H_{11}O_3N$	Gastrointestinal disturbances, nervousness, irritability	Meat, whole grains, vegetables
Vitamin B₁₂	$C_{63}H_{90}O_{14}N_{14}PCo$	Anemia	Milk, meat, poultry, eggs, fish
Folic acid (folacin)	$C_{19}H_{19}N_7O_6$	Anemia in pregnancy, sprue	Vegetables, meats, poultry, fish, eggs
Vitamin C (ascorbic acid)	$C_6H_8O_6$	Impairment of gums and teeth, hemorrhage, scurvy	Citrus fruits, other fruits, vegetables
Vitamin D	$C_{27}H_{44}O$	Impairment of bones, rickets	Fish-liver oils, fortified milk, exposure to sun
Vitamin E (α-tocopherol)	$C_{29}H_{50}O_2$	Uncertain (sterility?, more rapid aging?)	Salad oils, whole grains, nuts, vegetables
Vitamin K	$C_{31}H_{46}O_2$	Hemorrhage, impairment of blood clotting	Vegetables, soybeans

The addition of small amounts of vitamins to foods is common for preventing possible vitamin shortages. Flour, for example, is commonly fortified with vitamin B₁ (thiamine), vitamin B₂ (riboflavin), and niacin (another member of the B vitamin complex) to compensate for the vitamin losses during milling. (In many states this is required by law.) Vitamin A is commonly added to margarine to give it the same vitamin content as butter. Almost all fresh milk or canned evaporated milk is fortified with vitamin D. The fact that the deficiency disease rickets is rarely found in the United States is attributed to this added source of vitamin D in the diet of infants.

Few substances, however, have been involved in as much misunderstanding and misuse as the vitamins. Excessive and unproved claims for vitamin treatments are all too common, and some preparations are overpromoted. The use of vitamin tablets is justified in many cases, but it is best to be cautious and to be guided by professional advice.

11-8 WHAT MINERALS ARE ESSENTIAL IN FOODS?

The composition of the average adult human body in terms of the percent by weight of total atoms is given in Table 11-6. One of the many striking facts in this table is that about 96 percent by weight of the human body is due to only four different kinds of atoms, hydrogen, oxygen, carbon, and nitrogen. This is not surprising, however, when one remembers (Table 10-7) that 60 percent of the body is water (hydrogen and oxygen atoms), another 18 percent is fat (mostly carbon, hydrogen, and oxygen atoms), and 16 percent is protein (mostly carbon, hydrogen, oxygen, and nitrogen atoms). (We have not previously mentioned that a few fats contain small amounts of phosphorus and nitrogen. Proteins may contain small amounts of phosphorus and sulfur and smaller amounts of iron and copper.)

The remaining 4 percent of the weight of the human body is composed of some 20 different atoms. Many of the substances containing these atoms are referred to as **minerals,** a designation emphasizing their inorganic (noncarbon) nature. Even though these atoms are present in small amounts and some only in traces, most of them have critically important functions and must be present in foods for good health to prevail.

About 2 percent of the weight of the human body is due to calcium atoms, about 99 percent of which are found in the bones and teeth to which they give rigidity. About another 1.2 percent of the body weight is due to phosphorus atoms, of which between 80 and 90 percent are also part of the structural substances of bones and teeth. Table 11-7 lists the principal atoms in the mineral content of the body, together with their principal roles and dietary sources.

Recognition of the important role of iodine in synthesizing the thyroid hormone has led to the common practice of adding small amounts of potassium iodide to table salt (iodized salt) in order to offset the dietary shortage. (Iodine compounds are most common in foods from the sea.) The thyroid hormone, thyroxine, regulates important processes of the body. An iodine deficiency restricts the amount of this important hormone, causing a depression in the rate of many body reactions and a condition known as goiter, involving enlargement of the thyroid gland.

Table 11-6
Composition† of the Average Adult Human Body in Terms of the Percent by Weight of Atoms

Atom	Percent by weight
Oxygen	65
Carbon	18
Hydrogen	10
Nitrogen	3.0
Calcium	2.0
Phosphorus	1.2
Potassium	0.35
Sulfur	0.25
Sodium	0.15
Chlorine	0.15
Magnesium	0.05
Iron	0.004

† Plus traces of iodine, fluorine, silicon, manganese, cobalt, copper, zinc, boron, aluminum, vanadium, and molybdenum.

11-9 THE ROLE OF FERTILIZERS

The tremendous expansion of the use of concentrated fertilizers in the last 30 years has been a major factor in keeping food production ahead of population increases (Fig. 11-8). In many of the poorer countries, however, the application of the newer fertilizers is only beginning. It has been estimated that each ton of fertilizer used in the soil of many of the less developed countries increases the harvest by about 10 tons. The large-scale use of chemical fertilizers has been made possible by the tremendous strides achieved by the chemical industry in supplying fertilizing substances inexpensively and in quantity. The development of the Haber process for the synthesis of ammonia, for example, was an important turning point (Sec. 11-4), and most nitrogen-containing fertilizers now originate from this source (Fig. 11-9). Yet there is currently a worldwide shortage of fertilizers, worsened by the increasing energy short-

Mineral Substances in the Average Human Adult Body

	Percentage of body weight	Principal functions	Principal dietary sources
Calcium	2.0	Structural material of bones and teeth	Milk, cheese, cabbage, broccoli, shellfish
Phosphorus	1.2	Structural material of bones and teeth; enzymes; fat transport	Milk, cheese, meats, poultry, fish, whole grains
Potassium	0.35	Fluid of cells; water balance; nerve-impulse transmission	Meat, cereals, fruits and juices, vegetables
Sulfur	0.25	Protein substances	Meat, cereals, vegetables
Sodium	0.15	Fluid between cells; water balance; nerve stimulation	Table salt, baking powder, milk, meats, poultry, fish
Chlorine	0.15	Fluid between cells	Table salt, milk, meats, poultry
Magnesium	0.05	Structural material of bones and teeth; nerve-impulse transmission; enzymes	Vegetables
Iron	0.004	Hemoglobin (transport of oxygen); enzymes	Meats, fish, poultry, eggs, whole grains, vegetables
Iodine	0.00004	Thyroid gland	Shellfish, saltwater fish, iodized salt
Fluorine	Trace	Bones and teeth	Fluoridated water
Zinc	Trace	Enzymes	Meats, vegetables

age (Sec. 13-1C). This has made the improved fertilizers too expensive for many regions of the world which need them most. It has been estimated that the needed increase in production would cost $8 billion a year for 3 or 4 years and even more after that. But precautions are necessary; when fertilizers get into natural water systems, they can cause harmful pollution.

Growing plants require carbon dioxide, water, and sunlight for the process of photosynthesis. In addition, compounds containing at least a dozen other atoms are necessary. Those of nitrogen, phosphorus, and potassium are of major importance. Compounds of nitrogen are required for the proteins of plants. The most common sources of nitrogen in the soil are ammonium ions, NH_4^+, and nitrate ions, NO_3^-. Plants convert them into amino acids, with the aid of carbon compounds from the process of photosynthesis.

If left undisturbed, all the plant nutrients are eventually returned to the soil when the plants die and decay. This is true even if the plants are consumed by animals, for in time they also die and decay. Through crop harvesting and animal husbandry, however, plants and animals may be carried away from the place where they grew; then the source of valuable plant nutrients is lost as the natural replenishing processes are interrupted. Harvesting 1 ton of wheat, for example, removes

FIGURE 11-8
World fertilizer production, 1920 to 1975. The tonnage is calculated in terms of nitrogen as N, phosphorus as P_2O_5, and potassium as K_2O.

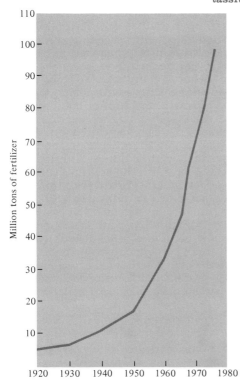

FIGURE 11-9
Gaseous ammonia being applied directly as a fertilizer. The ammonia is transported in pressurized tanks and released into the soil at a depth of 6 to 8 inches. (*Courtesy of the Department of Agriculture; photograph by Jens Jensen.*)

about 62 pounds of combined nitrogen, about 24 pounds of combined phosphorus, and about 38 pounds of combined potassium. Nutrient substances are also lost by erosion, which is appreciable in areas of high rainfall.

The nutrients lost from the soil must be replaced if the soil is to continue to furnish crops in high yield (Fig. 11-10). Organic fertilizers (manure, bones, dried blood) and some inorganic substances (calcium carbonate, sodium nitrate, potassium nitrate) have been used for centuries. Organic fertilizers continue to be a significant way of replenishing soil nutrients. They are a major source of fertilizing substances on small farms, but the pressures of rapid population growth and higher standards of living have demanded increased efficiency in agricultural methods and have led to the use of more concentrated fertilizers.

Since most plants consume larger quantities of combined nitrogen, phosphorus, and potassium than other soil nutrients, these are the most easily exhausted from the soil and the ones most in need of replacement by fertilizers. Fertilizers containing these combined elements are marketed in mixtures with their nitrogen-phosphorus-potassium content identified by a numerical system. A 14-10-8 mixture, for example, contains 14 percent combined nitrogen, 10 percent combined phosphorus (calculated as though it were P_2O_5), and 8 percent combined potassium (calculated as though it were K_2O). Fertilizers are generally applied in the form of solid granules and solutions. Table 11-8 lists some of the principal substances used in fertilizer mixtures.

FIGURE 11-10
The effect of fertilizer on the growth of wheat. The wheat on the left received no fertilizer and yielded grain at the rate of about 8 bushels per acre. The wheat on the right was fertilized with 500 pounds of 10-10-10 fertilizer per acre and yielded 52 bushels per acre. (*Courtesy of the Department of Agriculture.*)

Table 11-8
Some Substances Used in Fertilizer Mixtures

Nitrogen sources:	Calcium and phosphorus sources:
Ammonia, NH_3	Calcium dihydrogen phosphate, $Ca(H_2PO_4)_2$
Ammonium nitrate, NH_4NO_3	Calcium hydrogen phosphate, $CaHPO_4$
Urea, H_2NCONH_2	
Ammonium sulfate, $(NH_4)_2SO_4$	Potassium sources:
Sodium nitrate, $NaNO_3$	Potassium chloride, KCl
	Potassium sulfate, K_2SO_4
Nitrogen and phosphorus sources:	
Ammonium dihydrogen phosphate, $NH_4H_2PO_4$	
Ammonium hydrogen phosphate, $(NH_4)_2HPO_4$	

11-10 CAN WE DO WITHOUT CHEMICAL INSECTICIDES?

Some insects are beneficial and important to agriculture; e.g., bees pollinate flowering plants, and ladybugs eat smaller insects. Many insects, however, cause damage on a large scale (Figs. 11-11 and 11-12). It has been estimated that even with the intensive use of insecticides in the United States, insects destroy about 10 percent of all crops (Fig. 11-13). This means an annual financial loss between 5 and 6 billion dollars. Damage in many of the less developed countries is much greater. Government officials in Tanzania, for example, estimate that insects consume 25 percent of all crops, and the estimates in Kenya run as

FIGURE 11-12
Forest on Cape Cod, Massachusetts, devastated by an attack of gypsy moths, July 1970. The trees will revive, but birds and other wildlife have left. (*Courtesy of the Department of Agriculture; photograph by Larry Rana.*)

FIGURE 11-11
A gypsy moth larva chews on a leaf. (*United Press International.*)

high as 75 percent. At times insect populations reach appalling concentrations. A locust swarm in eastern Africa was reported so dense that it blacked out most daylight along a front 1 mile wide and took 9 hours to pass over.

Insect-borne diseases are a major cause of illness and death. It has been estimated that more people die annually from diseases carried by mosquitoes than from any other single cause. Despite the strong efforts to control this insect, mosquito-borne malaria alone kills more than 1 million people each year, and the disease afflicts about 100 million more. Sleeping sickness, carried by the tsetse fly, is so prevalent in a large area of central Africa that an estimated 1.7 billion acres of potential food-producing land is rendered relatively useless. (The role of insects in disease is discussed further in Sec. 12-3.)

Entomologists have identified almost 1 million different kinds of insects and estimate that there may be 3 or 4 million others. The combined weight of insects has been calculated to be about 12 times the weight of all the people on the earth. They have been around for about 400 million years and are well equipped for survival. Many lay hundreds of thousands of eggs after each mating, and the life-span of some is so short that dozens of generations can be produced in one season. Some species are able to stand temperatures as high as 120°F (49°C) and as low as −30°F (−34°C). The ability of most insects to fly and their extraordinary vision provide them with exceptional means of evading their enemies. (Tried to catch a fly lately?)

A. Insecticides

About 10,000 insect species are significantly harmful to agricultural productivity or to human health, and attempts to control them go back to the beginning of civilization. In ancient times locusts were fought by crushing them. Infusions prepared from tobacco were used as early as the seventeenth century to combat an insect pest of pear trees. The mixture contains the effective insecticide nicotine (Table 11-9), still in current use. A preparation obtained from the flowers of a species of chrysanthemums was used early in the nineteenth century as a flea and body louse powder. This is now known to contain several effective insecticides known collectively as pyrethrins (Table 11-9). These are still used, especially in sprays used on cattle and for household purposes. Rotenone (Table 11-9), another insecticide in current use, is present in plant preparations first used in 1848 to combat leaf-eating caterpillars.

Table 11-9
Some Representative Insecticides

Insecticide	Use
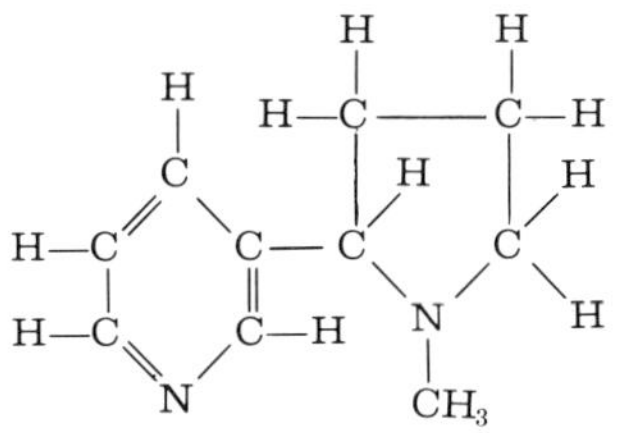 Nicotine, $C_{10}H_{14}N_2$	One of the principal insecticides in tobacco, from which some of earliest insecticides were obtained
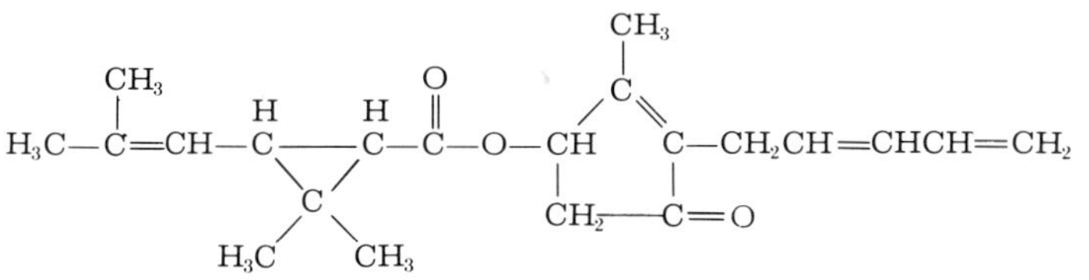 Pyrethrin I, $C_{21}H_{28}O_3$	One of the insecticides found in pyrethrum, obtained from a species of chrysanthemums

(Continued on next page)

Table 11-9
Some Representative Insecticides (*Continued*)

Insecticide	Use

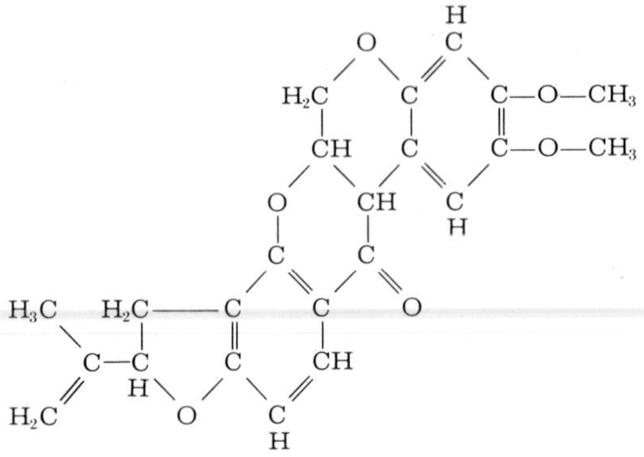

Rotenone, $C_{23}H_{22}O_6$

One of the insecticides obtained from derris root, used for over 100 years

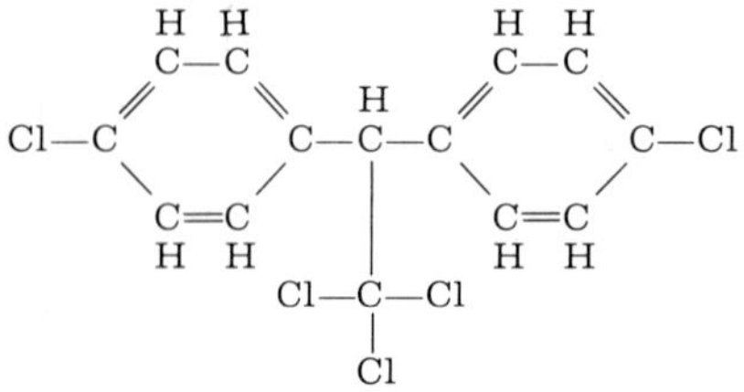

DDT (dichlorodiphenyltrichloroethane), $C_{14}H_9Cl_5$

An inexpensive insecticide effective against many different kinds of insects; synthesized in 1874 but not recognized as insecticide until 1939; use now heavily restricted because of potential harmful effects

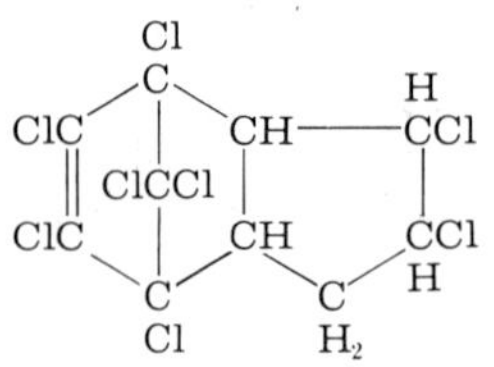

Chlordane, $C_{10}H_6Cl_8$

Effective against many different kinds of insects, including ants, cockroaches, termites and certain pests of agricultural products; synthesized in 1945; along with DDT one of the many "polychlorinated" insecticides, i.e., containing many chlorine atoms

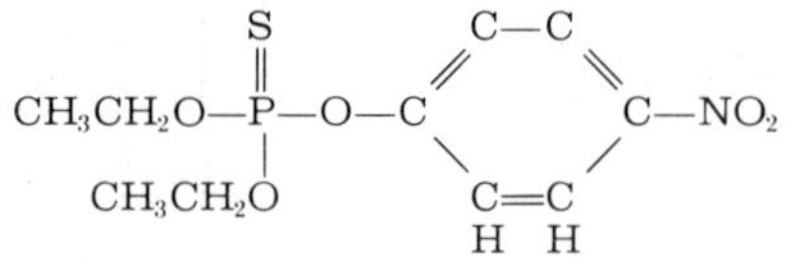

Parathion, $C_{10}H_{14}O_5NPS$

Effective against a wider variety of insects than any other insecticide; one of many organophosphorus insecticides, i.e., containing phosphorus atoms; synthesized in 1948

The introduction of DDT (*dichlorodiphenyltrichloroethane*) (Table 11-9) in 1944 had a great impact. In December 1943 an epidemic of typhus broke out in Naples in the midst of the World War II Allied invasion of Italy. Typhus is a serious, highly contagious disease spread by the body louse. An extensive DDT dusting program was initiated in January, and the epidemic was stopped by March. It has been estimated that the use of DDT in World War II operation areas against the body louse and the malaria-carrying anopheles mosquito prevented about as many American casualties as the total inflicted by our enemies.

The success of DDT led to the development and introduction of other chlorine-containing insecticides including chlordane (Table 11-9), toxaphene, aldrin, dieldrin, and mirex. The investigations that led to the development of nerve gases also led to a series of phosphorus-containing insecticides, including parathion (Table 11-9) and methyl parathion. The widespread use of these insecticides led to significant increases in crop yields and to promising developments in the war against insect-borne diseases, but serious problems began to surface. Insects began to develop resistance to many insecticides, and many of the agents, such as DDT, were found to be unusually resistant to natural forces which would decompose them. As a result they began to accumulate in soils and in the fatty tissues of fish, wildlife, and humans. It was also discovered that DDT causes the eggs of some species of birds to have very thin and easily broken shells. This was leading to a decline in the population of brown pelicans, bald eagles, peregrine falcons, ospreys, and others.

When laboratory tests showed in 1972 that under some circumstances DDT could cause cancer in small animals, the Environmental Protection Agency all but banned its use in the United States. The use of many other insecticides has subsequently been severely restricted. Many farmers feel that they are now handicapped in their efforts to control insects, but some experts hold that chemical insecticides have been used more than necessary and that some of the safer, older methods of pest control, e.g., the rotation and diversification of crops, have been neglected. Moreover, the rate at which many insects have been developing resistance to chemical insecticides is in itself a cause for great concern. This has required farmers to use increasingly greater amounts of insecticides, and the insecticides have been killing off many of the natural enemies of insects, adding to the problem. Some authorities think that insects are too adaptable to be controlled for long by chemical insecticides.

B. Insect Hormones

Hope is now being placed in approaches known as biological controls. One of these involves the use of **insect hormones,** substances secreted in extremely small amounts by insects which regulate their growth and development at various stages. Among the most important of these are the juvenile hormones. A **juvenile hormone** is a hormone secreted by insects which regulates the metamorphic changes involved in their development from egg to larva to pupa to adult. The hormone must be present at certain stages, but it is just as important for it to be absent in others. If the hormone comes in contact with the insect at the wrong time, it causes a derangement of further development and kills the insect (Fig. 11-14). Juvenile hormones therefore appear to be excellent candidates for insecticides. They seem to have no effect on other forms of life and can kill insects very selectively. Unfortunately, they have been found so far to be too unstable for practical applications. The formula of a juvenile hormone is given in Table 11-10.

FIGURE 11-14
The effect of a juvenile hormone on the development of an insect. (a) The normal pupa stage of the grain-eating mealworm. (b) The normal adult stage. (c) The juvenile hormone derails normal development of the adult stage; the head and thorax are those of an adult, but the abdomen is not. (*Courtesy of the Department of Agriculture.*)

In the course of determining the structures of the juvenile hormones chemists devised means of synthesizing their structures. This led directly to the synthesis of analogs of related structure but slightly different behavior. Some of these could prove to be effective insecticides. One such structural analog, methoprene (Table 11-10), is currently being marketed for use against floodwater mosquitoes and certain fleas. Others are under development although the use of the juvenile hormones and their structural variants may be limited by the short duration of the insect developmental period during which they are effective. Some laboratory tests have indicated that both houseflies and mosquitoes develop resistance to juvenile-hormone insecticides.

The discovery of the presence of anti-juvenile hormones in plants is an interesting recent development. An **anti-juvenile hormone** is a hormone antagonist which interferes with the production of juvenile hormones by insects. Two such substances have already been isolated from the flowering plant ageratum and have shown promise as insecticides in preliminary tests.

Table 11-10
Comparison of Structures of One of the Natural
Juvenile Hormones with a More Useful Structural Analog

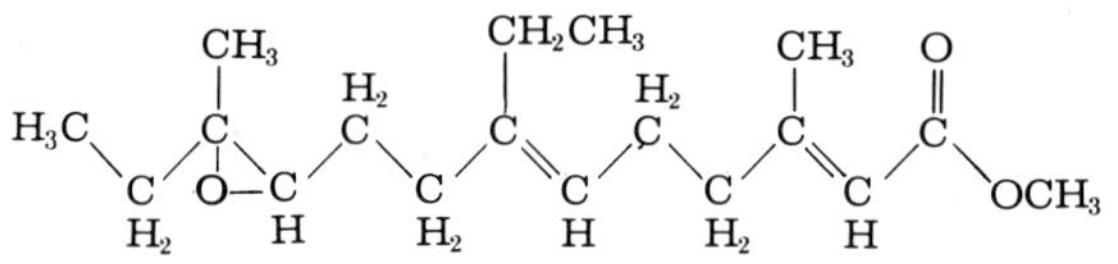

One of the three natural substances with juvenile hormone activity

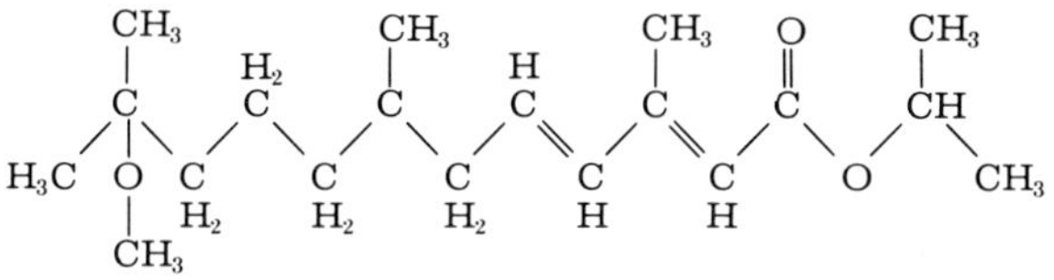

Methoprene, a synthetic structural analog, now available
as an insecticide for limited use

Note to Students

Although the discussion of functional groups in Sec. 8-5 called attention to only a relatively small number of the known groups, that discussion permits you to recognize may of the functional groups in the compounds of the tables in this chapter. As a sort of game, referring back to Table 8-3, see how many functional groups you can recognize. You should be able to name all the functional groups in the compounds of Table 11-10.

C. Insect Pheromones

Pheromones, another group of compounds with potential for controlling insects, are substances secreted by one insect which affect the behavior of another insect of the same species. They serve to announce food sources, signal alarm, or attract and excite members of the opposite sex. As agents of chemical communication they act over distances of many miles and, like the juvenile hormones, are effective in very small amounts.

The chemical characterization of pheromones is considerably complicated by the very small amounts found in an insect. It required the extraction of 500,000 female gypsy moths, for example, to obtain 0.020 gram of its sex attractant. Despite such problems, the pheromone sex attractants of many insects have been isolated and characterized, including those of the gypsy moth, the Mediterranean fruit fly, the American cockroach, and the oriental fruit fly. The formula for the sex attractant of the gypsy moth is given in Table 11-11.

Once the structures of sex attractants were determined, chemists set about to synthesize structural analogs in the hope of obtaining more useful or more easily obtained substances. The female gypsy moth secretes a sex pheromone which can be detected by the male gypsy moth miles away. The characterization of the structure of this sex pheromone led to the synthesis of the useful structural analog gyplure (Table 11-11). Gyplure has been used effectively as a bait to trap male gypsy moths. It is much easier to synthesize than to obtain the natural sex attractant either by synthesis or extraction from moths. It is hoped that insect sex pheromones or their structural analogs

Table 11-11
Comparison of the Structures of the Sex Pheromone of the
Gypsy Moth and an Active Synthetic Structural Analog

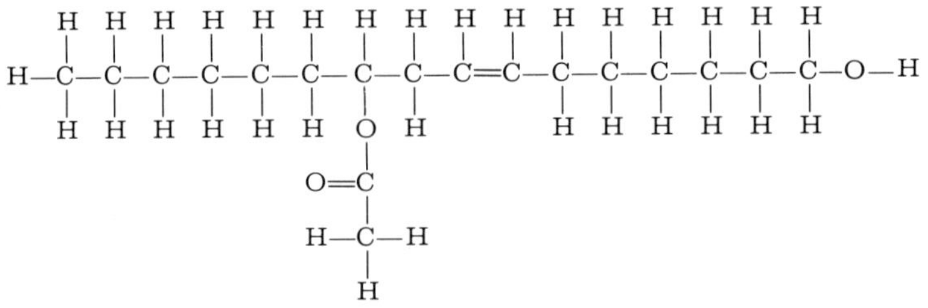

Natural sex pheromone of gypsy moth

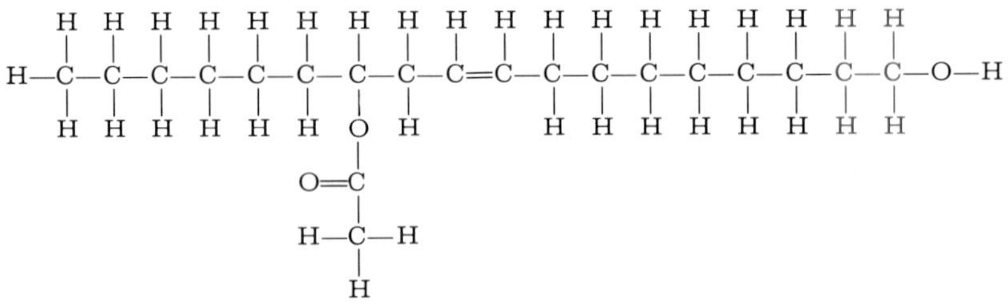

Gyplure, an active synthetic structural analog

will prove helpful in controlling harmful insects on a large scale. One of the most promising methods of application would probably be to spread them over an insect-infested area to confuse male insects and prevent them from finding females to mate with.

D. Insect Sterilization, Pest-resistant Plants, and Natural Predators

Other methods of controlling insects which hold some promise are sterilization by radiation, the development of pest-resistant plants, and the use of natural predators and parasites. Reduction in the populations of both the screwworm fly in Florida and the Mediterranean fruit fly in California has been achieved by releasing members of these species sterilized by radiation (Sec. 9-9). The subsequent insect mating involves many of the sterilized insects, and the birth rate drops.

Some plants have natural defenses against insect pests, and insect-resistant plant strains have been developed. An interesting example is a strain of cotton which matures faster than the standard varieties and can be harvested about a month earlier. Ordinarily, cotton is an easy prey for the very destructive boll weevil, but the new strain matures before the boll weevil population becomes large; the early harvest removes food and breeding grounds of the boll weevil and reduces their population the following season. Strains of about 100 food-producing plants are now resistant to many insect pests.

Natural predators have been used to control insects for over a century. A tiny wasp has been found, for example, which preys on a white fly harmful to poinsettia plants. The wasp destroys a fly by depositing its egg within it, and when the egg hatches it kills the fly. Predators for over 40 different insect pests have been found.

E. Future Prospects

A recent trend in insect control involves integrated pest management, an approach using such measures as crop rotation, natural predators, insect hormones and pheromones, insect-resistant plant strains, and chemical insecticides. Every effort is made to restrict the use of chemical insecticides in order to reduce the development of resistant insect strains and the large-scale destruction of helpful insects. Efforts are also being made to develop chemical insecticides which are readily biodegradable, i.e., easily destroyed by agents in the environment. Such insecticides are less persistent, more controllable, and less harmful.

Almost everyone agrees that our reliance on the present chemical insecticides should be lessened. It is very much to be hoped that some or all of the alternatives under development will be successful on a large scale; considerably more research and development, however, is necessary before this is assured. We are confronted with an awkward dilemma. Many present insecticides produce harmful effects, but we are not yet in a position to do without them, especially if the demands for increased supplies of low-cost foods are to be met.

A. The Race Between the Farmer and the Stork

The population of the world is estimated to be about 4 billion and to be growing at the rate of 80 million a year. This means that each day the population expands by over 200,000. Despite this phenomenal rate, the worldwide increase in food production has been keeping pace (Figs. 11-15 to 11-17). This extraordinary feat has been accomplished by remarkable improvements in many aspects of farming practice, including new varieties of crops, greater amounts of fertilizers, and more irrigation. In the race between the farmer and the stork, as it has been called, increases in food production, however, have not done much more than keep slightly ahead of population increases. Food reserves are now limited, and in view of the vulnerability of food production to uncertain weather patterns the situation is rather precarious. When famine strikes, there are severe limitations on the corrective measures available, both be-

FIGURE 11-15

World food production for 1963 to 1975, expressed on a relative scale where the average production during 1961 to 1965 is taken as 100. The colored line indicates the food production per person, and the white line indicates total food production. Although food production has been increasing significantly, the production per person has risen very little. (*Data from the "United Nations Statistical Yearbook," 1974, 1976. Copyright United Nations, 1974, 1976. Used by permission.*)

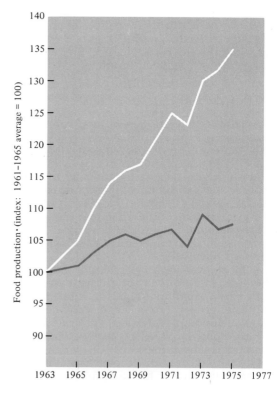

cause of limited supplies and because of poor distribution methods.

Estimates of future food supplies are usually affected more than they should be by the short-range supply situation prevailing at a given time. In 1972, for example, worldwide production of grain was rising faster than the population, and substantial reserves were in storage. New high-yield varieties of wheat and rice, heralded as the green revolution, were becoming established in many of the less developed countries such as India, Mexico, and the Philippines. Fish catches were rising spectacularly. In short, it seemed as though the world would at long last be able to come to terms with its food-supply problems. But by 1974 a food crisis was at hand. A reduction in harvests in several parts of the world due to bad weather almost exhausted all reserves. The worldwide fish catch dropped, and a sharp rise in oil prices made fertilizer much more expensive (Sec. 13-1C), jeopardizing the use of the new fertilizer-demanding crops of the green revolution in the very regions needing them the most. Grain prices were rising abruptly, and famine prevailed in regions of Africa, Asia, and Latin America.

FIGURE 11-16
Food production in the developed countries of the world for 1963 to 1975 expressed on a relative scale where the average production during 1961 to 1965 is taken as 100. The colored line indicates the food production per person and the white line indicates total food production. Both have been increasing. (*Data from "United Nations Statistical Yearbook," 1974, 1976. Copyright by United Nations, 1974, 1976. Used by permission.*)

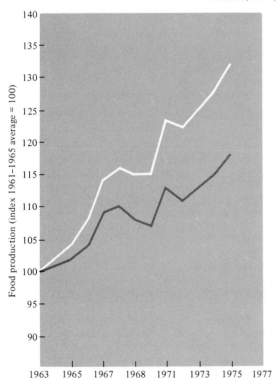

FIGURE 11-17
Food production in the less developed countries of the world for 1963 to 1975 expressed on a relative scale where the average production during 1961 to 1965 is taken as 100. The colored line indicates the food production per person, and the white line indicates the total food production. Total food production has been increasing as fast as in the developed countries, but the production per person has not increased. (*Data from the "United Nations Statistical Yearbook," 1974, 1976. Copyright by United Nations, 1974, 1976. Used by permission.*)

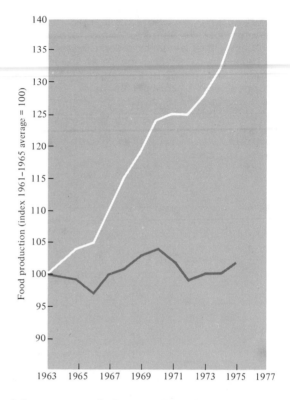

The situation had become much less serious by 1977, but the sobering events of 1974 emphasized the unmistakable need for a marked expansion of the food supply. In the food crisis of 1974 the ravages of famine descended on India, Bangladesh, northeastern Brazil, Ethiopia (Fig. 11-18) and the African countries just south of the Sahara, 10,000 people dying of starvation each week. Most adults who come close to starvation recover almost completely when food becomes available, but children are apt to be affected for life. A growing child requires more nutrients, pound for pound, than an adult. Without sufficient vitamin D, for example, legs become permanently bowed. Without sufficient protein brain development is permanently retarded.

The devastation of famine is appalling, but the effects of chronic malnutrition are more generally pervasive and affect a much greater number of people year in and year out. It is difficult to generalize about the minimum amount of food required for survival. Food experts estimate, however, that at

FIGURE 11-18
Hungry Ethiopian child being given a ration of corn at an aid
station. (*Courtesy of the Agency for International Development.*)

least 500 million people are suffering severely from inadequate
diets.

The adequacy of a diet is most commonly measured by the
energy released in the body by the food ingested. One common
unit of energy, the **calorie,** is a very small amount of energy
equal to the amount of heat energy required to raise the temperature of 1 gram of water 1°C. Food energy is expressed
in kilocalories, each kilocalorie containing 1000 calories.
(Nutritionists have given the kilocalorie the special and am-
biguous designation Calorie, written with a capital C, but we
shall continue to use kilocalories.) Broiled lean hamburger
is said to contain 288 kilocalories per 100 grams; that is, 288
kilocalories is released to the body as it is digested. Peanut
butter contains 594 kilocalories per 100 grams, and the same
weight of raw apples contains 47 kilocalories. The calorie content of some representative foodstuffs was given in Table 11-1.

The calorie content of a diet is not a full indication of its
quality, however; the proportion of proteins, fats, and carbohydrates is important and so is the presence of adequate supplies of vitamins and other essential nutrients. The calorie
content of a diet is, however, a convenient and briefly stated
measure of its sufficiency.

The minimum daily calorie content of an adequate diet
depends on the age and sex of the person, the climate, and
the amount of physical activity. The recommended calorie content of an adequate diet varies from about 1800 kilocalories
for an older woman leading a relatively sedentary life to about
3000 kilocalories for an active 22-year-old male. An average
value of about 2600 kilocalories is commonly used as a general
estimate for an adequate diet, but only about a third of the
world's population enjoys such a diet consistently. This third
includes the regions and countries where the overall supply
of food is usually plentiful, although there may be pockets of
severe malnutrition, namely, North America, Europe, the Soviet
Union, Argentina, South Africa, Australia, and New Zealand.
The calorie content of the diets in Japan, China, some of South
America, and some of southeast Asia approaches an adequate

level, but the average calorie intake is less than adequate in India, Bangladesh, Indonesia, Philippines, much of Central America, much of western South America, and most of Africa (Fig. 11-19).

Children are particularly harmed by malnutrition. A chronic protein deficiency retards growth. A severe deficiency causes kwashiorkor, a wasting disease characterized by a swollen body, a scaly rash, and copper-colored hair. A severe deficiency of both calories and protein causes marasmus, resulting in diarrhea and an extremely emaciated body. Undernourished people of all ages are vulnerable to infections and other illnesses, but the death rate of children can be especially high. In Zambia and Bolivia, for example, the death rate for infants less than 1 year old is reported to be about 250 in 1000; in India and Pakistan it is 140. In the United States the comparable number is 19, and in Sweden it is 12.

Malnutrition is often accompanied by deficiency diseases. Blindness caused by the lack of vitamin A occurs frequently in regions where the diets are most inadequate. In remote inland mountain areas goiter is common because of the deficiency of iodine (Sec. 11-8). Iron-deficiency anemia is also common.

B. What Can Be Done to Improve the World's Food Supply?

Improvement in the world's supply of food can be brought about if there is both planning and action at all levels: local, national, and global. There is an urgent need both for action that will bring immediate results and for research and development on a broad front to provide longer-range improvements. Some of the possibilities can be summarized briefly as follows.

Using more fertilizer As mentioned in Sec. 11-9, estimates suggest that in the less developed countries, where the soil has been undernourished for many years, harvests may be increased by 10 tons for each ton of fertilizer applied. Fertilizer, however, is currently in short supply and is much more expensive than before, mostly because of the shortage in energy (Sec. 13-1C). Experts estimate that fertilizer production should be increased threefold by the end of this century if demand is to be met. This will require a great amount of capital to build the plants and a lot of technological expertise to run them. Special efforts will be required to make ample fertilizer supplies available to the poorer countries despite the rising costs.

Cultivating new land areas Estimates suggest that only about half of the earth's potential agricultural acreage is now used. An area of almost 2 billion acres in central Africa could be brought into use with the solution of such problems as new supplies of water and the control of the malarial mosquito, the deadly tsetse fly (sleeping sickness), and other insects. The Amazon River basin in Brazil is another promising area. Large plains in Colombia, Ecuador, and Venezuela could be made available for livestock grazing if problems of the high acid soil could be overcome. There is also potential new acreage in some countries in Southeast Asia.

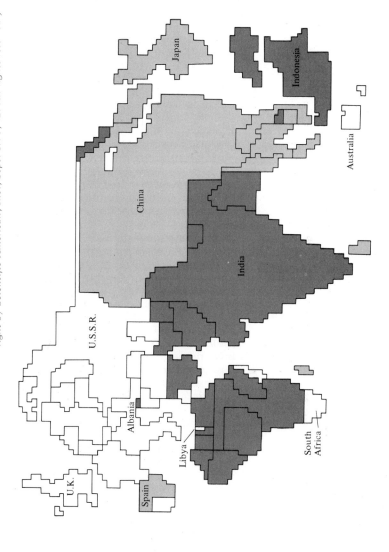
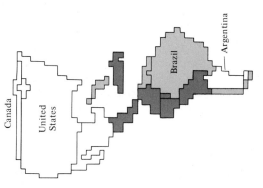

FIGURE 11-19
The level of calorie intake from food for the countries of the world. The area of each nation is represented in proportion to its population. Where the average intake is less than adequate, the country has been colored darkly. Where the average intake is adequate or as much as 10 percent above adequate, the country has been colored slightly. Where the average intake is higher than average by at least 10 percent, the country has been left uncolored. (*From Jean Mayer, The Dimensions of Human Hunger, Copyright by Scientific American, Inc., Sept. 1976, 42. All rights reserved.*)

The development of these areas will require considerable time, great amounts of energy, and huge capital expenditures for the construction of roads, irrigation canals, and facilities for food storage and distribution. It has been estimated that increasing the world's arable land by 10 percent would cost at least $400 billion dollars. Not surprisingly, large amounts of fertilizers, insecticides, and weed killers would also be needed. It is ironic that each year in the United States alone about 600,000 acres of fertile land are taken out of use by housing developments and highway construction.

Increasing the supply of water The Food and Agriculture Organization of the United Nations estimates that the world-wide demand for water will expand 2 or 3 times by the end of the century. A lack of water has placed severe limitations on the use of the new high-yielding crops, which require larger amounts of water than low-yielding crops. New dams and irrigation canals will require large capital expenditures.

Improving food storage and distribution In many regions of the world food is inadequately stored, and a large amount is lost to insects, rats, and molds. Transport systems in many areas are also woefully inadequate. It is estimated that better storage and transportation facilities could, in themselves, increase the world's food supply by 20 percent or more.

More intensive research and development The new high-yielding varieties of wheat and rice, introduced in the 1960s, have demonstrated the great possibilities of the development of new varieties of crops. Currently, only about 1 percent of the solar energy shining on an area is absorbed by photosynthesis. The detailed understanding of this process and the use of genetic manipulation may lead to greater efficiencies. A better understanding of the process of nitrogen fixation could lead to better use of nitrogen fertilizers and to other innovations. Improvements could also come from better management of tropical soils, more disease- and insect-resistant varieties of crops, more efficient processing of fish, and better insecticides, weed killers, and methods of preservation. Further research into nutrition could lead to a better use of available foodstuffs.

A worldwide early-warning system to spot incipient famine The Food and Agricultural Organization of the United Nations has proposed the establishment of an international agency to collect data concerning future weather, anticipated crop yields, expected imports and exports, and all other factors which determine regional food supplies. The agency would not only monitor every region but would develop contingency plans to prevent food-shortage disasters. Such an organization would not only need equipment such as weather sattelites and sophisticated computers but would also require a degree of international cooperation which has not yet been seen.

C. What Is the Role of Chemists and Chemistry?

The relationship between chemistry and the food supply has been the underlying theme throughout this chapter. Here we list the principal areas in which the activities of chemists relate to the food supply.

Fertilizers The importance of the production of fertilizers has been discussed in Sec. 11-9. The production of certain fertilizers is the subject of Sec. 13-1.

Insecticides The role of insecticides and the need for new ones were discussed in Sec. 11-10.

Plant-disease agents Crop losses caused by plant diseases are enormous. The diseases number in the thousands, and the current agents, although helpful, are not completely adequate. Chemicals for soil fumigation are among the agents presently used.

Weed killers Weeds are strong competitors of useful plants for water and soil nutrients. They can easily reduce the yield of corn, wheat, and rice by 20 percent. The selective chemical weed killers offer a means of control which is more practical in large-scale agriculture than the traditional mechanical weeding methods. Some weed killers, however, are toxic, and all must be used cautiously. It is hoped that more relatively safe agents will be developed.

Dietary supplements The chemical characterization of the vitamins and other essential nutrients has been an essential part of the understanding of their role in nutrition. The large-scale chemical synthesis of many and the extraction of others have made possible their low-cost use in treatment of deficiency diseases and in the fortification of foods to compensate for shortages.

Food-preservation additives The use of preservation additives, e.g., calcium propionate to reduce the formation of mold on bread, is important in combating the spoilage of foods. Such use must be carefully monitored, however, to avoid substances in any way harmful and to prevent their unwarranted use.

Food analysis The analytical methods developed by chemists and food technologists have played an important role in understanding the essentials of good nutrition. These methods are also of significance in monitoring the quality of food production and storage.

Veterinary medicines and animal-food supplements Recent developments in medicinal agents, such as antibiotics, have been as important in improving animal health as they have been in treating human diseases. The addition of vitamins and other dietary supplements to animal feeds has increased milk and egg production and the health of all domestic animals. As with food additives, however, careful monitoring is required.

Research and development Chemistry has a large potential role to play in the development of improvements in food production, processing, and preservation. Possible contributions include further understanding of such processes as photosynthesis and nitrogen fixation, improved fertilizers and fertilization techniques, and the production of safer and more effective insecticides, weed killers, and preservation additives. Chemists are also contributing to the development of possible low-cost desalting of seawater as a means of increasing the water available for food production. Large-scale chemical production should also play its traditional role of lowering the costs of all useful substances.

11-12 WHY FOOD ADDITIVES?

Optional The deterioration and spoilage of food during storage have presented problems since the human race lived in caves. Even today it is estimated that at least 25 percent of the world's food is lost between the farmer's field and the dinner table. Most fresh foods

---

WAS MALTHUS CORRECT AFTER ALL?

Thomas Malthus, the eighteenth-century British political economist, warned nearly two centuries ago that population increases, if unchecked, would eventually outrun our capacity to produce food. In 1798 he wrote in his "Essay on the Principle of Population,"

I think I may fairly make two postulata.

First, That food is necessary to the existence of man.

Second, That the passion between the sexes is necessary, and will remain nearly in its present state. These two laws ever since we have had any knowledge of mankind, appear to have been fixed laws of our nature. . . .

Assuming then, my postulata as granted, I say, that the power of population is indefinitely greater than the power in the earth to produce subsistence for man.

Population, when unchecked, increases in a geometrical ratio. Subsistence increases only in an arithmetical ratio. A slight acquaintance with numbers will shew the immensity of the first power as compared with the second.

By that law of our nature which makes food necessary to the life of man, the effects of these two unequal powers must be kept equal.

This implies a strong and constantly operating check on population from the difficulty of subsistence. . . . The race of plants, and the race of animals shrink under this great restrictive law. And the race of man cannot, by any efforts of reason, escape from it. Among plants and animals its effects are waste of seed, sickness,

and premature death. Among mankind, misery and vice. The former, misery, is an absolutely necessary consequence of it. Vice is a highly probable consequence; and we therefore see it abundantly prevail. . . . I see no way by which man can escape from the weight of this law which pervades all animated nature. . . .

Until now Malthus' pessimism has not been confirmed. Food supplies, on the whole, have managed to keep reasonable pace with the population even though there are over 4 times more people in the world now than there were when Malthus lived. And many competent experts hold the opinion that food supplies may be expanded to provide a good diet for 2 or 3 times the present population of the world. But they add some important provisos. For this to be achieved there must be intense application of present technological know-how and all the technological improvements which will be developed. There must be a will to meet the tremendous costs involved, and there must be a high order of international cooperation.

But the food supply cannot be expanded without limit. Warnings are already coming from our apparent finite energy resources and the environmental costs of such requirements as soil fertilization and the clearing and planting of new agricultural areas. On the basis of present evidence, population growth must slow down and eventually stop. Mankind apparently has before it a grave challenge. Will population growth be halted by starvation and the diseases which accompany it, or by a more reasonable course, befitting man's claim to rationality?

deteriorate on standing because of (1) the action of invading microorganisms such as molds, bacteria, and yeasts and (2) chemical changes catalyzed by enzymes present. The reactions lead, for example, to the loss of nutrients, loss of attractive flavors, and food softening. The action of bacteria may lead to illness and death.

Many different kinds of food-preservation methods have been developed. Heat (cooking, canning, pasteurization of milk) destroys microorganisms and enzymes; removal of water (dehydration), freezing, and refrigeration retard the growth of microorganisms and slow down the chemical reactions of spoilage. Minimizing exposure to air and light slows down the chemical reactions. The addition of certain substances can also retard spoilage. This began many centuries ago when it was discovered that the addition of table salt, sodium chloride, preserves meat. From this simple beginning the use of food additives has expanded enormously to include substances used not only to reduce deterioration and maintain nutritional quality but also to facilitate food processing, to enhance essential nutrients, and to improve the taste and attractiveness of foods (Table 11-12).

A **food additive** is any nutritive or nonnutritive substance added to food other than a basic foodstuff. Potassium iodide, added to enhance iodide ion in the diet and thereby prevent thryoid gland enlargement, is a food additive (Sec. 11-8). Table sugar, added to sweeten flavors, is not considered to be a food additive because sugar itself is a basic foodstuff. The use of food additives is sometimes made to appear sinister. Emphasis is placed on their chemical nature with the implication that anything chemical has no place in food. But the word chemical is after all only just a substitute for the word substance, and foods themselves, as well as the bodies of the people who eat them, are a mixture of chemicals. Moreover, some of the substances found in natural foodstuffs can be harmful. Rhubarb leaves, for example, contain a toxic concentration of the poison oxalic acid, $H_2C_2O_4$; the green part of sprouting potatoes contains solanine, which can cause pain, vomiting, and diarrhea; and the deadly poisons of some mushrooms are well known.

This is not to say that the use of food additives should not be severely restricted. Great care must be taken to confine their use to significant purposes and to make certain that they are not used to camouflage inferior ingredients or faulty processing or to deceive the consumer. Most important, food additives must constantly be tested to make certain that they are not harmful to the consumer in any way.

Perhaps the most important food additives are those which are used to reduce food spoilage. Calcium propionate is one of the compounds commonly used to retard the molding of food products such as bread and cheese. Butylated hydroxytoluene (BHT) and butylated hydroxyanisole (BHA) are among the substances called antioxidants, which retard the reactions responsible for causing fats and oils to become rancid. They are added to such products as cereals, nuts, potato chips, and crackers. Benzoic acid, sorbic acid, and their salts sodium benzoate and potassium sorbate are also used as spoilage inhibitors.

The use of sodium nitrate and sodium nitrite as inhibitors in the curing of meat is under investigation. These substances are

2
resentative Food Additives

Formula and name	Typical purpose
Preservatives	
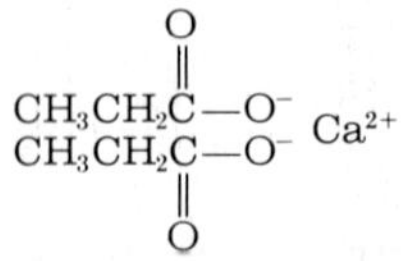 Calcium propionate	To retard molding of cheese, bread, and other bakery products
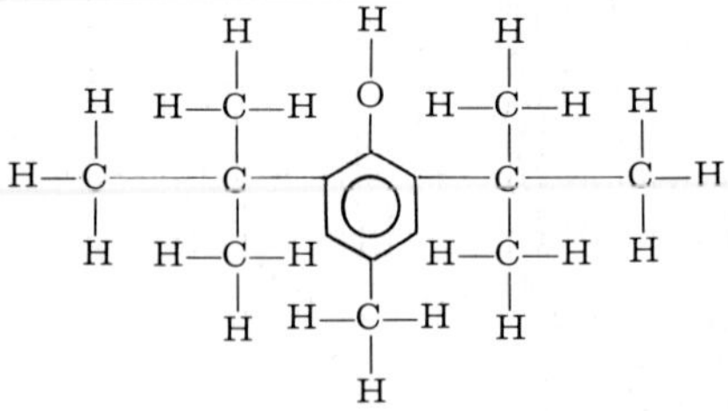 Butylated hydroxytoluene (BHT)	To retard the rate at which fats and oils become rancid; added to such products as cereals, nuts, potato chips, and crackers; safety has been questioned
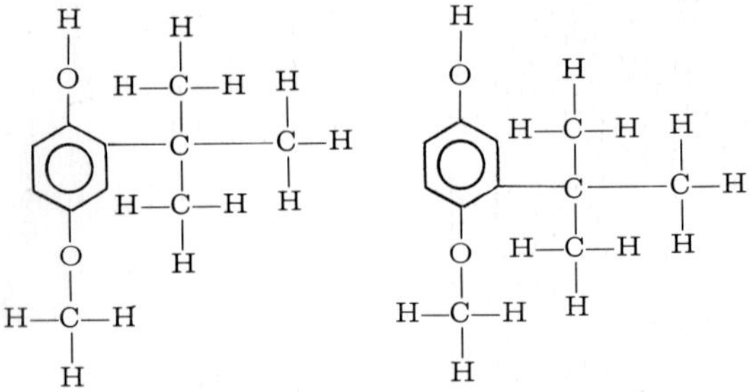 Butylated hydroxyanisole (BHA) (mixture of two compounds)	To retard the rate at which fats and oils become rancid; added to such products as cereals, nuts, potato chips, and crackers; safety has been questioned
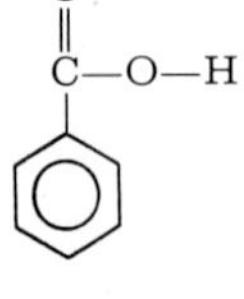 Benzoic acid	To retard spoilage of preserves, jams, jellies, dried vegetables and fruits, concentrated orange drinks
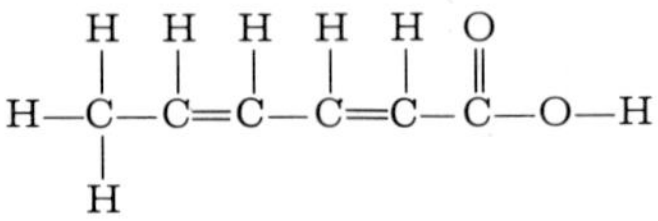 Sorbic acid	To retard spoilage of cheese products, pickles, fish products, carbonated beverages

Table 11-12
Some Representative Food Additives *(Continued)*

Formula and name	Typical purpose
Preservatives *(Continued)*	
$NaNO_3$ and $NaNO_2$ Sodium nitrate Sodium nitrite	Used in curing meats to inhibit bacteria causing botulism poisoning; to increase pink color and tangy flavor; safety has been questioned
Nutritional supplements	
$FeSO_4$ Ferrous sulfate	Used to increase iron content of foods to prevent anemia
KI Potassium iodide	Used in table salt to provide a source of iodide ion to prevent goiter

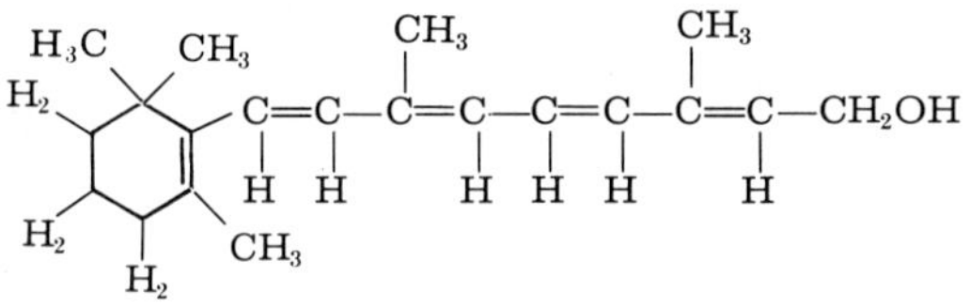

Vitamin A

Added to margarine to give it a vitamin content similar to that of butter

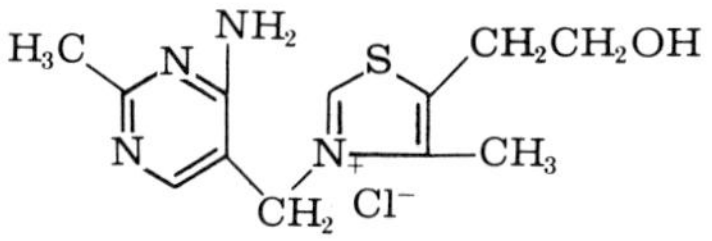

Vitamin B₁ (thiamine)

Added to milled flour to compensate for loss during milling

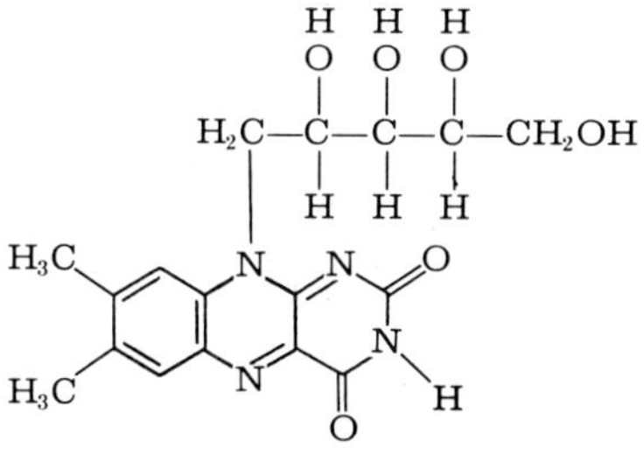

Vitamin B₂ (riboflavin)

Added to milled flour to compensate for loss during milling

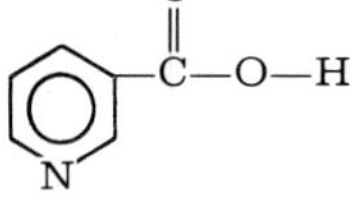

Niacin

Added to milled flour to compensate for loss during milling

(Continued on next page)

Table 11-12
Some Representative Food Additives (*Continued*)

Formula and name	Typical purpose
Nutritional supplements (*Continued*)	

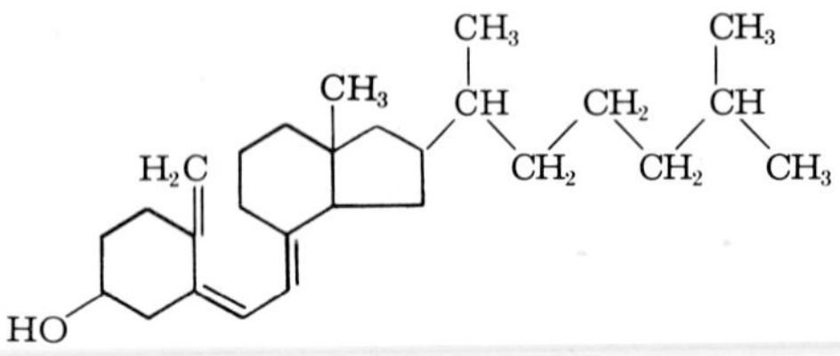

Vitamin D₃ — Added to milk to prevent rickets in children

Emulsifying agents	

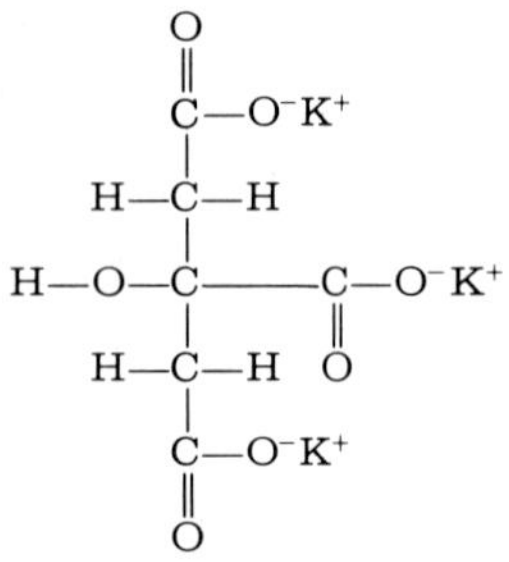

Potassium citrate — Added to process cheese to produce smooth texture

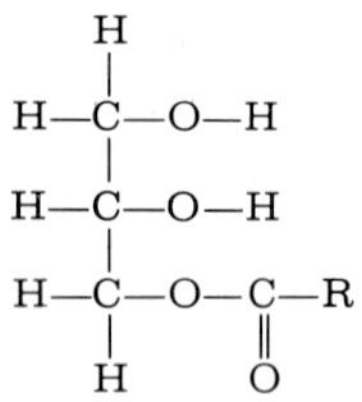

Monoglycerides (representative general formula) — Added to dairy products to produce smooth texture

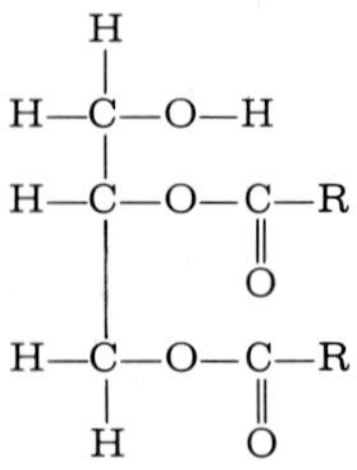

Diglycerides (representative general formula) — Added to dairy products to produce smooth texture

Table 11-12
Some Representative Food Additives (*Continued*)

Formula and name	Typical purpose

Emulsifying agents (*Continued*)

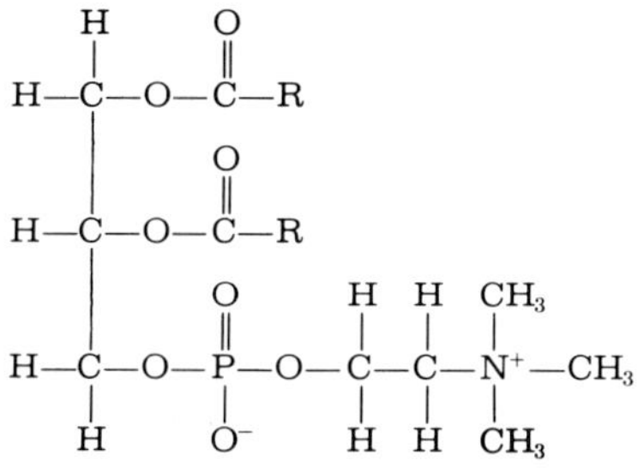

Lecithins (representative general formula)

Added to bakery goods and dairy products to improve texture

Flavoring agents and flavor enhancers

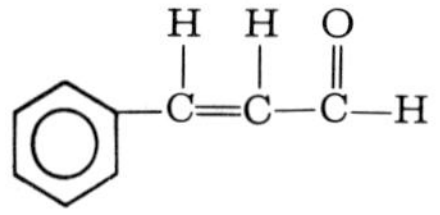

Cinnamaldehyde

Used in preparing artificial cinnamon flavoring

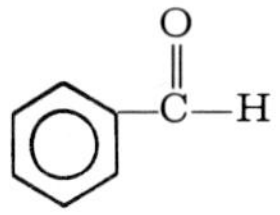

Benzaldehyde

Used in preparing artificial bitter almond and cherry flavoring

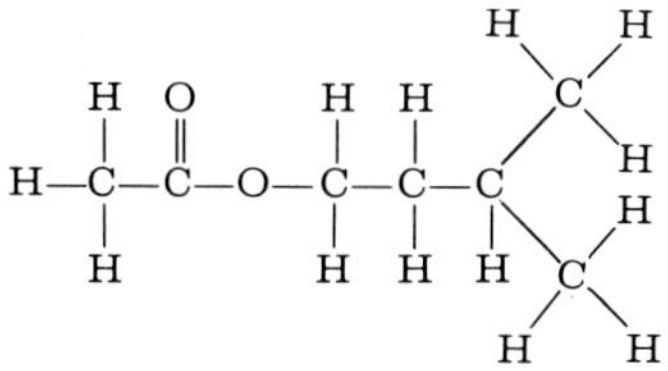

Isoamyl acetate

Used in preparing artificial banana flavoring

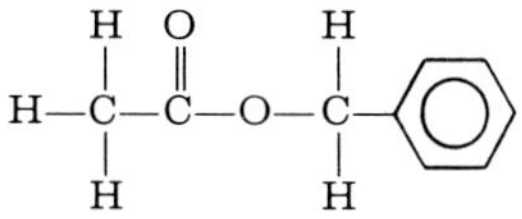

Benzyl acetate

Used in preparing artificial raspberry flavoring

(*Continued on next page*)

Table 11-12
Some Representative Food Additives (Continued)

Formula and name	Typical purpose
Flavoring agents and flavor enhancers (Continued)	

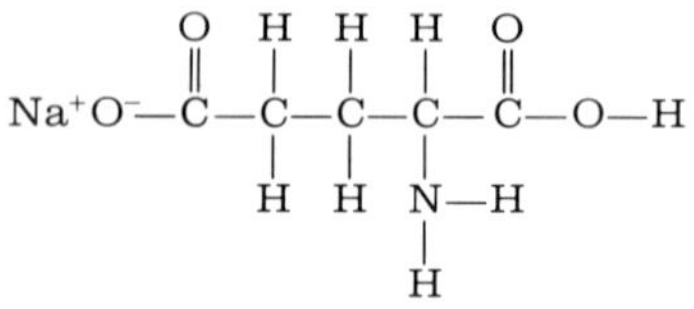 $$Na^+O^- - \overset{\overset{\displaystyle O}{\|\|}}{C} - \overset{\overset{\displaystyle H}{\|}}{\underset{\underset{\displaystyle H}{\|}}{C}} - \overset{\overset{\displaystyle H}{\|}}{\underset{\underset{\displaystyle H}{\|}}{C}} - \overset{\overset{\displaystyle H}{\|}}{\underset{\underset{\displaystyle N-H}{\|}}{C}} - \overset{\overset{\displaystyle O}{\|\|}}{C} - O - H$$ Monosodium glutamate (MSG) (sodium salt of glutamic acid, an amino acid)	Added to meat and vegetable products to enhance flavor
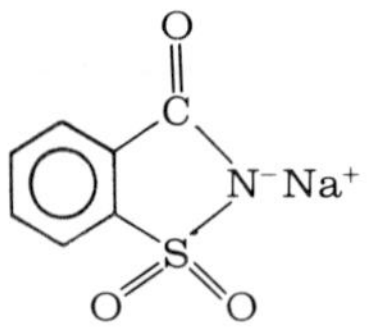 Saccharin (sodium saccharin)	Used as a nonnutritive sweetening agent in diet foods and beverages; safety has been questioned; use may be prohibited
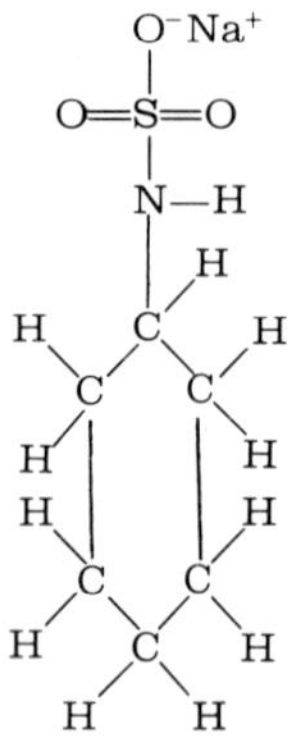 Sodium cyclamate	Formerly used as a nonnutritive sweetening agent in diet foods and beverages; use has been prohibited

effective in inhibiting the bacteria which causes the deadly botulism poisoning, but they are used also to add to the tangy flavor and pink color of meats such as bologna, hot dogs, and ham. Their use is being questioned because of the changes they undergo in the body. Sodium nitrate is converted into sodium nitrite, which reacts with the hydrochloric acid of the stomach in a proton-transfer reaction to produce nitrous acid, $H-O-N=O$:

$$\text{NaNO}_2 + \quad \text{HCl} \quad \rightarrow \text{H—O—N}\!\!=\!\!\text{O} + \text{NaCl}$$

Sodium nitrite	Hydrochloric acid in the stomach	Nitrous acid	Sodium chloride

Reaction which the additive sodium nitrite, NaNO_2, undergoes with the hydrochloric acid, HCl, of the stomach to produce nitrous acid, H—O—N$=$O

Nitrous acid is known to react with amines, R_2NH, to form nitroso compounds, $\text{R}_2\text{N—N}\!\!=\!\!\text{O}$, which have been found to be carcinogenic (cancer producing) in animals:

$$\begin{array}{c} \text{R—N—H} \\ | \\ \text{R} \end{array} + \text{H—O—N}\!\!=\!\!\text{O} \rightarrow \begin{array}{c} \text{R—N—N}\!\!=\!\!\text{O} \\ | \\ \text{R} \end{array} + \text{HOH}$$

General formula for a type of an amine	Nitrous acid from sodium nitrite	General formula for nitroso compounds, known to be carcinogenic	Water

Reaction of nitrous acid, H—O—N$=$O, with amines, R_2NH, to form carcinogenic nitroso compounds, $\text{R}_2\text{N—N}\!\!=\!\!\text{O}$

The question before the food authorities is whether or not the ingestion of sodium nitrate and sodium nitrite in the amounts found in foods can cause stomach cancer.

Some food additives are used to improve nutritional quality. Mention has already been made (Sec. 11-7) of the addition to flour of vitamin B_1 (thiamine), vitamin B_2 (riboflavin), and niacin to compensate for the vitamins lost in the milling process. Ferrous sulfate is also added to increase the content of iron, one of the essential minerals for the diet (Sec. 11-8). Previous mention has also been made (Sec. 11-7) of the addition of vitamin A to margarine, the addition of vitamin D to milk, and the addition of potassium iodide to salt (Sec. 11-8).

Coloring agents are frequently added to foods to improve their appearance and sometimes to compensate for colors lost in processing. They add nothing to the nutritive quality of food but often increase the consumer acceptance of food products. Margarine is colored yellow to make it more appealing, and even butter is often given more color by adding a coloring agent. Other examples abound; oranges are often made more orange, strawberry ice cream is often made a deeper pink.

Coloring agents were originally derived from natural substances: seeds of the annatto tree for yellows, extracts of sandalwood trees for reds, anthocyanins from blue flowers for blues. Coloring agents today, however, are usually the more intensely colored and less expensive synthetic dyes. The number of coloring additives considered to be safe is significantly smaller than it was some years ago because many have been found to be carcinogenic or hazardous in other ways. Since the effect of a coloring additive is only cosmetic, it is particularly important that only substances with no suspicion of risk be used.

Some additives are used as emulsifying agents to produce a smooth texture in such foods as bread and cake mixes, peanut butter, cheese, and ice cream. Representative emulsifying additives are potassium citrate, carboxylic esters of glycerol (mono-

and diglycerides), and lecithin, a glycerol derivative found in egg yolk and the fats of the spinal cord and brain.

Flavoring agents are among the most widely used food additives. Some have a natural origin, such as spices, fruit juices, and plant extracts. Others are synthetic. Compounding artificial flavoring agents may involve blending 20 or 30 substances. The preparation of a cinnamon flavoring agent involves the use of the aldehyde cinnamaldehyde (Table 11-12). Bitter almond and cherry flavors involve the use of the aldehyde benzaldehyde (Table 11-12). Banana-type flavors use isoamyl acetate (Table 11-12), and raspberry flavoring agents use benzyl acetate (Table 11-12) among other substances.

Other substances are added as flavor enhancers, which intensify flavors and in low concentrations contribute little or no flavor of their own. Table salt (sodium chloride) has served this purpose for many centuries. Monosodium glutamate (MSG) (Table 11-12), use of which has now become widespread, is the sodium salt of one of the amino acids, glutamic acid, the residue of which is found in the structures of many of the proteins (Sec. 11-2C) of natural foodstuffs. Glutamic acid has been used in oriental cooking for centuries, and the sodium salt is now added to hundreds of processed foods. Its use has been questioned because excessive amounts have been reported to bring on uncomfortable symptoms in some people. These include burning sensations, a general feeling of weakness, headaches, and sometimes chest pains, which can be confused with symptoms of heart attacks. Since susceptible people are reported to experience these symptoms after eating Chinese foods, the condition has been called the Chinese restaurant syndrome. Safety tests on animals are inconclusive, but the use of MSG in baby foods has been stopped until the effects of this flavor enhancer have been clarified.

The most common sweetening agent in foodstuffs is sucrose (common table sugar). It is not considered officially to be a food additive because it is a natural foodstuff, but most nutritionists are concerned about the effects of the high sugar content of the diet of most people in America and other highly industrialized, urbanized countries. The consumption of sugar and sugar syrups in the United States is in the neighborhood of 125 pounds per person each year. About 100 pounds of this is in the form of cane or beet sugar (sucrose), and the remainder is consumed as syrups containing glucose and fructose (Sec. 11-2). About three-quarters of this intake is hidden, i.e., contained in processed foods and beverages. Sugars are added to such products as soups, salad dressings, cured meats, and prepared entrees, as well as soft drinks, frozen desserts, and sweet baked goods. A recent survey showed that two-thirds of the breakfast cereals contain more than 10 percent sugar. Much of the concern of nutritionists centers on the fact that about half of our carbohydrate intake is now furnished by sugars, and they question whether sugars are nutritionally equivalent to the other carbohydrates which they have displaced from the more traditional diets of former generations.

The relationship between a high sugar diet and the degeneration of teeth is rather well documented. Numerous surveys on an international scope have shown that where sugar intake is high, tooth decay is prevalent. In the United States, for example, it has been estimated that over 95 percent of all children have some tooth decay and about half the population over 60 have no teeth at all. Excessive body weight is also a major health problem in the United States. With it is associated a more frequent occurrence of high blood pressure and certain chronic diseases. It can also cause diabetes, a major cause of death, in those who have a genetic predisposition to this disease.

In an effort to devise low-sugar diets for diabetics and the victims of certain other diseases and to provide sugar-free foods and beverages for the weight-conscious, synthetic sugar substitutes are being increasingly used. One of the oldest of these is saccharin (Table 11-12). Although this substance is about 400 times as sweet as sucrose, it has no nutritive content at all. It was synthesized by the American chemist Ira Remsen, and its sweetness was accidentally discovered in 1879 by his colleague Constantine Fahlberg. Another artificial sweetening agent, sodium cyclamate (Table 11-12) was developed in 1937 by the American chemist Michael Sweda. While it is not as sweet as saccharin, it lacks saccharin's bitter aftertaste. At one time it was a very popular sweetening agent. In the 1960s the annual consumption in the United States rose to 15 million pounds a year, about 70 percent of which was used in soft drinks. But then tests showed that an appreciable proportion of rats fed a diet very high in sodium cyclamate develop cancer of the bladder at the end of their life-span. As a result of this finding sodium cyclamate was banned as a food additive in 1969. The necessity of the ban has been questioned because the concentration of sodium cyclamate in the test rat diets was so high that it was equivalent in human terms to drinking 1000 glasses of sodium cyclamate–containing soft drinks each day.

The use of saccharin is also under attack. Like sodium cyclamate, saccharin has been found to produce bladder tumors in rats fed on very high saccharin diets. Experiments are now under way to determine whether the rat tumors are due to impurities in the saccharin and if saccharin is harmful to animals other than rats. About 70 percent of the saccharin consumed in the United States is used in diet foods and beverages. It is hoped that new synthetic sweetening agents in the process of development will be more acceptable.

The growing, processing, storage, and marketing of food in the United States involves thousands of processes and millions of people. It is obviously appropriate that the federal government be involved in monitoring the quality of the food supply. The first federal food and drug law was passed in 1906 and established the Food and Drug Administration (FDA). Since that time the law has been strengthened by amendments several times. The overall mandate of the FDA regarding food is to ensure that it is pure, wholesome, and honestly labeled. The amendment of 1958 required that food additives be tested for safety, but hundreds of substances were already in use, many, like baking soda (sodium bicarbonate) and table salt, for a long time. To simplify the enforcement of the new amendment at the start, the FDA asked several hundred food experts to draw up a list of food additives which could be assumed to be safe on the basis of use over a number of years. This list, first published in 1959, was called the *generally recognized as safe* (GRAS) list. The original list included 600 substances. The FDA can remove a substance from the list if testing suggests that it is potentially hazardous, and several substances on the original list have subsequently been banned, e.g., sodium cyclamate. Substances can be added to the list if tests by the manufacturer demonstrate their safety to the satisfaction of the FDA, and the list now includes over 1000 additives.

The 1958 amendment to the Food, Drug and Cosmetic Act contains a clause which requires banning from food and beverages any food additive found carcinogenic (cancer-causing) in man or animal. This clause is called the Delaney clause for its author and ardent champion, Representative James J. Delaney of New York. This firm clause was designed to provide the FDA with authority to cope with the strong commercial pressures which may oppose the agency's refusal to approve a given additive. Its enforcement, however, has become controversial. Many authorities question the clause's absolute approach, which leaves little or no opportunity to weigh benefits as well as risks. They also question the extrapolation of the effects of high-concentration doses on animals over short periods of time to the effects of low doses on human beings over 20 to 30 years. Almost any substance, they add, can be harmful if given in sufficiently large doses. Others maintain that the animal tests are relevant and that no food additive should be used if there is any potential risk of cancer, however small.

"If it's true that the world ant population is 10^{15}, then it's no wonder we never run into anyone we know."

KEY WORDS

1. Carbohydrate (Sec. 11-2A)
2. Sugar (Sec. 11-2A)
3. Monosaccharide (Sec. 11-2A)
4. Disaccharide (Sec. 11-2A)
5. Polysaccharide (Sec. 11-2A)
6. Fats (Sec. 11-2B)
7. Fatty acid (Sec. 11-2B)
8. Protein (Sec. 11-2C)
9. Amino acid (Sec. 11-2C)
10. Amino acid residue (Sec. 11-2C)
11. Peptide bond (Sec. 11-2C)
12. Polypeptide structure (Sec. 11-2C)
13. Prosthetic group (Sec. 11-2C)
14. Conjugated protein (Sec. 11-2C)
15. Simple protein (Sec. 11-2C)
16. Enzyme (Sec. 11-2C)
17. Photosynthesis (Sec. 11-3)
18. Nitrogen fixation (Sec. 11-4)
19. Symbiotic relationship (Sec. 11-4)
20. Digestion (Sec. 11-5)
21. Metabolism (Sec. 11-6)
22. Adipose tissue (Sec. 11-6B)
23. Essential fatty acid (Sec. 11-6B)
24. Essential amino acid (Sec. 11-6C)
25. Deficiency disease (Sec. 11-7)
26. Vitamin (Sec. 11-7)
27. Mineral (Sec. 11-8)
28. Insect hormone (Sec. 11-10B)
29. Juvenile hormone (Sec. 11-10B)
30. Anti-juvenile hormone (Sec. 11-10B)
31. Pheromone (Sec. 11-10C)
32. Calorie (Sec. 11-11)

Optional:

33. Food additive (Sec. 11-12)

SUMMARIZING QUESTIONS FOR SELF-STUDY

Sections 11-1 and 11-2

1. Q. What is the principal factor encouraging optimism about the long-term food prospects for the world's population?

A. Increases in food production are keeping ahead of population increases on a worldwide basis.

2. Q. What inclines some people to pessimism?

A. The large proportion of the world's population currently undernourished, the rapid increase in the world's population, and the frequency of severe food shortages in certain regions.

3. Q. What are carbohydrates?

A. Substances containing carbon, hydrogen, and oxygen atoms ranging from sugars to starch and cellulose. (A more chemical definition is found in Sec. 10-2G.)

4. Q. What is a sugar?

A. A carbohydrate of simple structure, such as glucose and sucrose (table sugar).

5. Q. What is a polysaccharide?

A. A carbohydrate which contains many structural units, e.g., starch and cellulose.

6. Q. What is the name and molecular formula of the product of the hydrolysis of starch?

A. Glucose, $C_6H_{12}O_6$.

7. Q. What is a fat?

A. A carboxylic ester of the alcohol glycerol and carboxylic acids containing from 12 to 18 carbon atoms.

8. Q. What are the products of the complete hydrolysis of a fat?

A. Glycerol and a mixture of carboxylic acids.

9. Q. What is a fatty acid?

A. A carboxylic acid obtained from the hydrolysis of a fat.

10. Q. What is a protein?

A. A large polymeric molecule made up of amino acid residues joined by amide groups (peptide groups).

11. Q. What is an amino acid?

A. A substance containing an amine and a carboxylic acid group.

12. Q. Give a general formula for an amino acid.

A.

$$H-\underset{\underset{H}{|}}{N}-\underset{\underset{H}{|}}{\overset{\overset{R}{|}}{C}}-\overset{\overset{O}{\|}}{C}-O-H$$

13. Q. What is an amino acid residue?

A. The part of an amino acid molecule actually found in a protein molecule.

14. Q. Give a general formula for an amino acid residue.

A.

$$-\underset{\underset{H}{|}}{N}-\underset{\underset{H}{|}}{\overset{\overset{R}{|}}{C}}-\overset{\overset{O}{\|}}{C}-$$

15. Q. What is a peptide bond?

A. The amide bond of a protein molecule.

16. Q. What is a polypeptide structure?

A. The polyamino acid residue structure of a protein.

17. Q. What is an enzyme?

A. A protein which serves as a catalyst for a reaction in a living organism.

Section 11-3

18. Q. What is meant by photosynthesis?

A. The process in which light energy from

the sun is converted into chemical energy, as glucose is synthesized from carbon dioxide and water.

19. Q. Where does photosynthesis occur?

A. In microscopic algae and bacteria of lakes and oceans and in the green plants of land areas.

20. Q. What is a product of photosynthesis in addition to glucose?

A. Oxygen.

21. Q. What is important about the production of oxygen?

A. It is the only significant means of replenishing the oxygen content of the atmosphere.

22. Q. Give a balanced equation for the overall process of photosynthesis.

A. $6CO_2 + 6H_2O + \text{solar energy} \rightarrow$
$$C_6H_{12}O_6 + 6O_2$$

23. Q. What is the function of the chlorophylls in the process of photosynthesis?

A. The two chlorophylls are the major sunlight-absorbing substances of green plants.

24. Q. What is the role of photosynthesis in the body energy of human beings?

A. In photosynthesis solar energy is absorbed while carbon dioxide and water are converted in plants into glucose and then other carbohydrates. We obtain the body energy we need by subjecting the carbohydrates of foods to a series of reactions which release the energy.

Section 11-4

25. Q. What is meant by nitrogen fixation?

A. A process of converting elemental nitrogen to a nitrogen-containing compound.

26. Q. What is one of the routes of nitrogen fixation in nature?

A. A kind of bacteria which lives in the roots of a class of plants (legumes) and is able to convert elemental nitrogen into ammonia.

27. Q. What is the significance of this route?

A. All living organisms depend on the nitrogen-fixing ability of these bacteria (and that of certain other microorganisms). They convert elemental nitrogen, which most living organisms cannot use directly, into nitrogen-containing substances which living organisms need.

28. Q. What is the principal synthetic method of nitrogen fixation? Give an equation for the overall reaction involved.

A. The Haber process: $N_2 + 3H_2 \rightarrow 2NH_3$.

Section 11-5

29. Q. What is meant by digestion?

A. The name given to a series of changes whereby the complex substances of foods are converted into substances of simple structure, which can be absorbed by the body.

30. Q. What is the principal type of reaction which takes place during the digestion of carbohydrates, fats, and proteins?

A. Hydrolysis.

31. Q. What is the principal product of the digestion of carbohydrates?

A. Glucose.

32. Q. What are the principal products of the digestion of fats?

A. Partially hydrolyzed esters of glycerol, carboxylic acids, and glycerol.

33. Q. What are the principal products of the digestion of proteins?

A. Amino acids.

Section 11-6

34. Q. What is meant by metabolism?

A. All the chemical processes which food substances undergo in a living organism.

35. Q. What are the principal roles of glucose in the metabolism of animals?

A. The generation of energy and the synthesis of carbohydrates, fats, and proteins.

36. Q. What are the principal roles of fats in the metabolism of animals?

A. They are the richest source of energy in the body, and fatty acids obtained from them are the starting points for the synthesis of substances with important special functions.

37. Q. What other purposes do fats serve?

A. Body insulation and protection of body organs from physical injury.

38. Q. What are the principal roles of amino acids in the metabolism of animals?

A. They serve as a starting point in the synthesis of proteins, fats, and carbohydrates. They may also serve as a source of energy.

Sections 11-7 and 11-8

39. Q. What is a vitamin?

A. A substance known to be necessary in the diet for normal health, in addition to the usual carbohydrates, fats, and proteins.

40. Q. What is a deficiency disease? Give an example and the cause.

A. An ailment caused by the absence of an essential dietary substance, rather than by a poison or an invading parasite, e.g., scurvy, caused by a lack of vitamin C; beriberi, caused by a lack of vitamin B_1; or rickets, caused by a lack of vitamin D. (See Table 11-5 for other examples.)

41. Q. From a dietary viewpoint, what is meant by a mineral?

A. A substance which is inorganic (does not contain carbon atoms).

42. Q. What are the two most abundant atoms in the human body which are common components of substances classified as minerals?

A. Calcium and phosphorus.

43. Q. What are the principal functions in the human body of substances containing these atoms?

A. Structural material of bones and teeth.

Section 11-9

44. Q. What is the role of fertilizers?

A. To replace the nutrients of the soil lost principally by growing and harvesting crops.

45. Q. What three elements (in the combined form) are taken from the soil in the greatest amounts by plants?

A. Nitrogen, phosphorus, and potassium.

46. Q. What is the role of the chemical industry in the use of fertilizers?

A. The inexpensive production of large amounts of concentrated fertilizers.

Section 11-10

47. Q. What is the basic dilemma associated with the use of insecticides?

A. Many present insecticides produce harmful effects, but in order to control insects we do not seem to be able to do without them.

48. Q. Why did DDT attract so much attention when it was introduced in the 1940s?

A. It played such a significant role in controlling disease-carrying insects in many of the disease-infected regions occupied by Allied troops during World War II.

49. Q. Why is its use now severely restricted?

A. It has been found to accumulate in the fatty tissues of animals and humans, and laboratory tests have indicated that under some circumstances it can cause cancer in small animals.

50. Q. What is a principal limitation to its effectiveness?

A. Insects develop resistance to it.

51. Q. What are juvenile hormones?

A. Hormones secreted by insects which regulate the changes involved in their development from eggs to adults.

52. Q. What properties do juvenile hormones have which make them appear to be excellent candidates for insecticides?

A. They can kill insects selectively without harm to other forms of life.

53. Q. What practical application has resulted from investigations of juvenile hormones?

A. One of the synthetic structural variants of the hormones has been marketed for use against certain mosquitoes and flies.

54. Q. What is a pheromone?

A. A substance secreted by one insect which can act over distances of many miles to affect the behavior of another insect of the same species.

55. Q. How do insects use pheromones?

A. To announce food sources, signal alarm, or attract and excite members of the opposite sex.

56. Q. In what way could pheromones be useful in the control of insects?

A. As a bait to trap insects or to confuse insects in their search for members of the opposite sex.

57. Q. In what way could structural variants of pheromones be useful?

A. They could prove to be much more inexpensive to obtain in quantity, and they might turn out to be more effective than the natural pheromones.

Section 11-11

58. Q. Over the last 15 years food production has been increasing at about the same rate in both the developed and the less developed countries. Why, on the whole, has the food situation not improved in the less developed countries?

A. Because of the faster increase in population among the less developed countries.

59. Q. What is a calorie?

A. It is an amount of heat energy equal to that required to raise the temperature of 1 gram of water 1°C. (It is a very small amount of energy.)

60. Q. What is a Calorie?

A. A kilocalorie, containing 1000 calories; a unit used by nutritionists.

61. Q. Why is the calorie content of a diet not a full indication of its quality?

A. Because it does not specify such important factors as the proportion of proteins, fats, and carbohydrates and the presence of vitamins and other essential nutrients.

62. Q. Why are children especially susceptible to malnutrition and famines?

A. Growing children require more nutrients, pound for pound, than adults. They usually do not recover as completely as adults from a state of near-starvation.

63. Q. List six possible means of improving the world's food supply.

A. The use of more fertilizer, the cultivation of new land areas, an increase in the supply of water, the improvement of food storage and distribution, more intensive research and development of food production, processing, and preservation, and a worldwide early-warning system to spot incipient famine.

64. Q. List nine areas in which chemistry can be influential in improving the food supply.

A. Fertilizers, insecticides, plant-disease agents, weed killers, dietary supplements, food-preservation additives, food analysis, veterinary medicines and animal-food supplements, research and development to improve all the above.

Section 11-12 (Optional)

65. Q. What is a food additive?

A. Any nutritive or nonnutritive substance added to food other than a basic foodstuff.

66. Q. Sugar is one of the substances most frequently added in the processing of foodstuffs. Is it considered to be a food additive? Why?

A. No, because it is a basic foodstuff.

67. Q. List six of the classes of food additives.

A. Preservatives, nutritional supplements, emulsifying agents, flavoring agents, flavor enhancers, and coloring agents.

68. Q. What are two mechanisms whereby foods deteriorate?

A. The action of invading microorganisms, such as molds, bacteria, and yeasts, and chemical changes catalyzed by enzymes.

69. Q. What is the oldest chemical food preservative?

A. Table salt (sodium chloride).

70. Q. Why is the use of sodium nitrate and sodium nitrite in the processing of meats under investigation?

A. Because of the possibility that they react in the body to produce cancer-causing agents.

71. Q. What food additives are required by law in some states (Sec. 11-7)?

A. The addition of vitamin B_1, vitamin B_2, and niacin to flour to compensate for the vitamin losses during milling.

72. Q. What agency of the federal government has been established by Congress to monitor the quality of the food supply?

A. The Food and Drug Administration (FDA).

73. Q. What is the GRAS list?

A. A list of substances, generally recognized *as safe*, issued by the FDA when it was mandated by Congress in 1958 to approve the safety of food additives. This list included substances already in use for some years when the safety check began that were tentatively considered to be safe without testing. Substances have been subtracted and added to the list since 1958 as testing has proceeded.

PRACTICE EXERCISES

1. Match each definition or other statement with the numbered term above with which it is most closely associated. A numbered term may be used only once.

   1. Polypeptide
   2. Fat
   3. Conjugated protein
   4. Maltose
   5. Amino acid residue
   6. Fatty acid
   7. Carbohydrates
   8. Hexane
   9. Enzyme
   10. Simple protein

   (a) A carboxylic acid obtained from the hydrolysis of fats
   (b) The part of an amino acid molecule actually found in a protein molecule
   (c) Substances which contain carbon, hydrogen, and oxygen atoms ranging from sugars to starch and cellulose
   (d) A protein without a prosthetic group
   (e) A disaccharide
   (f) The poly-amino acid residue structure of a protein
   (g) A protein with a prosthetic group
   (h) A catalyst for the reaction of a living organism

2. Match each definition or other statement with the numbered term above with which it is most closely associated. A numbered term may be used only once.

   1. Glucose
   2. An amide bond
   3. Fatty acid
   4. A prosthetic group
   5. Methane
   6. Fats
   7. Starch
   8. Protein
   9. Sucrose
   10. Amino acid

   (a) Carboxylic esters of carboxylic acids containing from 12 to 18 carbon atoms and glycerol
   (b) A carboxylic acid which also contains an amino group
   (c) Table sugar
   (d) A large polymeric molecule made up of structural units known as amino acid residues
   (e) The nonpolypeptide unit of a protein
   (f) Peptide bond
   (g) Polysaccharide
   (h) A monosaccharide

3. Match each definition or other statement with the numbered term above with which it is most closely associated. A numbered term may be used only once.

   1. Essential amino acid
   2. Vitamin
   3. Symbiotic relationship
   4. Photosynthesis
   5. Anti-juvenile hormone
   6. Metabolism
   7. Calorie
   8. Adipose tissue
   9. Watt
   10. Digestion

   (a) An amount of energy required to raise the temperature of 1 gram of water 1°C
   (b) The process in which light energy is converted into chemical energy as glucose is synthesized from carbon dioxide and water
   (c) A depot of reserve fat in the body
   (d) One of a group of substances necessary in the diet for normal health, in addition to carbohydrates, fats, and proteins
   (e) An amino acid which is needed by the body but which it cannot synthesize or cannot synthesize fast enough to meet bodily needs
   (f) A way two species of organisms live together which is advantageous to both

(g) A hormone antagonist which interferes with the production of the juvenile hormone

(h) A series of changes whereby the complex substances of foods are converted into substances of simpler structure which can be absorbed by the body

4. Match each definition or other statement with the numbered term above with which it is most closely associated. A numbered term may be used only once.
   1. Deficiency disease 2. Fats
   3. Metabolism   4. Kilogram
   5. Juvenile hormone   6. Nitrogen fixation
   7. Essential fatty acids 8. Mineral
   9. Kilocalorie   10. Pheromone

(a) A substance secreted by an insect which regulates the growth and developmental changes at the various stages of its life

(b) A process of converting elemental nitrogen into a nitrogen-containing compound

(c) Fatty acids which the human body must obtain from food

(d) A substance containing phosphorus or calcium found in the human body

(e) The chemical changes which the products of digestion undergo during the synthesis of the substances needed for bodily functions and the generation of energy

(f) An ailment caused by the absence of a dietary substance

(g) A substance secreted by one insect which affects the behavior of another insect of the same species

(h) A Calorie

5. Write a structural formula for one of the ring forms of glucose.

6. What is the product of the complete hydrolysis of starch?

7. What is the most characteristic functional group found in fat molecules?

8. (a) Give the names of the classes of substances obtained in the complete hydrolysis of a fat molecule. (b) Give a specific or a general structural formula for each.

9. (a) What is the difference between an amino acid and an amino acid residue? (b) Represent one of each by a structural formula.

10. What is a fatty acid?

11. What class of products is obtained from the complete hydrolysis of a simple protein?

12. (a) What gaseous product is formed in the process of photosynthesis? (b) From what reactant do the atoms of the molecule of this gas come?

13. In the process of photosynthesis from what reactant do (a) the hydrogen atoms and (b) the oxygen atoms in the glucose product come?

14. (a) To what products is glucose ultimately converted in the body? (b) What substances are the ultimate sources of the glucose used as food? (c) What is the ultimate source of the energy evolved as glucose reacts with oxygen in the body?

15. (a) What is nitrogen fixation? (b) What is the initial product of one such process? (c) What is one of the reactants? (d) What is the significance of nitrogen fixation?

16. Give an equation for the process used in one type of industrial nitrogen fixation.

17. (a) What are the principal products of the human digestion of carbohydrates, fats, and proteins? (b) What is the name of the principal type of reaction involved in each case? (c) What single reactant is involved in all cases?

18. What is meant by metabolism?

19. What are two principal roles of (a) glucose and (b) fats in the human body? (c) What are four roles in the human body of the amino acids obtained from proteins?

20. (a) What is a deficiency disease? (b) Name two.

21. (a) What is a vitamin? (b) Give both names for three vitamins which have two commonly used names. (c) For what purposes are vitamins legitimately used?

22. (a) Identify the functional groups in the molecule of vitamin C. (b) Identify two of the functional groups in the molecule of vitamin B_1 (the other functional groups present have not been discussed).

23. (a) What is the most abundant substance in the human body? (b) Of carbohydrates, fats, and proteins, which class is not present in substantial amounts in the human body?

24. Where is (a) calcium and (b) phosphorus principally found in the human body?

25. What is iodized salt and what is its purpose?

26. (a) What are the sources of carbon atoms, oxygen atoms, and hydrogen atoms required for plant growth? (b) What three other kinds of atoms are of major importance to plant growth?

27. (a) Fertilizers most commonly contain the atoms of what three elements? Why? (b) Give the name and formula of one of the compounds commonly used to furnish each of the three atoms (one compound for each kind of atom).

28. What are two harmful effects of certain insects from an overall point of view?

29. (a) What currently used insecticide was also used (as part of a natural mixture) in the seventeenth century? (b) What is the full name of DDT, and why is its use for the most part banned in the United States?

30. (a) What is a juvenile hormone and what is the principle of its use to control insects? (b) What possible role could an anti-juvenile hormone play? (c) What possible role could structural

variants of juvenile hormones play in the control of insects?

31. (*a*) What are pheromones, and what role could they possibly play in the control of insects? (*b*) What role could structural variants of pheromones play in the control of insects?

32. What measures are included in integrated pest management?

33. Food production has increased in the less developed countries during the last 15 years at about the same rate as in the developed countries. Yet the amount of food available per person in the less developed countries has not been growing, whereas in the developed countries it has. Explain.

34. (*a*) What is the Calorie unit used by nutritionists? (*b*) Why is the calorie content of a diet not a full indication of its quality?

35. (*a*) What fraction of the world's population consistently enjoys what is generally considered to be an adequate diet? (*b*) Why are children harmed more by malnutrition and near-starvation than adults?

36. (*a*) Name six of the possible approaches to improving the world's food supply. (*b*) What are nine of the areas in which chemists can be of help?

Section 11-12 (Optional)

37. (*a*) What is a food additive? (*b*) What was the first food additive and what was its purpose? (*c*) For what reasons are food additives used today?

38. Why are the food additives sodium nitrate and sodium nitrite under investigation?

39. Why is the number of approved coloring additives smaller now than some years ago?

40. What is MSG, and why has its use in baby foods been stopped?

41. Why are nutritionists concerned about the high sugar content of the diet of most Americans?

42. Why is the use of saccharin under investigation?

43. (*a*) What is the FDA and the GRAS list? (*b*) What is the Delaney clause?

SUGGESTIONS FOR FURTHER READING

The following articles are of an introductory nature.

Benditt, Earl P.: The Origin of Artherosclerosis, **Scientific American**, February 1977, p. 74.

Boerma, Addeke H.: A World Agricultural Plan, **Scientific American**, August 1970, p. 54.

Boffey, Philip M.: Color Additives: Botched Experiment Leads to Banning of Red Dye No. 2, **Science, 191**:450 (1976).

Boffey, Philip M.: Color Additives: Is Successor to Red Dye No. 2 Any Safer?, **Science, 191**:832 (1976).

Brown, Lester R.: The World Food Prospect, **Science, 190**:1053 (1975).

Frejka, Tomas: The Prospects for a Stationary World Population, **Scientific American**, March 1973, p. 15.

Haggin, Joseph H. S.: The Candy Chemist, **Chemistry**, April 1967, p. 9.

Holden, Constance: Mirex: Persistent Pesticide on Its Way Out, **Science, 194**:301 (1976).

Jacobson, Martin, and Morton Beroza: Insect Attractants, **Scientific American**, August 1964, p. 20.

Kermode, G. O.: Food Additives, **Scientific American**, March 1972, p. 15.

Majtenyi, Joan Z.: Food Additives: Food for Thought, **Chemistry**, May 1974, p. 6.

Marx, Jean L.: Insect Control, I: Use of Pheromones, **Science, 181**:736 (1973).

Marx, Jean L.: Applied Ecology: Showing the Way to Better Insect Control, **Science, 195**:860 (1977).

Mayer, Jean: The Bitter Truth about Sugar, **The New York Times Magazine**, June 20, 1976, p. 26.

Mayer, Jean: The Dimensions of Human Hunger, **Scientific American**, September 1975, p. 40.

Peakall, David B.: Pesticides and the Reproduction of Birds, **Scientific American**, April 1970, p. 72.

Poleman, T. T.: World Food: A Perspective, **Science, 188**:510 (1975).

Pratt, Christopher J.: Chemical Fertilizers, **Scientific American**, August 1965, p. 62.

Raw, Isaias, and Gerald W. Holleman: Water: Energy for Life, **Chemistry**, May 1973, p. 6.

Revelle, Roger: Food and Population, **Scientific American**, September 1974, p. 160.

Sanders, Howard J.: New Weapons against Insects, **Chemical and Engineering News**, July 28, 1975, p. 18.

Sanderson, F. H.: The Great Food Fumble, **Science, 188**:503 (1975).

Schrauzer, G. N.: Biological Nitrogen Fixation, **Chemistry**, March 1977, p. 13.

Scrimshaw, Nevin S., and Vernon R. Young: The Requirements of Human Nutrition, **Scientific American**, September 1976, p. 50.

Shapley, Deborah: Nitrosamines: Scientists on the Trail of Prime Suspect of Urban Cancer, **Science, 191**: 268 (1976).

Vollmer, John J., and Susan A. Gordon: Chemical Communication, II: Between Plants and Insects and among Social Insects, **Chemistry**, April 1975, p. 6.

Wade, Nicholas: Green Revolution, I: A Just Technology, Often Unjust in Use, **Science, 186**:1093 (1974).

Wade, Nicholas: Green Revolution, II: Problems of Adapting a Western Technology, **Science, 186:** 1186 (1974).

Williams, Carroll M.: Third-Generation Pesticides, **Scientific American,** July 1967, p. 13.

Wilson, Edward O.: Pheromones, **Scientific American,** May 1963, p. 100.

Wortman, Sterling: Food and Agriculture, **Scientific American,** September 1976, p. 30.

The following articles are at a more advanced level.

Bowers, William S., Tomihusa Ohta, Jeanne S. Cleve, and Patricia A. Morsella: Discovery of Insect Antijuvenile Hormones in Plants, **Science, 193**:542 (1976).

Davenport, Horace W.: Why the Stomach Does Not Digest Itself, **Scientific American,** January 1972, p. 86.

Djerassi, Carl, Christina Shih-Coleman, and John Diekman: Insect Control of the Future: Operational and Policy Aspects, **Science, 186**:596 (1974).

Levine, R. P.: The Mechanism of Photosynthesis, **Scientific American,** December 1969, p. 58.

Marx, Jean L.: Nitrogen Fixation: Prospects for Genetic Manipulation, **Science, 196**:638 (1977).

Safrany, David R.: Nitrogen Fixation, **Scientific American,** October 1974, p. 64.

Skinner, Karen Joy: Nitrogen Fixation, **Chemical and Engineering News,** Oct. 4, 1976, p. 22.

Wolff, I. A., and A. E. Wasserman: Nitrates, Nitrites, and Nitrosamines, **Science, 177**:15 (1972).

12

THE RELATIONSHIP BETWEEN CHEMISTRY AND MEDICINE

12-1 THE ROLE OF CHEMISTRY IN PUBLIC HEALTH

We who are fortunate enough to live in parts of the world with a strong program of public health are apt to take its great advantages for granted. We drink water directly from the faucet, enjoy milk without fear of infection, and dine without hesitation at public restaurants. While occasional lapses occur in health-protecting procedures, by and large the record is very good. For the most part we are not aware of the tremendous amount of research, applied medicine, and engineering which has been required to make these benefits commonly available, and we tend to forget that there are many parts of the world where these conditions do not prevail. Moreover, the first significant steps toward making sanitation and hygiene part of public policy in the United States were taken only a little more than 100 years ago.

The desolation and human suffering caused by the great epidemics of the past are all but impossible to comprehend.

Bubonic plague (the Black Death) is a deadly disease caused by a bacterium which attacks rats and other animals and is transmitted to people by the bites of fleas which have become infected from diseased rats. One of the major epidemics of bubonic plague broke out near Constantinople in 1348 and spread throughout Europe. Records are incomplete, but the death toll was probably around one-half of the entire population. Specific accounts report, for example, that four-fifths of the population of Marseilles were killed. Major epidemics of bubonic plague have occurred all over the world (Fig. 12-1). More than 12 million died of an outbreak in India around the turn of the present century. Areas of infection continue to exist, for the most part in Asia.

Typhus fever is a highly infectious disease transmitted by the bites of body lice. It has flourished particularly in time of war, when malnutrition and personal uncleanliness are common. A major epidemic swept through Europe during World War I, when there were an estimated 30 million cases and 3 million deaths in European Russia alone. In 1914 more than 150,000 men of the Serbian army were reported to have died of the disease in less than 6 months.

Typhoid fever is a contagious disease known since ancient times. Epidemics have frequently ravaged many parts of the world, and it continues to be a menace in many countries. Polluted drinking water is commonly responsible for spreading the disease. One of the first epidemics in the United States traceable to a specific cause broke out during April 1885 in a small town in Pennsylvania with a population of about 8000. Within 2 months there were over 1100 cases and 114 deaths. Investigation showed that there had been a case of typhoid fever the preceding winter in one of the farmhouses along the stream supplying the town's water. Excrement of the patient had been thrown on the snow and was washed into the stream on the first thaw.

The control of such diseases as bubonic plague, typhus fever, and typhoid fever has been brought about by a combination of measures related to public sanitation and personal hygiene in which chemistry has played significant roles. Chemical poisons, for example, along with reducing refuse and access to food, are of help in checking the population of the rats which spread bubonic plague. Insecticides are important in killing the body lice which spread typhus fever. The introduction of DDT (dichlorodiphenyltrichloroethane) in 1944 happened to coincide with an outbreak of typhus in Naples as the Allied armies of World War II were invading Italy. Its use abruptly stopped an epidemic which would almost certainly have become very serious (Sec. 11-10). It is a matter of record that since disinfection of public water supplies has been practiced in the United States (1908), the incidence of typhoid fever and other water-borne diseases has decreased steadily. Today such diseases are rare and confined to areas with undisinfected water supplies. The most common disinfecting agent used is chlorine.

The activities of chemists and the maintenance of good health are related on a broad front. Nowhere are the contributions of chemistry to the general welfare more striking than in the relief of disease and discomfort. Many of the principal areas in which chemistry supports the practice of medicine are summarized in Table 12-1. Although this chapter cannot elaborate on all these

topics, the table illustrates the range and significance of the
relationship between chemistry and the practice of medicine.

FIGURE 12-1
Scenes in London during the Great (bubonic) Plague. (*By permission
of the Master and Fellows, Magdalene College, Cambridge.*)

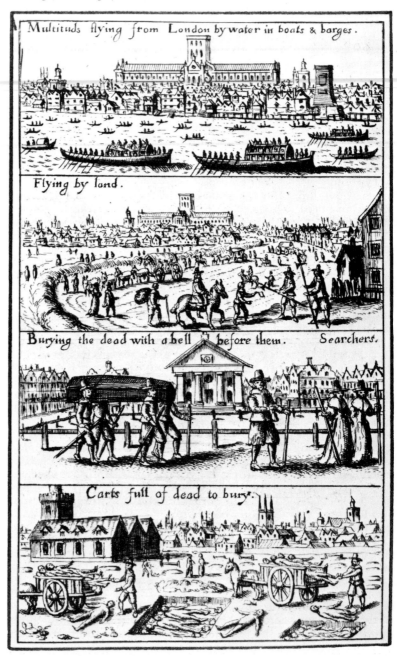

Table 12-1
Representative Areas Where
Chemistry Supports the Practice of Medicine

Area	Description and example
Chemotherapeutic agents	Substances effective against infectious diseases and malignant growths, e.g., penicillin and other antibiotics, chloroquine and other antimalarials, methotrexate and other anti-cancer drugs
Disinfectants and antiseptics	Agents used to destroy most types of harmful microorganisms, e.g., chlorine, used to disinfect drinking water supplies; iodine, used to treat wounds; formaldehyde, used to disinfect instruments
Analgesics	Agents used to relieve pain regardless of origin, e.g., morphine, aspirin
Anesthetics	General anesthetics used in surgery to render the entire body insensitive to pain and usually produce loss of consciousness, e.g., halothane, ethyl ether, nitrous oxide; local anesthetics render particular parts of the body insensitive to pain without loss of consciousness, e.g., lidocaine, used in dental drilling and extraction
Agents to treat anxiety and psychiatric disorders	Substances acting on the central nervous system to relieve anxiety and symptoms of psychiatric abnormalities, e.g., Valium, used principally to relieve anxiety; chlorpromazine, used to treat schizophrenia and manic depression
Insecticides	Agents used to control insect-borne diseases such as malaria and typhus, e.g., parathion, chlordane
Vitamins	Substances used to treat and prevent dietary deficiency diseases, e.g., vitamin B_1 (thiamine), for treatment of beriberi; vitamin D for prevention of rickets; vitamin B_{12} for treatment of pernicious anemia
Investigation of mechanism of drug action	One of the many current research areas of chemistry directly related to the practice of medicine; the molecular basis for the action of the various medicinal agents currently in use is only incompletely known; better understanding could lead to more effective agents with fewer side effects and perhaps whole new approaches to the treatment of human ills

12-2 CHEMOTHERAPEUTIC AGENTS

Before the germ theory was established in the latter part of the nineteenth century, the causes of disease were enveloped in all sorts of superstitious notions. Beginning with a study of the diseases of animals it gradually became clear that microorganisms are a principal agent of disease. The French chemist Louis Pasteur established that bacteria are the cause of anthrax, a deadly disease of animals, especially cattle. In 1878 the German physician Robert Koch demonstrated that six different infectious diseases of animals are caused by invading bacteria. The microorganisms responsible for many infectious diseases were soon isolated and characterized.

The term **chemotherapeutic agent** is usually restricted to substances used to treat diseases of microbial origin without significant harm to the animal or person afflicted, but it is also applied to chemical agents used for other ailments. Common examples are penicillin, active against a wide range of infectious diseases, and isoniazid, one of the principal drugs effective in the treatment of tuberculosis. Crude extracts from plant, animal, and mineral sources have been used in the practice of medicine since ancient times, but widely used and effective disease-combating chemotherapeutic agents have been available for only about 40 years. The term **drug** is defined as any chemical agent which somehow affects the processes of living systems. Chemotherapeutic agents are one class of drugs. **Pharmacology** is that branch of science which is concerned with the development of drugs, their use, and the mechanism of their action.

A. Sulfa Drugs

An important step toward modern chemotherapy was taken around 1880, when the German physician and chemist Paul Ehrlich discovered that some of the synthetic dyes then available were able to prevent the growth and multiplication of disease-producing microorganisms. His long and patient search for specific germ-killing agents led him to the synthesis of salvarsan, the first chemotherapeutic agent reasonably effective in the treatment of syphilis. The continued examination of dyestuffs as possible medicinal agents led to the development in 1933, principally by the German pharmacologist Gerhard Domagk of the drug Prontosil, a red dye which for the first time provided a means of combating bacterial infections. Its dramatic cure of a deadly infection in the bloodstream of a 10-month-old infant initiated a revolution in the treatment of infectious diseases.

Prontosil, a red dye with remarkable properties of curing
certain infectious diseases, discovered in 1933

It was soon discovered that Prontosil is converted into the substance sulfanilamide in the body tissues and that sulfanilamide is the therapeutically active agent. This substance had been synthesized in 1908 in the course of some research related to dyes. It was found to be astoundingly successful in treating many diseases caused by microorganisms, including pneumonia, diphtheria, and gonorrhea. Widespread use soon revealed, however, that sulfanilamide is moderately toxic and that it is ineffective against some types of disease-causing microorganisms. Variations of the sulfanilamide molecule were soon synthesized in an attempt to find a better chemotherapeutic agent. This rather blind manipulation of molecular structure is a common method (Sec. 1-8) of searching for useful drugs. Even though it is usually carried out in the absence of any guiding explanatory hypothesis, it has led to the development of many of the most useful agents in medicine today. In the case of sulfanilamide over 5400 structural variants were synthesized, out of which about a dozen proved useful. They are known collectively as the sulfa drugs. Most, however, were quickly replaced by the less toxic antibiotics (Sec. 12-2B), and very few remain in use. One still prescribed is sulfamethoxazole (Table 12-2), used in the treatment of general bodily infections, especially those of the urinary tract.

Sulfanilamide, one of the first chemotherapeutic agents used to combat bacterial infections in people (1935)

B Antibiotics

It was the famous French chemist Louis Pasteur who tellingly observed that "chance favors the prepared mind." There is no better example of this than the discovery of penicillin by the English bacteriologist Alexander Fleming in 1928. While he was working on some bacterial cultures, the protective cover of one of the dishes in which they were being grown was accidentally left off. A few days later he noticed that a spot of mold had formed and had begun to grow. He also noticed that the bacteria around the spot of mold were no longer there but those some distance from it had continued to increase. Since as a bacteriologist he had been on the lookout for antibiotics, he investigated the reason for the action of the mold. He found that the broth in which the mold had grown was promisingly active against a wide range of bacteria and was almost completely nontoxic to animals and people. He named the active broth penicillin, from the mold which produced it, *Penicillium notatum*. (Figure 12-2 shows a related species, *P. chrysogenum*, now used to manufacture penicillin.)

The chemical characterization of a very important substance seemed to be at hand, but Fleming and subsequent investigators were unable to isolate it from the broth because of its instability and extremely low concentration. Moreover, safe, nontoxic agents effective against internal bacterial infections were considered so improbable as to be hardly worth pursuing. But then the development of the sulfa drugs during the 1930s demonstrated that such substances were well within the realm of possibility. Encouraged by this, the British biochemists Howard Florey and Ernest Chain reexamined Fleming's promising mold product. In 1939 they successfully isolated an active solid extract and demonstrated its spectacular antibacterial properties.

Agent	Description
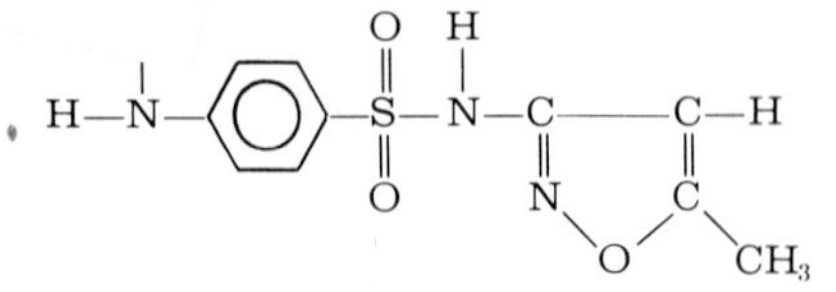 **Sulfamethoxazole**	One of the sulfa drugs in current use; effective in the treatment of general bodily infections, especially those of the urinary tract; administered orally
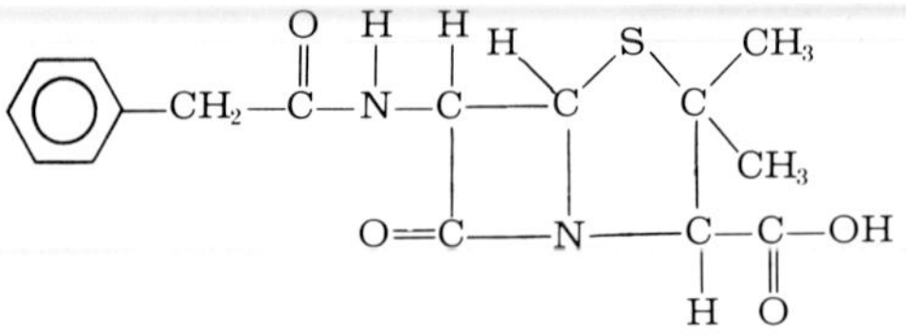 **Penicillin G**	An antibiotic; most useful of the natural penicillins; effective in the treatment of many (but not all) infectious diseases, e.g., pneumonia, diphtheria, syphilis, gonorrhea, and meningitis; administered by injection
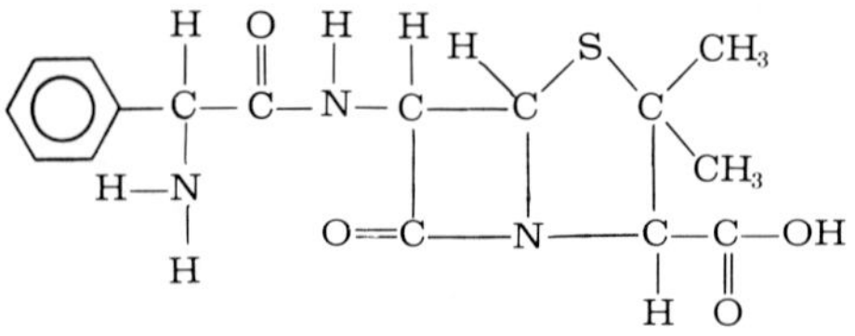 **Ampicillin**	A synthetic structural variant of penicillin G; administered orally; one of the broad-spectrum antibiotics, effective against a wider range of infections than penicillin G, e.g., salmonella infections, types of meningitis, and urinary tract infections
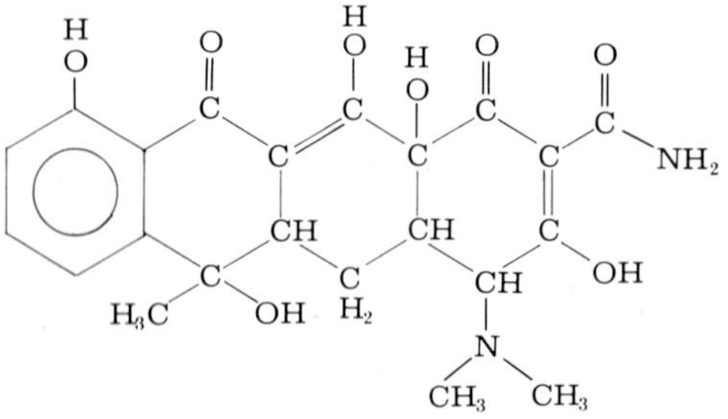 **Tetracycline**	Broad-spectrum natural antibiotic found in the soil, effective against a broad range of harmful microorganisms, e.g., those which cause typhus fever, Rocky Mountain spotted fever, psittacosis, and trachoma; administered orally; frequently used when patients are sensitive to penicillin G

The work of Florey and Chain coincided with World War II demands for agents effective against infections and led to the creation of a special joint British-American crash program to follow through on their findings. This work resulted in the isolation of a group of penicillins and the determination of their structure. The antibacterial activity was found to be concen-

FIGURE 12-2
The mold *Penicillium chrysogenum*, the species now used to produce
the antibiotic penicillin, growing in a nutrient medium. (*Courtesy
of Pfizer Incorporated.*)

trated mostly in the substance designated penicillin G (Fig.
12-3), the penicillin now in common use (Table 12-2).

Once again chemists immediately used the structural-variant
approach in an attempt to develop agents better or more easily
available than penicillin. These investigations have led to more
than a dozen useful agents, one of which is ampicillin (Table
12-2). It belongs to a group of agents that are effective against a
broader range of microorganisms than penicillin G, so-called
broad-spectrum penicillin derivatives.

An **antibiotic** is a substance produced by a microorganism
capable of destroying or inhibiting the growth of other microorganisms. The production of such substances is one of the
means used by microorganisms to compete with other microorganisms for survival in their highly competitive environment.
Penicillin G was the first antibiotic chemists were able to extract from the microorganism which produced it (the mold
Penicillium notatum) and to use on a large scale against other microorganisms. The second antibiotic to be isolated and identified
was streptomycin (1944), due to the work of the American
biochemist Selman Waksman and his colleagues. The medical

profession now has at its disposal a large assortment of antibiotics. One of the most frequently used is tetracycline (Table
12-2), effective against typhus fever and a broad spectrum of
other infectious diseases. These substances are able to cure a
wide variety of diseases which were once fatal and have brought
about a virtual revolution in the treatment of disease (Fig. 12-4).
One of their limitations, however, is that microorganisms can
develop resistance to their action, a disadvantage which the
antibiotics share with many other antibacterial agents. This is
one reason why such agents should be used sparingly.

FIGURE 12-4
The major causes of death in the United States in 1900, 1940, and
1970. At the same time the average life expectancy at birth increased from 47 years in 1900 to 71 years in 1970. Note the reduction in deaths due to pneumonia, tuberculosis, and diphtheria,
due partly to the introduction of the sulfa drugs and (later)
antibiotics. Note also the accompanying increase in deaths due
to heart ailments and cancer. (*From John H. Dingle, The Ills of Man.
Copyright by Scientific American, Inc. All rights reserved.*)

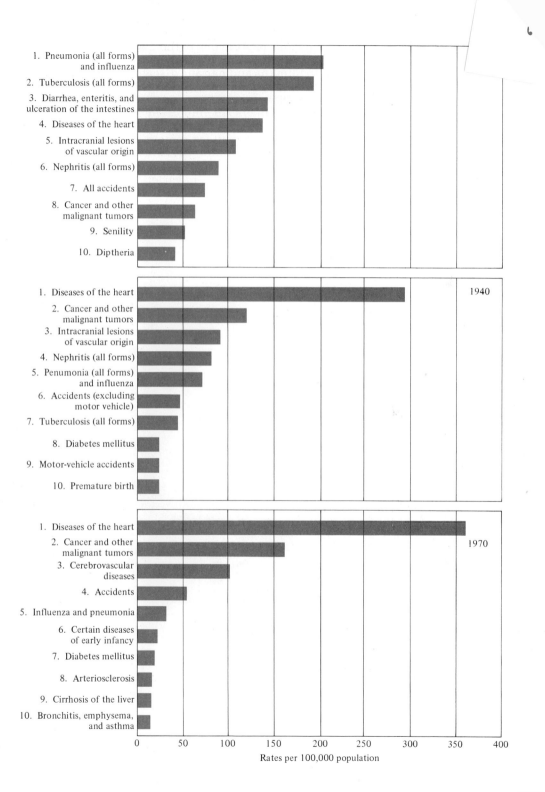

1. Pneumonia (all forms) and influenza
2. Tuberculosis (all forms)
3. Diarrhea, enteritis, and ulceration of the intestines
4. Diseases of the heart
5. Intracranial lesions of vascular origin
6. Nephritis (all forms)
7. All accidents
8. Cancer and other malignant tumors
9. Senility
10. Diptheria

1940

1. Diseases of the heart
2. Cancer and other malignant tumors
3. Intracranial lesions of vascular origin
4. Nephritis (all forms)
5. Penumonia (all forms) and influenza
6. Accidents (excluding motor vehicle)
7. Tuberculosis (all forms)
8. Diabetes mellitus
9. Motor-vehicle accidents
10. Premature birth

1970

1. Diseases of the heart
2. Cancer and other malignant tumors
3. Cerebrovascular diseases
4. Accidents
5. Influenza and pneumonia
6. Certain diseases of early infancy
7. Diabetes mellitus
8. Arteriosclerosis
9. Cirrhosis of the liver
10. Bronchitis, emphysema, and asthma

0   50   100   150   200   250   300   350   400
Rates per 100,000 population

12-3 THE MALARIA PROBLEM

Insect-borne diseases are one of the major scourges of mankind. The biting tsetse fly (Fig. 12-5), which transmits fatal sleeping sickness to humans and nagana disease to horses and cattle, has for centuries hampered the agricultural development of a region in central Africa larger than the entire United States. The oriental rat flea plays an important role in the transmission of the deadly bubonic plague, and the body louse carries typhus fever. But mosquitoes remain the most harmful insects of all. More deaths around the world are due each year to mosquito-borne diseases than from any other single cause. There are almost 3000 different varieties of these pests, and they are found almost everywhere, including the Arctic. They are the bearers of yellow fever, which remains a major health

FIGURE 12-5
The tsetse fly, transmitter of fatal sleeping sickness. (*Courtesy of the Department of Agriculture.*)

problem. Between 15,000 and 30,000 people died from an epidemic in Ethiopia in the 1960s. There have been serious outbreaks in the United States of encephalitis, a mosquito-borne disease accompanied by inflammation of the brain.

But mosquitoes do their greatest harm in transmitting malaria. The World Health Association estimates that there are more than 100 million cases worldwide. Even though malaria is more debilitating than deadly, between 1 and 2 million people die from the disease each year. In the South Pacific during World War II it removed 5 times as many troops from combat as the Japanese did. The disease is caused by microorganisms which spend part of their complicated life cycle in the human bloodstream and the other part in the female anopheles mosquito (Fig. 12-6), mosquito bites serving as the method of transmission. Measures to combat malaria are therefore two-pronged: control of mosquitoes and treatment of the human infection.

Mosquitoes are flying animals in their adult stage and aquatic animals in their preadult larval stage. They can be attacked by draining swamps and spraying breeding grounds with oil to reduce the population of larvae and by using chemical insecticides to kill the adults on contact. With the introduction of inexpensive insecticides such as DDT in the 1940s, considerable success was obtained with the insecticide approach, and the older methods of attacking breeding grounds were deemphasized. The incidence of malaria dropped dramatically in many parts of the world, and it appeared that it was only a matter of time before it would be brought under control. In Ceylon, for example, where malaria had been heavily entrenched, cases fell to only about 100 a year by 1963. But the mosquitoes began to build up resistance to the insecticides, and as questions arose about their environmental costs and toxicity,

FIGURE 12-6
A malaria mosquito, one of humanity's worst enemies (*Courtesy of the American Museum of Natural History.*)

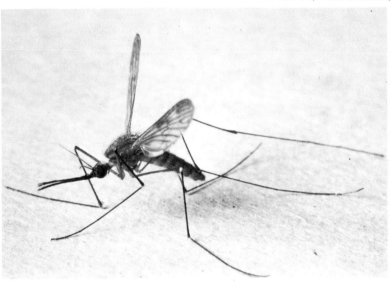

the use of insecticides was reduced. In the meantime, use of older methods had been relaxed. As a consequence malaria cases have risen. Although deaths from malaria on a worldwide basis are now only half of what they were before the antimalarial campaign was initiated in 1955, malaria remains the biggest killer of human beings and the world's principal medical problem. Adequate mosquito control presents a tremendous challenge. It is hoped that greater understanding of the life cycle of the mosquito and the invading microorganism will lead to new approaches. In the meantime, there is a great need for the older methods of attack on breeding grounds and for safer and more effective insecticides.

Malaria can also be attacked by curing the infection in people. When Spanish armies entered Peru in the sixteenth century, they found the native Indians using the bark of the cinchona tree to treat malaria, and this practice was subsequently introduced into Europe. In 1820 the French chemists P. J. Pelletier and J. B. Caventou succeeded in isolating from cinchona bark the substance quinine, responsible for its antimalarial activity. The details of quinine's molecular structure were unraveled by 1908, mostly through the efforts of the German chemist Paul Rabe (Table 12-3).

Although quinine suppresses malaria and is still useful in the treatment of the disease, it is not a cure. There has been active interest for many years, therefore, in developing a more effective agent. The structural-variant approach (Sec. 1-8) has been used here, as in the case of many medicinal chemicals. Attention was focused not only on the structure of quinine but on the structure of methylene blue, the dye in which Paul Ehrlich had discovered a low level of antimalarial activity (1891). This antimalarial work, mostly by German chemists, led to the synthesis of some useful drugs, including atabrine, which is similar to quinine in its action.

With the advent of World War II the United States and its allies were faced with military campaigns in many malaria-stricken areas, especially in the South Pacific. With the invasion of the world's principal quinine plantations in the East Indies by the Japanese early in the war, supplies of quinine were cut off. Atabrine was synthesized on a large scale, and its wartime availability was of paramount military importance. Meanwhile a crash program was initiated to develop better antimalarial drugs, and over 12,000 different substances were tested. Several superior agents emerged, including the principal drugs used to treat malaria today. Two of these are chloroquine and primaquine, structural variants of quinine (Table 12-3), which are very effective against several malarial strains. There is a need for better agents, however, and the search continues.

12-4 PAIN-RELIEVING SUBSTANCES

It was observed in Sec. 1-5 that pain is a common denominator of all poor health and disease. The management of pain, however, is a very challenging assignment, since a completely satisfactory means of control remains to be developed. Pain-relieving agents are of three general types. A **centrally acting**

Table 12-3
Three Representative Antimalarial Agents

Agent	Description
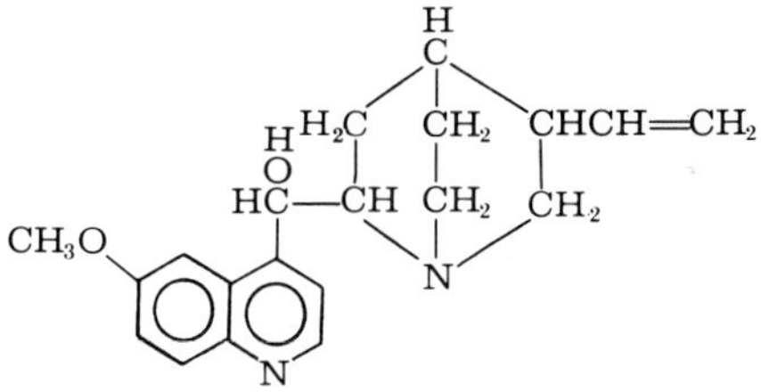 Quinine	The original antimalarial agent, obtained from the bark of the cinchona tree; now largely replaced by other agents but sometimes used in conjunction with newer antimalarials
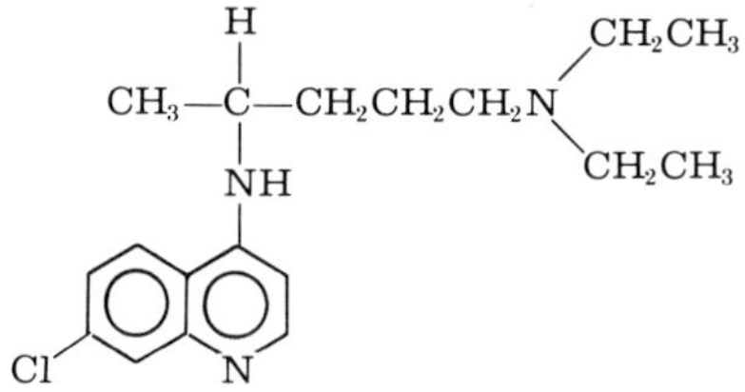 Chloroquine	More powerful and less toxic than quinine; an effective suppressive agent for most types of malaria; frequently serves as a cure for one type
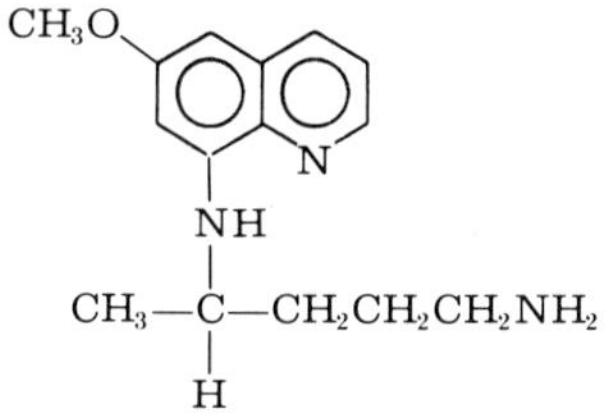 Primaquine	Frequently serves as a cure for a common type of malaria

analgesic is a pain-relieving substance which relieves most forms of pain throughout the body, usually without loss of consciousness. Common examples are morphine and aspirin. A **general anesthetic** is a substance which relieves pain throughout the body but usually brings about a loss of consciousness. Common examples are the surgical anesthetics ethyl ether and nitrous oxide. A **local anesthetic** is a substance which is capable of relieving pain in specific parts of the body without loss of consciousness. A common example is novocain (procaine hydrochloride) used in dental practice.

A. Centrally Acting Analgesics

For over 170 years morphine (Table 12-4) has played an important role in the practice of medicine as the standard agent for the relief of severe pain. The use of opium, the dried exudate of unripe poppy seeds in which morphine is the principal active ingredient, goes back thousands of years. Morphine

Table 12-4
Some Powerful Pain-relieving Agents

Agent	Description
Morphine	Isolated from opium, obtained from unripe poppy seeds; despite its addiction liability and other adverse effects, remains standard agent for relief of severe pain
Meperidine	Synthetic substance, widely used as an alternative to morphine for the relief of severe pain; addicting
Methadone	Synthetic substance; effective pain-reliever but addicting; largely restricted to treatment of heroin addiction
Pentazocine	Synthetic substance, introduced in 1964; less powerful than morphine and meperidine but with very low addiction liability; useful in chronic pain

Note to Students	With an assist from Table 8-3 you should be able to recognize all of the functional groups in meperidine.

plays a double role, alleviating pain from any cause and often relieving anxiety and promoting a feeling of well-being. But it has many undesirable side effects; it depresses the respiratory center in the central nervous system, and large overdoses may cause death because breathing stops. In the smaller doses usually used it may also cause nausea and constipation. Moreover, its effects wane as doses are repeated, and worst of all it becomes addicting on prolonged use.

Consequently there has been a continuous search for a better pain-relieving agent, one which has morphine's beneficial properties without its harmful side effects, especially addiction (Sec. 1-8). As soon as the molecular structure of morphine was determined, chemists began to put together molecules of similar structure to see if they could find a better agent. In this work over 150 structural variants of the morphine molecule have been synthesized, along with some 300 other substances. Several useful pain-relieving substances have been developed in these investigations, including pentazocine, the agent with very low addiction liability, but none is more effective in relieving pain than morphine (Secs. 1-10 and 1-11).

The terms **opiate** and **narcotic analgesic** are used to describe morphine and other substances which in use behave like morphine. The physician now has at his disposal several useful narcotic analgesics in addition to morphine. One of these is meperidine, also called Demerol, a synthetic drug first made by German chemists in 1939. This substance is better than morphine when administered orally (as opposed to intravenous injections) (Table 12-4), but it is addicting and has many of the other disadvantages of morphine.

Methadone, another useful synthetic narcotic analgesic (Table 12-4), was developed by German chemists and introduced during World War II. It is now used in the treatment of heroin addiction (Sec. 1-13) even though addicting itself, and its use is now largely restricted to this treatment. Pentazocine is a pain-relieving substance synthesized by American chemists and introduced in 1964 (Sec. 1-11 and Table 12-4). Because of its very low addiction liability, it is especially useful in relieving the pain of chronic illnesses.

Aspirin is the common household pain-relieving agent. While it is effective only against pain of low intensity, continued use does not lead to addiction or diminution of its effectiveness. Aspirin is valued not only as a pain-relieving agent but is also effective in the relief of fever and in combating the inflammation associated with rheumatism and arthritis. It is the most widely used medicinal agent. An estimated 20 billion tablets of aspirin or drug combinations containing aspirin are used in the United States each year. Although aspirin is one of the safest drugs known, the potential toxic effects of large doses are sometimes overlooked.

The chemical name for aspirin is acetylsalicylic acid. The substance was first synthesized by the Alsatian chemist Charles Gerhardt in 1853, but its effectiveness in treating arthritis and fevers was not discovered until 1899, when the German chemist Felix Hoffman used it for this purpose. Aspirin's ability to relieve pain became apparent shortly thereafter. The development of the drug can be traced back to the use of willow bark in the herbal remedies of tribal peoples. The ability of willow bark to relieve fever was first carefully characterized

Salicylic acid

Acetylsalicylic acid (aspirin)

Comparison of the structures of salicylic acid and acetylsalicylic acid; salicylic acid, somewhat effective against rheumatism but relatively toxic, was obtained from substances in the fever-reducing willow bark; acetylsalicylic acid is the much more useful structural variant

in 1763 by the English physician and clergyman Edward Stone. Chemical investigations of willow bark led eventually to the recognition in 1838 that salicylic acid, prepared from substances isolated from the bark, had the ability to relieve fever and the symptoms of rheumatism. Unfortunately, however, it proved to have toxic side effects: nausea, vomiting spasms, and in extreme cases, prolonged unconsciousness. Felix Hoffman set about to find a better agent, and guided by the molecular structure of salicylic acid, he turned to one of its structural variants, acetylsalicylic acid, already recorded in the chemical literature. Once again the structural-variant approach paid impressive dividends.

Aspirin was developed without any knowledge of how salicylic acid acts to exert its physiological effects, but recently clues to the mechanism of aspirin's action have come from an unexpected source. In the 1950s a group of Swedish scientists isolated from human semen representatives of a class of substances called prostaglandins, which have interesting hormonelike activity. Individual prostaglandins show promise as possible agents for the regulation of blood pressure, relief of bronchial asthma, and fertility control. Some of the prostaglandins or structurally related substances are apparently able to induce fever and inflammation. Even more interesting is that aspirin has been found capable of blocking the synthesis of prostaglandins in animals. This finding has led investigators to the provocative hypothesis that aspirin's ability to counter the inflammation of rheumatoid arthritis and to lower fever is due to its ability to block the body's synthesis of the fever- and inflammation-causing prostaglandins. Should this work lead to a confirmed mechanism of aspirin's physiological activity, a systematic approach will be opened for the development of improved replacements for aspirin. It will be very interesting to follow the progress in this area.

Prostaglandin E$_2$, which shows some promise in the treatment of bronchial asthma; this is one of a group of prostaglandins, some of which induce fever and inflammation

B. General Anesthetics

The dramatic introduction of ethyl ether as a general anesthetic over 130 years ago was discussed in Chap. 8. This made surgery a humane form of treatment and began a new era in the practice of medicine. The development of surgical anesthesia stands as one of the truly great contributions which science has made to human welfare (Fig. 12-7).

The surgeon has available today a sophisticated array of anesthetic agents. Indeed anesthesiology is a field of medicine in itself, and for the most part it is the anesthesiologists who administer the anesthetics during surgery. Table 12-5 lists some modern agents used for general anesthesia. Ethyl ether is no longer used very much, mostly because of its flammability and the nausea, vomiting, and other side effects which follow its use. One of the other early anesthetics, nitrous oxide, still makes an important contribution. It is of interest that while several hypotheses have been advanced to explain how anesthetics exert their effect, there is no established theory which explains the fundamental mechanism satisfactorily.

C. Local Anesthetics

Local anesthetics are usually administered by injection or by spreading them on the skin or membrane surface. They act by blocking the conduction of nerve impulses and do not cause loss of consciousness. A small region, e.g., one side of the jaw, can be rendered insensitive to pain (as in dental practice), or larger areas can be anesthetized. Spinal injection of a local anesthetic, for example, can be used to desensitize all the lower extremities of the body and is often used in major surgery for this purpose.

FIGURE12-7
An anesthesiologist monitors the administration of a general anesthetic during a major surgical operation. It is interesting to compare this operating-room scene with the first public demonstration of ethyl ether as a general anesthetic in Fig. 8-2. (*Courtesy of Hillcrest Hospital, Pittsfield, Massachusetts.*)

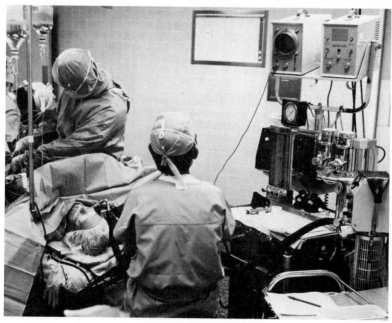

Anesthetic	Description
N≡N—O Nitrous oxide, a gas	Administered by inhalation; nonflammable, non-irritating, pleasant-smelling; because of low potency often supplemented with more powerful anesthetics and muscle relaxants; one of earliest anesthetics
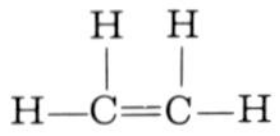 H—C=C—H (with H's) Ethylene, a gas	Administered by inhalation; not much used now because of explosive nature of ethylene-oxygen mixtures; little more powerful than nitrous oxide
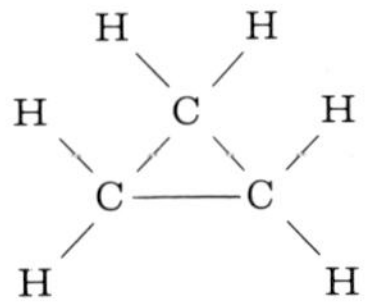 Cyclopropane, a gas	Administered by inhalation; very powerful and controllable anesthetic, relatively free of adverse side effects but is flammable and explosive
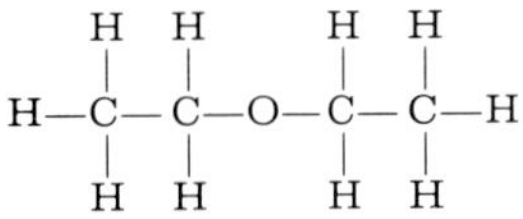 H—C—C—O—C—C—H (with H's) Ethyl ether, a liquid	Administered by inhalation; relatively free of serious side effects and simple to administer, not much used because flammable and explosive and because of nausea, vomiting, and other consequences
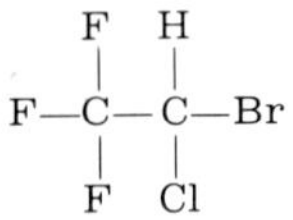 F—C—C—Br (with F, H, Cl) Halothane (Fluothane), a liquid	Administered by inhalation; one of the popular newer anesthetics; nonflammable, nonirritating, and powerful; usually used in conjunction with nitrous oxide and muscle relaxants
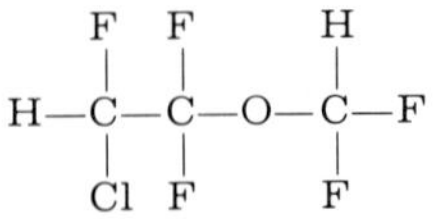 H—C—C—O—C—F (with F, F, H, Cl, F, F) Enflurane (Ethrane), a liquid	Administered by inhalation; one of popular newer anesthetics; nonflammable, nonirritating, and powerful
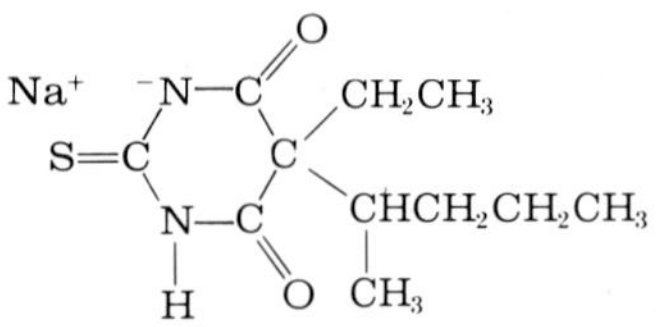 Sodium thiopental (Pentothal), a solid	Administered by intravenous injection of an aqueous solution; very rapid in action, with a minimum of aftereffects; not very strong in pain relief, and generally used with an inhalation anesthetic; often used to induce anesthesia, because of its quick, comfortable action, before use of inhalation anesthetic

Local anesthetics, like general anesthetics, are among the major contributions of science to humanity, and chemists have played a major role in their development. Modern local anesthetics are descendants of investigations of the chemical composition of the leaves of the coca plant which grows in the Andes mountains of South America. For centuries these leaves have been chewed for their stimulating effect by the people of the region. The active substance itself, cocaine, was first isolated late in the nineteenth century, and the young Viennese physicians Sigmund Freud and Karl Koller were principally responsible for its introduction as a local anesthetic around 1884. Although useful, cocaine has many deficiencies as an anesthetic agent. It causes allergic reactions, it irritates tissues, and it is strongly addicting. It has been largely replaced by the newer, improved agents.

The synthesis of structural variants of cocaine led to the development of the first generally effective local anesthetic, novocain (procaine hydrochloride), synthesized by the German chemist Alfred Einhorn in 1909. Further synthesis and testing have produced many useful agents. One of the most widely used currently is lidocaine (Xylocaine), introduced in 1948.

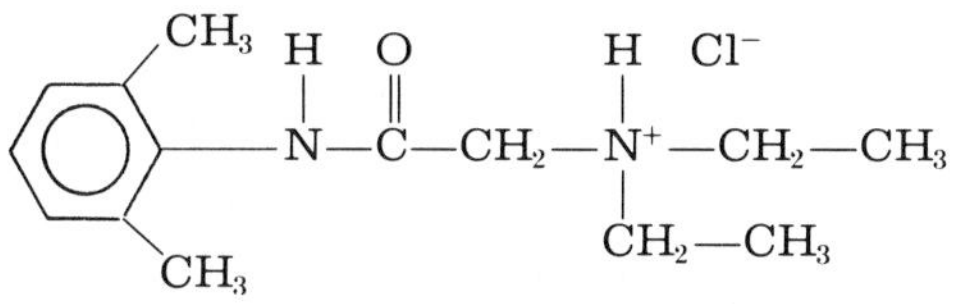

Lidocaine hydrochloride (Xylocaine hydrochloride), one of the most widely used local anesthetics, has only a remote structural resemblance to cocaine

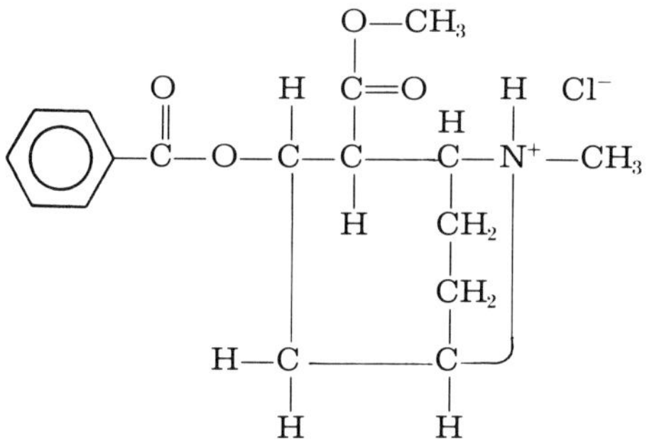

Cocaine hydrochloride

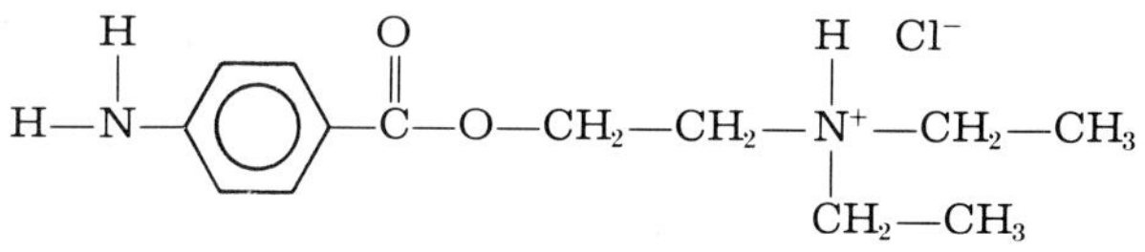

Novocain (procaine hydrochloride)

Comparison of the structures of the local anesthetics cocaine hydrochloride and procaine hydrochloride; both are represented as hydrochloride salts, the form in which they are frequently used; although the structures are quite different, some similarities prevail

Note to Students

Can you spot the two carboxylic ester groups in cocaine hydrochloride? The carboxylic ester and amine groups in novocain? (Consulting Table 8-3 is allowed.)

12-5 THE ANTIMETABOLITE THEORY OF DRUG ACTION

The development of sulfanilamide was an extraordinary event not only because of the drug's life-saving antibacterial action and the encouragement its success gave to the subsequent development of penicillin; the investigation of sulfanilamide's properties led directly to an important theory of drug action. This theory is concerned with the various chemical reactions which occur within disease-causing microorganisms and within humans. A **metabolite** is a substance which is converted into other substances by the metabolism of living organisms. The chemical reactions of humans and many kinds of disease-causing microorganisms require the metabolite folic acid. Humans must obtain this substance from their food, but the microorganisms are able to synthesize it from a metabolite of simpler structure, p-aminobenzoic acid. This chemical resourcefulness of microorganisms proves, however, to be an Achilles' heel. For if sulfanilamide is made available to the microorganisms, they will mistake it for p-aminobenzoic acid. The error is understandable since the overall structure of the two substances is similar, as the comparison of their structural formulas in Fig. 12-8 suggests.

The sulfanilamide is able to behave like p-aminobenzoic acid, but only to a small extent. The net effect of the sulfanilamide is to deprive the microorganism of the p-aminobenzoic acid and thus folic acid essential for its well-being. Since humans do not synthesize folic acid but obtain it directly from food, they

FIGURE 12-8
Comparison of the structures of folic acid, p-aminobenzoic acid, and sulfanilamide. Folic acid is an important substance in the normal chemical reactions of many microorganisms, which synthesize it from p-aminobenzoic acid. If you look closely at the structures, focusing particularly on the region of the benzene nucleus, you will be able to see how the structure of p-aminobenzoic acid is incorporated into the structure of folic acid. Since sulfanilamide has a structure similar to that of p-aminobenzoic acid, it can partially replace it, thus preventing the microorganism from synthesizing folic acid.

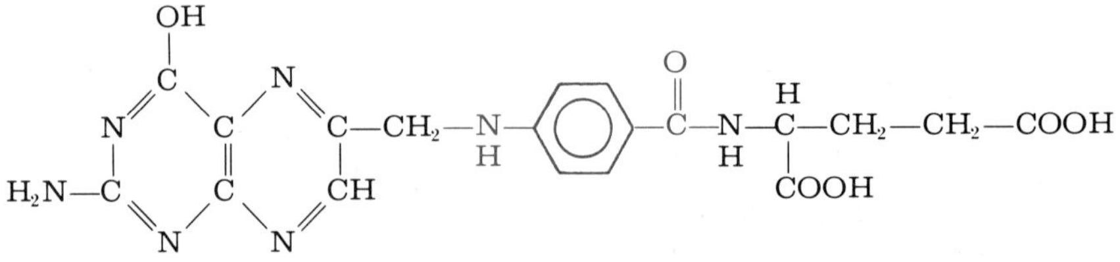

Folic acid

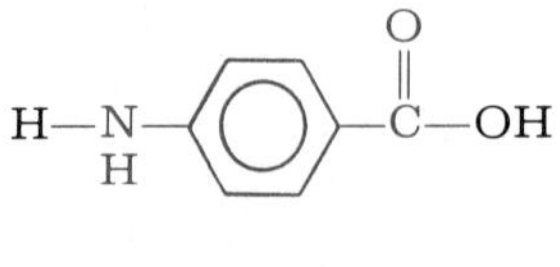

p-Aminobenzoic acid

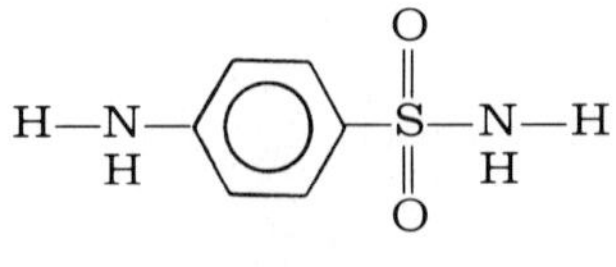

Sulfanilamide

are not bothered by the sulfanilamide's jamming action. Hence sulfanilamide is able to interfere selectively with the chemistry of only the microorganism.

Sulfanilamide acts as an **antimetabolite,** i.e., a substance of structure similar to a normal metabolite (in this case p-amino-benzoic acid) which is capable of interfering with the reactions of the normal metabolite. (The exact mechanism of this interference is discussed in the optional Sec. 12-11.) The term **antimetabolite theory of drug action** is applied to the provocative idea that a substance may serve as a chemotherapeutic agent if it has a structure similar to a normal metabolite of a disease-causing microorganism. Many antibiotics are known to function as antimetabolites.

This theory can be used in the development of useful chemotherapeutic agents in the laboratory (Fig. 12-9) as follows.

1. Determine the structure of an important metabolite of a disease-causing microorganism which is not also a metabolite of human beings.
2. Synthesize structural variants of the metabolite.
3. Test these substances for useful drug action.

FIGURE 12-9
General view of a research laboratory devoted to pharmacological **research.** (*Courtesy of Merck and Co., Inc.*)

This approach has not yet been as fruitful as hoped, since the chemistry of human beings and microorganisms is closer than one might expect. Moreover, in order for a substance to be a useful chemotherapeutic agent it must be able to reach the correct site in the microorganism intact and must have other necessary properties, requirements that are not easily fulfilled. The antimetabolite approach, however, has had a limited success in the treatment of some forms of cancer (Sec. 12-6).

12-6 HOW IS CANCER TREATED?

As Fig. 12-4 indicates, cancer is the second most frequent cause of death after diseases of the heart. One of the reasons the treatment of cancer is such a difficult problem is that there are over 100 different kinds of malignancies. Certain types respond to the treatments currently available much better than others. Cancerous tumors are composed of cells lacking the normal controls on growth and reproduction. They are apt to grow quickly and to spread out beyond the tissue in which they develop. The most destructive characteristic of cancerous tumors is their ability to release malignant cells which migrate through the body fluids and establish new cancerous pockets in distant parts of the body. This process is called metastasis.

Present forms of treatment include the three approaches of surgery, nuclear radiation (Sec. 9-9A), and chemotherapy (the use of chemical agents). A fourth approach in the early stages of development is immunotherapy (the use of agents which increase a patient's resistance against a cancer attack). Surgery and radiation therapy are the principal forms of treatment, but they are for the most part limited to use against localized tumors. Once the cancer has metastasized, it is usually beyond the reach of these methods. Chemotherapy, at least in principle, is effective against malignant cells wherever they are distributed.

There are 40 different chemical agents which are currently being used in the treatment of cancer, and another 10 are being used on a limited basis. Possible new agents are being screened at the rate of more than 40,000 a year. The agents used in the pioneering work of about 20 years ago led to many disappointments. Progress has been slow, but the combinations of drugs now in use are bringing about more promising results. Frequently surgery, radiation, and chemotherapy are used together. About 20 years ago the lives of only about one-fourth of the cancer victims could be extended 5 years or more by treatment. Today the fraction has been increased to one-third. While this increase is only a beginning, it represents the saving of about 50,000 additional lives each year. Although chemical agents have contributed significantly to this improvement, they still leave much to be desired. The body usually recovers rapidly from their side effects, but these include loss of hair, nausea, vomiting, and increased susceptibility to infection. New developments, however, are yielding some encouraging improvements.

The chemical agents in use today are of many different types and have been developed from several different approaches. The antimetabolite theory of drug action has led to some useful forms of medication. We shall mention two. The first is methotrexate, a synthetic structural variant of folic acid. It acts as an antimetabolite of folic acid and interferes with the replication of tumor cells. It is used against many types of cancer and is one of the drugs which has had some promising results in the treatment of childhood leukemia.

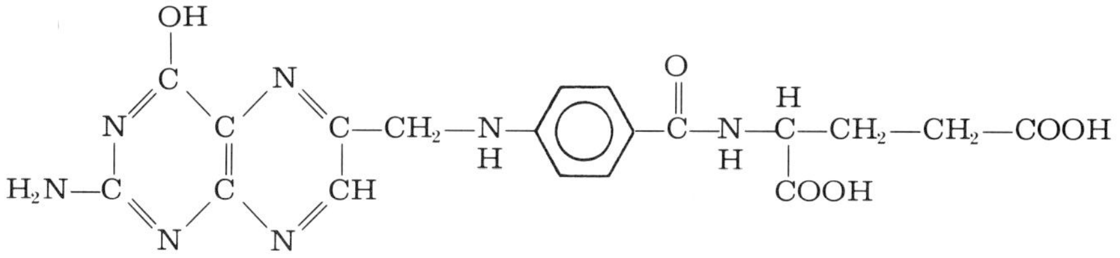

Folic acid

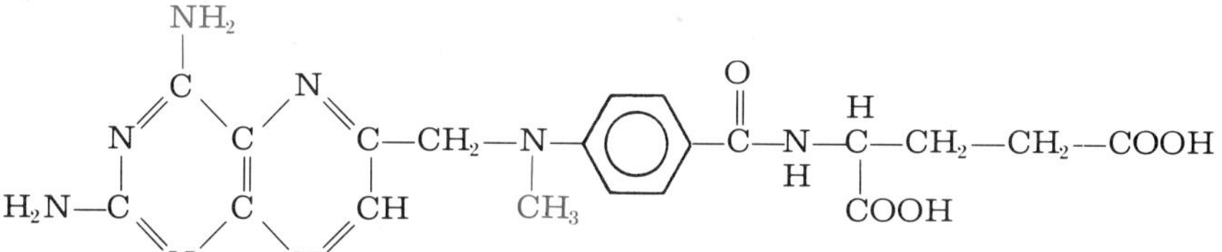

Methotrexate

Comparison of the structure of folic acid, an important substance in the chemistry of human beings, with the structure of methotrexate, a synthetic antimetabolite used in the treatment of childhood leukemia and other forms of cancer

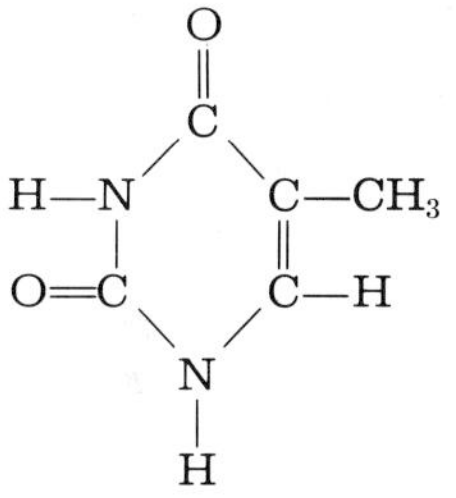

The substance 5-fluorouracil has proved valuable in treating cancers of the breast and intestinal tract. It is a structural variant of thymine, one of the structural components of the nucleic acids (Sec. 10-8D) and serves to interfere with the replication of tumor cells. Unfortunately the agents which obstruct the growth of cancer cells also disturb many normal cells, causing unfortunate side effects and limiting the use of the drugs.

Thymine

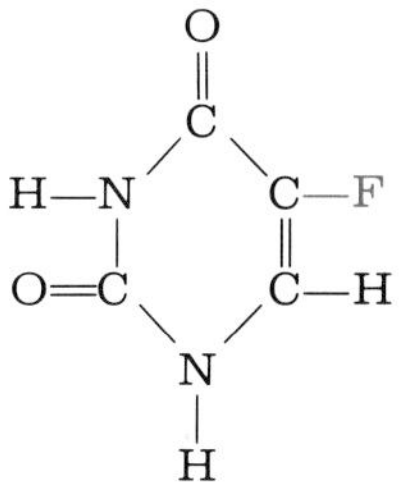

5-Fluorouracil

Comparison of the structures of thymine, a structural component of the nucleic acids, and 5-fluorouracil, a synthetic antimetabolite used in the treatment of tumors of the breast and intestinal tract

12-7 THE DEVELOPMENT OF ORAL CONTRACEPTIVES

Hormones are substances secreted by specific body organs which exert important controlling influences on body chemistry. Insulin, for example, a hormone secreted by the pancreas, regulates the reactions involving the use of glucose in mammals. Testosterone, a male sex hormone secreted by the testes, influences body chemistry to bring about the male sexual characteristics.

One of the most important hormones is the female hormone progesterone, which plays an important role in maintaining pregnancy. Since it can prevent ovulation, it is a natural contraceptive. Unfortunately, when administered orally it is too weak in its activity to be practical. In their usual manner chemists set about preparing structural variants in the search of a more powerful substance. This work led to the synthesis of norethynodrel in 1952 by Frank Colton of the laboratories of

FIGURE 12-10
Comparison of the structures of progesterone, a natural female sex hormone with mild oral contraceptive properties, and norethynodrel, a synthetic structural variant, which is much more active and is the principal component of the original oral contraceptive, Enovid.

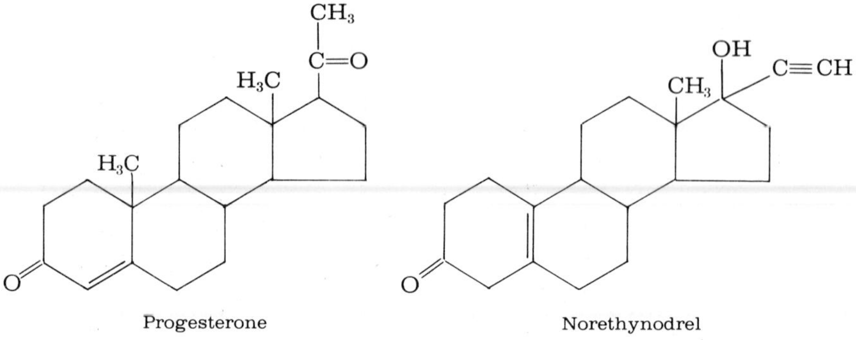

Progesterone   Norethynodrel

G. D. Searle and Company (Fig. 12-10). Clinical trials showed that this substance is very effective as a female oral contraceptive and that its activity is enhanced when mixed with a small amount of another structural variant called mestranol. The mixture was marketed under the trade name Enovid beginning in 1960. Other effective structural variants of progesterone have subsequently been synthesized, and today more than 30 formulations of oral contraceptives are available. Collectively known as "the pill," they represent a landmark development in contraceptive methods. It has been estimated they are used by more than 11 million women in the United States and by another 40 million in the world at large.

12-8 DRUGS USED IN THE TREATMENT OF ANXIETY AND PSYCHIATRIC DISORDERS

The use of drugs to alter mood, emotion, and behavior is as old as civilization itself. The effective use of chemical substances in the treatment of psychiatric disorders, however, began only about 25 years ago. Today drugs are one of the principal means of treating mental illness, and their successes have been largely responsible for the substantial decrease in the number of these patients hospitalized. The agents used have emerged from many different types of investigations, and the drugs themselves exert a variety of actions. The search for more effective drugs with fewer annoying and harmful side effects is being urgently pursued, and considerably more needs to be known about the molecular mechanism of such agents. One of the many limitations in the use of chemical substances for the treatment of psychiatric disorders is due to the difficulties of diagnosing the various forms of mental illness.

One of the very first agents found effective in treatment of mental disorders was chlorpromazine (Thorazine), first synthesized in 1950 and introduced clinically in 1952 (Table 12-6). It is used today in the treatment of schizophrenia and manic depression. Meprobamate (Miltown, Equanil), one of the early tranquilizers (1954), is still used to treat anxiety. Another antianxiety drug, diazepam (Valium), is the most frequently prescribed drug in the United States. The use of lithium carbonate in the treatment of manic depression was discussed in Chap. 7. Since, for the most part, this agent acts to remove the hyperelated state of manic depression, it is often used along with a drug which alleviates depression. Imipramine (Presamine) is one of the drugs currently used for this purpose (Table 12-6).

Table 12-6
Representative Agents Used in Treating Anxiety and Psychiatric Disorders

Agent	Description
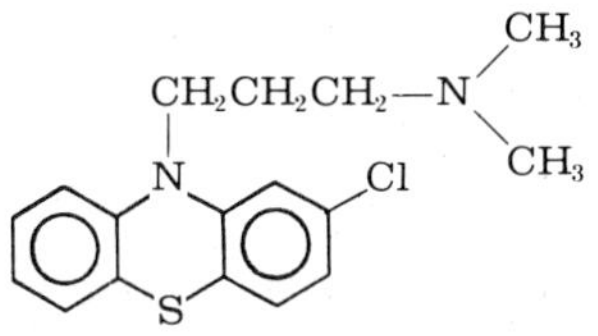 Chlorpromazine (Thorazine)	Used in treating schizophrenia and manic depression; one of the first effective agents (1952)
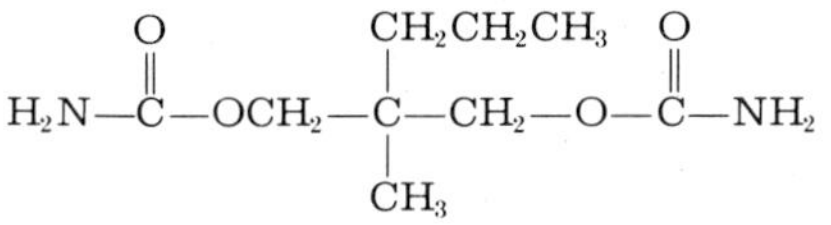 Meprobamate (Miltown, Equanil)	One of the first tranquilizers (1954) used to relieve anxiety
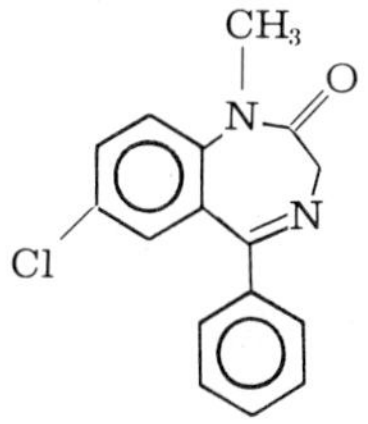 Diazepam (Valium)	Used mainly to relieve anxiety; most frequently prescribed drug in United States
Li_2CO_3 Lithium carbonate	Used to treat the hyperelated state of manic depression
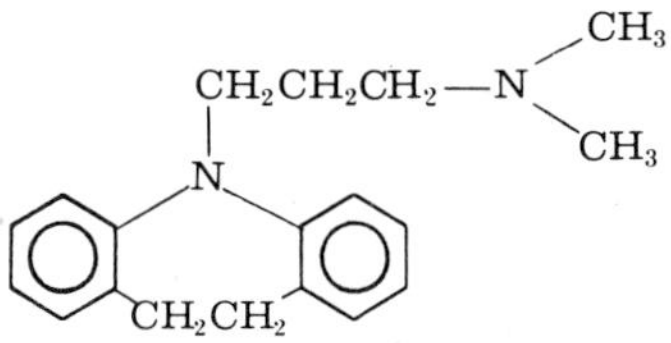 Imipramine (Presamine)	Used to treat the depressed state of manic depression

12-9 THE PROBLEM OF DRUG ABUSE

The nonmedical use of drugs, especially those having pronounced effects on mood, emotion, and thought, appears always to have been a part of human behavior. The distinction between socially acceptable use and use which is not acceptable has generally been determined by the customs of a particular culture at a particular time. Moreover, the laws governing drug use have often had some arbitrary and capricious features. There is no doubt, however, that the self-administration of drugs often carries with it severe damage to body and mind. All too frequently compulsive drug use leads to great personal tragedy and very high social cost. It is ironic that the use of drugs, with all their potential benefits, also reveals the darker side of human nature. Heroin addicts in the United States were estimated in the early 1970s to number about 250,000, and at that time drug abuse was reported to be the leading cause of death in New York City in the age group 15 to 35. Some authorities estimate that the alcoholics in the United States number about 5 million. The National Institute of Mental Health estimates those with a serious alcohol problem to number about 9 million. Alcohol was reported to be a factor in more than 25,000 of the 50,000 fatal automobile accidents in the United States in 1970.

We have previously used the word drug to designate any chemical agent which somehow affects the processes of living systems (Sec. 12-2). This is a broad definition and includes agents like ethyl alcohol and the nicotine of cigarettes, as well as pharmaceutical agents such as penicillin and morphine. The term **drug abuse** refers in general to the use of a drug, usually by self-administration, in a way which departs from the prevailing medical or social conventions. For the most part, however, it implies the compulsive use of a substance capable of causing harm (Fig. 12-11). **Drug addiction** is a condition characterized by a physiological and psychological dependence which results in disagreeable and dangerous symptoms when the use of the drug is stopped (Sec. 1-5).

Table 12-7 lists some representative examples of the substances which are frequently involved in drug abuse. Some of these find beneficial use in the medical profession. Amphetamine is used to treat mild depression, obesity, and narcolepsy, a pathological tendency to fall asleep. Seconal is one of the six leading derivatives of barbituric acid, known as barbiturates, used in sleeping pills, in the treatment of epilepsy, and in surgical anesthesia (Sec. 12-4B). Nicotine is used as an insecticide (Sec. 11-10).

12-10 MONITORING THE SAFETY OF MEDICINAL AGENTS

It is obviously essential to have the safety of all medicinal agents maintained at a very high level. The use of any drug, even the most beneficial, however, is always accompanied by some risk. People vary, sometimes greatly, in their reaction to drugs. Some agents which are well tolerated by the large majority of people can cause seriously harmful reactions in a few who happen to have an inborn susceptibility. Sometimes, as is frequently the case in the treatment of cancer, for example, the extension of life involves using drugs with severe side reactions because no other form of treatment is available. What is of utmost importance, of course, is to make certain that the ratio of the potential benefit to the potential risk is as high as possible. This requires the judgment and diagnostic skill of an experienced physician. Notwithstanding the disease-combating and life-saving capacities of many medicinal agents, prudence dictates that they be used with caution and no more frequently than necessary.

In 1906 the Congress established the Food and Drug Administration to watch over the quality and safety of food supplies and medicinal agents. Its mandate with drugs includes clearing all new agents for both safety and effectiveness before they are marketed. This is a difficult assignment. If the agency is not sufficiently strict, drugs with unnecessarily harmful side effects may become available. If it is too stringent, beneficial drugs could be delayed or made unavailable.

A drug has already been defined (Sec. 12-2) as a chemical agent which somehow affects the processes of living systems. An **ethical drug** is an agent which requires a physician's prescription, e.g., penicillin and morphine. An **over-the-counter drug** is one which is sold freely without a prescription, e.g., aspirin and tincture of iodine. Drugs often have more than one name. The **generic name** is a name which usually designates a chemical relationship and does not depend on the company which supplies it. A **trade name** is a name given to a drug or a

FIGURE 12-11
A branch of the marihuana plant together with derivatives. Prepared marihuana is at the lower left. (*Courtesy of Carolina Biological Supply Company.*)

Substance	Description
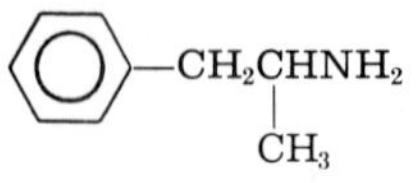 **Amphetamine**	Stimulant which increases alertness, reduces fatigue, and gives an illusion of well-being; used medically to treat mild depression, obesity, and other conditions; causes insomnia and loss of appetite; excessive amounts toxic and addicting and cause psychotic symptoms
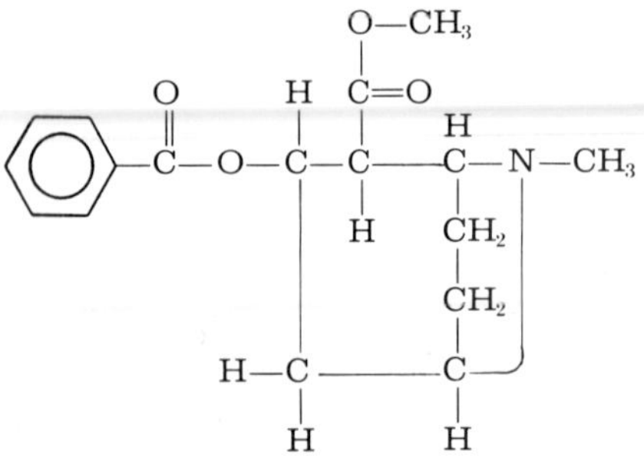 **Cocaine**	Stimulant which gives an illusion of well-being; formerly used in medicine as a local anesthetic (Sec. 12-4C); toxic and addicting; large doses cause psychotic symptoms
CH_3CH_2OH **Ethyl alcohol**	Can act as depressant, stimulant, tranquilizer, or sleep producer in the form of beer, wine, whiskey, gin, rum, and many other alcoholic beverages; formerly used medically as a tonic; most people unaffected by small amounts; larger amounts cause drowsiness and impaired coordination and judgment; can be addicting; alcoholics in United States, estimated to number at least 5 million, are major social and economic problem
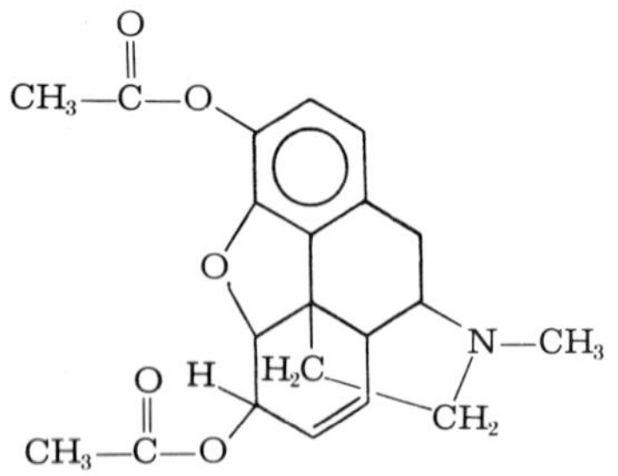 **Heroin**	Produces loss of pain, sedation, deadened sensibility, and illusion of well-being; now being cautiously revived for medical use after having been outlawed; very dangerously addicting; addicts in the United States estimated at around 250,000
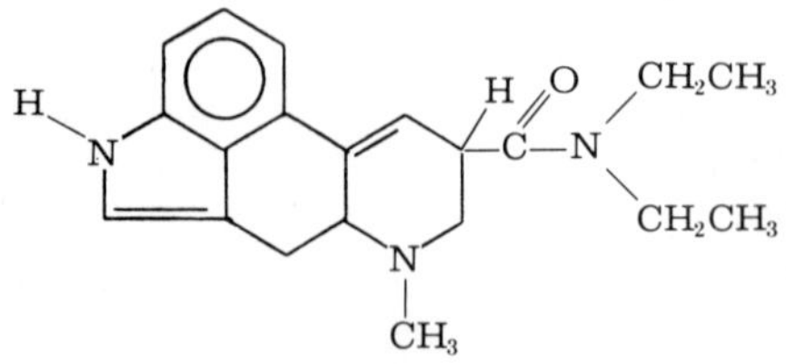 **LSD (lysergic acid diethylamide)**	One of several hallucination-producing agents (hallucinogens); causes periods of altered mood, consciousness and perception; formerly used to create model psychoses resembling schizophrenia; active in unusually small amounts

Table 12-7
Representative Substances Involved in Drug Abuse (*Continued*)

Substance	Description
Tetrahydrocannabinol (the most active of the substances in marihuana)	Marihuana is hemp plant and the drug prepared from its dried leaves and flowering tops; contains a mixture of cannabinols; in small amounts produces an illusion of well-being and impairment of coordination and short-time memory; higher doses induce hallucinations, anxiety, and possible permanent psychological changes
Nicotine	Can act as tranquilizer, depressant, or stimulant; present in tobacco in form of cigarettes, cigars, pipes, snuff, and chews; one of the most toxic and addicting drugs known; about 800 billion cigarettes consumed annually in United States; nicotine-containing products cause an estimated 350,000 deaths a year; considered by health authorities to be greatest public health hazard in United States
Seconal	Acts as sedative and sleep-producer; one of the class of substances known as barbiturates, used in medicine as sleeping pills and in general anesthesia; causes drowsiness and impairment of coordination and judgment; addicting; withdrawal symptoms can be severe after prolonged use; large doses can cause death
Caffeine	Stimulant which increases alertness and reduction of fatigue; used in pain remedies in combination with aspirin and other agents; commonly used in form of coffee, tea, and cola drinks; habit-forming; very heavy use can produce toxic and addictionlike reactions

specific drug preparation by a specific manufacturer. Methadone is a generic name. The trade names which have been given to methadone by suppliers include Amidone, Butolgin, Dolophine, Methadon, Miadone, and others. Note that trade names are capitalized. Generic names and trade names are frequently used interchangeably. In many states when a physician uses a trade name in a prescription the pharmacist who fills the prescription must use the specific preparation of the specified supplier. If the physician uses a generic name, the pharmacist may dispense any of the available preparations.

Distrastrous incidents within recent memory have been painful reminders of the need for careful drug regulation. In 1937 a small pharmaceutical company marketed a solution for oral dosage of the newly discovered sulfanilamide, which had already become famous as a "miracle" medicine. The company had difficulty getting the sulfanilamide to dissolve in any of the common medicinal solvents, but a mixture of water and diethyleneglycol, $HOCH_2CH_2OCH_2CH_2OH$, worked, and the company's "Elixir Sulfanilamide" was distributed all over the United States. Unfortunately, no tests were made for the safety of the solvent, and no heed was paid to published reports that diethyleneglycol was extremely poisonous to animals.

Within a month reports of deaths reached the American Medical Association, which quickly scheduled a news conference to warn the public. Agents of the Food and Drug Administration organized an elaborate countrywide search for all 1920 pints of the elixir which had been distributed. Diethyleneglycol proved to be a cumulative poison with no known antidote. Once substantial amounts had been swallowed, there was no possible treatment. Although the heroic search tracked down practically all the elixir, it proved to be too late for many. A total of 108 people lost their lives, including the formulator of the solvent mixture, who committed suicide.

Ironically, despite the tragic consequences, the "Elixir Sulfanilamide" failed to meet the drug regulations as they were then formulated only in a technical sense. The law in force did not require that ingredients be listed on the label, but it gave the government the power to act if the label was untruthful in some way. The word "elixir" in the name officially implied a nonpoisonous solvent containing ethyl alcohol. "Elixir Sulfanilamide" obviously did not fulfill these requirements. This disaster became a principal reason, however, for amending the Food and Drug Act in 1938 to require drug manufacterers to furnish the Food and Drug Administration with proof of the safety of a medicinal agent before permission for marketing could be granted.

The vigilance of a staff member of the Food and Drug Administration prevented residents of the United States from being among the victims of one of the most appalling drug catastrophies of all time. The drug was thalidomide, a popular, inexpensive sedative used between 1959 and 1961 in Europe, Australia, New Zealand, and Canada. Tests indicated that the drug was safe. This was further substantiated when people intending suicide survived very large doses without harm. It was used especially by pregnant women because of its ability to relieve the nausea of pregnancy. But by late 1961 it had become clear that if it was used some time between the third and sixth week of pregnancy, before many women know they are pregnant, it could cause gross deformities in the developing child. These were distinguished principally by useless short arms and legs, characteristic of a rare condition called phocomelia. The drug was withdrawn in late 1961 but not before it had cripped about 8000 children, mostly in West Germany and England. The developers had failed to anticipate its toxicity to unborn infants.

In September 1960 one of the companies which had marketed it in Canada applied to the Food and Drug Administration (FDA)

Thalidomide, the sedative which caused the birth of thousands of children with severe deformities because of its effect during pregnancy; it was never marketed in the United States but was popular in Europe

for permission to distribute the drug in the United States. At this time the drug's toxic effects had not become apparent, but the application was held to be incomplete by the FDA. While the company was assembling further information, complaints of tingling hands and motor disturbances after the use of thalidomide were reported in West Germany. Frances Kelsey, a pharmacologist at the FDA with an awareness of the special drug susceptibilities of unborn children, became concerned both about the reports of nervous disorders and the recommendation of the pharmaceutical company that the drug be used for relieving the nausea of pregnancy. She requested evidence that the drug was safe for use during pregnancy. Before the company had time to respond, the dreadful reports of deformed babies in Europe were confirmed many times over. Because of Frances Kelsey's watchfulness, Americans were spared this calamity. Congress, aware of how close to this disaster we had come, strengthened the Food and Drug Law.

12-11 HOW ANTIMETABOLITES INTERFERE WITH THE NORMAL CHEMICAL REACTIONS OF LIVING ORGANISMS

The discussion of Sec. 10-5† described the role of enzymes as the catalysts of the reactions of living organisms. The all-important formation of the transitory enzyme-substrate complex was summarized as follows:

Optional

Extremely slow
(no enzyme present):   Substrate → products

Extremely fast
(enzyme present):   Substrate + enzyme ⇌ enzyme-substrate complex

Extremely fast: Enzyme-substrate complex → products + enzyme

General representation of the role of enzyme-substrate formation in the catalytic action of enzymes; the temporary association between enzyme and substrate facilitates the reaction, enormously increasing its rate; the enzyme is regenerated for repeated use

The formation of the enzyme-substrate complex requires a close three-dimensional fit between the active site of the enzyme and the reaction site of the substrate. This lock-and-key fit is represented diagrammatically in Fig. 10-5.

The conversion of p-aminobenzoic acid into folic acid in microorganisms was mentioned in Sec. 12-5. The first step in this conversion involves an enzyme-catalyzed reaction. As this enzyme functions, it forms an enzyme-substrate complex with the p-aminobenzoic acid, which is the substrate. Normally the enzyme is quickly regenerated and thus can repeat its catalytic effect with many p-aminobenzoic acid molecules in turn.

As sulfanilamide molecules function as antimetabolites, they compete with the p-aminobenzoic acid molecules for the active site of the enzyme. The sulfanilamide molecules are able to do this because of the similarity between their structures and the structures of p-aminobenzoic acid molecules (Fig. 12-8).

† You should read that section now if you have not already done so.

Although able to form a complex with the enzyme, they cannot replace the *p*-aminobenzoic acid in the synthesis of folic acid. The sulfanilamide serves, however, to jam the enzyme, making it unavailable to the *p*-aminobenzoic acid. As a result the microorganism is denied the folic acid it needs for normal functioning.

"This new drug works on streptococci, pneumococci, and staphylococci. Now here's where you come in . . ."

© 1971 American Scientist; reprinted by permission of Sidney Harris.

KEY WORDS

1. Drug (Sec. 12-2)
2. Chemotherapeutic agent (Sec. 12-2)
3. Pharmacology (Sec. 12-2)
4. Antibiotic (Sec. 12-2B)
5. Centrally acting analgesic (Sec. 12-4)
6. General anesthetic (Sec. 12-4)
7. Local anesthetic (Sec. 12-4)
8. Opiate (Sec. 12-4A)
9. Narcotic analgesic (Sec. 12-4A)
10. Metabolite (Sec. 12-5)
11. Antimetabolite (Sec. 12-5)
12. Antimetabolite theory of drug action (Sec. 12-5)
13. Hormone (Sec. 12-7)
14. Drug abuse (Sec. 12-9)
15. Drug addiction (Sec. 12-9)
16. Ethical drug (Sec. 12-10)
17. Over-the-counter drug (Sec. 12-10)
18. Generic name (Sec. 12-10)
19. Trade name (Sec. 12-10)

SUMMARIZING QUESTIONS FOR SELF-STUDY

Section 12-1

1. **Q.** Name eight areas in which chemistry supports the practice of medicine, and give the names of two examples of the classes of substances involved.

   A. Chemotherapeutic agents (penicillin, chloroquine), disinfectants and antiseptics (chlorine, iodine), analgesics (morphine, aspirin), anesthetics (nitrous oxide, lidocaine), agents for anxiety and psychiatric disorders (Valium, chlorpromazine), insecticides (parathion, chlordane), vitamins [vitamin B_1 (thiamine), vitamin B_{12}], and investigation of mechanism of drug action.

2. **Q.** What is a drug?

A. Any chemical agent which somehow affects the processes of living organisms.

3. **Q.** What is a chemotherapeutic agent?

A. The term is usually restricted to substances used to treat diseases of microbial origin without significant harm to the animal or person afflicted, but it is also applied to chemical agents used for other ailments.

4. **Q.** What is pharmacology?

A. The branch of science concerned with the development of drugs, their use, and the mechanism of their action.

5. **Q.** What is a sulfa drug?

A. The chemotherapeutic agent sulfanilamide or one of its structural variants.

6. **Q.** What is the structural formula for sulfanilamide?

A.

7. **Q.** What is the name of the chemotherapeutic agent whose development led to the recognition of the chemotherapeutic action of sulfanilamide? To what general class of substance does it belong?

A. Prontosil, a synthetic dye.

8. **Q.** Sulfamethoxazole is a sulfa drug in current use. How was it developed?

A. By synthesizing structural variants of the sulfanilamide molecule.

9. **Q.** What is an antibiotic?

A. A substance produced by a microorganism capable of destroying or inhibiting the growth of other microorganisms.

10. **Q.** What is the name of the first widely used antibiotic?

A. Penicillin (more specifically, penicillin G).

11. **Q.** For what, in general, is penicillin used?

A. To treat (very effectively) many infectious diseases.

12. **Q.** Why were most of the sulfa drugs replaced rather quickly by antibiotics?

A. Because the antibiotics have many fewer toxic effects.

13. **Q.** Ampicillin is a chemotherapeutic agent in current use. How was it developed?

A. By synthesizing structural variants of the penicillin G molecule.

14. **Q.** What is one of the limitations of antibiotics?

A. The ability of microorganisms to develop resistance to their action.

Section 12-3

15. **Q.** Name four insect-borne diseases and name the insect which transmits each.

A. Sleeping sickness (tsetse fly), bubonic plague (oriental rat flea), typhus fever (body louse), yellow fever (mosquito), encephalitis (mosquito), and malaria (mosquito).

16. **Q.** About how many people suffer from malaria?

A. About 100 million.

17. **Q.** About how many deaths are caused by malaria each year?

A. Between 1 and 2 million.

18. **Q.** Name three phases of the attempts to control malaria.

A. Draining swamps and spraying breeding grounds with oil to reduce the population of mosquito larvae; using insecticides to kill adult mosquitoes; and treating the human infection with antimalarial chemotherapeutic agents.

19. **Q.** Name the original drug used widely to treat malaria.

A. Quinine.

20. **Q.** How was quinine discovered?

A. It is the active ingredient isolated from the cinchona bark used to treat malaria by South Americans for centuries.

21. **Q.** Atabrine is an antimalarial widely used by the military in World War II. How was it developed?

A. By synthesizing structural variants of quinine and other antimalarials.

22. **Q.** Chloroquine is a German-developed antimalarial currently in use. Its existence was made known to the Allies in World War II when the Anglo-American forces captured Tunis (North Africa) from the Germans and came into possession of samples of the drug. How had the Germans developed the drug?

A. By synthesizing structural variants of quinine.

Section 12-4

23. **Q.** What is a centrally acting analgesic? Name two examples.

A. A pain-relieving substance which relieves most forms of pain throughout the body, usually without loss of consciousness, e.g., morphine, meperidine, pentazocine, aspirin.

24. **Q.** What are some of the disadvantages of morphine as a pain-relieving agent?

A. It is addicting, patients develop tolerance, and excessive doses may cause death.

25. **Q.** Pentazocine is a pain-relieving agent of low addiction liability. How was it developed?

A. By synthesizing structural variants of morphine (and variants of other structural variants of morphine).

26. **Q.** What is a narcotic analgesic?

A. Morphine or a morphinelike substance.

27. **Q.** What is an opiate?

A. Same as a narcotic analgesic.

28. **Q.** Give the structural formula for aspirin.

A.

```
            O   H
            ‖   |
      O—C—C—H
      |       |
  (benzene)   H
      |
      C—O—H
      ‖
      O
```

29. Q. Aspirin is used to stop pain of low intensity. What are its other principal physiological properties?

A. It relieves fever and reduces somewhat the inflammation of rheumatoid arthritis.

30. Q. What is aspirin's chemical name?

A. Acetylsalicylic acid.

31. Q. How was aspirin developed?

A. It is a structural variant of salicylic acid, obtained from a substance isolated from willow bark, an old household remedy for fevers.

32. Q. Give the structural formula for salicylic acid.

A.

```
  (benzene)—O—H
      |
      C—O—H
      ‖
      O
```

33. Q. What is a general anesthetic?

A. A substance which relieves pain throughout the body but usually brings about loss of consciousness.

34. Q. Why is ethyl ether no longer used very much as a surgical anesthetic?

A. Because of its flammability and because of the nausea, vomiting, and other after effects.

35. Q. What one of the early general anesthetics is still in use?

A. Nitrous oxide, N_2O.

36. Q. Give the structural formula for the popular, nonflammable anesthetic halothane.

A.

```
      F   H
      |   |
  F—C—C—Br
      |   |
      F   Cl
```

37. Q. What is a local anesthetic?

A. A substance which is capable of relieving pain in specific parts of the body without loss of consciousness.

38. Q. The first widely used local anesthetic was novocaine. How was it developed?

A. By synthesizing structural variants of cocaine, a naturally occurring local anesthetic found in the leaves of the coca plant.

Section 12-5

39. Q. What is a metabolite?

A. A substance which is converted into another substance by the reactions of living organisms.

40. Q. What is an antimetabolite?

A. A substance of structure similar to a normal metabolite which is capable of interfering with the reactions of the normal metabolite.

41. Q. What is the antimetabolite theory of a drug action?

A. It proposes that substances may serve as chemotherapeutic agents if they have structures similar to the normal metabolites of disease-causing microorganisms and are able to act as antimetabolites.

Section 12-6

42. Q. What three approaches are currently being used in cancer treatment?

A. Surgery, nuclear radiation, and chemotherapy.

43. Q. What, in principle, is an advantage of chemotherapy compared with the other forms of the treatment of cancer?

A. It is effective, in principle, against malignant cells after the original growth has metastasized.

44. Q. What is meant by metastasis?

A. The ability of cancerous tumors to release malignant cells, which migrate through the body fluids to establish new cancerous pockets in distant parts of the body.

45. Q. What are some of the unfortunate side effects of the chemical agents used in the treatment of cancer?

A. Increased susceptibility to infection, loss of hair, nausea, and vomiting.

46. Q. 5-Fluorouracil is one of the principal chemical agents used in the treatment of cancer. Write its structural formula.

A.

```
                O
                ‖
                C
      H—N           C—F
       |            ‖
      O=C           C—H
                N
                |
                H
```

47. Q. What general principle led to the investigation of 5-fluorouracil as a treatment of cancer?

A. The antimetabolite theory. 5-Fluorouracil is an antimetabolite of thymine, a component of the nucleic acids.

48. Q. Give the structural formula of thymine.

A.

```
                O
                ‖
                C
      H—N           C—CH_3
       |            ‖
      O=C           C—H
                N
                |
                H
```

49. Q. What is a hormone? Give the names and principal functions of two examples.

A. A substance secreted by specific body organs which exert an important controlling influence on body chemistry, e.g., insulin (regulates use of glucose), testosterone (responsible for male sexual characteristics), and progesterone (helps maintain pregnancy).

50. Q. Norethynodrel is one of the synthetic substances used in a version of the oral contraceptive pill. How was it developed?

A. By synthesizing structural variants of the female sex hormone progesterone, which prevents ovulation.

Section 12-8

51. Q. Valium is the trade name for the most frequently prescribed drug in the United States. What is its principal use?

A. As a tranquilizer to treat anxiety.

52. Q. Imipramine is one of the drugs used to alleviate the depressing phase of manic depression. What drug is commonly used to relieve the hyperelation phase, and what is its formula?

A. Lithium carbonate, Li_2CO_3.

Section 12-9

53. Q. What is meant by drug abuse?

A. The use of a drug, usually by self-administration, in a way which departs from the prevailing medical or social conventions. It implies the compulsive use of a substance capable of causing harm.

54. Q. What is meant by drug addiction?

A. A condition characterized by a physiological and psychological dependence which results in disagreeable and dangerous symptoms when the use of a drug is stopped.

55. Q. Amphetamine is a drug frequently involved in drug abuse. What is it used for by the medical profession?

A. As a stimulant to treat mild depression and in the treatment of obesity and narcolepsy.

56. Q. The barbiturates are also frequently involved in drug abuse. What are their constructive uses?

A. Used to induce sleep, in the treatment of epilepsy, and in surgical anesthesia.

Section 12-10

57. Q. The use of any drug, even the most beneficial, is always accompanied by some risk. What,

then, should be two prevailing principles governing the use of a drug?

A. Keep the ratio of the potential benefit to the potential risk as high as possible and use it no more frequently than necessary.

58. Q. What mandate has Congress given the FDA with regard to drugs?

A. To clear all medicinal agents for safety and effectiveness before they are marketed.

59. Q. What is the difference between an ethical drug and an over-the-counter drug? Name one example of each.

A. An ethical drug requires a physician's prescription (penicillin, morphine). An over-the-counter drug is sold without a prescription (aspirin, iodine).

60. Q. What is the difference between a generic name for a drug and a trade name?

A. A generic name usually designates a chemical relationship and does not depend on the company which supplies the drug. A trade name is a special name given to a drug by a specific manufacturer.

61. Q. The lethal "Elixir Sulfanilamide" was widely distributed in 1937 and caused over 100 deaths. What was the poisonous component and why was it used?

A. Diethyleneglycol, $HOCH_2CH_2OCH_2CH_2OH$, used as a solvent for the sulfanilamide.

62. Q. What was the thalidomide disaster and how could it have been avoided?

A. Thalidomide, a widely used sedative (mostly used in Europe), led to the birth of thousands of deformed children in the early 1960s. It could have been avoided if the safety of its use by pregnant women had been checked (especially the effect on the developing child).

Section 12-11 (Optional)

63. Q. How do antimetabolites interfere with the normal chemical reactions of living organisms?

A. They compete with a natural metabolite for the active site of the enzyme which serves as the catalyst for the reaction of the natural metabolite. In so doing they jam the enzyme and render it incapable of performing its normal catalytic function. The living organism is denied the normal product of the reaction which the enzyme catalyzes.

64. Q. Why is sulfanilamide able to act as an antimetabolite of the metabolite p-aminobenzoic acid?

A. Because its structure is sufficiently similar to form an enzyme-substrate complex with the enzyme.

PRACTICE EXERCISES

1. Match each definition or other statement with the numbered term above with which it is most closely associated. Each numbered term may be used only once.

1. Metabolism	2. Antibiotic
3. Local anesthetic	4. Drug abuse
5. Chemotherapeutic agent	6. Generic name

7. A centrally acting analgesic
8. Antimetabolite
9. Trade name
10. Carbohydrate

(a) A pain-relieving substance which relieves most forms of pain throughout the body, usually without loss of consciousness

(b) A substance of structure similar to a normal metabolite capable of interfering with the reactions of the normal metabolite

(c) A substance used to treat diseases of microbial origin and other ailments without significant harm to the person or animal afflicted

(d) A name for a drug which usually designates a chemical relationship and does not depend on the company which supplies it

(e) The compulsive use of a drug, usually by self-administration, in a way which departs from the prevailing medical or social conventions

(f) A substance produced by a microorganism capable of destroying or inhibiting the growth of other microorganisms

(g) A substance capable of relieving pain in specific parts of the body without loss of consciousness

(h) A name given to a drug or a specific drug preparation by a specific manufacturer

2. Match each definition or other statement with the numbered term above with which it is most closely associated. Each numbered term may be used only once.

1. General anesthetic
2. Metabolite
3. Hormone
4. Pharmacology
5. Over-the-counter drug
6. Monoglyceride
7. Ethical drug
8. Narcotic analgesic
9. Drug
10. Digestion
11. Drug addiction

(a) A drug sold freely without a prescription

(b) A chemical agent which somehow affects the processes of living systems

(c) A substance which behaves like morphine

(d) A substance which is converted into other substances by the reactions of living organisms

(e) The branch of science which is concerned with the development of drugs, their use, and the mechanism of their action

(f) A substance which relieves pain throughout the body and usually brings about loss of consciousness

(g) A condition characterized by a physiological and psychological dependence which results in disagreeable and dangerous symptoms when use of the drug is stopped

(h) A drug which requires a physician's prescription

3. The sulfa drugs were the first generally available chemotherapeutic agents which had the ability to cure many diseases caused by microorganisms. What did their development have to do with dyes?

4. (a) What is an antibiotic? (b) What was the first generally useful antibiotic to be discovered? (c) What is the relationship between ampicillin and this first antibiotic? (d) What is a broad-spectrum antibiotic?

5. While Chap. 8 did not introduce all the functional groups in the ampicillin molecule, you should be able to recognize all except the one involving the sulfur atom (S). Identify the other functional groups present. (Use Table 8-3.)

6. Compare the number of nitrogen atoms in a molecule of penicillin G with the number in a molecule of ampicillin.

7. (a) What kind of insect is the most harmful in carrying disease? (b) Name three diseases it spreads.

8. (a) What method of controlling insects had a dramatically helpful effect in reducing the incidence of malaria in the 1960s? (b) Why has the incidence of malaria risen recently?

9. (a) What was the first drug widely used in the treatment of malaria? (b) How did it come to be used? (c) What is the relationship between this first drug and the currently used improved drugs chloroquine and primoquine?

10. What are three kinds of functional groups in a molecule of quinine that are not present in a molecule of chloroquine? (Use Table 8-3.)

11. (a) What is the difference between a centrally acting analgesic and a general anesthetic? (b) Name one example of each.

12. (a) What is the difference between a local anesthetic and a general anesthetic? (b) Name one example of each.

13. (a) Why have over 150 structural variants of morphine been synthesized? (b) Has the purpose of these syntheses been fulfilled?

14. What three different physiological effects does aspirin have?

15. What two functional groups are found in a molecule of aspirin?

16. What did the use of willow bark in the herbal remedies of folk medicine have to do with the development of aspirin?

17. (a) Why is ethyl ether no longer used very much as a surgical anesthetic? (b) Give the name and formula of one of its popular recent replacements. (Consult Table 12-5.)

18. (a) From what naturally occurring substance are modern local anesthetics descended? (b) Why is the natural substance no longer used?

19. What is a metabolite and an antimetabolite?

366

20. (a) Give structural formulas for sulfanilamide and p-aminobenzoic acid. (b) Which is a metabolite and which an antimetabolite?

21. Using the substances sulfanilamide and p-aminobenzoic acid, describe what is meant by the antimetabolite theory of drug action.

22. (a) In what principal way do cancerous cells differ from normal body cells? (b) What are the three principal current forms of treating cancer?

23. What is the principle of the use of the anticancer drug methotrexate?

24. What is the principle of the use of 5-fluorouracil in the treatment of cancer?

25. (a) What is a hormone? (b) Name two examples and describe their function briefly.

26. One of the substances used in oral contraceptives is norethynodrel. How did this substance come to be synthesized?

27. Lithium carbonate and imipramine are frequently used together in the treatment of manic depression. (a) What is the effect of the lithium carbonate and why is imipramine often used with it? (b) Consult Table 12-6 and identify the functional groups present in the molecule of imipramine. (c) Imipramine is one of the newer drugs used in the treatment of mental illness. Consult Table 12-6 and identify the drug whose structure inspired the synthesis of imipramine.

28. What is meant by (a) drug abuse and (b) drug addiction?

29. (a) Name two drugs frequently abused which are used beneficially by the medical profession. (b) Give one beneficial use for each.

30. Consult Table 12-7 and give the name of a drug which is (a) a stimulant, (b) causes hallucinations, (c) is present in tobacco, (d) is present in marihuana, (e) is present in a cup of coffee.

31. Consult Table 12-7 and identify the functional groups in (a) the heroin molecule, (b) the amphetamine molecule, (c) the ethyl alcohol molecule, (d) the cocaine molecule.

32. What is the difference between (a) an ethical drug and an over-the-counter drug and (b) the generic name for a drug and the trade name?

33. What briefly was the "Elixir Sulfanilamide" disaster, and to what change in the Food and Drug Act did it lead?

34. (a) For what was the drug thalidomide used? (b) Why is it no longer used?

Section 12-11 (Optional)

35. What is the mechanism of the action of sulfanilamide as an antimetabolite?

36. The presence of a substantial amount of p-aminobenzoic acid is able to prevent the antimetabolite action of sulfanilamide. Can you explain why?

SUGGESTIONS FOR FURTHER READING

The following articles are of an introductory nature.

American Chemical Society: "Chemistry in Medicine: The Legacy and the Responsibility," Washington, D.C., 1977. A report of a committee of authorities on the relationship between chemistry and medicine. The current status of many chemical agents is discussed.

Aspirin Action Linked to Prostaglandins, **Chemical and Engineering News,** Aug. 7, 1972, p. 4.

Barron, Frank, Murray E. Jarvik, and Sterling Bunnell, Jr.: The Hallucinogenic Drugs, **Scientific American,** April 1964, p. 29.

Burchenal, J. H., and J. R. Burchenal: Chemotherapy of Cancer, **Chemistry,** July-August 1977, p. 1.

Burger, Alfred: Behind the Decline in New Drugs, **Chemical and Engineering News,** Sept. 22, 1975, p. 37.

Cairns, John: The Cancer Problem, **Scientific American,** November 1975, p. 64.

Collier, H. O. J.: Aspirin, **Scientific American,** November 1963, p. 96.

Dingle, John H.: The Ills of Man, **Scientific American,** September 1973, p. 76.

Drugs for Treating Narcotic Addicts, **Chemical and Engineering News,** Mar. 28, 1977, p. 30.

Frederickson, Donald S.: Cancer: The Outlaw Cell, **Chemistry,** January-February 1977, p. 9.

Gates, Marshall: Analgesic Drugs, **Scientific American,** November 1966, p. 131.

Gillett, J. D.: The Mosquito: Still Man's Worst Enemy, **American Scientist, 61:**430 (1973).

Grinspoon, Lester: Marihuana, **Scientific American,** December 1969, p. 17.

Hammond, Allen L.: Aspirin: New Perspective on Everybody's Medicine, **Science, 174:**48 (1971).

Hansch, Corwin: Drug Research or the Luck of the Draw, **Journal of Chemical Education, 51:**360 (1974).

Holden, Constance: Amphetamines: Tighter Controls on the Horizon, **Science, 194:**1027 (1976).

Holden, Constance: New Look at Heroin Could Spur Better Medical Use of Narcotics, **Science, 198:**807 (1977).

Hollister, Leo F.: Marihuana in Man: Three Years Later, **Science, 172:**21 (1971).

Jones, R. G.: Antibiotics of the Penicillin and Cephalosporin Family, **American Scientist, 58:**404 (1970).

Majtenyi, Joan Z.: Antibiotics: Drugs from the Soil, I, **Chemistry**, January 1975, p. 6.

Majtenyi, Joan Z.: Antibiotics: Drugs from the Soil, II, **Chemistry**, March 1975, p. 15.

Maugh, Thomas H., II: Marihuana: The Grass May No Longer Be Greener, **Science, 185**:683 (1974).

Mellinkoff, Sherman M.: Chemical Intervention, **Scientific American**, September 1973, p. 102.

Morrison, Bayard H.: A Cancer Researcher Looks at Laetrile, **Chemistry**, July-August, 1977, p. 21.

Naves, Renee G., and Barbara Strickland: Barbiturates, **Chemistry**, March 1974, p. 11.

Pike, John E.: Prostaglandins, **Scientific American**, November 1971, p. 84.

Pilot, Henry C.: Cancer: An Overview, **Chemistry**, January-February 1977, p. 11.

Rose, Anthony H.: New Penicillins, **Scientific American**, March 1961, p. 66.

Stokes, Jimmy C., and W. Glenn Esslinger: Consumer Drug Index, **Journal of Chemical Education, 52**:784 (1975).

Synthetic Drugs Used and Abused, **Chemical and Engineering News**, Nov. 2, 1970, p. 26.

Taussig, Helen B.: The Thalidomide Syndrome, **Scientific American**, August 1962, p. 29.

Wade, Nicholas: Anabolic Steroids: Doctors Denounce Them, But Athletes Aren't Listening, **Science, 176**:1399 (1972).

The following articles are at a more advanced level.

Axelrod, Julius: Neurotransmitters, **Scientific American**, June 1974, p. 58.

Eddy, Nathan B., and Everette L. May: The Search for a Better Analgesic, **Science, 181**:407 (1973).

Magliulo, Anthony: Prostaglandins, **Journal of Chemical Education, 50**:602 (1973).

Marx, Jean L.: Prostaglandins: Mediators of Inflammation?, **Science, 177**:780 (1972).

Marx, Jean L.: Opiate Receptors: Implications and Applications, **Science, 189**:708 (1975).

Marx, Jean L.: Neurobiology: Researchers High on Endogennous Opiates, **Science, 193**:1227 (1976).

Marx, Jean L.: Analgesia: How the Body Inhibits Pain Perception, **Science, 195**:471 (1977).

Maugh, Thomas H., II: Cancer Chemotherapy: Now a Promising Weapon, **Science, 184**:970 (1974).

Polypeptide from Pituitary Acts as Analgesic, **Chemical and Engineering News**, Nov. 15, 1976, p. 26.

Snyder, Solomon H.: Opiate Receptors and Internal Opiates, **Scientific American**, March 1977, p. 44.

Snyder, Solomon, H., Shailesh P. Banerjee, Henry Y. Yamamura, and David Greenberg: Drugs, Neuro-Transmitters and Schizophrenia, **Science, 184**:1243 (1974).

13

WHAT IS THE ROLE OF THE CHEMICAL INDUSTRY?

13-1 THE DISTINGUISHING FEATURES OF THE CHEMICAL INDUSTRY

A. Large-Scale Synthetic Nitrogen Fixation

One of the most striking characteristics of the chemical industry is its capacity for innovation. Nowhere has this been more evident than in the support the industry has provided the large-scale production, processing, and preservation of food. It is remarkable that although the population of the world has more than doubled in the last 50 years, the supply of food by and large has been able, up to now at least, to keep pace with this tremendous growth. Among the most important factors underlying this achievement have been the introduction of concentrated fertilizers and their low-cost availability in very large quantities. The chemical industry has been almost wholly responsible for this accomplishment. The further expansion of food supplies to take care of the 200,000 or more new members of the human race born each day and the millions of undernourished in the present population continues, of course, to be a matter of extreme concern. But increases in the effectiveness and quantity of fertilizers can be expected to remain important approaches toward solution of this problem.

Bones, manures, and other natural forms of fertilizers dominated agricultural practice until about 60 years ago. The use of more concentrated fertilizing mixtures began around 1830 with the importation of sodium nitrate from the natural deposits in Chile. With the accelerating growth rate of the population, it soon became clear that the fertilizing mixtures available from organic sources and Chilean sodium nitrate would not be sufficient to meet the expanding demand for food. Attention was turned toward even more concentrated fertilizers.

The nitrogen-containing fertilizers used in the United States reveal the trend. In 1900 about 90 percent of the 72,000 tons of commercially prepared nitrogen-containing fertilizers used were organic in origin; 50 years later the annual consumption of commercially prepared nitrogen fertilizers had risen to more than 1 million tons, but only 30,000 tons were of organic origin. The major portion contained substances lower in cost and much higher in nitrogen content. Small farms, of course, continue to rely heavily on animal manures for fertilizing soils, but these account for only a small part of the total fertilizer used nationwide.

Elemental nitrogen is, of course, plentiful and easily available, constituting as it does about 78 percent of the atmosphere. Unfortunately, plants are unable to make direct use of elemental nitrogen, despite their great need for nitrogen-containing nutrients. For plant consumption the nitrogen must be converted into simple substances, such as ammonia, NH_3, or salts of the nitrate ion, NO_3^-. **Nitrogen fixation** is any process of converting elemental nitrogen into a nitrogen-containing compound. The most significant route of nitrogen fixation in nature is carried out by bacteria in the roots of plants like peas and beans (legumes). These bacteria can convert elemental nitrogen into ammonia. Their combined action is estimated to consume about 190 million tons of nitrogen a year.

Chemists have long sought to develop methods of nitrogen fixation to supplement the processes of nature. The first highly successful method was developed in Germany before World War I by Fritz Haber, Karl Bosch, and their coworkers. It has come to be known as the Haber-Bosch process or, more frequently, the Haber process. It involves the direct combination of elemental nitrogen and elemental hydrogen at a high temperature (range, 400 to 650°C) and high pressure (range 200 to 1000 times atmospheric) in the presence of a catalyst to yield ammonia. Today, worldwide production of ammonia is estimated to be about 60 million tons. In 1977 it was almost 17 million tons in the United States alone.

The availability of ammonia at low cost and in high volume from the Haber process has made nitrogen-containing fertilizers much more accessible and has made it possible to use mixtures more concentrated in nitrogen content. Currently the two most popular nitrogen fertilizers in the United States are ammonium nitrate (33.5 percent nitrogen content) and ammonia itself (82 percent content), which is applied in aqueous solution or in the pure gaseous form. The impact of the development of the Haber process on the food supply is a notable example of the extra-

N_2

Nitrogen

+

$3H_2$

Hydrogen

high press. | cat.
high temp.

$2NH_3$

Ammonia

Balanced equation for the overall reaction of the Haber (or Haber-Bosch) method of nitrogen fixation

ordinary ability of the chemical industry to bring about new and far-reaching solutions to pressing problems.

B. What Are the Raw Materials of Synthetic Ammonia Production?

It is of interest to trace the ammonia produced by the Haber process back to the natural resources from which it is synthesized (Fig. 13-1). One of the advantages of the Haber process is the use it makes of nitrogen directly from the atmosphere, an inexpensive and almost inexhaustible source. The source of the hydrogen is of special interest. Up until about 1940 at least 90 percent was produced from water or from coal. The **destructive distillation of coal** is the process which involves the decomposition of bituminous (soft) coal in high-temperature ovens (900 to 1200°C) sealed from air (alternate source II in Fig. 13-1). Bituminous coal contains a mixture of hydrocarbons and oxygen-containing carbon compounds. The treatment yields an array of products. The solid residue, called coke, is mostly carbon. The gas, called coke-oven gas, is mostly hydrogen, H_2, and methane, CH_4. Benzene and a wide variety of other volatile liquid and solid substances are also obtained:

$$CH_xO_x \xrightarrow[\substack{\text{sealed ovens} \\ (900-1200°C)}]{\text{decomposed in}} C + H_2 + CH_4 + \text{ benzene and other carbon-containing products}$$

Bituminous   Coke   Coke-oven
soft coal   gas

Outline of the production of coke, C, and coke-oven gas, H_2 and CH_4,
from bituminous coal, CH_xO_x, in coke ovens by destructive
distillation

The second hydrogen source (alternate source I in Fig. 13-1) is the water-gas process, in which steam is passed over hot

FIGURE 13-1
Outline tracing ammonia, NH_3, produced by the Haber process back to the natural resources that provide the raw materials. Names of natural resources are in boldface. Bituminous coal, which contains hydrocarbons and oxygen-containing carbon compounds, is represented as CH_xO_x.

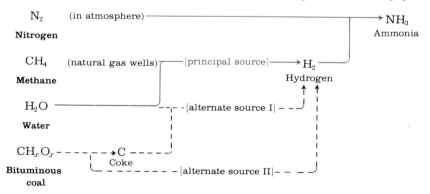

H$_2$O

Steam

+

C(coke)

Carbon

| high | temp. |

↓

CO

Carbon
monoxide

+

H$_2$

Hydroxen

CO

Carbon
monoxide

+

H$_2$O

Steam

| cat., | high temp. |

↓

CO$_2$

Carbon
dioxide

+

H$_2$

Hydrogen

Overall reactions of the water-gas process for producing hydrogen, H$_2$

coke. The overall reactions are summarized in the margin. The mixture of carbon monoxide, CO, and hydrogen, H$_2$, obtained in the first reaction in known as water gas. The carbon dioxide, CO$_2$, produced in the second reaction is removed from the hydrogen by dissolving it in water.

Today hydrogen is obtained principally from methane, CH$_4$, (and secondarily from other hydrocarbons) by treatment with steam at high temperatures in the presence of a catalyst. A typical overall reaction is shown. Methane became the principal

$$CH_4 \;+\; 2H_2O \xrightarrow[\text{high temp.}]{\text{cat.,}} CO_2 \;+\; 4H_2$$

Methane   Steam   Carbon   Hydrogen
                                              dioxide

Typical overall reaction for the production of hydrogen by treating methane, CH$_4$, with steam at a high temperature

source of hydrogen because of the high purity and quality of the supplies, its ease of transportation (pipeline), and (until recently) its low cost and abundant supply. Now that the cost of methane (natural gas) is rising and the competition for the available supply is increasing a switch back to versions of the water-gas and coke-oven processes could begin. This might prove to be especially desirable in view of the large supplies of coal in the United States. The alternate sources of hydrogen for ammonia synthesis underscore the flexibility which is an important feature of the chemical industry. The industry's ability to substitute one raw-material source for another is one of its most distinctive characteristics.

C. The Production of Ammonium Nitrate, NH$_4$NO$_3$

The most frequently used fertilizer, ammonium nitrate, NH$_4$NO$_3$, is produced from ammonia, NH$_3$, and nitric acid, HNO$_3$, which in turn is made from ammonia. Its derivation from naturally occurring substances is outlined in Fig. 13-2. The ammonia, produced principally by the Haber process, is converted into nitric acid, HNO$_3$, by the Ostwald process, developed by the German chemist Wilhelm Ostwald at the turn of the century. The steps involved are carried out in an uninterrupted sequence and may be represented as follows:

$$4NH_3 \;+\; 5O_2 \xrightarrow[\substack{\text{high temp.,}\\\text{high press.}}]{\text{cat.,}} 4NO + 6H_2O \qquad (1)$$

Ammonia   Oxygen   Nitric   Water
                                              oxide

Reaction of ammonia, NH$_3$, with oxygen, O$_2$, of the atmosphere, to form nitric oxide, NO

$$2NO + \quad O_2 \;\rightarrow\; 2NO_2 \qquad (2)$$

Nitric   Oxygen   Nitrogen
oxide   dioxide

Reaction of the nitric oxide, NO, with oxygen, O$_2$, of the atmosphere, to form nitrogen dioxide, NO$_2$

$$3NO_2 + H_2O \rightarrow 2HNO_3 + NO \qquad (3)$$

Nitrogen   Water   Nitric   Nitric
dioxide   acid   oxide

Reaction of the nitrogen dioxide, NO$_2$, with water, H$_2$O, to form nitric acid, HNO$_3$, and (unavoidably) some nitric oxide, NO

Step (4) (margin) involves a proton-transfer reaction (Sec. 7-4) between ammonia and nitric acid.

Ammonium nitrate is produced in huge quantities. Over 7 millions tons are synthesized each year in the United States alone. Most of this is used as a fertilizer, but ammonium nitrate is also used in explosive mixtures. One of the disadvantages of ammonium nitrate as a fertilizer is its explosive nature, especially when mixed with carbon-containing compounds. A combination of ammonium nitrate and hydrocarbon oils is often used as an explosive in mining operations to replace dynamite (Fig. 13-3).

D. The Production of the General Anesthetic Nitrous Oxide, N$_2$O

A significant use of ammonium nitrate is the preparation of the useful general anesthetic nitrous oxide, N$_2$O (Sec. 7-8 and 12-4B). One of the very first general anesthetics, it still finds important uses, especially in conjunction with other surgical anesthetics. When dry ammonium nitrate is heated to 200 to 250°C, it undergoes a spontaneous decomposition to form nitrous oxide and water (margin).

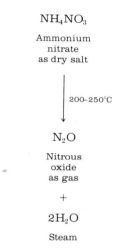

$$NH_3 + HNO_3 \rightarrow NH_4NO_3$$

Ammonia   Nitric   Ammonium
acid   nitrate

The proton-transfer reaction of step (4) in the production of ammonium nitrate, NH$_4$NO$_3$

$$NH_4NO_3$$

Ammonium
nitrate
as dry salt

200–250°C

$$N_2O$$

Nitrous
oxide
as gas

+

$$2H_2O$$

Steam

Decomposition of ammonium nitrate in the commercial preparation of nitrous oxide, N$_2$O

FIGURE 13-2
Outline tracing the fertilizer and explosive ammonium nitrate, NH$_4$NO$_3$, to the natural resources (boldface) providing the raw materials.

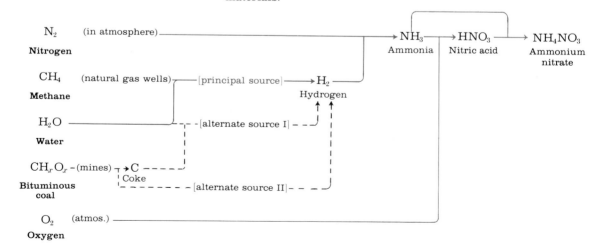

FIGURE 13-3
A major disaster resulted from the explosion of ammonium nitrate in Texas City, Texas, in 1947. A fire aboard a ship while the fertilizer-grade ammonium nitrate was being loaded triggered explosions, which in turn caused fires in a chemical plant on shore and another explosion in a nearby ship, also being loaded with ammonium nitrate. In all 576 people were killed, and damages totaled $67 million. (*United Press International.*)

E. The Special Importance of Nitric Acid, HNO_3

The nitric acid used in the synthesis of ammonium nitrate has a variety of other important uses. Indeed it is the second most important acid commercially (after sulfuric acid, H_2SO_4). Over 7 million tons are consumed each year by the chemical industry in the United States. The largest single use is in the synthesis of ammonium nitrate, but it is also very important in the synthesis of nylon and many other carbon-containing substances, especially dyes, pharmaceuticals, and explosives. A **chemical intermediate** is a substance (like nitric acid) which is derived by chemical reactions from raw materials or other chemical intermediates and which undergoes further chemical treatment along the pathway toward the production of an end product.

F. What Are the Distinguishing Features of the Chemical Industry?

What do the various aspects of the production and use of ammonium nitrate tell us about the chemical industry? One of the first characteristics we notice is that the operations of industrial chemistry convert naturally occurring raw materials into products of greater economic usefulness. In the case of ammonium nitrate, natural components of the atmosphere (nitrogen, oxygen) and the earth (coal, natural gas, water) are converted into a substance which can be used as a fertilizer, an explosive, and as a reactant in the preparation of a widely used general anesthetic.

Another feature of the chemical industry which becomes evident is its ability to use radically new approaches to solve problems of supply. The expansion of nitrogen-containing fertilizer production, essential to keeping food supplies abreast of population increases, was accomplished by developing a revolutionary, low-cost method of producing ammonia from abundant atmospheric nitrogen.

The three alternate sources of hydrogen for the Haber process (natural gas, water gas, coke-oven gas) illustrate the flexibility which is typical of many of the processes used by the chemical industry. This built-in capacity for interchanging raw materials engenders efficiency, keeps costs down, and often permits conservation of natural resources. The Haber process itself, replacing as it does Chilean sodium nitrate by abundant atmospheric nitrogen, exemplifies the capacity of chemical processes to conserve raw materials. While the Chilean deposits are extensive, they are not limitless and probably would not long meet the current demand for nitrogen fertilizer. The capacity of the chemical industry to produce products which serve as replacements for others also provides some relief from the vulnerability of imported raw materials.

While some products of the chemical industry are of a trivial nature (lipstick coloring agents, plastic wrappers), many are of substantial significance. Ammonium nitrate typifies the products of importance, serving, as it does, as an agricultural fertilizer, a mining explosive, and a precursor to a surgical anesthetic. The diversity in the uses of ammonium nitrate also suggests the pervasive influence of the chemical industry's products. It is rather common for a given substance to perform a variety of functions in industry and in meeting human needs.

An interesting aspect of the chemical industry is the anonymous nature of most of its products. The principal reason is that many products exert their effects indirectly. Farmers are aware of ammonia and ammonium nitrate fertilizer, but the consumer of the agricultural products probably is not. For the most part, the activities and products of the chemical industry are far removed from the public's general knowledge, even though the scale of operations is often huge and the social and economic impact substantial.

The production of ammonium nitrate is a reminder of the potential hazards which abound in chemical operations and in their products. Ammonium nitrate itself is explosive. Nitric acid is dangerously corrosive, and the gases which go into its production are highly toxic. Great care must be taken in the production, shipment, and use of many chemical products. On the whole the safety record is good, especially in view of the severe hazards, but through carelessness or expediency errors are sometimes made, and the costs can be high.

Special vigilance is required to make certain that chemical processes, their products and their waste products, are not harmful to the environment. Long-term effects, in particular, must be evaluated with care. The chemical industry produces an amazing array of products of great value both to the economy and the general welfare, but obviously this work must be carried out in harmonious balance with environmental considerations. Considering the huge scale of operations and their many built-in hazards, the record is moderately good. But problems remain, and satisfactory solutions will require the full capacities of the chemical industry for innovation. The distinguishing features of the chemical industry are summarized in Table 13-1.

13-2 THE SCOPE AND PATTERN OF CHEMICAL PRODUCTION

Products of the chemical industry range so widely that it is difficult to define the scope of the industry. In a broad sense the chemical industry includes all manufacturing processes which produce products by altering molecular structure, but this criterion embraces many processes not usually included as part of the chemical industry, e.g., petroleum refining and the production of steel. The statistics-gathering agencies of the federal government have a classification called "chemicals and allied products" which includes such products as basic inorganic and organic chemicals, fertilizers and other agricultural chemicals, drugs, soaps and detergents, dyes, paints, explosives, and plastics. The classification "chemical process industries" adds petroleum products, rubber products, food processing, metals, pulp and paper, and stone-clay-glass products. By either classification the chemical industry ranks among the largest in the country. The annual contribution of chemical process industries to the national income was estimated in 1970 at $79 billion.

Many industries, such as the aerospace industry, produce relatively few products from a great number of starting materials. The chemical industry is characterized both by the variety of products which can be prepared from a small group of raw materials and by the diversity of uses of many of its products. Products as different as synthetic rubber and rocket fuels can be produced from petroleum. A single product, sulfuric acid, is essential to the preparation of an extraordinary number of other products: fertilizers, dyes, reinforced concrete, metals, paper, textiles, explosives, plastics, and many drugs. Through these products it plays a critical role in the construction of all automobiles, airplanes, roads, railroads, telephones, and television sets.

The production of most chemical end products relies on a sequence of steps involving several intermediates. The manufacture of ammonium nitrate is simpler than most, but it involves the production of hydrogen from natural gas, the subsequent synthesis of ammonia and nitric acid, as well as the formation of the final product. Most intermediates are used in many different processes, and they are frequently produced as separate products and shipped from plant to plant. The products of one company may be the starting substances of another. Table 13-2 lists some representative raw materials, intermediates, and end products of the chemical industry. Some substances serve as both intermediates and end products. Ethylene glycol, listed as an end product because of its use as a permanent automobile radiator antifreeze, also serves as an intermediate in the production of polyester textile fibers and many other products.

Table 13-1
Summary of the Distinguishing Features of the Chemical Industry

1. Large-scale conversion of naturally occurring raw materials into products of greater economic usefulness, usually by way of several intermediate substances

2. Substantial capacity for innovation

3. Products frequently able to replace materials of limited supply, providing opportunity for conservation of raw materials and relief from vulnerability of imports

4. Processes often very flexible, with alternative raw materials and intermediates possible

5. Importance of products ranges from trivial to essential, often critically indispensable

6. Products usually often of anonymous quality reaching consumers in altered form

7. Products tend to be versatile and to be used throughout a wide range of services and other industries

8. Products and processes frequently hazardous, sometimes extremely so, requiring special safety precautions to prevent harm to the production workers, consumers, and environment; continuing need for improvement

Table 13-2
Representative Raw Materials, Intermediates, and End Products of the Chemical Industry

Raw materials	Intermediates	End products
Air:	Acetylene	Aluminum
Nitrogen	Adipic acid	Ampicillin and other synthetic antibiotics
Oxygen	Ammonia	Ammonium nitrate and other fertilizers
	Benzene	Aspirin
Earth's crust:	Butadiene	Copper
Coal	Calcium carbide	Meperidine and other pain relievers
Limestone	Carbon monoxide	Dynamite and other explosives
Metal ores	Chlorine	Ethylene glycol
Natural gas	Cyclohexane	Gasoline
Petroleum	Ethyl alcohol	Lidocaine and other local anesthetics
Sodium chloride	Ethylene	Magnesium
Sulfur	Ethylene oxide	Nitrous oxide and other general anesthetics
	Hexamethylenetetramine	Nylon fibers
Seawater:	Hydrochloric acid	Parathion and other insecticides
Magnesium salts	Hydrogen	Polyester and other synthetic fibers
	Lime	Soap
Agriculture:	Methyl alcohol	Synthetic detergents
Animal fat	Nitric acid	Synthetic rubber
Coconut oil	Sodium hydroxide	
Grain	Styrene	
Sugar cane	Sulfuric acid	
	Terephthalic acid	
Water	p-Xylene	

Chlorine, listed as an intermediate, is used directly to disinfect public drinking-water supplies.

13-3 THE PRODUCTION OF POLYMERS

The introduction of nylon stockings for women in 1939 (Fig. 13-4) made of the synthetic fiber just developed by the Du Pont Company, probably did more than any other single event to correct the impression that "synthetic" implies something inferior. The strength and toughness of nylon textile fibers were greatly superior to those of natural silk, and the women's stockings made from them immediately became famous for their durability, washability, and sheerness. Over 15 million pairs were sold the first year, and "nylons" became a synonym for women's stockings.

The introduction of nylon also made industry and the public at large much more aware of synthetic polymers in general. Such materials had been in use ever since the introduction of celluloid in 1870, which in turn made possible the development of the motion picture industry. But for the most part the public image of the early commercial polymers was influenced more by the shortcomings of the products (the extreme flammability, for example, of celluloid motion picture film) than by their advantages. The introduction of nylon (Fig. 13-5) played a large role in improving the status of synthetic polymers.

Today synthetic polymers have displaced traditional materials in thousands of end products. Synthetic rubber has largely replaced natural rubber in tires, synthetic fibers have revolu-

FIGURE 13-4
The public first saw nylon stockings when they were worn by these models at the New York World's Fair in 1939. (*Courtesy of E. I. du Pont de Nemours & Co.*)

FIGURE 13-5
The development of nylon played an important role in World War II since supplies of silk, the original material of parachutes, were cut off. Fortunately not only was nylon able to replace silk but parachutes made of nylon were stronger and longer-lasting. Nylon was also used for cord in military airplane tires. (*Courtesy of E. I. du Pont de Nemours & Co.*)

tionized the textile industry, water-based latex paints have dramatically changed the coatings industry, and synthetic polymers have greatly improved electrical insulation. Many of these products are known as plastics. A **plastic** is a polymeric material capable of being molded into such objects as eyeglass frames and buttons. Scores of polymeric products are produced in huge amounts. Over 2.5 million tons of synthetic rubber, 2.5 million tons of synthetic textile fibers, and 10 billion tons of molded and related synthetic polymers are produced in the United States each year.

Many synthetic polymers are highly valued for their inertness to water, oxygen, sunlight, and microorganisms, but these very properties complicate the plastics litter problem. A great number of polymeric materials persist annoyingly in the environment or present other disposal problems. Efforts are now under way to devise polymeric materials so that many of these problems will be eliminated, e.g., polymers which will decompose under the influence of sunlight or microorganisms.

A **polymer** was defined in Sec. 7-9 as a substance whose molecules are made up of many structural units held together by interatomic bonds. More than 100 polymeric substances are produced in large quantities. Facts about some common synthetic polymers are summarized in Table 13-3. Three examples have been chosen for brief discussion here: representative types of polyethylene, synthetic rubber, and nylon fiber.

13-4 THE PRODUCTION OF POLYETHYLENE

Polyethylene is produced by a **polymerization reaction,** defined in Sec. 7-9 as a reaction in which many molecules of one or a small number of substances of relatively simple structure are

Table 13-3
Some Representative Synthetic Polymers

Starting substance	Polymer	Trade names	Representative uses
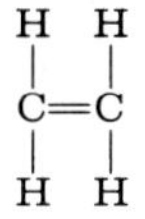 $\underset{\text{Ethylene}}{\overset{\displaystyle \underset{H}{\overset{H}{C}}=\underset{H}{\overset{H}{C}}}{}}$	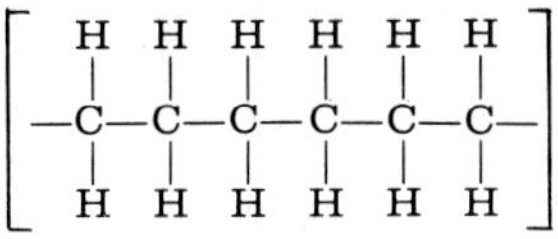 Polyethylene: $\left[-\text{CH}_2-\text{CH}_2-\cdots\right]_n$	Polythene	Wire and cable insulation, pipe, packaging, bottles and other molded products
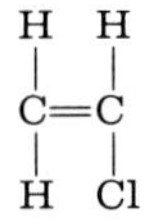 Vinyl chloride: $\overset{H}{\underset{H}{C}}=\overset{H}{\underset{Cl}{C}}$	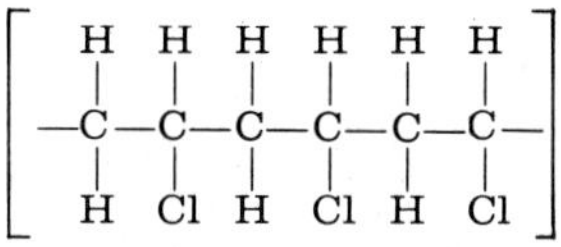 Polyvinyl chloride	Vinyon, Dynel, Koroseal	Floor tile, wire insulation, phonograph records, rainwear
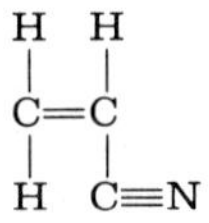 Acrylonitrile: $\overset{H}{\underset{H}{C}}=\overset{H}{\underset{C\equiv N}{C}}$	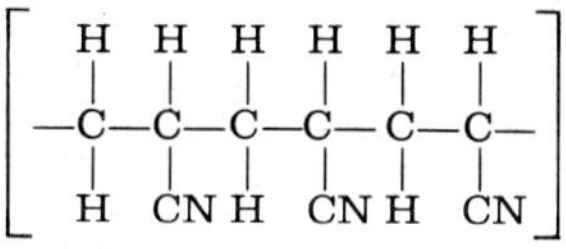 Polyacrylonitrile	Acrilan, Creslan, Orlon, Dynel	Rugs, textiles
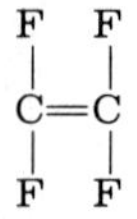 Tetrafluoroethylene: $\overset{F}{\underset{F}{C}}=\overset{F}{\underset{F}{C}}$	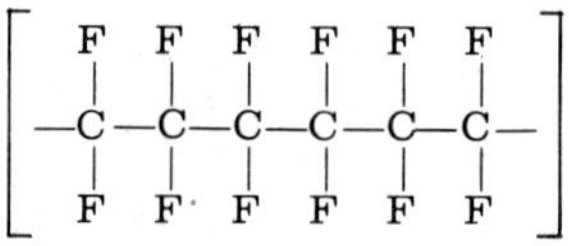 Polytetrafluoroethylene	Teflon	Electrical insulation, gaskets, cooking-pan coatings
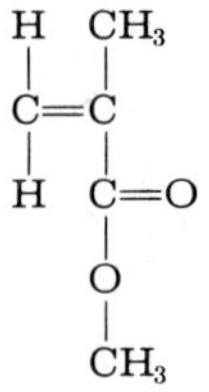 Methyl methacrylate	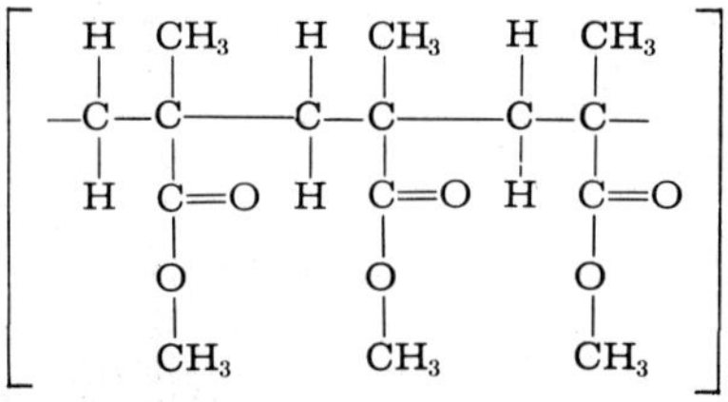 Polymethyl methacrylate	Lucite, Plexiglas	Windows, latex paints, molded products
$\text{CH}_2=\text{CH}-\text{CH}=\text{CH}_2$ Butadiene; 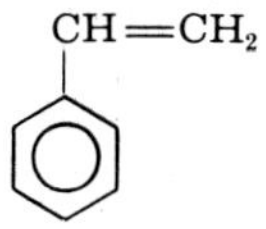 $\text{CH}=\text{CH}_2$ Styrene	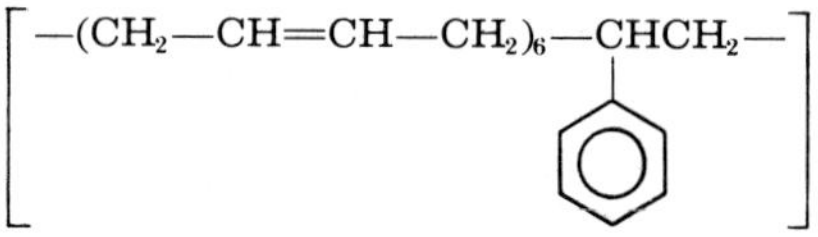 $\left[-(\text{CH}_2-\text{CH}=\text{CH}-\text{CH}_2)_6-\text{CHCH}_2-\right]_n$ Styrene-butadiene synthetic rubber	SBR, GR-S	Tires, garden hose
$\text{HOOC}(\text{CH}_2)_4\text{COOH}$ Adipic acid; $\text{H}_2\text{N}(\text{CH}_2)_6\text{NH}_2$ Hexamethylamine-diamine	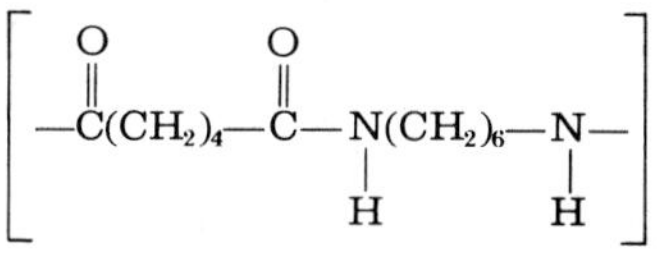 $\left[-\overset{O}{\overset{\|}{C}}(\text{CH}_2)_4-\overset{O}{\overset{\|}{C}}-\underset{H}{N}(\text{CH}_2)_6-\underset{H}{N}-\right]_n$ Nylon 66	Nylon	Textiles, rugs, molded products

(Continued on next page)

ng ince	Polymer	Trade names	Representative uses
$HOCH_2CH_2OH$ Ethylene glycol 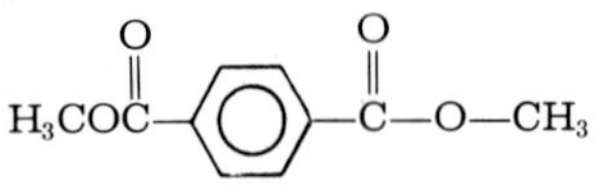 Dimethylterephthalate	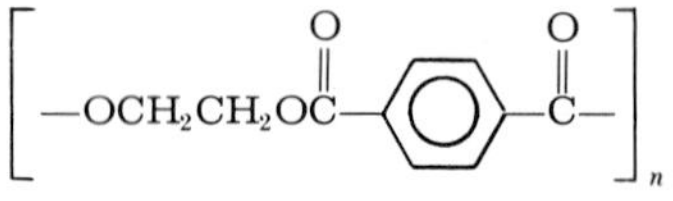 Polyester (one type)	Dacron, Terylene	Textiles, rugs

linked together to make a large molecule. The polymerization of ethylene to form polyethylene was represented as

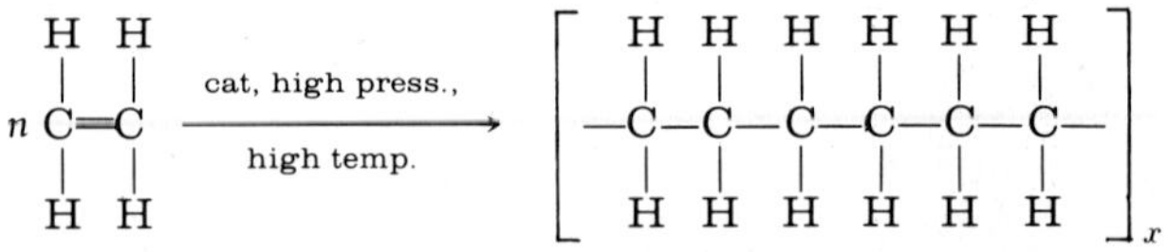

n units of ethylene molecules, typically 500 to 1000

A polyethylene molecule, formed with a mixture of molecules of a varying number of units; x = typically 170 to 340

Representation of the polymerization reaction for the formation of a polyethylene molecule

Actually there are several different kinds of polyethylene; in some the chains of —CH_2— units contain branches, leading to a less rigid product. The reaction above, however, may be taken as representative. Polyethylene polymers are valued for their corrosion resistance, toughness, excellent electrical-insulation properties, flexibility, ease of fabrication, and low cost. Very early in polyethylene's development, just before World War II, it proved to be an excellent shielding for microwave cable in radar equipment. In this capacity it made an important contribution to the Allied cause in the war. Annual worldwide production of polyethylene polymers now amounts to almost 9 billion tons. They are used for wire and cable insulation, pipe, packaging, and the fabrication of bottles and many other molded products.

The raw material of polyethylene is ethylene, $CH_2=CH_2$, readily obtained from petroleum (Fig. 13-6). When crude petroleum, essentially a mixture of hydrocarbons (Sec. 13-11A), is heated as it is refined to obtain gasoline and other fuels, a gaseous mixture called refinery gas is obtained. This contains a significant amount of the gases ethylene, $CH_2=CH_2$, and ethane, CH_3CH_3. The yield of ethylene is increased substantially by decomposing the ethane by heating it to about 800°C in the presence of a catalyst:

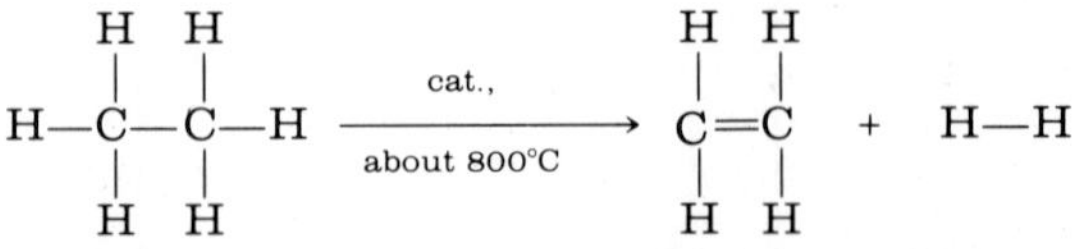

Reaction for the decomposition of ethane in the production of ethylene; the ethane is obtained from petroleum

The production of polyethylene from petroleum can be outlined as follows:

Crude petroleum $\longrightarrow$ $CH_2{=}CH_2$ $\longrightarrow$ $(CH_2CH_2)_x$
Essentially a   Ethylene   Polyethylene
mixture of
hydrocarbons

Outline tracing the polymer polyethylene, $(CH_2CH_2)_x$, to the natural crude petroleum

13-5 THE PRODUCTION OF SYNTHETIC RUBBER

Natural rubber substances are hydrocarbon polymers obtained from rubber trees. The repeating unit of the polymeric molecule is a residue of a simple alkene hydrocarbon, isoprene. The number of residues in the rubber molecules varies from about 1500 to 40,000 and beyond. Raw rubber is made more useful by **vulcanization,** which involves heating raw rubber with elemental sulfur. At the molecular level the principal change is the formation of covalently bonded sulfur-containing cross-links between chains of isoprene residues. The rubber becomes less tacky, more elastic, and more resistant to abrasion. The production of rubber products involves sophisticated compounding with a variety of agents, which vary with the product desired. Making tires involves compounding with substantial amounts of elemental carbon (carbon black) as a reinforcing agent to increase strength and resistance to wear.

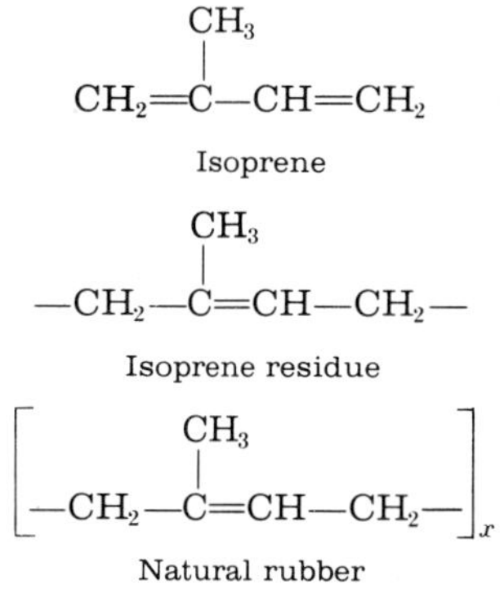

Many kinds of synthetic rubber substances have been developed. The most widely used type is a polymer of styrene and

$$CH_2{=}\overset{\displaystyle CH_3}{\underset{\displaystyle |}{C}}{-}CH{=}CH_2$$

Isoprene

$$-CH_2{-}\overset{\displaystyle CH_3}{\underset{\displaystyle |}{C}}{=}CH{-}CH_2-$$

Isoprene residue

$$\left[-CH_2{-}\overset{\displaystyle CH_3}{\underset{\displaystyle |}{C}}{=}CH{-}CH_2-\right]_x$$

Natural rubber

Comparison of the isoprene molecule, an isoprene residue, and rubber molecules; x varies over a wide range, from about 1500 to 40,000 and beyond

381

butadiene residues in a ratio of 1:6. It is called SBR (styrene-butadiene rubber).

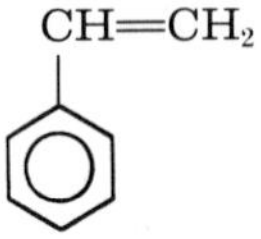

$$CH=CH_2$$

Styrene

$$CH_2=CH-CH=CH_2$$

Butadiene
(1,3-butadiene)

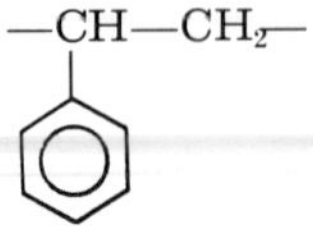

$$-CH-CH_2-$$

Styrene residue

$$-CH_2-CH=CH-CH_2-$$

Butadiene residue

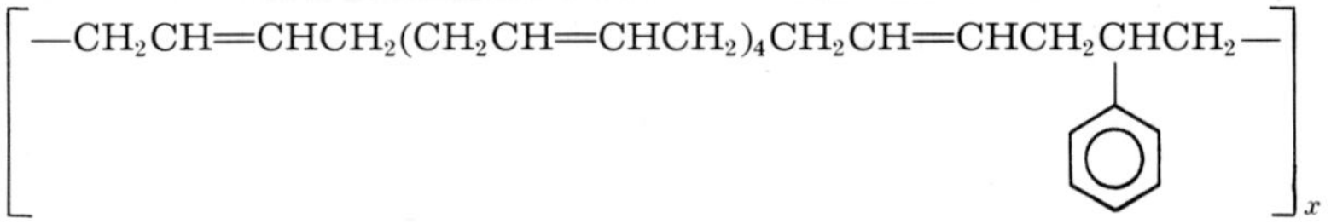

$$\left[-CH_2CH=CHCH_2(CH_2CH=CHCH_2)_4CH_2CH=CHCH_2CHCH_2- \right]_x$$

SBR has about six butadiene residues to one styrene
residue; the structure is actually more complicated
and includes some chain branching and covalently
bonded cross-links; x varies from a few hundred to
thousands

**Comparison of structures of SBR with those of the component
styrene and butadiene residues**

SBR has better wearing properties in tires than natural rubber.

SBR was the synthetic rubber which rescued the United States and its allies from what might have been a disastrous rubber shortage in World War II. At the opening of the war in the Pacific, the Japanese swept over the principal sources of rubber. The stockpiled natural rubber supplies were far short of the amount needed for the automobile tires, airplane tires, and tank tracks for the war. Fortunately a high-priority crash program had been organized in 1940 to develop and produce synthetic rubber modeled on a process developed in Germany. Two years later production was under way, and it increased to the remarkable level of almost 1 million tons a year by 1945.

The styrene is obtained from benzene, which in turn is obtained mostly from petroleum but can also be obtained from coal. Relatively small amounts of benzene are found in crude petroleum, but it can be made from the petroleum hydrocarbon hexane by the use of high temperatures and pressures in the presence of catalysts. In practice a mixture of alkane hydrocarbons obtained from petroleum is treated together to yield benzene and some simple benzene derivatives (the benzene is later isolated by distillation):

$$CH_3CH_2CH_2CH_2CH_2CH_3 \xrightarrow[\substack{\text{high temp.,} \\ \text{high press.}}]{\text{cat.,}} \bigcirc + 4H_2$$

Hexane (and similar alkane hydrocarbons containing more carbon atoms)

Benzene (and some benzene derivatives containing one or two more carbon atoms)

Hydrogen

Overall reaction for the conversion of hexane into benzene; in practice mixtures of hydrocarbons serve as reactants and are obtained as products

WHY IS RUBBER ELASTIC?

The word rubber suggests stretchability more than anything else (although the word was invented by Joseph Priestley because of the material's ability to rub out pencil marks). What in the structure of rubber causes it to snap back to its original shape after being stretched as much as 8 times its length? Many polymeric fibers can be stretched. In the fabrication of nylon fiber, for example, filaments of the polymer are stretched to 400 to 600 percent of their original length; but once stretched, they remain that way. They do not snap back because the stretching process brings about an alignment of the long, thin polymer chains which increases the forces operating between them. These forces (hydrogen bonds and polar forces) are strong enough to hold the fiber in the stretched condition.

The individual polymer chains of natural and synthetic rubber, on the other hand, are held to each other only by weak intermolecular forces such as van der Waals forces. Although the chains are long, thin, and flexible, like the polymer chains of fibers, the structural conditions for polar forces and hydrogen bonds are absent. The few cross-links involving covalent bonds are increased during vulcanization (Sec. 13-5). As a piece of rubber is stretched, the covalently bonded cross-links prevent the polymer chains from slipping past each other and hold the material in one piece (Fig. 13-7).

But the intermolecular forces operating between the stretched polymer chains are too weak to hold them in the extended position, and the polymer chains return to their original shape. For a polymer to have the characteristics of a rubber, therefore, its chains must be long and flexible with occasional covalently bonded cross-links, but only weak intermolecular forces must operate between chains.

FIGURE 13-7
Schematic representation of (a) unstretched and (b) stretched vulcanized-rubber fibers. The cross-links are those of the rubber itself and those, containing sulfur atoms, formed by vulcanization. The cross-links prevent the polymer chains from slipping and hold the material together. The absence of strong intermolecular forces between the stretched chains allows them to snap back into their original shape.

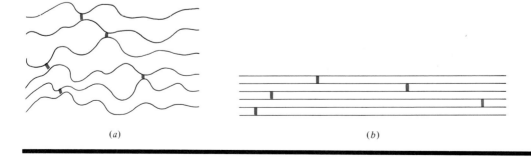

(a)   (b)

Benzene is also obtainable from coal. When bituminous coal is heated at 500 to 750°C in the absence of air, a wide variety of products is obtained from which benzene is isolated by a series of distillations (Sec. 13-1B).

Benzene is converted into ethylbenzene by reaction with ethylene in the presence of a catalyst. The ethylene is obtained from petroleum, as mentioned in Sec. 13-4. Styrene is formed from ethylbenzene through a high-temperature catalyzed decomposition:

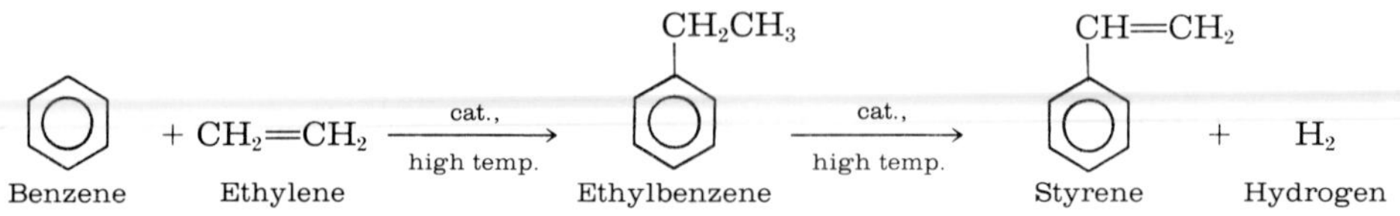

Conversion of benzene into styrene; the benzene is first converted into ethylbenzene, and then styrene is obtained from the ethylbenzene by decomposition

Butadiene can be obtained either from petroleum hydrocarbons or from grain through ethyl alcohol. In the petroleum process a mixture of hydrocarbons containing butane and closely related substances is decomposed by heat in the presence of a catalyst:

$$\underset{\substack{\text{Butane}\\\text{(together with}\\\text{related hydrocarbons)}}}{CH_3CH_2CH_2CH_3} \xrightarrow[\text{high temp.}]{\text{cat.,}} \underset{\substack{\text{Butadiene}\\\text{(plus other products)}}}{CH_2{=}CH{-}CH{=}CH_2} + \underset{\text{Hydrogen}}{2H_2}$$

The final step in the formation of the SBR polymer is the polymerization of a 1:6 mixture of styrene and butadiene in the presence of a catalyst. The overall production of SBR synthetic rubber from naturally occurring raw materials is outlined in Fig. 13-8. While today most SBR synthetic rubber is derived from petroleum-based hydrocarbons because they are currently inexpensive and available, alternate sources (now more costly and less convenient) of the intermediates are available. Benzene can be obtained from coal and butadiene from grain via ethyl alcohol.

13-6 HOW IS NYLON PRODUCED?

As has been true for many chemical products, the practical large-scale production of nylon depended on the development of inexpensive routes to the intermediates required by the synthesis. The two immediate precursors of the nylon polymer are

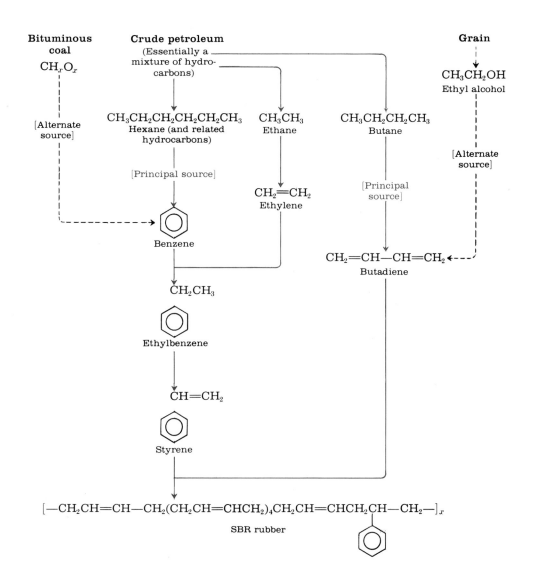

FIGURE 13-8
Outline tracing SBR synthetic rubber to the natural resources (boldface) providing the raw materials.

Note to Students   The production of styrene-butadiene synthetic rubber is an interesting example of the chemical industry's conversion of naturally occurring raw materials into products of greater usefulness.

---

an acid with two carboxylic acid groups, adipic acid, and an amine with two amino groups, hexamethylenediamine:

$$HO-\overset{\overset{\displaystyle O}{\|}}{C}-CH_2CH_2CH_2CH_2-\overset{\overset{\displaystyle O}{\|}}{C}-OH \qquad\qquad H_2N-CH_2CH_2CH_2CH_2CH_2CH_2-NH_2$$

$$HO-\overset{\overset{\displaystyle O}{\|}}{C}-(CH_2)_4-\overset{\overset{\displaystyle O}{\|}}{C}-OH \qquad\qquad\qquad H_2N-(CH_2)_6-NH_2$$

Abbreviated formula $\qquad\qquad\qquad\qquad\qquad$ Abbreviated formula

Formulas for adipic acid $\qquad\qquad$ Formulas for hexamethylenediamine

"Nylon salt"

200–300°C

$$\left[-\overset{\overset{\displaystyle O}{\|}}{C}-(CH_2)_4-\overset{\overset{\displaystyle O}{\|}}{C}-\underset{\underset{\displaystyle H}{|}}{N}-(CH_2)_6-\underset{\underset{\displaystyle H}{|}}{N}-\overset{\overset{\displaystyle O}{\|}}{C}-(CH_2)_4-\overset{\overset{\displaystyle O}{\|}}{C}-\underset{\underset{\displaystyle H}{|}}{N}-(CH_2)_6-\underset{\underset{\displaystyle H}{|}}{N}-\right]_n$$

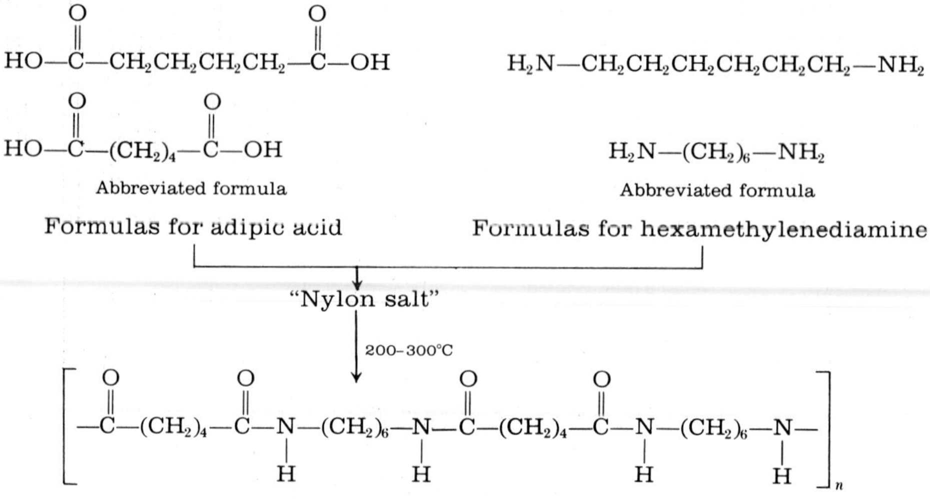

Nylon 66 polymer; n is about 25 to 40

The last stage in the formation of nylon polymer; adipic acid and hexamethylenediamine react to form a "nylon salt," from which the nylon polymer is formed on heating at 200 to 300°C

Adipic acid is formed from cyclohexane, which is in turn obtained from benzene:

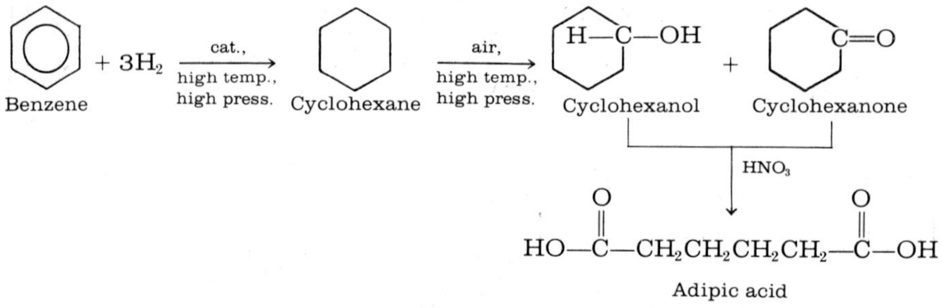

Benzene $+ 3H_2 \xrightarrow[\text{high press.}]{\text{cat.,} \atop \text{high temp.,}}$ Cyclohexane $\xrightarrow[\text{high press.}]{\text{air,} \atop \text{high temp.,}}$ Cyclohexanol $+$ Cyclohexanone

$\xrightarrow{HNO_3}$

$$HO-\overset{\overset{\displaystyle O}{\|}}{C}-CH_2CH_2CH_2CH_2-\overset{\overset{\displaystyle O}{\|}}{C}-OH$$

Adipic acid

Note that in the final step the nitric acid treatment breaks a carbon-carbon bond and opens up the cyclohexanol and cyclohexanone rings of six carbon atoms to form the open-chain six-carbon structure of adipic acid. Hexamethylenediamine is obtained from the adipic acid:

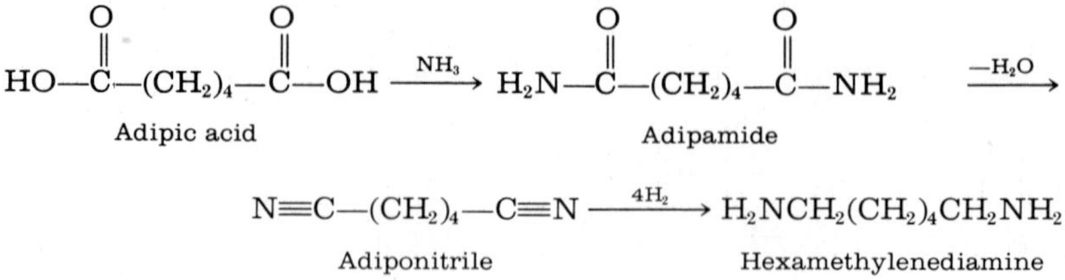

$$HO-\overset{\overset{\displaystyle O}{\|}}{C}-(CH_2)_4-\overset{\overset{\displaystyle O}{\|}}{C}-OH \xrightarrow{NH_3} H_2N-\overset{\overset{\displaystyle O}{\|}}{C}-(CH_2)_4-\overset{\overset{\displaystyle O}{\|}}{C}-NH_2 \xrightarrow{-H_2O}$$

Adipic acid $\qquad\qquad\qquad\qquad\qquad$ Adipamide

$$N\equiv C-(CH_2)_4-C\equiv N \xrightarrow{4H_2} H_2NCH_2(CH_2)_4CH_2NH_2$$

Adiponitrile $\qquad\qquad\qquad\qquad$ Hexamethylenediamine

Note that the conversion of the amide adipamide to adiponitrile involves the removal of the atoms of two molecules of water. The details have been omitted.

The production of nylon from naturally occurring raw materials is outlined in Fig. 13-9.

There are several different nylon polymers. The one we have discussed is known as nylon 66 because it is made from adipic acid, which contains six carbon atoms, and hexamethylenediamine, which also contains six carbon atoms. The individual units of all nylons are joined by amide groups, thus resembling proteins (Sec. 10-4). Natural silk is a protein, and the resemblance between nylon and natural silk is traceable to similarities in molecular structure. Nylons are valued for their strength, toughness, and resistance to many reagents (except acids and

FIGURE 13-9
Outline tracing nylon to the natural resources (boldface) providing the raw materials. Note that nylon can be made either from coal, air, and water or from petroleum, air, natural gas, and water.

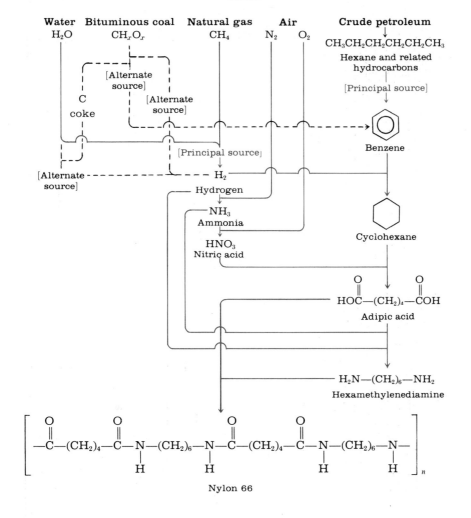

Nylon 66

bases). They are used as textile fibers, rug fibers, and to make many molded products, such as gears.

The development of nylon resulted from the thoroughgoing pioneering study of Wallace Carothers into the nature of polymers and polymerization, beginning in 1928. This work was carried out in the Du Pont laboratories. It led far beyond the creation of nylon, although that would have been achievement enough. The research established the foundation for the later production of the polyester polymers and polymers with rubberlike characteristics.

13-7 HOW IS METALLIC MAGNESIUM OBTAINED FROM THE SEA?

A technological society consumes large quantities of metals. Table 13-4 lists the approximate annual consumption of the most widely used. The extraction of metals from their ores is basically a chemical operation, and a large part of the total chemical activity is devoted to this work. The production of magnesium will be discussed not only because the metal is important in the aerospace industry but because of its interesting source: about 90 percent of the United States' supply is obtained from the sea (Fig. 13-13). As metallic ores become less plentiful, increasing attention will be paid to the oceans, which are vast dilute solutions of the ions of many different elements (Table 15-7). Though the concentration of magnesium ion in seawater is only a low 0.13 percent, a cubic mile contains about 6 million tons. The supply, therefore, is practically inexhaustible.

Table 13-4
Approximate Annual Consumption
of Metals, 1973, Millions of Tons

	United States	World[†]
Iron	102	552
Aluminum	6.5	13.4
Copper	2.4	7.9
Zinc	1.9	6.3
Lead	1.5	3.9
Chromium	0.64	3.4
Nickel	0.20	0.58
Magnesium	0.12	0.27
Tin	0.084	0.26

[†] Based on production.

Magnesium metal is the lightest of all structural metals and it has a high strength-to-weight ratio. It is valued also for its dimensional stability, its ease of fabrication, and its low cost. Magnesium occurs in the sea as the positive ion, Mg^{2+}. Since the predominant negative ion in the sea is chloride ion, Cl^-, we shall represent the presence of magnesium ion in the sea as a dilute solution of magnesium chloride, $MgCl_2$. The first step toward obtaining the metal, of course, is to get the magnesium ion in a concentrated form. This is accomplished by converting it into a substance which precipitates from water, magnesium hydroxide, Mg(OH)$_2$ (Figs. 13-10 and 13-11). The conversion is made by treatment with calcium hydroxide, Ca(OH)$_2$:

$$MgCl_2 \quad + \quad Ca(OH)_2 \rightarrow Mg(OH)_2 \downarrow + \quad CaCl_2 \tag{1}$$

| Magnesium chloride (dilute, impure in seawater) | Calcium hydroxide | Magnesium hydroxide | Calcium chloride (remains in solution) |

The reaction which concentrates the dilute magnesium ion, Mg^{2+}, in seawater as a precipitate of solid magnesium hydroxide, Mg(OH)$_2$

The calcium hydroxide, Ca(OH)$_2$ also comes from the sea as calcium carbonate, CaCO$_3$, present in shells of oysters found close to the seaside plant:

$$CaCO_3 \xrightarrow{\text{heat}} CaO + CO_2 \tag{2}$$

| Calcium carbonate (from oyster shells) | Calcium oxide (lime) | Carbon dioxide |

$$CaO + H_2O \rightarrow Ca(OH)_2 \tag{3}$$

| Calcium oxide | Water | Calcium hydroxide |

Reactions in the production of calcium hydroxide, Ca(OH)$_2$, from the calcium carbonate, CaCO$_3$, of oyster shells

FIGURE 13-10
The intake of seawater for the extraction of magnesium. (*Courtesy of Dow Chemical U.S.A.*)

The solid magnesium hydroxide, $Mg(OH)_2$, is removed by filtration and converted into magnesium chloride with hydrochloric acid, HCl:

$$Mg(OH)_2 \ + \quad 2HCl \quad \rightarrow \quad MgCl_2 \ + 2H_2O \qquad (4)$$

Magnesium hydroxide	Hydrochloric acid	Magnesium chloride	Water

**Reaction for the conversion of magnesium
hydroxide, $Mg(OH)_2$, into magnesium chloride, $MgCl_2$**

The hydrochloric acid, HCl, for this conversion is made at the plant from hydrogen and chlorine, Cl_2:

$$H_2 \quad + \quad Cl_2 \quad \rightarrow \quad\quad 2HCl \qquad (5)$$

Hydrogen   Chlorine   Hydrochloric acid
(hydrogen chloride
in aqueous solution)

**Reaction for the production of the hydrochloric acid
used in the isolation of magnesium**

The hydrogen comes from methane, CH_4, of natural gas (Sec. 13-1B) or directly from the destructive distillation of coal by the water-gas process (Sec. 13-1B). The chlorine is obtained from the electrolysis in reaction (6), below. The water is evaporated from the magnesium chloride, and the magnesium is obtained in the elemental state by decomposing the pure magnesium chloride using an electric current, in a process called electrolysis (Sec. 7-14):

$$MgCl_2 \ \xrightarrow[\text{current}]{\text{electric}} \ Mg \quad + \quad Cl_2 \qquad (6)$$

Magnesium   Magnesium   Chlorine
chloride
(melted)

**The decomposition of molten magnesium
chloride, $MgCl_2$, using an electric current in
a process called electrolysis**

In the process of electrolysis, the magnesium ions, Mg^{2+}, of the melted magnesium chloride, $MgCl_2$, are converted into magnesium atoms by the electrons of the electric current:

$$Mg^{2+} \quad + \quad 2e^- \quad \rightarrow \quad Mg$$

Magnesium   Electrons   Magnesium
ions

**The conversion of magnesium
ions, Mg^{2+}, into magnesium, Mg,
by the electrons of the electric
current during the process
of electrolysis**

A plant for extracting magnesium from seawater is shown in Figs. 13-11 and 13-12. The production of magnesium from naturally occurring substances is outlined in Fig. 13-13.

FIGURE 13-11
The settling tanks in a magnesium plant containing precipitated magnesium hydroxide, $Mg(OH)_2$. (*Courtesy of Dow Chemical U.S.A.*)

FIGURE 13-12
General view of a plant for the extraction of magnesium from seawater. The circular tanks at the near right contain precipitated magnesium hydroxide, $Mg(OH)_2$. The buildings containing the electrolytic cells for the production of the metallic magnesium are on the left. (*Courtesy of Dow Chemical U.S.A.*)

H
|
H—C—O—H
|
H—C—O—H
|
H—C—O—H
|
H

Glycerol, an alcohol with three
hydroxyl groups

$$
\begin{array}{c}
O \\
\parallel \\
H—O—C—R
\end{array}
$$

General formula for a
carboxylic acid

H   O
| ∥
H—C—O—C—R
|
| O
| ∥
H—C—O—C—R
|
| O
| ∥
H—C—O—C—R
|
H

General formula for a fat
molecule; the R groups may
be the same or different and
usually contain from 12 to
18 carbon atoms

**Comparison of general
formulas of a fat molecule
with a carboxylic acid and
glycerol**

13-8 HOW IS SOAP MADE?

Soap has been made from fatty substances for thousands of years (a soap factory was uncovered in the excavations at Pompeii). In the course of time the chemistry of the structure and synthesis of soap has been clarified and the process improved. **Soaps,** for the most part, are sodium salts of carboxylic acids containing 12 to 18 carbon atoms. The substance sodium palmitate is representative:

$$CH_3CH_2CH_2CH_2CH_2CH_2CH_2CH_2CH_2CH_2CH_2CH_2CH_2CH_2CH_2COO^-Na^+$$

Full formula

$$CH_3(CH_2)_{14}COO^-Na^+$$

Abbreviated formula

Formulas for sodium palmitate, a representative component of soaps

Fats, the starting materials in the preparation of soap, are for the most part carboxylic esters of carboxylic acids and the alcohol glycerol, which contains three hydroxyl groups (Sec. 10-3). If we use the general formula of a carboxylic acid (Sec. 8-5), the general formula of a fat molecule is as shown. The carboxylic acid residue parts of a given fat molecule are usually different from each another, but for simplicity we shall consider them to be alike. On this basis we take as a representative fat molecule one in which the carboxylic residues are derived from palmitic acid, an acid with 16 carbon atoms. (The structure and chemical behavior of fats are discussed in Sec. 10-3.)

For our immediate purposes the most important chemical property of fats is the readiness with which they undergo hydrolysis (reaction with water). The focus of action is the carbox-

FIGURE 13-13
Outline tracing the production of elemental magnesium, Mg, back to natural resources (boldface). Note that while the hydrogen is obtained from natural gas, it could be obtained from bituminous coal (Sec. 13-1B).

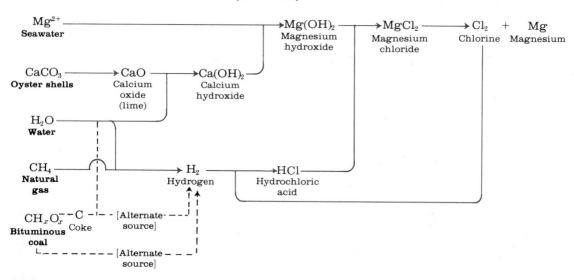

392

ylic ester group, which in general reacts with water to give a carboxylic acid and an alcohol:

$$R—O—\overset{\overset{\displaystyle O}{\|}}{C}—R + H—O—H \rightarrow R—O—H + H—O—\overset{\overset{\displaystyle O}{\|}}{C}—R$$

Bond breaks   Bond breaks   Bond forms   Bond forms

Carboxylic ester   Water   Alcohol   Carboxylic acid

Representation by general formulas of the hydrolysis of an ester to form an alcohol and a carboxylic acid

$$H—\overset{\overset{\displaystyle H}{|}}{C}—O—\overset{\overset{\displaystyle O}{\|}}{C}—(CH_2)_{14}CH_3$$
$$H—\overset{|}{C}—O—\overset{\overset{\displaystyle O}{\|}}{C}—(CH_2)_{14}CH_3$$
$$H—\overset{|}{C}—O—\overset{\overset{\displaystyle O}{\|}}{C}—(CH_2)_{14}CH_3$$
$$\overset{\overset{\displaystyle |}{}}{H}$$

The structure of an idealized fat in which all the carboxylic acid residues are derived from palmitic acid; the carboxylic acid residues in naturally occurring fat molecules usually differ from each other

In the conversion of fats into soap the ester groups are hydrolyzed. Sodium hydroxide is used as the catalyst for the reaction, and it in turn reacts with the carboxylic acids obtained as products to form sodium salts in a proton-transfer reaction (Secs. 7-4 and 7-5).

$$H—\overset{\overset{\displaystyle H}{|}}{C}—O—\overset{\overset{\displaystyle O}{\|}}{C}—(CH_2)_{14}CH_3$$
$$H—\overset{|}{C}—O—\overset{\overset{\displaystyle O}{\|}}{C}—(CH_2)_{14}CH_3 + 3H_2O \xrightarrow[\text{cat.}]{\text{NaOH}} \begin{matrix} H—\overset{\overset{\displaystyle H}{|}}{C}—O—H \\ H—\overset{|}{C}—O—H \\ H—\overset{|}{\underset{\overset{|}{H}}{C}}—O—H \end{matrix} + 3H—O—\overset{\overset{\displaystyle O}{\|}}{C}(CH_2)_{14}CH_3 \quad (1)$$
$$H—\overset{\overset{|}{\underset{}{}}{C}}{C}—O—\overset{\overset{\displaystyle O}{\|}}{C}—(CH_2)_{14}CH_3$$

Representative fat molecule   Water   Glycerol   Palmitic acid

$$Na^+OH^- + H—O—\overset{\overset{\displaystyle O}{\|}}{C}—(CH_2)_{14}CH_3 \rightarrow Na^+O—\overset{\overset{\displaystyle O}{\|}}{C}(CH_2)_{14}CH_3 + H—O—H \quad (2)$$

Sodium hydroxide   Palmitic acid from fat hydrolysis   Sodium palmitate, a representative soap molecule   Water

The preparation of soap from fats; both steps are frequently represented as one since they take place in the same reaction mixture

The sodium hydroxide is obtained from sodium chloride through the use of electric energy in an electrolysis reaction (Sec. 7-14). The naturally occurring substances used in the production of soap are therefore sodium chloride (salt), water, and fats from animal and plant sources.

13-9 WHAT ARE SYNTHETIC DETERGENTS?

One of the disadvantages of soaps is that their negative ions form insoluble substances with such ions as calcium ions, Ca^{2+}, and magnesium ions, Mg^{2+}, found in most freshwater supplies. Hard water is water that contains appreciable concentrations of such ions. The reaction with calcium ion, Ca^{2+}, is typical:

$$2CH_3(CH_2)_{14}\overset{O}{\overset{\|}{C}}-O^- \quad + \quad Ca^{2+} \quad \rightarrow \quad Ca(CH_3(CH_2)_{14}\overset{O}{\overset{\|}{C}}-O)_2 \downarrow$$

Negative ion of sodium palmitate, a typical component of soaps	Calcium ion, found in most freshwater supplies	Calcium palmitate, insoluble and precipitates from solution

Typical reaction between the negative ion of a soap and calcium ions, Ca^{2+}, one of the ions frequently found in freshwater supplies

The insoluble substances formed from the soap are the film which "dulls your hair" and forms the "ring around the bathtub." The formation of the precipitates with the calcium ions, Ca^{2+}, and other ions, consumes soap and impairs its cleaning efficiency.

Many substances besides soaps have a cleaning action. One is sodium dodecylbenzenesulfonate, also called linear alkylbenzenesulfate (LAS):

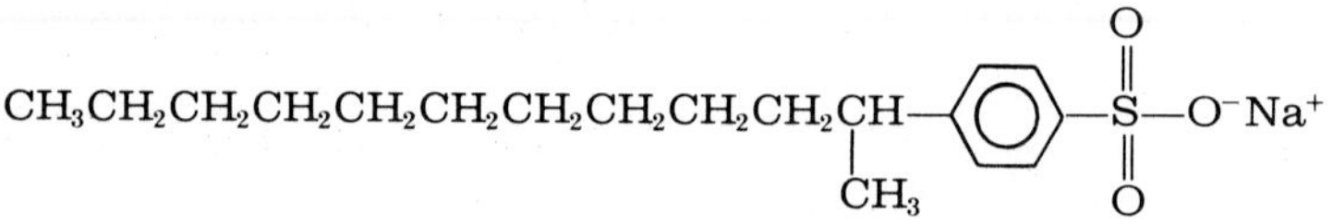

$$CH_3CH_2CH_2CH_2CH_2CH_2CH_2CH_2CH_2CH_2CH-\underset{CH_3}{\underset{|}{\bigcirc}}-\overset{O}{\underset{O}{\overset{\|}{S}}}-O^-Na^+$$

The structure of sodium dodecylbenzenesulfonate, a representative synthetic detergent

A **synthetic detergent** is any of the synthetic substances, other than soaps, which are capable of soaplike cleaning action. Soaps, of course, are also synthetic, but through an accident of history they are usually not labeled as such. (The "natural" soaps of the television commercials are a falsification.) One of the advantages of synthetic detergents is that they do not form insoluble precipitates with calcium ions, Ca^{2+}, and the other ions commonly found in water supplies.

There are many types of synthetic detergents. Before 1965 the most widely used was a sodium alkylbenzene sulfonate (ABS), with the formula

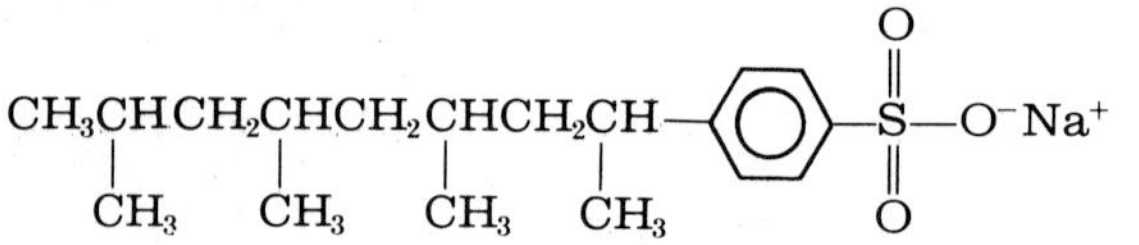

$$CH_3CHCH_2CHCH_2CHCH_2CH-\underset{\underset{CH_3}{|}}{\overset{}{\bigcirc}}-\overset{O}{\underset{O}{\overset{\|}{S}}}-O^-Na^+$$
$$\underset{CH_3}{|} \quad \underset{CH_3}{|} \quad \underset{CH_3}{|} \quad \underset{CH_3}{|}$$

The structure of sodium alkylbenzene sulfonate (ABS), the most widely used synthetic detergent up to 1965

Unfortunately, the highly branched carbon chain makes the substance resistant to microorganisms. **Biodegradable** means that large molecules or the substances they constitute can be degraded by the biological or chemical changes of natural processes into small molecules readily assimilated by the environment. The highly branched carbon chain of sodium alkylbenzenesulfonate makes it resistant to such processes, or nonbiodegradable. With the widespread use of nonbiodegradable substances as detergents, they formed an undesirable accumulation in groundwaters and rivers. Beginning in 1965, however, a large-scale switch was made to such biodegradable detergents

HOW DO SOAPS AND DETERGENTS EXERT THEIR CLEANING ACTION?

Sodium palmitate is an example of a polar molecule, described in Sec. 4-11 as one in which the centers of positive and negative charge are separated. The polar characteristics of sodium palmitate are concentrated at the end containing the oxygen atoms. The opposite end of the molecule is nonpolar; i.e., the center of positive charge and the center of negative charge overlap (Sec. 4-11):

$$CH_3CH_2CH_2CH_2CH_2CH_2CH_2CH_2CH_2CH_2CH_2CH_2CH_2CH_2CH_2\overset{\overset{\text{O}}{\|}}{C}{-}O^-Na^+$$

Nonpolar end   Polar end

The polar and nonpolar ends of sodium palmitate,
a representative component of soap

Most dirt is held on textile fabrics and other surfaces by a thin film of oil or grease. Oil and grease are nonpolar, and water by itself, being polar, has only limited ability to remove them (Sec. 5-3). If soap is present, however, the nonpolar ends of molecules like sodium palmitate can dissolve in the nonpolar oils and greases while their polar ends remain in the surrounding water (Fig. 13-14). Aided by the action of the soap, mechanical action breaks up the oil and grease into tiny droplets. The polar ends of the soap ions project out of the surfaces of the oil droplets, and the charges on their negative ends prevent the oil droplets from coalescing. The dispersed droplets of oil (and dirt) are suspended in the water and are mechanically washed away.

The essential structural features a substance must have to exert this kind of cleaning action is a relatively long structural unit with polar and nonpolar ends. Synthetic detergents have such structures. One example is sodium dodecylbenzenesulfonate, also called linear alkylbenzenesulfonate (LAS):

$$CH_3CH_2CH_2CH_2CH_2CH_2CH_2CH_2CH_2CH_2\underset{\underset{CH_3}{|}}{CH}{-}\bigcirc{-}\overset{\overset{\text{O}}{\|}}{\underset{\underset{\text{O}}{\|}}{S}}{-}O^-Na^+$$

Nonpolar end CH_3 O   Polar end

The polar and nonpolar ends of sodium
dodecylbenzenesulfonate, a representative
component of synthetic detergents

FIGURE 13-14
Two droplets of oil covered with the negative ions of soap substances. The nonpolar ends of the soap ions dissolve in the nonpolar oil droplets while their polar ionic ends stay in the surrounding water. The negative charges on each droplet prevent them from coalescing. The dispersed droplets remain suspended in the water and are washed away.

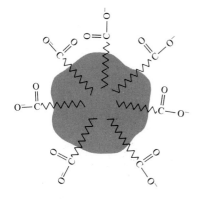
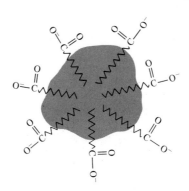

as sodium dodecylbenzenesulfonate. The carbon chains of these substances, like the carbon chains of soaps, contain no (or almost no) branches, which makes them readily biodegradable by microorganisms. The original branched-carbon-chain detergents were made because they were easily produced at low cost. Unfortunately, their long-term disadvantages were not anticipated.

Synthetic detergents are now made in considerably greater amounts than soap. The starting substances for most detergents are derived principally from petroleum hydrocarbons, in contrast to the vegetable and animal fat sources for soaps.

WHY PHOSPHATES IN DETERGENTS?

Soap and detergent products usually are mixtures of substances in which the soap or detergent itself is supplemented by other substances added to enhance its cleaning action. Appreciable amounts of sodium carbonate, Na_2CO_3, for example, are often added to powdered and beaded soaps to remove from water supplies ions like calcium, Ca^{2+}, and magnesium, Mg^{2+}, which consume some of the soap and form annoying precipitates.

$$CO_3^{2-} \quad + \quad Ca^{2+} \quad \rightarrow \quad CaCO_3 \downarrow$$

Carbonate ion   Calcium ion   Calcium carbonate
(from sodium   (present in
carbonate,   water supplies)
Na_2CO_3)

The action of sodium carbonate, Na_2CO_3, in reducing the concentration of calcium ions, Ca^{2+}, in water supplies (similar precipitates are formed with Mg^{2+} ions)

The cleaning action of synthetic detergents is greatly increased by the addition of pyrophosphates and tripolyphosphates, which are derivatives of phosphoric acid, H_3PO_4. Ionic derivatives of phosphoric acid, such as sodium phosphate, Na_3PO_4, contain the phosphate ion, PO_4^{3-}. The pyrophosphate and tripolyphosphate ions have structures related to the phosphate ion, as the comparison below reveals.

The sodium pyrophosphates and tripolyphosphates increase the detergent action of synthetic detergents, and they also remove the ions responsible for the hardness of water by reacting with them to form soluble products which do not interfere with the action of detergents. This incorporation of ions into more elaborate soluble molecular species is known as sequestering. The pyrophosphate- and tripolyphosphate–synthetic detergent mixtures have been successful in solving the tough cleaning problems presented by automatic washers for clothes and dishes.

The trouble with phosphates as detergent enhancers is that they find their way into rivers and lakes, where they fertilize the excessive growth of algae and other plant life. This depletes the oxygen in the water, destroying fish and encouraging foul-smelling bacterial action. Eutrophication is the process of enriching waters with such nutrients as the nitrates, phosphates, and sulfates of potassium, calcium, and magnesium. In order to reduce the eutrophication of lakes by detergent phosphates the phosphate content of many detergent mixtures is being reduced, and some contain no phosphates at all. Certain states have banned sales of most detergents containing phosphates.

Phosphate ion   Pyrophosphate ion   Tripolyphosphate ion
(also called
dipolyphosphate ion)

Comparison of the phosphate, pyrophosphate, and tripolyphosphate ions; many salts, e.g., sodium salts, of these ions can be prepared; they are known collectively as phosphates

In 1976 the production of soap totaled 1 billion pounds, and the production of detergents reached almost 6 billion pounds. People in the United States probably consume more soap and detergents than necessary, but the extraordinary availability of these products helps to maintain high levels of personal cleanliness and high standards of sanitation in the food and beverage industry. Large amounts of detergents are also used in finishing textiles and in other industrial operations.

13-10 HOW ARE CHEMICALS OBTAINED FROM COAL?

A. Acetylene, a Very Versatile Gas

Acetylene is a colorless, odorless gas which reacts with oxygen in an extremely exothermic reaction:

$$2HC\equiv CH + 5O_2 \rightarrow 4CO_2 + 2H_2O + energy$$

Acetylene　　Oxygen　　Carbon　　Water
　　　　　　　　　　　　dioxide

**The highly exothermic reaction between
acetylene and oxygen**

The flame produced by this reaction reaches about 2800°C and is widely used in welding and cutting iron and steel. Acetylene can also serve as the starting point in the synthesis of an extraordinary variety of carbon-containing compounds.

Although acetylene can now be prepared from substances obtained from petroleum, it became easily and economically available at the end of the nineteenth century from the raw materials coal and limestone (essentially calcium carbonate, $CaCO_3$). The first step is the preparation of calcium carbide in the high temperatures of an electric furnace (2500°C). Acetylene is subsequently generated when the calcium carbide is treated with water at room temperature.

$$CaCO_3 \xrightarrow[\text{temp.}]{\text{high}} CaO + CO_2 \tag{1}$$

Calcium　　　　　Calcium　Carbon
carbonate　　　　oxide　　dioxide
(limestone)　　　(lime)

$$CaO + 3C \xrightarrow[\text{2500°C}]{\text{electric furnace}} CaC_2 + CO \tag{2}$$

Calcium　　Coke　　　　　　Calcium　　Carbon
oxide　(from coal)　　　　carbide　　monoxide
(lime)

$$CaC_2 + 2H_2O \xrightarrow[\text{temp.}]{\text{room}} HC\equiv CH + Ca(OH)_2 \tag{3}$$

Calcium　　Water　　　　Acetylene　　Calcium
carbide　　　　　　　　　　　　　　hydroxide

**Steps in the production of acetylene from coal
(via coke) and limestone; since acetylene can serve
as the starting substance for a great variety of
carbon-containing substances, these reactions
provide a source of important compounds
directly from coal**

The many commercially important reactions of acetylene include the combination with hydrogen chloride, HCl, to form vinyl chloride, CH_2=CHCl. This is the starting substance in the production of the widely used polymer polyvinyl chloride, important in the fabrication of such products as wire insulation, floor tile, phonograph records, and rainwear (Fig. 13-15).

$$HC{\equiv}CH + HCl \xrightarrow[200°C]{cat.} HC{=}CCl \xrightarrow{cat.} \left[\begin{array}{cccccc} H & H & H & H & H & H \\ | & | & | & | & | & | \\ -C- & C- & C- & C- & C- & C- \\ | & | & | & | & | & | \\ H & Cl & H & Cl & H & Cl \end{array} \right]_n$$

Acetylene   Hydrogen   Vinyl   Polyvinyl chloride
chloride   chloride

**Preparation on polyvinylchloride
from acetylene and hydrogen chloride**

Acetylene can also react with itself in a variety of ways which serve as the first steps in the preparation of a vast array of useful substances. Under present market conditions,

FIGURE 13-15
Familiar products made of polyvinylchloride (Table 13-2).
(*Courtesy of Union Carbide Corporation.*)

however, many of the products can be obtained less expensively from other sources, especially petroleum:

$$HC\equiv CH + HC\equiv CH \xrightarrow{cat.} HC\equiv C-CH=CH_2$$

Acetylene   Vinylacetylene

$$HC\equiv CH + HC\equiv CH + HC\equiv CH \xrightarrow{cat.}$$

Acetylene

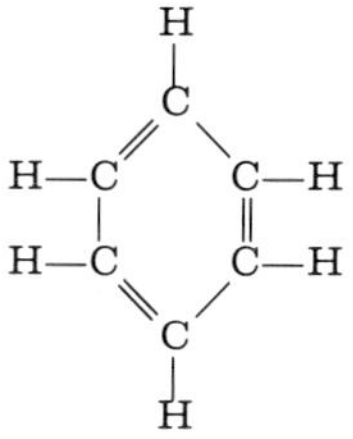

Benzene

Reactions illustrating acetylene's versatility in the production of carbon-containing substances

B. The Destructive Distillation of Coal; the Water-Gas Process

In the destructive distillation of coal (Sec. 13-1B) heating coal in the absence of air produces a mixture of hydrogen, H_2, and methane, CH_4 (called coke-oven gas), as well as substantial amounts of benzene and other volatile carbon-containing products. The nonvolatile residue, mostly carbon, is coke:

$$CH_xO_x \xrightarrow[900-1200°C]{\substack{\text{decomposed in} \\ \text{sealed ovens}}} C \;+\; H_2 + CH_4 \;+ \text{benzene} + \text{other carbon-containing products}$$

Bituminous coal   Coke   Coke-oven gas

Outline of destructive distillation of coal

In the **water-gas process** (Sec. 13-1B) a mixture of carbon monoxide, CO, and hydrogen, H_2, is obtained by passing steam over hot coke. The carbon monoxide and hydrogen obtained can be used to synthesize a wide variety of carbon-containing substances (See Sec. 13-10C).

C. Gasoline from Coal?

Gasoline is primarily a carefully blended mixture of hydrocarbon molecules containing from 5 to 12 carbon atoms in greatest concentration. Other substances are often added to improve its performance as a fuel, including tetraethyllead, $Pb(CH_2CH_3)_4$, and agents to reduce engine corrosion, gum formation, and carbon deposits. The basic hydrocarbon mixture is most readily obtained from the components of petroleum (Sec. 13-11). Several processes have been developed, however, for obtaining the hydrocarbons from coal. One of the most successful is the Fischer-Tropsch synthesis, named for the German chemists Franz Fischer and Hans Tropsch who developed the original version.

H_2O
Steam

+

C
Coke

high | temp.

CO
Carbon
monoxide

+

H_2
Hydrogen

The water-gas process for producing a mixture of carbon monoxide, CO, and hydrogen, H_2

Coal is used in one version of this process without first being converted into coke. It is treated at a high temperature with a mixture of steam and oxygen under pressure. Benzene and a number of other carbon-containing liquids and solid substances are obtained, along with a mixture of gases, typically hydrogen, H_2 (40 percent), carbon monoxide, CO (15 percent), methane, CH_4 (15 percent), and carbon dioxide, CO_2 (30 percent). After removal of the carbon dioxide and purification, the gas is led directly into reactors, where at elevated temperature and pressure with the aid of a catalyst, reactions occur which yield a complex mixture of hydrocarbons. The following reactions are representative:

$$CO \;+\; 3H_2 \;\rightarrow\; CH_4 \;+\; H_2O$$

Carbon   Hydrogen   Methane   Water
monoxide

$$7CO \;+\; 15H_2 \;\rightarrow\; CH_3CH_2CH_2CH_2CH_2CH_2CH_3 \;+\; 7H_2O$$

Carbon   Hydrogen   Heptane   Water
monoxide

**Typical reactions occurring during production of
hydrocarbons in the Fischer-Tropsch process**

The mixture of products depends on conditions but 60 percent yields of gasoline-type hydrocarbons are easily achieved. Large quantities of motor fuel were produced in Germany during World War II by mixing the products of processes like this with benzene, also obtained from coal. Usable gasoline can therefore be produced from coal, but under present market conditions and at the present stage of development, the cost is substantially higher than that of gasoline obtained from petroleum.

Note to Students

The production of gasoline from coal is one of the better examples of the chemical industry's potential for opening up alternate sources of important materials.

13-11 CHEMICALS FROM PETROLEUM AND NATURAL GAS

A. Petroleum Refining

Crude petroleum is, for the most part, a complex mixture of hydrocarbons, but it also contains small amounts of sulfur, nitrogen, and oxygen-containing compounds. It formed over geologic periods of time from buried organic material of living organisms. The substances range in complexity from methane, with a single carbon atom, to substances containing 100 or more carbon atoms in their molecules. They include open-chain alkanes, cycloalkanes, and aromatic substances, some containing multiple ring systems. Even an incomplete analysis of a sample of crude petroleum yields several hundred different hydrocarbons.

Petroleum refining involves the separation of the crude oil into various fractions according to volatility and the conversion by chemical reactions of the products of distillation into other substances of greater usefulness (Fig. 13-16). The reactions decompose the large molecules into smaller ones (cracking), rearrange the atoms of medium-sized molecules (reforming), and convert small molecules into larger ones (polymerization). The aim is to get high yields of substances which make up gasoline and other useful products and to combine mixtures of molecules in each of the products to ensure a high level of performance. Hydrocarbons with branched carbon chains perform better as fuels in automobile engines than hydrocarbonds with straight carbon chains. **Reforming** rearranges molecular structures to increase the branching of the carbon chains. At high temperatures and pressures in the presence of a catalyst heptane, for example, can be converted into methylhexane:

$$CH_3CH_2CH_2CH_2CH_2CH_2CH_3 \xrightarrow[\substack{\text{high temp.,} \\ \text{high press.}}]{\text{cat.,}} \overset{\overset{\displaystyle CH_3}{\displaystyle |}}{CH_3CHCH_2CH_2CH_2CH_3}$$

Heptane   2-Methylhexane

Typical reforming conversion

FIGURE 13-16
A production unit of a petroleum refinery. Many of the tall columns carry out distillations. (*Courtesy of Exxon Corporation.*)

BLACK GOLD

Crude petroleum has been known since ancient times. Moses is quoted in the Bible as describing priests "who had fire on the altar which they took from the deep pit without water." The Phoenicians used a petroleum product to caulk their ships. Oil obtained from natural seepages was used for medicinal purposes by the American Indians. The nuisance of the pollution of underground salt

...s was frequently reported in the early nineteenth century. Samuel Kier, an owner of salt wells, marketed crude petroleum in 1849 as a treatment for rheumatism, gout, neuralgia and toothache. Obviously this was before medicines were monitored for their effectiveness, since petroleum is worthless for these purposes.

The scarcity and expense of fish and whale oil, burned to produce light, led to recognition of petroleum's commercial possibilities. Bituminous coal had first been investigated as a source of illuminants. Heated in the absence of air, it produced an oil which was distilled into "lamp oil" (close to the present kerosine). The experience gained in distilling coal oil was soon applied to petroleum, and the kerosine produced became a popular illuminant. Distillation of crude petroleum also yielded considerable quantities of a lower-boiling, highly flammable mixture called gasoline, which was discarded as a nuisance. The full potentialities of crude petroleum as a source of useful substances, however, soon began to be appreciated. Professor Benjamin Silliman, Jr., of Yale, was asked to investigate petroleum by a group of businessmen. He concluded his 1854 report: "It appears to me that there is ground for encouragement in the belief that your company have in their possession a raw material from which, by simple and not expensive processes, they may manufacture very valuable products."

In 1855 the Seneca Oil Company decided to undertake the first search for underground oil by drilling, with "Colonel" Edwin L. Drake in charge of operations. A site at Titusville, in northwest Pennsylvania, was chosen because petroleum seeped naturally to the surface in that area. Oil was found in 1859 at a depth of 70 feet, and the well produced nine gallons a day (Fig. 13-17). Although "Drake's folly" ran dry rather quickly, thousands of people rushed to the scene. Within a few years over 200 successful wells were drilled in the area, and oil deposits were soon discovered in other parts of the world.

With the development of the internal combustion engine and the arrival of the automobile, a demand was created for gasoline which has increased steadily ever since. Today, petroleum is the source of about 43 percent of the nation's energy needs, in the form of furnace oil, diesel oil, jet fuel, and gasoline. It is also the source of almost 90 percent of the raw materials for synthetic rubber, fibers, plastics, detergents, and hundreds of other products. Crude petroleum has often been called "black gold," but in usefulness and versatility it far surpasses any precious metal.

FIGURE 13-17
The derrick and engine house of the first successful oil-well drilling, Titusville, Pennsylvania, August 1859. "Colonel" Drake, in stovepipe hat, stands nearby. The well produced about 9 gallons of crude petroleum a day, causing a rush of prospectors to the area. (*Courtesy of Exxon Corporation.*)

Figure 13-18 outlines the principal processes of a petroleum refinery. The operations are usually carried out on a huge scale. Processing crude petroleum at the rate of 50,000 barrels a day is common, and some plants have a capacity of 400,000 barrels a day. The distillation columns are several stories high. The outline shows some of the special efforts made to increase the quantity and quality of gasoline.

Straight-run gasoline is the fraction of crude petroleum which is directly obtainable as gasoline by distillation. For the average crude petroleum this amounts to a yield of about 25 percent. In modern refineries the gasoline yield can be increased to almost 50 percent by (1) polymerizing the small molecules of lower-boiling distillation fractions and (2) decomposing (cracking) the large molecules of the higher-boiling distillation fractions. The quality of the gasoline is improved by increasing the

FIGURE 13-18
Simplified outline of the operation of a petroleum refinery. The hydrocarbons of the crude petroleum are separated by distillation into fractions with a series of boiling-point ranges. The components of several of these fractions are converted by chemical reactions into more useful substances. The processes are usually carried out on a huge scale; the distillation columns, for example, are several stories high. Many of the products serve as starting materials for important chemical processes.

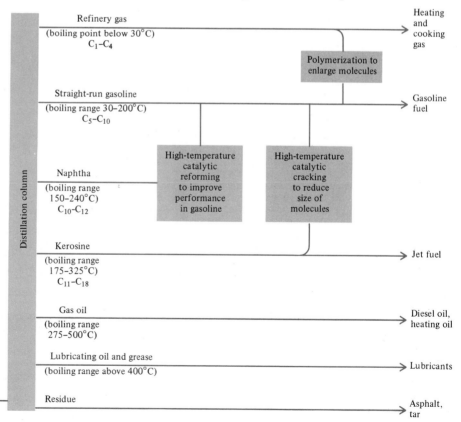

amount of branching in the carbon chains of the hydrocarbons (reforming).

B. Petrochemicals

We have mentioned several examples of important chemical intermediates obtained from the substances of crude petroleum and natural gas:

1. Hydrogen, H_2, for the production of the fertilizer ammonium nitrate (Sec. 13-1B) and extraction of magnesium from the sea (Sec. 13-7)

2. Ethylene, $CH_2=CH_2$, for the production of the polymer polyethylene (Sec. 13-4)

3. Benzene, C_6H_6, styrene, $C_6H_5CH=CH_2$, and butadiene, $CH_2=CH—CH=CH_2$, for the production of synthetic rubber (Sec. 13-5)

4. Adipic acid, $HOOC(CH_2)_4COOH$, and hexamethylenediamine, $H_2N(CH_2)_6NH_2$, for the production of nylon 66 (Sec. 13-6)

Petroleum, in fact, is the present source of almost 90 percent of all of the carbon-containing products of the chemical industry. A **petrochemical** is any chemical intermediate derived from petroleum or natural gas. The production of petrochemicals consumes about 10 percent of the entire natural gas and petroleum-refinery production. Most of these chemicals are also obtainable from coal, but since the 1940s there has been a changeover to petroleum-based chemicals, largely because they are cheaper, cleaner, and more adaptable to large-scale production processes. With the rising cost of petroleum, the increasing concern about possible shortages, and the large amount which has to be imported from foreign sources, a return to our plentiful coal reserves appears likely in the future. The flexibility of the processes of the chemical industry should be able to accommodate the conversion.

KEY WORDS

1. Nitrogen fixation (Sec. 13-1A)
2. Destructive distillation of coal (Sec. 13-1B)
3. Chemical intermediate (Sec. 13-1E)
4. Polymer (Sec. 13-3)
5. Plastic (Sec. 13-3)
6. Polymerization reaction (Sec. 13-4)
7. Vulcanization (Sec. 13-5)
8. Soap (Sec. 13-8)
9. Synthetic detergent (Sec. 13-9)
10. Biodegradable (Sec. 13-9)
11. Water-gas process (Sec. 13-10B)
12. Gasoline (Sec. 13-10C)
13. Reforming (Sec. 13-11A)
14. Straight-run gasoline (Sec. 13-11A)
15. Petrochemical (Sec. 13-11B)

"AH, HA!"

SUMMARIZING QUESTIONS FOR SELF-STUDY

Section 13-1

1. Q. What is meant by nitrogen fixation?

A. Any process converting elemental nitrogen into a nitrogen-containing compound.

2. Q. What is the most significant route of natural nitrogen fixation?

A. The action of certain bacteria which live in the roots of a class of plants (legumes) and are capable of converting nitrogen into ammonia.

3. Q. What is the special significance of the processes of nitrogen fixation for the growth of plants?

A. The processes convert elemental nitrogen, which does not serve as a plant nutrient, into simple compounds of nitrogen, which do serve as plant nutrients.

4. Q. Give the name and overall reaction for the leading process of synthetic nitrogen fixation.

A. The Haber process:

$$N_2 + 3H_2 \xrightarrow[\substack{\text{high temp.,} \\ \text{high press.}}]{\text{cat.,}} 2NH_3$$

5. Q. What is the special significance of this process of synthetic nitrogen fixation?

405

A. It provides large amounts of ammonia at low cost for direct use as a fertilizer and to make other fertilizers. It serves as an important supplement to the natural processes of nitrogen fixation.

6. Q. What are the sources for the hydrogen required for the production of ammonia? Give the reactions involved.

A. From bituminous coal:

$$CH_rO_r \xrightarrow[\substack{\text{sealed ovens} \\ (900-1200°C)}]{\text{decomposed in}}$$

$$C + H_2 + CH_4 + \text{benzene} + \text{other compounds}$$
Coke

From coke:

$$H_2O + C \xrightarrow[\text{temp.}]{\text{high}} CO + H_2$$
$$\text{Coke}$$

$$CO + H_2O \xrightarrow[\text{high temp.}]{\text{cat.,}} CO_2 + H_2$$

From methane (natural gas):

$$CH_4 + 2H_2O \xrightarrow[\text{high temp.}]{\text{cat.,}} CO_2 + 4H_2$$

7. Q. Which of these is now the principal source?
A. Methane (natural gas).

8. Q. Why?
A. The high purity and quality of the supplies, its ease of transportation (pipeline), and (until recently) its low cost and abundant supply.

9. Q. Why may this source not be the leading one in the future?
A. Dwindling supplies and probable higher cost.

10. Q. What are the natural sources for the starting substances used now in the production of ammonia by the Haber process?
A. The atmosphere (nitrogen) and natural gas (methane).

11. Q. Name three principal uses of ammonium nitrate.
A. Fertilizers, explosives, and source of the surgical anesthetic nitrous oxide.

12. Q. Give an equation for the reaction of the last step in the production of ammonium nitrate. To what general class does this reaction belong?
A. $NH_3 + HNO_3 \rightarrow NH_4NO_3$. This is a proton-transfer reaction. The proton is transferred from the nitric acid, HNO_3, to the ammonia, NH_3.

13. Q. What is meant by a chemical intermediate?
A. A substance which is derived in one or more chemical reactions from raw materials or other chemical intermediates and which undergoes further chemical treatment along the pathway toward the production of an end product.

14. Q. Give the names and formulas of three chemical intermediates involved in the production of ammonium nitrate.
A. Hydrogen, H_2; ammonia, NH_3; nitric acid, HNO_3.

15. Q. List briefly eight distinguishing features of the chemical industry.

A. Converts on a large scale naturally occurring raw materials to products of greater economic usefulness; is highly innovative; provides replacements for important materials, sometimes conserving natural resources and/or reducing imports; has flexibility in raw materials; turns out many important products; the industry and its products tend to be anonymous; products often versatile in their uses; products and processes frequently hazardous and must be made and used with care.

Section 13-2

16. Q. Broadly defined, what types of production does the chemical industry embrace?
A. All manufacturing processes which make products by altering molecular structure.

17. Q. Name five products which come under this heading.
A. Five from fertilizers, drugs, soaps, detergents, paints, explosives, plastics, petroleum products, rubber products, metals, pulp and paper (without exhausting the possibilities).

18. Q. Name five products the manufacture of which requires the intermediate sulfuric acid.
A. Five products from fertilizers, dyes, reinforced concrete, metals, paper, textiles, explosives, plastics, many drugs.

19. Q. The raw materials of the chemical industry come from (*a*) the earth's crust, (*b*) the atmosphere, (*c*) seawater, and (*d*) agriculture. Name one raw material from each source.
A. (*a*) Petroleum, (*b*) nitrogen, (*c*) magnesium salts, (*d*) grain.

Sections 13-3 and 13-4

20. Q. What is meant by a polymer?
A. A substance whose molecules are made up of many structural units held together by interatomic bonds.

21. Q. What is the fundamental structural unit of polyethylene, and what is the starting substance in its production?
A. The $—CH_2—$ group (but since the starting substance for polyethylene is ethylene, $CH_2{=}CH_2$, the fundamental structural unit may be taken as $—CH_2—CH_2—$).

22. Q. What is a plastic?
A. A polymeric material capable of being molded into such objects as eyeglass frames and buttons.

23. Q. What is a polymerization reaction?
A. A reaction in which many molecules of one or a small number of substances of relatively simple structure are linked together to make a large molecule.

24. Q. Name two of the principal uses of polyethylene.

A. Wire and cable insulation, pipe, packaging materials, bottles, and other molded products.

25. Q. From what raw material is polyethylene made?

A. Petroleum.

Section 13-5

26. Q. What is the name and formula for the structural unit of natural rubber?

A. An isoprene residue

$$-CH_2-\underset{\underset{CH_3}{|}}{C}=CH-CH_2-$$

27. Q. About how many of these structural units are found in individual rubber molecules?

A. Somewhere between 1500 and 40,000 (and beyond).

28. Q. What are the names and formulas of the two structural units of SBR synthetic rubber?

A. Styrene residue

$$-CH-CH_2-$$

and butadiene residue, $-CH_2-CH=CH-CH_2-$.

29. Q. What is the ratio of these structural units in the SBR molecules?

A. Six of the butadiene residues to one of the styrene residues.

30. Q. What is the principal raw material source of styrene?

A. Petroleum.

31. Q. What is an alternate raw material source of the benzene used to produce styrene?

A. Bituminous coal.

32. Q. What is the principal raw material source of butadiene?

A. Petroleum.

33. Q. What is an alternate raw material source of butadiene?

A. Grain, by way of ethyl alcohol.

34. Q. What is the process of vulcanization, and what change occurs during the treatment at the molecular level?

A. Vulcanization involves heating raw rubber with elemental sulfur. The treatment forms covalently bonded sulfur-containing cross-links between polymeric chains.

35. Q. How does the process of vulcanization change the rubber?

A. It becomes less tacky, more elastic, and more resistant to abrasion.

Section 13-6

36. Q. What are the names and formulas of the two immediate precursors of nylon 66 in the process for its production?

A. Adipic acid, $HOOCCH_2CH_2CH_2CH_2COOH$, and hexamethylenediamine,

$$H_2NCH_2CH_2CH_2CH_2CH_2CH_2NH_2$$

37. Q. Both these substances are obtained from the intermediate benzene. What are two raw material sources of benzene?

A. The current principal source is petroleum, by way of hexane. An alternate source is bituminous coal.

38. Q. What functional group serves as the connecting link between the structural units of nylon 66?

A. An amide group.

39. Q. When nylon was introduced, it was announced that it was made from coal, air, and water. Petroleum and natural gas have since replaced the coal for the most part, but how did the air and water enter into the process?

A. The air furnished the nitrogen for the synthesis of ammonia and the oxygen used in the conversion of ammonia to nitric acid. The water served as a source of hydrogen for the synthesis of ammonia.

Section 13-7

40. Q. In what form does magnesium occur in the sea?

A. Magnesium ion, Mg^{2+}.

41. Q. Why is the magnesium in seawater frequently represented as the salt magnesium chloride, $MgCl_2$?

A. Because the predominant negative ion in seawater is the chloride ion, Cl^-.

42. Q. How is the magnesium obtained in a concentrated form? Give an equation for the overall reaction.

A. The magnesium is concentrated by treating the seawater with calcium hydroxide, $Ca(OH)_2$, to obtain a precipitate of magnesium hydroxide, $Mg(OH)_2$:

$$MgCl_2 + Ca(OH)_2 \rightarrow Mg(OH)_2 \downarrow + CaCl_2$$

43. Q. What is the source of the calcium hydroxide?

A. The calcium carbonate, $CaCO_3$, of oyster shells.

44. Q. How is elemental magnesium obtained from the magnesium hydroxide, $Mg(OH)_2$?

A. The magnesium hydroxide, $Mg(OH)_2$, is converted into magnesium chloride, $MgCl_2$, which is converted into a dry salt. The magnesium ion, Mg^{2+}, of the magnesium chloride is converted into elemental magnesium by an electric current, which furnishes the necessary electrons in the process of electrolysis.

$$Mg(OH)_2 + 2HCl \rightarrow MgCl_2 + 2H_2O$$

$$\underset{\underset{MgCl_2}{From}}{Mg^{2+}} + \underset{\underset{\substack{electric \\ current}}{From}}{2e^-} \rightarrow Mg$$

Overall reaction for last step:

$$MgCl_2 \xrightarrow[\text{current}]{\text{electric}} Mg + Cl_2$$

45. Q. The hydrochloric acid is made by the reaction of hydrogen and chlorine. What are the sources of these intermediates?

A. The chlorine is obtained from the electrolysis of the magnesium chloride. The hydrogen may be obtained from methane (natural gas), bituminous coal, or coke and water.

46. Q. Assuming that the hydrogen is obtained from the methane of natural gas, what additional raw materials are needed for the production of magnesium?

A. Seawater and calcium carbonate.

Sections 13-8 and 13-9

47. Q. What is the composition of soaps?

A. They are sodium salts of carboxylic acids containing between 12 and 18 carbon atoms.

48. Q. Give the formula of a representative substance of soap.

A. $CH_3(CH_2)_4COO^-Na^+$.

49. Q. What is the source of the carboxylic acids of soaps?

A. Vegetable and animal fats.

50. Q. What type of reaction is involved in the preparation of the carboxylic acids from fats?

A. Hydrolysis.

51. Q. How are the soap salts formed?

A. The sodium hydroxide, NaOH, used as the catalyst in the hydrolysis of the fats, reacts with the carboxylic acids as soon as they are formed to produce the soap salts.

$$Na^+OH^- + H\text{—}O\text{—}\overset{\overset{\displaystyle O}{\|}}{C}\text{—}(CH_2)_{14}CH_3 \rightarrow$$

$$Na^+O\text{—}\overset{\overset{\displaystyle O}{\|}}{C}(CH_2)_{14}CH_3 + H\text{—}O\text{—}H$$

52. Q. What is a synthetic detergent?

A. Any synthetic substance other than a soap which is capable of soaplike cleaning action.

53. Q. What does biodegradable mean?

A. Large molecules or the substances they constitute are capable of being degraded by the biological or chemical changes of natural processes into small molecules readily assimilated by the environment.

54. Q. Give the formula for a substance representing a biodegradable detergent and one for a nonbiodegradable detergent.

A. Biodegradable

$$CH_3(CH_2)_7CH_2CH_2\underset{\overset{|}{CH_3}}{CH}\text{—}\langle O \rangle\text{—}\overset{\overset{\displaystyle O}{\|}}{\underset{\underset{\displaystyle O}{\|}}{S}}\text{—}O^-Na^+$$

Nonbiodegradable

$$CH_3\underset{\overset{|}{CH_3}}{CH}CH_2\underset{\overset{|}{CH_3}}{CH}CH_2\underset{\overset{|}{CH_3}}{CH}CH_2\underset{\overset{|}{CH_3}}{CH}\text{—}\langle O \rangle\text{—}\overset{\overset{\displaystyle O}{\|}}{\underset{\underset{\displaystyle O}{\|}}{S}}\text{—}O^-Na^+$$

55. Q. Why are most synthetic detergents today of the biodegradable type?

A. Because the nonbiodegradable detergents produced undesirable accumulations in groundwaters and rivers.

56. Q. What is one of the principal advantages of synthetic detergents over soaps?

A. The negative ions of the synthetic detergents do not form precipitates with the ions of hard water, such as calcium ion, Ca^{2+}, as the negative ions of soaps do:

$$Ca^{2+} + 2CH_3(CH_2)_{14}COO^- \rightarrow Ca[CH_3(CH_2)_{14}COO]_2 \downarrow$$

The formation of these precipitates consumes soap and interferes with its cleaning action.

Section 13-10

57. Q. Why is acetylene a valuable substance? Give equations for reactions where appropriate.

A. It burns in oxygen to produce a very hot flame used in welding. $2CH{\equiv}CH + 5O_2 \rightarrow 4CO_2 + 2H_2O$ + energy, and is a chemical intermediate in the production of many useful substances; e.g., polyvinylchloride:

$$HC{\equiv}CH + HCl \xrightarrow[200°C]{cat.,} \underset{\overset{|}{H}}{\overset{\overset{\displaystyle H}{|}}{HC}}{=}\underset{\overset{|}{Cl}}{\overset{\overset{\displaystyle H}{|}}{C}} \xrightarrow{cat.} \left[\text{—}\underset{\overset{|}{H}}{\overset{\overset{\displaystyle H}{|}}{C}}\text{—}\underset{\overset{|}{Cl}}{\overset{\overset{\displaystyle H}{|}}{C}}\text{—}\right]_x$$

58. Q. What is the destructive distillation of coal, and what kinds of substances are produced in the process?

A. The process of heating coal in the absence of air. It produces a mixture of hydrogen and methane, substantial amounts of benzene and other volatile organic compounds, and coke, as a nonvolatile residue.

59. Q. The two substances which serve as the starting compounds of the Fischer-Tropsch synthesis are carbon monoxide, CO, and hydrogen, H_2. What are two sources of these substances?

A. The water-gas process, in which steam is passed over hot coke:

$$H_2O + \underset{\text{Coke}}{C} \xrightarrow[\text{temp.}]{\text{high}} CO + H_2$$

and the treatment of coal directly with a mixture of steam and oxygen under pressure. A mixture of CO, H_2, CH_4, and CO_2 is obtained.

60. Q. Give a balanced equation for the formation of pentane, $CH_3CH_2CH_2CH_2CH_3$ (as a representative hydrocarbon), in the Fischer-Tropsch synthesis.

A. $5CO + 11H_2 \rightarrow CH_3CH_2CH_2CH_2CH_3 + 5H_2O$.

61. Q. What is the significance of the Fischer-Tropsch synthesis?

A. It is a means of producing the hydrocarbons for gasoline and many other useful carbon-containing substances from coal rather than from petroleum.

62. Q. What two types of processes are used in refining crude petroleum?

A. The separation of the various components into groups, or fractions, by distillation and the conversion by chemical reactions of the products of distillation into other substances of greater usefulness.

63. Q. The petroleum-refining process of reforming brings about rearrangements of molecular structures which increase the branching of carbon chains, as illustrated in the reaction

$$CH_3CH_2CH_2CH_2CH_2CH_2CH_3 \xrightarrow[\substack{\text{high temp.,}\\ \text{high press.}}]{\text{cat.,}}$$

$$\underset{\overset{|}{CH_3}}{CH_3CHCH_2CH_2CH_2CH_3}$$

What is the point of such rearrangements?

A. Hydrocarbons with branched carbon chains perform better as fuels in automobile engines than hydrocarbons with straight carbon chains.

64. Q. What is gasoline?

A. Primarily a mixture of hydrocarbon molecules containing from 5 to 12 carbon atoms in greatest concentration.

65. Q. What is straight-run gasoline?

A. The fraction of crude petroleum directly obtainable as gasoline by distillation.

66. Q. What two processes are used to increase the yield of gasoline from crude petroleum?

A. Polymerizing the small molecules of lower-boiling distillation fractions and decomposing (cracking) the large molecules of the higher-boiling distillation fractions.

67. Q. Gasoline is only one of the principal products of a petroleum refinery. Name five others.

A. Heating and cooking gas, jet fuel, diesel oil, heating oil, lubricants, asphalt, and intermediates for chemical synthesis.

68. Q. What is a petrochemical?

A. Any chemical intermediate derived from petroleum or natural gas.

69. Q. Name six petrochemicals mentioned in the production processes of this chapter.

A. Hydrogen, e.g., for ammonium nitrate and magnesium; ethylene, e.g., for polyethylene; benzene, e.g., for synthetic rubber and nylon; styrene, e.g., for synthetic rubber; butadiene, e.g., for synthetic rubber; adipic acid, e.g., for nylon 66; hexamethylenediamine, e.g., for nylon 66.

70. Q. What is the economic significance of petrochemicals?

A. They are the current source of about 90 percent of all the carbon-containing products of the chemical industry, and their production consumes about 10 percent of the entire natural-gas and petroleum-refinery production.

71. Q. As the cost of petroleum rises and the supply of petroleum decreases, what alternate raw material for carbon-containing substances is available?

A. Bituminous coal, which supplied most of such substances four decades ago.

PRACTICE EXERCISES

1. Match each definition or other statement with the numbered term above with which it is most closely associated. Each numbered term may be used only once.

   1. Reforming
   2. Soap
   3. Straight-run gasoline
   4. Cracking
   5. Nitrogen fixation
   6. Schultz-Brown process
   7. Biodegradable substance
   8. Polymer
   9. Water-gas process
   10. Vulcanization

   (a) The production of carbon monoxide and hydrogen by passing steam over hot coke
   (b) A substance capable of being degraded into substances readily assimilated by the environment through the biological or chemical changes of natural processes
   (c) A process of converting elemental nitrogen into a nitrogen-containing compound
   (d) A process in which raw rubber is heated with elemental sulfur
   (e) A process which brings about rearrangement of the molecular structure of hydrocarbons to increase the branching of the carbon chains
   (f) A substance whose molecules are made up of many structural units held together by interatomic bonds
   (g) A sodium salt of a carboxylic acid containing 16 carbon atoms
   (h) A fraction of crude petroleum

2. Match each definition or other statement with the numbered term above with which it is most closely associated. Each numbered term may be used only once.

   1. Synthetic detergent
   2. Petrochemical
   3. Polymerization
   4. Heavy water
   5. Destructive distillation
   6. Plastic
   7. Chemical intermediate
   8. Soap
   9. Quinine
   10. Hard water
   11. Gasoline

   (a) Heating coal in the absence of air
   (b) A substance which is derived by chemical reactions from raw materials and which

undergoes further chemical treatment in the production of an end product

(c) A polymeric material capable of being molded into such objects as eyeglass frames and buttons

(d) Water containing appreciable concentrations of calcium and magnesium ions

(e) A reaction in which many molecules of a substance of relatively simple structure are linked together to make a large molecule

(f) A synthetic substance which is not a soap but is capable of soaplike cleaning action

(g) A blended mixture of hydrocarbons containing from 5 to 12 carbon atoms in greatest concentration

(h) A chemical intermediate derived from petroleum or natural gas

3. (a) Give a balanced equation for the Haber method of nitrogen fixation. (b) What is the source of the nitrogen? (c) What is the current principal source of the hydrogen? (d) Give an equation for the overall process.

4. (a) What alternate sources of hydrogen for the Haber process are available? Give equations where appropriate. (b) Why may the principal ultimate source of hydrogen change in the future?

5. (a) Give the names and formulas for the naturally occurring substances used in the current principal method of synthesizing the fertilizer and explosive ammonium nitrate, NH_4NO_3. (b) What naturally occurring material is involved in alternate sources and what naturally occurring substance would it replace?

6. Give balanced equations for the Ostwald process of synthesizing nitric acid, HNO_3, from ammonia, NH_3.

7. (a) Give a balanced equation for the last step in the synthesis of ammonium nitrate, NH_4NO_3. (b) What type of reaction is involved?

8. (a) What is one of the disadvantages of ammonium nitrate as a fertilizer? (b) What use is made of this disadvantage? (c) What further large-scale use is made of ammonium nitrate?

9. (a) Give a balanced equation for the production of the general anesthetic nitrous oxide, N_2O. (b) What is the ultimate source of the oxygen used in the process?

10. Nitric acid is described as an important chemical intermediate. What does this mean?

11. (a) List eight distinguishing features of the chemical industry. (b) Give one example of each.

12. Give the name of one representative raw material used by the chemical industry obtained from each of the following: the atmosphere, the earth's crust, seawater, agriculture.

13. (a) What is a polymer? (b) What is a plastic?

14. Using Table 13-2 give the name and formula for the starting substance (or substances) in the production of (a) polyvinylchloride (Dynel), (b) polytetrafluoroethylene (Teflon), and (c) polyester (Dacron).

15. (a) Give a general formula for polyethylene plastic. (b) What is the starting substance used in the production? (c) What is the ultimate naturally occurring source of this substance?

16. Give an equation for the reaction for the preparation of ethylene.

17. (a) What are the principal useful properties of polyethylene? (b) What are some of the principal uses of polyethylene?

18. (a) To what class of carbon-containing substances does ethylene belong? (b) Assuming the structure of polyethylene given in the text, to what class of carbon-containing substances does it belong?

19. (a) What is the difference between isoprene and an isoprene residue? (b) What is the relationship between an isoprene residue and the structure of natural rubber?

20. (a) What is vulcanization? (b) From a structural viewpoint what does it do? (c) What does it do to the properties of rubber?

21. (a) To what class of carbon-containing substances does isoprene belong? (b) Assuming the formula for natural rubber given in the text, to what class of carbon-containing substances does it belong?

22. (a) What do the letters SBR stand for? (b) What was the significance of SBR to the United States in World War II?

23. (a) What is the ultimate naturally occurring raw material used in making SBR? (b) What is an alternate ultimate source for the intermediate benzene? (c) What is an alternate ultimate source for the intermediate butadiene?

24. (a) Give the names and formulas for the two intermediates used in the last step of the production of SBR. (b) Give the names and formulas of the other intermediates used in the production steps leading to the preparation of the final two intermediates.

25. (a) To what class of carbon-containing substances does butadiene belong? (b) To what family of carbon-containing substances does styrene belong (see Sec. 8-6)?

26. To what family of carbon-containing substances does (a) hexane and (b) benzene belong (see Sec. 8-6)?

27. Give a formula representing the structure of SBR.

28. (a) What are the naturally occurring raw materials used in the principal current production route of nylon 66? (b) What alternate

ultimate source of benzene is available? (c) What alternate ultimate source of hydrogen is available?

29. Give the names and formulas for the two intermediates used in the last steps of the production of nylon 66.

30. What is the significance of 66 in nylon 66?

31. Name the functional groups in (a) adipic acid, (b) hexamethylenediamine, and (c) nylon 66.

32. Give the formula for a general representation of nylon 66.

33. Give the names and formulas of the principal intermediates involved in the production of the intermediates used in the last steps of the production of nylon 66.

34. (a) Why is magnesium a widely used metal? (b) Why is the raw material from which it is produced practically inexhaustible?

35. (a) What are the ultimate naturally occurring sources of substances used in the current production of magnesium from the sea? (b) What alternate source of hydrogen is available and what raw material would it replace?

36. (a) Give an equation for the reaction which concentrates the dilute magnesium ion, Mg^{2+}, in seawater. (b) To what class does the reaction belong?

37. Trace the calcium hydroxide, $Ca(OH)_2$, used in the production of magnesium from seawater back to its ultimate natural source, giving names and formulas for the intermediates involved.

38. (a) Give an equation for the reaction in the last step of the production of magnesium from the sea. (b) What type of reaction is involved?

39. (a) What is the source of the chlorine used in the production of magnesium from the sea? (b) Give an equation for the reaction involved.

40. What is the distinctive functional group of fat molecules?

41. Give a typical reaction for the preparation of the carboxylic acids used in the production of soap.

42. Give the formula of a representative substance of soap and give an equation for its formation from the corresponding carboxylic acid.

43. Give an equation for a typical reaction of a soap with one of the distinctive components of hard water.

44. Compare the structure of a typical soap molecule with the structure of a typical synthetic detergent molecule by giving a formula for each.

45. What is one advantage of synthetic detergents over soaps?

46. Would you expect a soap to be biodegradable? Explain. Hint: Examine the structures of a soap, a biodegradable detergent, and a non-biodegradable detergent.

47. Give equations for the reactions involved in the production of acetylene from coal.

48. What is the special significance of the fact that acetylene can be produced from coal and limestone?

49. How can acetylene be used to produce the aromatic hydrocarbon benzene? Give the equation.

50. What are two different processes for obtaining benzene from coal?

51. Using equations for typical reactions show how gasoline can be obtained from coal and water.

52. (a) In general terms, what is the composition of crude petroleum? (b) Name seven of the principal products obtained from petroleum.

53. What two principal types of processes are used in the refining of petroleum?

54. What is meant by straight-run rasoline?

55. What are the names of two processes used in petroleum refining to increase the yield of gasoline?

56. (a) What is meant by the process of reforming, and what is its purpose? (b) Use formulas to represent a typical reforming conversion.

57. (a) What is meant by a petrochemical? (b) Give the names and formulas of six petrochemicals.

58. Compare the processes for obtaining carbon-containing products from petroeum with the processes of obtaining them from bituminous coal.

The following exercises assume a knowledge of Supplement 4.

59. How many grams of ammonium nitrate, NH_4NO_3, would be formed from 1000 g of ammonia, NH_3, assuming that there is an excess of nitric acid, HNO_3, and that all the ammonia is converted into ammonium nitrate?

60. How many grams of ammonium nitrate, NH_4NO_3, would be required to form 800.0 g of nitrous oxide, N_2O, assuming that all the ammonium nitrate is converted?

61. How many grams of ethylene would be required to form 10.00 kg of polyethylene assuming no loss in the process?

62. How many grams of calcium hydroxide, $Ca(OH)_2$, will be required to form 2.000 kg of magnesium hydroxide, $Mg(OH)_2$, assuming an excess of magnesium ion and no loss in the process?

63. (a) How many grams of chlorine, Cl_2, will be formed in the electrolysis of 3.000 kg of magnesium chloride, $MgCl_2$, assuming no loss in the process? (b) How many grams of magnesium will be formed at the same time?

SUGGESTIONS FOR FURTHER READING

The following articles are of an introductory nature.

American Chemical Society, "Chemistry in the Economy," Washington, D.C., 1973. A study by a panel of distinguished experts of the relationship of chemical production to the economy. It covers many of the significant contributions of the chemical industry.

Fisher, Harry L.: Rubber, **Scientific American,** November 1956, p. 75.

Flory, Paul J.: Understanding Unruly Molecules, **Chemistry,** May 1964, p. 6.

Hall, Dana, and Earle Allen: The Many Faces of Rubber, **Chemistry,** June 1972, p. 6.

Hammond, A. L.: Phosphate Replacements: Problems with the Washday Miracle, **Science, 172:** 361 (1971).

Keller, Eugenia: Nylon: From Test Tube to Counter, **Chemistry,** September 1964, p. 8.

Kushner, Lawrence M., and James I. Hoffman: Synthetic Detergents, **Scientific American,** October 1951, p. 26.

Mark, H. F.: Giant Molecules, **Scientific American,** September 1957, p. 204.

Mark, H. F.: The Nature of Polymeric Materials, **Scientific American,** September 1967, p. 148.

Meloan, Clifton E.: Detergents: Soaps and Syndets, **Chemistry,** September 1976, p. 6.

Meyerhoff, Arthur A.: Economic Impact and Geopolitical Implications of Giant Petroleum Fields, **American Scientist,** 64:536 (1976).

Morton, Maurice: Big Molecules, **Chemistry,** January 1964, p. 12.

Morton, Maurice: Polymers: Ten Years Later, **Chemistry,** October 1974, p. 6.

Natta, Guilio: How Giant Molecules Are Made, **Scientific American,** September 1957, p. 98.

Oster, Gerald: Polyethylene, **Scientific American,** September 1957, p. 139.

Smith, Cyril Stanley: Materials, **Scientific American,** September 1967, p. 68.

The following articles are at a more advanced level.

Frazer, A. H.: High Temperature Plastics, **Scientific American,** July 1969, p. 96.

Probstein, Ronald F.: Desalination, **American Scientist, 61:** 280 (1973).

14

WHAT WILL BE THE ENERGY SOURCES IN THE FUTURE?

14-1 WHY THE CONCERN ABOUT ENERGY SUPPLY?

The United States, with about 6 percent of the world's population, is consuming about 35 percent of the energy used in the world. Moreover, 93 percent of its energy is derived from the nonrenewable fossil sources of petroleum, natural gas, and coal (Table 14-1). It is a matter of growing concern that 75 percent is obtained from petroleum and natural gas, supplies of which we are consuming faster than we are producing. The situation with regard to crude petroleum is the most disturbing. Although it is the source of over 40 percent of our energy needs, we are finding it necessary to import almost half of what we are using.

Moreover, our consumption of energy has been increasing rapidly (Fig. 14-1), 4 times faster than the population. An intensive use of energy is characteristic of any highly technological society. Whereas primitive people required only about 2000 kilocalories a day to sustain themselves, people in the United States are consuming over 230,000 kilocalories a day.

Table 14-1
Principal Sources of Energy
in the United States (1976)

Source	Percentage of current total United States energy sources
Petroleum	43
Natural gas	32
Coal	18
Water power	4
Nuclear energy	3

Energy is indispensable to the characteristic high output of goods and services. Not only is it required for transportation, industrial production, space heating, space cooling, and lighting, but it is also essential for such basic needs as building homes, fabricating clothing, and growing, preparing, and storing food. The United States currently spends about $125 billion a year on ferreting out energy sources and on the refining, distribution, and consumption of fuels. This represents about 10 percent of the nation's entire economic activity.

Given the critical importance of an adequate amount of energy and the uncertainties which becloud energy sources, it is understandable that questions about what is in store for the nation are widespread, especially in view of the temporary fuel shortages which have already occurred. How is energy produced? What are the present and potential future sources of energy? How is energy used in the United States, and how can reductions in its consumption be made? This chapter is concerned with these questions. The various sources of energy are described and the pros and cons of each source described. Attention is then turned to questions of supply in the future.

FIGURE 14-1
Total annual energy consumption in the United States from all
sources during 1860 to 1976. (*Data from Department of Energy.*)

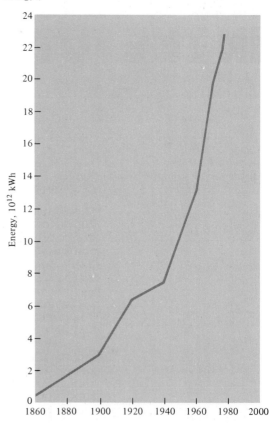

14-2 WHAT ARE SOME OF THE
UNITS USED TO EXPRESS AMOUNTS OF ENERGY?

415

14-3   HOW IS ENERGY OBTAINED
FROM FOSSIL FUELS?

Energy has been defined as the ability to do work (Sec. 2-3). One of the units of energy is the **calorie,** the amount of heat energy required to raise the temperature of 1 gram of water 1°C. A kilocalorie (the Calorie of nutritionists) contains 1000 calories. A person weighing about 150 pounds expends about 5 kilocalories of energy walking at an average pace for 1 minute. A 100-watt light bulb consumes 86 kilocalories of electric energy in 1 hour. Amounts of energy may also be expressed as **British thermal units** (Btu); one Btu contains the quantity of heat energy consumed in raising the temperature of 1 pound of water 1°F; 1 Btu is equivalent to 252 calories. **Power** is the rate of expending energy. The unit of power is the **watt,** which is equivalent to the expenditure of about 860 calories of energy in an hour. A person weighing about 150 pounds uses energy at the rate of about 350 watts in walking at an average pace. In running, energy is consumed at about 4 times this rate, or 1400 watts. Amounts of electric energy are ordinarily calculated by utility companies as kilowatthours, a **kilowatthour** being an amount equivalent to the energy consumed in 1 hour at the rate of 1000 watts. A person weighing about 150 pounds has to run at about 12 miles per hour for about $\frac{3}{4}$ hour to use an amount of energy equivalent to 1 kilowatthour. A 100-watt light bulb consumes 1 kilowatthour of electricity in 10 hours. One kilowatthour of electric energy is equivalent to 3410 Btu.

14-3 HOW IS ENERGY OBTAINED FROM FOSSIL FUELS?

Fossil fuels are derived from petroleum, natural gas, and coal (fossil means formed from the remains of living organisms). Energy is obtained from these fuels through **combustion,** a rapid reaction with oxygen, which generates heat and light. In the combustion of methane, CH_4, the principal component of natural gas, the exothermic reaction is easily described in terms of a balanced equation (margin).

The discussion of Sec. 6-7 related the evolution or absorption of energy in a reaction to the energy involved in breaking and forming interatomic bonds during the reaction. In applying that analysis to the combustion of methane, let us look at the bonds broken and bonds formed during the reaction.

CH_4

Methane

+

$2O_2$

Oxygen

↓

CO_2

Carbon
dioxide

+

$2H_2O$

Water

The exothermic reaction for the combustion of methane, CH_4, the principal component of natural gas

Methane;
all bonds
broken

Oxygen;
bonds
broken

Carbon
dioxide;
bonds
formed

Water;
bonds
formed

$$H-\overset{\overset{H}{|}}{\underset{\underset{H}{|}}{C}}-H + O{=}O + O{=}O \rightarrow O{=}C{=}O + H-O-H + H-O-H + energy$$

Breaking these bonds
absorbs energy

Forming these bonds evolves
energy

During the combustion of methane the energy required to break the bonds of the reactants is less than the energy generated in forming the bonds of the products. The extra energy evolved represents the energy obtainable on combustion.

Gasoline is primarily a carefully blended mixture of hydrocarbon molecules containing from 5 to 12 carbon atoms. The combustion of a typical component of a gasoline can be represented as

$$CH_3CHCH_2CH_2CH_2CH_3 + 11O_2 \rightarrow 7CO_2 + 8H_2O + energy$$
$$|$$
$$CH_3$$

2-Methylhexane, a   Oxygen   Carbon   Water
typical component   dioxide
of gasoline

The exothermic reaction for the complete combustion of a typical component of gasoline; the reaction evolves energy because the energy generated in forming the bonds of the products is greater than the energy required to break the bonds of the reactants

C

Coal
(simplified)

+

O_2

Oxygen

↓

CO_2

Carbon
dioxide

+

energy

Simplified representation of the complete combustion of coal; coal is not pure carbon but contains other substances as well; again energy is evolved because more energy is generated in forming the bonds of the product than is absorbed in breaking the bonds of the reactants

The composition of coal varies with its origin, but the typical coal may be considered to be a mixture of carbon and compounds containing carbon. **Anthracite, or hard coal,** contains a relatively low proportion of hydrogen and oxygen-containing substances. **Bituminous, or soft coal,** contains a more substantial proportion of hydrogen and oxygen-containing substances. In a simplified fashion the combustion of coal can be represented by the reaction shown, in which coal is represented as though it were pure carbon.

The combustion processes represented by the preceding reactions are all **complete combustion** reactions; i.e., all of the carbon atoms are converted into carbon dioxide molecules and all the hydrogen atoms into water molecules. In the actual combustion of these substances the products obtained depend on the conditions. Frequently the oxygen supply is limited, and **incomplete combustion** results; i.e., carbon monoxide, CO, may be formed instead of, or along with, the carbon dioxide, and other incompletely oxidized products may result:

$$CH_3CHCH_2CH_2CH_2CH_3 + 10O_2 \rightarrow 5CO_2 + 2CO + 8H_2O + energy$$
$$|$$
$$CH_3$$

2-Methylhexane, a   Oxygen   Carbon   Carbon   Water
typical component   (limited   dioxide   monoxide
of gasoline   supply)

A representative reaction for the incomplete combustion of a typical component of gasoline, which occurs when the hydrocarbon is burned in insufficient oxygen; carbon monoxide is among the products

The toxic effects of carbon monoxide are discussed briefly in Sec. 15-2A.

14-4 A POSSIBLE GREENHOUSE EFFECT FROM THE COMBUSTION OF FOSSIL FUELS

Carbon dioxide is a normal constituent of the atmosphere at the small concentration of about 0.03 percent by volume. Its presence is essential to the life processes of plants since it is one of the raw materials required by photosynthesis. As discussed in Sec. 11-3, photosynthesis is a complex multistep process, but the role of carbon dioxide is suggested by the overall reaction:

$$6CO_2 \quad + \quad 6H_2O \quad \xrightarrow{\text{solar energy}} \quad C_6H_{12}O_6 \quad + \quad 6O_2$$

| Carbon dioxide from atmosphere | Water from environment | Glucose, a carbohydrate, converted in turn into more complex carbohydrates | Oxygen, to atmosphere |

The overall reaction of photosynthesis

It is estimated that each year 60×10^9 tons of carbon dioxide are consumed by plants as they synthesize carbon-containing compounds.

Carbon dioxide is released to the atmosphere by the processes of respiration and the decay of animals and plants (Sec. 11-3). Normally the amount of carbon dioxide generated from these sources is about equal to the amount of carbon dioxide consumed in the photosynthesis process. Although the concentration of carbon dioxide remaining in the atmosphere is very small, its presence could have a significant effect on the temperature of the earth. Molecules of carbon dioxide do not block the energy of the sun from reaching the earth's surface, but they are capable of absorbing some of the heat energy radiated back from the surface. The **greenhouse effect** is the insulating action of an atmospheric substance such as carbon dioxide; like the glass of a greenhouse, it permits the energy from the sun to pass through but prevents the heat from leaving (Fig. 14-2). The temperature of the earth, however, depends upon several factors and it is difficult on the basis of present information to judge exactly what the specific effect of the carbon dioxide is. Water vapor in the atmosphere, for example, has a greenhouse effect of its own. Moreover, fluctuations in the intensity of the sun's radiation also affect the earth's temperature. In addition, tiny particles in the atmosphere can cause a turbidity which cools by preventing some of the sun's energy from reaching the earth.

FIGURE 14-2
Diagrammatic representation of the role of carbon dioxide in the greenhouse effect: (*a*) normal CO_2 content; (*b*) increased CO_2 content.

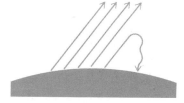 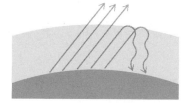

The combustion of fossil fuels releases carbon dioxide to the atmosphere in enormous amounts, about 6×10^9 tons a year. Moreover, extensive land clearing also adds to the carbon dioxide content by reducing the amount consumed by photosynthesis. The question therefore arises: How does this extra carbon dioxide affect the natural processes dependent on carbon dioxide?

Unfortunately, conclusions about the impact of the added carbon dioxide on the environment are not easy to derive from what is presently known about the various factors involved. Since 1958 the concentration of carbon dioxide in the atmosphere has been carefully monitored at a measuring station in Hawaii and has been showing a steady trend upward (Fig. 14-3). In 1958 the concentration was 314 ppm (parts per million parts). In 1976 it was 330 ppm, a 5 percent rise. It is not yet clear, however, whether this rise is part of a natural fluctuation or part of a more prolonged rising trend. It could be caused, for example, by a natural warming trend in the ocean water brought on by other forces. Such a warming trend would cause aquatic microorganisms to release added amounts of carbon dioxide.

The mean temperature of the earth's surface appears to have risen about 0.4°C between 1880 and 1940, but it seems to have declined slightly since then. The increasing turbidity of the atmosphere may have been a factor in this fluctuation. It has been suggested that the increase in the carbon dioxide concentration could have accounted for the earlier increase in the global temperature and the increased turbidity could have

FIGURE 14-3
The increase in the concentration of carbon dioxide in the atmosphere as determined at the Mauna Loa Observatory in Hawaii. Investigations are now under way to determine to what extent this represents a continuing trend and is not part of a temporary fluctuation. (*From Wil Lepkowski, Carbon Dioxide: A Problem of Producing Usable Data, Chemical and Engineering News, Oct. 17, 1977, p. 26.*)

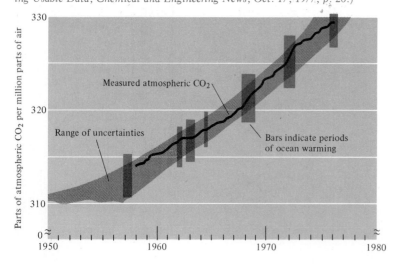

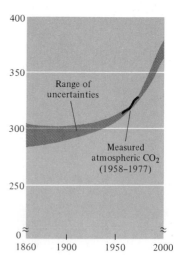

caused the subsequent decline. Research is now under way to gather additional information in an attempt to pin down the cause-and-effect relationships. If it should turn out that the carbon dioxide concentration in the atmosphere is actually increasing and that as a result the global temperature can be expected to rise, the earth could be in for some severe changes in climate. One study has projected that by 2100 the concentration of carbon dioxide may have risen to about 400 ppm and the earth's mean surface temperature may have increased by almost 9°C. This would melt much of the polar ice, causing periodic flooding of many low coastal areas, and a change in ocean currents, altering weather patterns. Considerable migration of people and disruption of agriculture could result, a possibility that has concerned many investigators. Should the combustion of fossil fuels be found to be the principal cause, we would be faced with some very difficult decisions about our sources of energy.

14-5 THE PROSPECTS OF OBTAINING GASEOUS AND LIQUID FUELS FROM COAL†

With petroleum reserves in decline more and more attention has been paid to the possibility of large-scale production of synthetic gaseous and liquid fuels from our extensive coal reserves. The chemical principles of such production have been known for some time. The production of gaseous fuel from coal for illumination and cooking was carried out at the gas works in cities throughout the Middle West and Northeast for more than a century. In later years the principal components of this gas were carbon monoxide, CO, and hydrogen, H_2. This mixture was replaced by natural gas (methane, CH_4) when pipelines to transport it made it generally available shortly after World War II.

During World War II Germany produced significant amounts of synthetic gasoline from coal; production reached about 500,000 gallons a day. South Africa has been producing synthetic gasoline from coal at a substantial rate for more than 20 years. Several processes are now under study in the United States designed to synthesize gasoline and methane on an even larger scale. A few of the more promising processes are variations of the Fischer-Tropsch synthesis, named for its originators. Coal is used in one variation of this process without first being converted into coke. It is treated at a high temperature with a mixture of steam and oxygen under pressure. Benzene and a number of other carbon-containing liquids and solid substances are obtained, along with a mixture of gases, typically hydrogen, H_2 (40 percent), carbon monoxide, CO (15 percent), methane, CH_4 (15 percent), and carbon dioxide, CO_2 (30 percent). After removal of the carbon dioxide and purification, the gas is led directly into reactors, where at elevated temperatures and pressures with the aid of a catalyst reactions

† Methods of obtaining useful chemicals from coal were also discussed in Sec. 13-10.

yield a complex mixture of hydrocarbons. The following reactions are representative of those taking place:

$$CO + 3H_2 \rightarrow CH_4 + H_2O$$

Carbon   Hydrogen   Methane   Water
monoxide

$$7CO + 15H_2 \rightarrow CH_3CH_2CH_2CH_2CH_2CH_2CH_3 + 7H_2O$$

Carbon   Hydrogen   Heptane   Water
monoxide

Typical reactions occurring during the Fischer-Tropsch process, in which gases obtained from coal react to form hydrocarbons suitable for use as gaseous and liquid fuels

The hydrocarbons synthesized by this process can serve as the basic components of gaseous and liquid fuels. The methane, CH_4, can be used as a gaseous fuel equivalent to natural gas and is curiously labeled synthetic natural gas. The liquid hydrocarbons of gasoline are readily obtained from this process, and the larger hydrocarbon molecules which are the components of jet fuel and heating oil can also be produced.

A production capacity of about 400 million gallons of synthetic fuel a year by this or similar processes is reputed to be well within the reach of current technology, but it would require a huge, expensive construction program and considerable expansion of coal production. It is hoped that further development will lower costs, since with the present technology the products would be substantially more expensive than equivalent fuels currently obtained from petroleum and natural gas.

14-6 ADVANTAGES AND DISADVANTAGES OF USING PETROLEUM AND NATURAL GAS AS SOURCES OF ENERGY

A. Advantages

Petroleum and natural gas have become popular sources of energy since the fuels derived from them are in a convenient liquid or gaseous form. The technology has been developed to make them readily available at low cost, encouraging a great expansion in the use of energy. Although in recent years petroleum and natural-gas costs have been rising, they remain well below other sources of liquid and gaseous fuels. Natural gas burns very cleanly, and oil is readily obtainable with a low sulfur content, reducing pollution by sulfur oxides.

B. Disadvantages

The principal disadvantage in the use of petroleum and natural gas is that supplies of these resources are limited and when they have been exhausted there will be no more. The known reserves of natural gas in the United States have been estimated to be sufficient for only about 11 years at the current rates of consumption. Undiscovered resources are estimated to be equivalent to about a 30 or 40 years' supply. The known reserves of petroleum in the United States (not including shale and sand oil) are equivalent to a meager 11-year supply, assuming current import and consumption rates. Undiscovered resources are estimated to be equivalent to a 32-year supply. World resources are greater but still finite (Fig. 14-20). The United States is currently importing almost half of the crude petroleum it is using and an increasing amount of natural gas. This produces a strain on the economy and subjects both the supply and prices to sudden changes.

The use of petroleum and natural gas as sources of energy competes with their use as sources of starting materials for making drugs, dyes, polymers, and a wide range of other carbon-containing products. Moreover, the use of both petroleum products and natural gas is dangerous because of their inflammable and explosive nature. Over 150 people were killed during 1975 in the United States in accidents due to the use of natural gas and petroleum-derived fuel gases. Property damage from fires and explosions due to the use of gas exceeded \$20 million. The importation of liquefied natural gas by ship necessitates special containers in which the substance is kept in a liquid state with pressure and refrigeration. Such ships involve great potential hazards. Offshore oil wells and the transport of imported crude petroleum by ship are possible sources of oil spills, which cause serious pollution of the seawater and coastal areas.

The use of gasoline in motor vehicles generates the toxic gases carbon monoxide, CO, sulfur dioxide, SO_2, nitric oxide, NO, and nitrogen dioxide, NO_2. Engine exhaust may also contain unburned hydrocarbons. The use of catalytic converters in automobiles has reduced the amounts of carbon monoxide and hydrocarbons in car exhaust (Sec. 15-3). In addition, the combustion of natural gas and petroleum fuels produces carbon dioxide, CO_2, which may possibly lead to a disruptive greenhouse effect (Sec. 14-4).

14-7 ADVANTAGES AND DISADVANTAGES OF COAL AS A SOURCE OF ENERGY

A. Advantages

The leading advantage in the use of coal as a source of energy is the abundance of resources, especially compared with those of petroleum and natural gas. The location of major deposits is given in Fig. 14-4. Known reserves economically recoverable with current mining techniques are judged to be sufficient for 300 years at the current rate of consumption and at least 200 years at the accelerated rates possible. Undiscovered resources are estimated to be equivalent to a 5000-year supply. Such abundance eliminates the need for any imports and gives the nation plenty of time to develop renewable sources of energy. Coal is a relatively low-cost fuel, and the techniques for using it are already in place. The technology for deriving more convenient (but more expensive) liquid and gaseous fuels from coal is almost fully developed.

FIGURE 14-4
The locations of the major coalfields and mining areas in the
United States. [*From John Walsh,* Problems of Expanding Coal Production,
Science: **184**:337 (*1974*). *Copyright by the American Association for the
Advancement of Science.*]

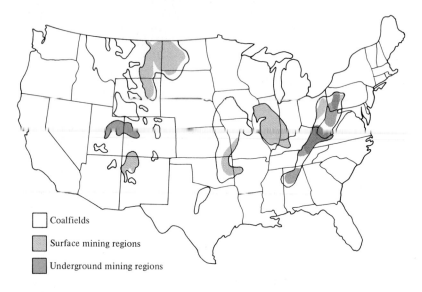

Coalfields

Surface mining regions

Underground mining regions

B. Disadvantages

Unfortunately, the use of coal has a number of disadvantages.

Expensive to transport   Coal can be transported either by railroad or by pipeline in slurry form. The railroads are the principal carriers at present; slurry pipelines have yet to be constructed. Both methods involve considerable cost and limit the distances over which coal can be transported economically (Fig. 14-5). The construction of electric-power-generating stations close to coal mines will probably become more common.

Dangerous to mine underground   About two-thirds of the 140,000 coal miners are currently working underground (Figs. 14-6 and 14-7), and the rest are operating open-pit surface mines. It has been estimated that if the optimum use is to be made of the nation's coal reserves, the mining work force will have to be doubled. With improvements in underground mining techniques and more attention to safety precautions casualties have declined markedly in the last 50 years, but cave-ins and explosions occur. An average of 1 underground miner per 1000 is killed each year, and about 1 in 10 suffers a disabling injury which requires time off for recovery. Moreover, many miners succumb to coal workers' pneumoconiosis (black lung disease). The lung disorder occurs in persons exposed to coal-mine dust and is reported to be particularly debilitating if the miner is also a cigarette smoker. A recent study reported that about 30 percent of the long-time bituminous miners are afflicted. In 1973 alone the federal government paid about $1

Coal being hauled from mines to electric power plants. The solid, bulky nature of coal makes it expensive to transport. (*Courtesy of Norfolk and Western Railway.*)

FIGURE 14-6
View in an underground coal mine. (*Courtesy of Exxon Corporation.*)

FIGURE 14-7
A personnel carrier in a Monterey Coal Company's mine in south-
ern Illinois. (*Courtesy of Exxon Corporation.*)

billion dollars to miners and their dependents in compensation
for this ailment.

Disruption of surface strip mining The aboveground mining
of coal is much less hazardous than underground mining, and
the development of new and efficient equipment has promoted
a great expansion in surface mining, or strip mining, as it is
called (Figs. 14-8 and 14-9). Unfortunately, this has led to an
offensive scarring of large areas of land. Strip mining involves
removing topsoil and soil overburden to get at the coal seams.
After the coal has been removed, vast depressions alternate
with banks of dislodged soil in a bleak landscape. Frequently
deposits of iron pyrite (iron disulfide) remain at or near the
surface. As these react with the oxygen and moisture of the
atmosphere, sulfuric acid is produced. The acid earth and lack
of topsoil discourages plant growth, and strip-mined areas may
remain barren for years.

Strip mining is often carried out on a huge scale. The power
shovels can be big enough to scoop up 200 tons at a time
and have control cabs more than 40 feet in the air. Land in the
United States is currently being strip-mined at the rate of about
1000 acres a week, and somewhere between 1 and 2 million
acres have already been stripped. While about one-third of this
area has been reclaimed by man or nature, most of it remains
in a deplorable conditon.

In 1977 Congress passed a law which requires the reclama-
tion of the strip-mined land by the miners and prohibits surface
mining unless the land can be restored to "a condition capable
of supporting the uses it was capable of supporting prior to
any mining." This will require saving the topsoil and controlling the formation of acid soils. The government has also

FIGURE 14-9
The strip mining of coal near Cadiz, Ohio. Some power shovels
are big enough to scoop up 200 tons at a time. (*Calonius,
EPA-DOCUMERICA.*)

$$S \;+\; O_2 \;\rightarrow\; SO_2$$

Sulfur,	Oxygen	Sulfur
present in	from	dioxide
coal as	air	
compounds		

$$2SO_2 \;+\; O_2 \;\rightarrow\; 2SO_3$$

Sulfur	Oxygen	Sulfur
dioxide	from	trioxide
	air	

**Reactions occurring during
the combustion of sulfur-
containing coal which
generate the toxic gases
sulfur dioxide and sulfur
trioxide**

$$SO_3 \;+\; H_2O \;\rightarrow\; H_2SO_4$$

Sulfur	Water	Sulfuric acid,
trioxide		often as a mist

**Reaction of the sulfur
trioxide from the combustion
of coal with water droplets
in the atmosphere to form
sulfuric acid, often as a mist**

$$CaCO_3 \xrightarrow{\text{heat}} CaO \;+\; CO_2$$

Calcium	Calcium	Carbon
carbonate,	oxide	dioxide
representative	(lime)	
of limestone		

$$CaO \;+\; SO_2 \;\rightarrow\; CaSO_3$$

Calcium	Sulfur	Calcium
oxide	dioxide	sulfite

$$CaO \;+\; SO_3 \;\rightarrow\; CaSO_4$$

Calcium	Sulfur	Calcium
oxide	trioxide	sulfate

**Simplified representation
of the removal of sulfur
compounds by adding
limestone during the com-
bustion of coal; these
reactions also illustrate
the chemistry of one of the
methods under develop-
ment for removing sulfur
oxides from stack gases
after coal combustion**

enacted a law to collect about $4 billion during the next 15 years from coal-mining operations to reclaim the land already laid waste. It is hoped that this will bring the formation of barren wastelands due to strip mining to a halt. The new law could prove to be particularly significant in view of the expected expansion in coal production to meet energy needs.

The formation of sulfuric acid from the sulfur compounds found in coal deposits is a problem with both strip mining and underground mining, especially in the eastern part of the United States. Surveys have led to estimates that the acidic drainage from mining operations has ruined between 6000 and 10,000 miles of streams in the United States.

Hazardous to use   One of the greatest deterrents to the use of coal is the production of the toxic gases sulfur dioxide, SO_2, and sulfur trioxide, SO_3, during combustion. They are formed from the sulfur compounds present. Much of the sulfur dioxide is subsequently converted into sulfur trioxide. Most of the sulfur trioxide, SO_3, reacts with water droplets to form sulfuric acid, H_2SO_4, which may remain suspended as a mist. The annoying and harmful smog generated in many highly industrialized areas is made up of sulfur dioxide, sulfuric acid mist, and airborne particles such as smoke and ash. From time to time it has caused acute air pollution resulting in many deaths (Sec. 15-1).

The new federal air quality and emission standards can generally be met by burning coal with 1 percent sulfur, but coal often has sulfur contents of 3 to 6 percent. It is possible, in principle, to remove the sulfur-containing compounds before, during, or after combustion. The problem has been to find ways of doing it which are low in cost and reliable. Crushing coal fine and washing it before combustion can remove up to half the sulfur compounds. Adding limestone (essentially calcium carbonate, $CaCO_3$) during combustion also has possibilities. The chemistry is complex, but in simplified terms the calcium carbonate, $CaCO_3$, decomposes, and the resulting calcium oxide, CaO (lime), combines with the oxides, SO_2 and SO_3 (margin).

At the present stage of development the removal of the sulfur oxides from the stack gases after combustion appears to provide the most practical solution. Of the various methods of accomplishing this, the use of wet lime, CaO, or limestone is one of the most thoroughly studied. A simplified view of the chemical reactions involved is given by the last two reactions shown for the use of limestone during the combustion process. For effective removal of the gaseous sulfur oxides, large volumes of a slurry of very finely ground lime or limestone are required. As much as 170 pounds of lime (or 300 pounds of limestone) is required to remove about 90 percent of the sulfur oxides produced by the combustion of 1 ton of high-sulfur (4 percent) coal. The great amount of sludge formed, containing the spent calcium compounds, presents a major disposal problem. It has been estimated that for the modern power plant of average size, sludge will form at the rate of 80,000 cubic feet per day. Other methods of removing sulfur oxides from stack gases are under development, and it is probable that several will be put in operation. The cost of the sulfur oxide removal will increase the expense of the generation of electricity by an estimated 3 to 6 percent and in some instances may be as high as 15 percent.

The stack gas from an electric power plant using a coal with 2.4 percent sulfur contains approximately 1500 ppm of sulfur oxides, 400 ppm of nitrogen oxides, and 8.5 grams of particles per cubic meter. The problems of the nitrogen oxides and particulate matter are discussed in Sec. 15-2. The possibility of a greenhouse effect from the carbon dioxide produced during coal combustion was discussed in Sec. 14-4.

$$^{1}_{0}n$$

Neutron

+

$$^{235}_{92}U$$

Uranium 235

↓

$$^{142}_{56}Ba$$

Barium 142

+

$$^{91}_{36}Kr$$

Krypton 91

+

$$3^{1}_{0}n$$

Neutrons

Typical nuclear reaction occurring during the fission of uranium 235; a large amount of energy is evolved

14-8 HOW IS ENERGY GENERATED BY NUCLEAR FISSION? †

When a few of the larger, more complex varieties of atoms are subjected to bombardment with neutrons they undergo **nuclear fission;** i.e., their nuclei are split into two nuclei of substantial size, accompanied by the evolution of very large amounts of energy. A typical reaction occurring during the fission of uranium 235 is represented by the nuclear equation in the margin. The terms **nuclear energy** and **atomic energy** are applied to the energy released during the fission of the nucleus of an atom. (The representation and naming of isotopes were discussed in Sec. 2-5, and the nature of nuclear reactions was discussed in Chap. 9.) About 2 million times more energy is obtained from the fission of 1 gram of uranium 235 than from burning 1 gram of natural gas.

A **nuclear chain reaction** is the self-sustaining sequence of nuclear fission reactions which is made possible by the ability of the neutrons produced in one nuclear fission reaction to bring about a subsequent nuclear fission. A **nuclear reactor** is an apparatus for carrying out a controlled self-sustaining nuclear chain reaction. The reactors most commonly in use are called light-water reactors or uranium 235 reactors. The hearts of these reactors are long thin tubes fueled with pellets of uranium oxide, UO_2, in which the uranium 235 content is about 3 percent. The tubes are immersed in ordinary "light" water, which serves as a neutron moderator to increase the probability of nuclear fissions. (Nuclear reactors are discussed in more detail in Sec. 9-8.)

A **breeder reactor** is a nuclear reactor which produces more nuclear fuel than it consumes. This is accomplished by the synthesis of a fissionable isotope, using the neutrons generated during a nuclear chain reaction. The neutrons produced during the fission of uranium 235, for example, can be used to produce fissionable plutonium 239 from uranium 238. Since uranium 238 is present in the fuel material of nuclear reactors along

† The generation of energy by nuclear fission has been discussed in greater detail in Secs. 9-7 to 9-9. Those who have studied those sections may go directly to Sec. 14-9 without loss of continuity.

with the uranium 235, the production of some plutonium 239 inevitably occurs.

$$^{1}_{0}n \quad + \quad ^{238}_{92}U \quad \rightarrow \quad ^{239}_{92}U$$

Neutron   Uranium 238   Uranium 239

$$^{239}_{92}U \quad \rightarrow \quad ^{239}_{93}Np \quad + \quad ^{0}_{-1}e$$

Uranium 239   Neptunium 239   Beta particle

$$^{239}_{93}Np \quad \rightarrow \quad ^{239}_{94}Pu \quad + \quad ^{0}_{-1}e$$

Neptunium 239   Plutonium 239   Beta particle

Sequence of steps in the synthesis of fissionable plutonium 239 in a breeder reactor using the neutrons of a nuclear chain reaction

A breeder reactor can also be constructed to synthesize fissionable uranium 233 by the neutron bombardment of thorium 232 in the sequence of reactions shown.

$$^{1}_{0}n \quad + \quad ^{232}_{90}Th \quad \rightarrow \quad ^{233}_{90}Th$$

Neutron   Thorium 232   Thorium 233

$$^{233}_{90}Th \quad \rightarrow \quad ^{233}_{91}Pa \quad + \quad ^{0}_{-1}e$$

Thorium 233   Protactinium 233   Beta particle

$$^{233}_{91}Pa \quad \rightarrow \quad ^{233}_{92}U \quad + \quad ^{0}_{-1}e$$

Protactinium 233   Uranium 233   Beta particle

Sequence of steps in the synthesis of fissionable uranium 233 in a breeder reactor using the neutrons of a nuclear chain reaction

For a nuclear reactor to function as a breeder reactor the production of either plutonium 239 from uranium 238 or uranium 233 from thorium 232 must exceed the consumption of uranium 235 (or other fissionable material) during the nuclear chain reaction. This is possible because on the average more than three neutrons are produced in the fission of an atom of uranium 235. Only one of these is required to carry on the subsequent fission of uranium 235 in the nuclear chain reaction. The extra neutrons are available for the production of plutonium 239 from uranium 238 or uranium 233 from thorium 232. If the loss of neutrons can be controlled, new fissionable isotopes can be produced faster than the uranium 235 is consumed. Breeder reactors could make use of both uranium 238 and thorium 232, supplies of which are substantial. Moreover, by synthesizing more nuclear fuel than they consume, breeder reactors could, in principle, provide the means of generating an unlimited supply of energy.

As a controlled nuclear chain reaction takes place in a nuclear reactor, considerable heat is generated. Provision must be made to remove this heat by circulating a coolant through the reactor. When nuclear reactors are used as a source of energy to generate electricity, the heat removed by the coolant is typically used to produce steam, which in turn is used to generate electricity by driving a conventional turbine. The principal parts of one type of nuclear power plant were represented diagrammatically in Fig. 9-7.

14-9 ADVANTAGES AND DISADVANTAGES
OF USING NUCLEAR FISSION AS A SOURCE OF
ENERGY

429

14-9 ADVANTAGES AND
DISADVANTAGES OF USING
NUCLEAR FISSION AS A SOURCE OF
ENERGY

A. Advantages

One of the advantages of nuclear energy is that it replaces fossil fuels, allowing them to be saved for their unique use as a source of essential starting substances in the production of drugs, dyes, polymers, and a wide range of other carbon-containing products. Moreover, nuclear power plants do not pollute the air with toxic sulfur oxides, nitrogen oxides, and particles. Further, they do not generate carbon dioxide, avoiding a possible greenhouse effect.

The use of nuclear energy also avoids the need of imported energy sources. The use of breeder reactors, when fully developed, could be an almost unlimited energy source, avoiding the problems of mining coal and obtaining crude petroleum. It would also head off any possible future shortages in the uranium 235 required by nonbreeder reactors. Nuclear reactors, moreover, use a concentrated fuel. For example, 1 kilogram of plutonium 239 can produce as much energy as about 3 million kilograms of coal. Among other things, transportation costs are reduced.

B. Disadvantages

The most controversial aspect of the use of nuclear energy is its safety. To clarify uncertainties about previous evaluations of nuclear-energy safety the Atomic Energy Commission, an agency of the government formerly in charge of the development of nuclear energy, appointed a special committee a few years ago to make a thorough study of nuclear-reactor safety. The study took into consideration all aspects of nuclear-reactor operation, including the probabilities and consequences of catastrophic accidents. Its 1975 report is highly detailed and not readily reduced to a brief summary, but the overall conclusion was that nuclear-reactor risks are low and the probability of major accidents is very small. A more recent study of the feasibility of nuclear energy sponsored by the Ford Foundation has challenged some details of the report. Although this study found the health and accident hazards of nuclear energy to be about in the range of, or less than, the effects of generating electricity with coal, it considered the Atomic Energy Commission report to have underestimated the accident hazards associated with nuclear reactors.

Although the safety record of the 64 commercial nuclear power plants has been good, the chances of major accidents due to technological failures, earthquakes, and other natural disasters are considered by some authorities to be a possible hazard, despite the many safeguards. Any large release of radioactive substances to the environment could be extremely hazardous. Plutonium 239 retains its radioactivity for thousands of years and could cause contamination of an area for a very long time in the event of a major accident. Failures of reactor cooling systems are a matter of concern since the radioactivity of fission products in a reactor produces appreciable heat even after it is shut down. An extremely high order of construction standards and operating skill is mandatory.

Accumulation of fission products in reactors necessitates periodic removal and reprocessing of radioactive products to recover fuels. Waste products contain long-lived radioactive substances which must be stored safely over a very long period of time. The problems of storage have not yet been resolved to the satisfaction of everyone (Sec. 9-8). Further, radioactive substances must be shipped from power plants to reprocessing stations, giving rise to the possibility of shipping accidents or diversion into unauthorized hands. Some people believe that power plants and fuel reprocessing facilities are both vulnerable to sabotage. Plutonium 239 is worth about $10,000 a kilogram, and a considerable amount will be used in a plutonium fast breeder reactor. The construction details of crude nuclear bombs are generally known, although their construction from stolen plutonium is reported to be difficult and extremely dangerous. Moreover, nuclear bomb materials could also be obtained from military sources not connected with power plants.

Another disadvantage of nuclear energy is that the uranium 235 light-water reactors now in use release more waste heat than modern plants using fossil fuels. Further, unlike the breeder reactors, the fuel supply for the uranium 235 reactors is not unlimited.

^2_1H

Deuterium

+

^3_1H

Tritium

$\downarrow$

^4_2He

Helium

+

^1_0n

Neutron

An example of a nuclear fusion reaction in which the nuclei of two atoms combine to form a somewhat more complex nucleus of higher atomic number; a large amount of energy is evolved

14-10 ADVANTAGES OF CONTROLLED NUCLEAR FUSION

Nuclear fusion is a type of nuclear change in which the nuclei of two atoms combine to form a somewhat more complex nucleus of higher atomic number. One such reaction is the union of a deuterium nucleus, ^2_1H, and a tritium nucleus, ^3_1H, to form a helium nucleus, ^4_2He, accompanied by the evolution of a large amount of energy. Extremely high temperatures, of the order of 100 million degrees, are required for such a fusion reaction to occur. (Fusion nuclear reactions are discussed in Secs. 9-6, 9-11, and 9-12.) Fusion reactions are the basis for the energy of the sun and other stars, as well as for the nuclear fusion (hydrogen) bombs. Unfortunately, the means of carrying out fusion reactions in a controlled manner which would generate a sustained source of energy have not yet been developed, despite considerable effort. Although encouraging discoveries have been made, some authorities have suggested that at least another 50 years of investigation will be required before a practical process is developed.

One of the potential advantages of nuclear fusion as a source of energy is found in the amount of energy obtainable from a given amount of fuel: it is about 4 times greater than the huge amount produced from nuclear fission. Further, there is a good possibility that the radiation and pollution effects will be very low. Most important, since deuterium can be obtained from seawater and tritium can be made from deuterium, the sources of fuel supplies are virtually unlimited. Controlled nuclear fusion, if perfected, could be a major source of energy in the future.

14-11 HOW CAN SUNLIGHT BE
USED DIRECTLY AS AN ENERGY SOURCE?

431

14-11 HOW CAN SUNLIGHT BE
USED DIRECTLY AS AN ENERGY
SOURCE?

A. Heating and Cooling Buildings

The heart of a solar heating system is the unit collecting the sun's energy. It is typically a large shallow tank with a black metal base to absorb the energy. It is covered with transparent glass, which permits the sunlight to enter the tank but prevents most of the absorbed heat from escaping, creating a greenhouse effect. The heat is collected by repeatedly circulating water through the tank. The circulation of the water heated to a temperature close to its boiling point through a building is similar to that in present hot-water heating systems, using an insulated hot-water reservoir, which serves as storage for use at night and at other times when the sun is not shining (Figs. 14-10 and 14-11).

Solar energy for air conditioning can be used, for example, by utilizing the heat from the sun to vaporize ammonia, NH_3, from its aqueous solution. The ammonia is collected and condensed into liquid ammonia. The subsequent expansion of the ammonia to the gaseous state produces a cooling effect which is used as a basis for an air conditioner. The gaseous ammonia is redissolved in water for repeated use.

B. The Generation of Electricity at Central Power Stations

The solar-energy collecting unit of a large-scale central power station could be an array of glass-covered shallow tanks similar to those used in individual building units. Or, as now favored, it could be a field of **heliostats,** sun-following mirrors which would reflect sunlight toward a central unit on a tower above

FIGURE 14-10
Diagram of a system for heating a house with solar energy.

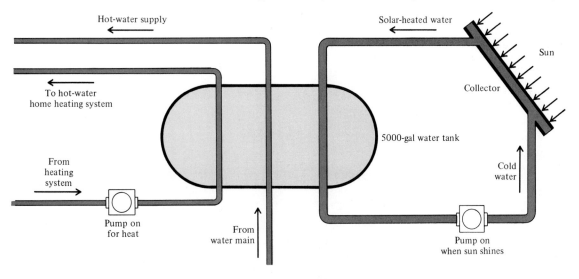

Hot-water supply

Solar-heated water

Sun

To hot-water
home heating system

Collector

5000-gal water tank

From
heating
system

Cold
water

Pump on
for heat

From
water main

Pump on
when sun shines

FIGURE 14-11
House in West Virginia equipped to use solar energy for hot water
and space heating. (*Courtesy of Grumman Energy Systems, Inc.*)

them (Fig. 14-12). The collected energy could then be used to
form steam from water. The steam, in turn, would drive conventional steam turbines like those which generate electricity
in fuel-burning power plants. An experimental plant using
such an approach is under construction in California's Mohave

FIGURE 14-12
Sketch of an experimental central electric generating plant operated by solar energy. A field of sun-following mirrors is focused
on a steam-generating facility on top of a tower. (*Courtesy of
Department of Energy.*)

Desert (Sec. 14-17). The Department of Energy foresees many such central generating stations in operation in about 25 years. At the present state of technical development electric energy generated by such a method costs somewhere between 4 and 8 times as much as electric energy generated by fossil fuel or nuclear plants.

C. The Direct Conversion of Solar Energy into Electric Energy

Specially treated silicon and certain other materials can be arranged so that an electric current is generated by the action of light. A **photon** is a pulse of light energy. When photons from the sun with the appropriate amount of energy impinge on the specially treated silicon, they dislodge electrons, which can be collected to generate a flow of electrons, in other words, an electric current (Sec. 2-3). A **solar cell** or **photovoltaic cell** is a device which converts light energy directly into an electric current (Fig. 14-13). Solar cells have provided a reliable source of electricity in many space vehicles, manned and unmanned (Fig. 14-14).

14-12 ADVANTAGES AND DISADVANTAGES OF THE DIRECT USE OF SOLAR ENERGY

A. Advantages

The overwhelming advantage of solar energy is its huge and unlimited supply. The sunlight shining on less than 1 percent

FIGURE 14-13
An array of solar cells at an experimental irrigation project near Mead, Nebraska. The cells convert solar energy directly into electricity, which is used to drive the motors of the water pumps. *(Courtesy of Department of Energy.)*

FIGURE 14-14
Arrays of solar cells extending from a spacecraft will supply electric energy for its operations. The craft, a Landsat-C, contains special cameras to measure natural resources like water, arable land, and mineral deposits both in daylight and in total darkness. (*Courtesy of NASA.*)

of the land area of the United States could supply more than the nation's estimated energy needs for many years to come. Furthermore, solar energy is relatively free from environmental pollution; it avoids the problems of mining coal, importing and otherwise transporting crude oil, the pollution caused by the combustion of fossil fuels, and the disadvantages of nuclear energy. Although central power stations are only now entering the developmental stage, direct solar energy is already available for individual use in building heating.

The development of photovoltaic cells could provide a means of converting solar energy directly into electric energy, a very convenient and useful form of energy. It has been proposed that photovoltaic cells in orbiting space stations could avoid the intermittent availability of sunlight which impairs other methods of using solar energy. The orbiting stations could conceivably transmit the energy to the earth if the technology could be developed.

435

14-13 HOW CAN THE INDIRECT EFFECTS OF THE SUN'S ENERGY BE USED AS SOURCES OF ENERGY?

B. Disadvantages

The sunlight reaching any given area of the earth is of low intensity. It has been estimated that a central power station with a capacity equivalent to a modern fossil fuel plant of average size would require from 7 to 10 square miles of solar collectors. Sunlight also is intermittent, requiring collected energy to be stored. Adequate technology for doing this on a large scale remains to be developed. Moreover, the amount of solar energy reaching the United States varies markedly with the region. The Southwest, for example, receives almost twice as much as the Northeast.

Photovoltaic cells could be an extremely attractive method of trapping solar energy, but much further development is required. The solar cells in current use are inefficient and have very limited lifetimes. Furthermore, they cost about 40 times more than the equivalent electrical generating capacity of a fuel-burning power station.

14-13 HOW CAN THE INDIRECT EFFECTS OF THE SUN'S ENERGY BE USED AS SOURCES OF ENERGY?

A. Water Power

The flow of water from the mountains to the sea is an indirect result of the action of the sun. Solar energy causes the evaporation of water from the ocean and land areas. The water vapor is ultimately returned to the earth as rain, snow, or sleet. Much of this precipitation ends up in streams and rivers, the power of which has been used as a source of energy since ancient times. In recent years large dams have been built to supply the power for the generation of electricity at relatively modest cost. About 12 percent of the electric energy currently produced in the United States (4 percent of the total energy supply) comes from water power. It may be expected that this source will be further developed, but it has been estimated that the ultimate maximum water-power capacity of stream flow in the nation is only about 3 times the current level of production.

Hydroelectric power generation is pollution-free and relatively safe once construction has been completed. Moreover, there is no need for a continuous supply of fuel. The formation of lakes, however, destroys agricultural lands and wilderness areas, and the failure of a major dam can lead to catastrophe.

B. Wind Power

The winds, like the stream flow of water, are indirect results of solar energy. Windmills for the generation of energy have

FIGURE 14-15
An experimental windmill
used in the Wind Energy
Program. The blades span
125 feet, and the windmill is
designed to generate 100
kilowatts of electricity.
(*Courtesy of NASA.*)

been used for centuries and were common in the rural landscape
in the United States until the electric lines of central power
plants began to reach the farm areas in the 1930s. Growing
concern about energy supplies and environmental considerations have reawakened interest in wind power as a means of
generating electricity.

Experimental wind generators (Fig. 14-15) are now under
study. One test unit uses propellors with blades spanning
125 feet mounted on a tower 125 feet tall. It has been estimated,
however, that in order to generate an amount of energy equivalent to a modern central generating plant, about 10,000 such
installations would be required. The intermittent nature of the
wind presents special problems of energy storage for times
when no wind is blowing. Windmill construction to withstand
sudden gusts, vibration, and in some areas ice formation are
among the other problems. Although windmills may be expected to find practical use in the future, they seem to be most
readily adaptable to small installations of supplemental energy
sources. One estimate, however, anticipates windmills supplying from 5 to 10 percent of the electric-energy needs of the
nation by the year 2000.

C. Biomass Sources

The **biomass** is the total amount of mass in all living organisms, plants and animals. Plants, in particular, may serve
as a source of energy. One of the most important ongoing
processes making use of solar energy takes place in plants,
ranging from microscopically small algae and bacteria of lakes
and oceans to the trees and other plants of the land areas.
Photosynthesis is the process whereby solar energy is converted into chemical energy as these plants synthesize the
carbohydrate glucose from carbon dioxide and water. This
conversion has been discussed in Sec. 11-3. It is sufficient
here to describe the overall reaction (the actual process involves
an intricate sequence of many reactions):

$$6CO_2 \quad + \quad 6H_2O \quad + \quad \text{solar energy} \quad \rightarrow \quad C_6H_{12}O_6 \quad + \quad 6O_2$$

| Carbon dioxide from atmosphere | Water from environment | | Glucose, a carbohydrate, which is converted into more complex carbohydrates | Oxygen, to atmosphere |

**The overall reaction of photosynthesis; the actual process involves
an intricate sequence of many reactions; the solar energy absorbed is
converted into chemical energy of the glucose molecules**

The glucose molecules formed by photosynthesis are combined in the plants to form the larger molecules of the carbohydrates starch and cellulose, among other substances. Starch
and cellulose are polymers of glucose and can be represented
by the general formula $(C_6H_{10}O_5)_n$. Starch and many of the other
products of photosynthesis are important as foodstuffs.
Cellulose is an important structural material of plants; it
accounts, for example, for about half the solid material in most
forms of wood.

Coal, petroleum, natural gas, and other fossil fuels are the remains of photosynthesis processes which took place many millions of years ago, and photosynthesis products of currently growing plants also have great potential as sources of energy. Indeed, the combustion of wood was a very important source of energy up until the end of the nineteenth century, when it was replaced by coal (Table 14-4).

The direct combustion of wood is one way of obtaining energy from the products of plant life. It turns out, however, that wood is not a very good fuel. There are more efficient ways of obtaining energy from the products of photosynthesis. Wood, however, can be used as a source of methyl alcohol, CH_3OH, also called wood alcohol. The **destructive distillation of wood** is the process of heating wood in the absence of air. Methyl alcohol is the principal component of the liquid products obtained. It can be used as a fuel in the pure form or blended into gasoline.

Another way of obtaining energy from the products of photosynthesis is to use sugar cane as a source of ethyl alcohol, CH_3CH_2OH. Sugar cane contains about 15 to 20 percent by weight of the sugar sucrose, $C_{12}H_{22}O_{11}$. Ethyl alcohol can be obtained from sucrose by fermentation brought about by the enzymes of yeasts. The sucrose is hydrolyzed into two simpler sugars, glucose, $C_6H_{12}O_6$, and fructose, $C_6H_{12}O_6$:

$$C_{12}H_{22}O_{11} + H_2O \xrightarrow{\text{enzyme}} C_6H_{12}O_6 + C_6H_{12}O_6$$

| Sucrose | Water | Glucose | Fructose |

The hydrolysis of sucrose into two simpler sugars before fermentation

The glucose and fructose then undergo fermentation to ethyl alcohol and carbon dioxide:

$$C_6H_{12}O_6 \xrightarrow{\text{enzyme}} 2CH_3CH_2OH + 2CO_2$$

Glucose or fructose   Ethyl alcohol   Carbon dioxide

Fermentation of glucose or fructose to ethyl alcohol and carbon dioxide

One use of ethyl alcohol is as an automobile fuel, either by itself or blended with gasoline. Brazil, with its large capacity for growing sugar cane, has launched a bold program to replace 20 percent of its gasoline consumption with ethyl alcohol by 1980.

Some plants are capable of synthesizing hydrocarbons. The rubber tree, for example, produces natural rubber, a hydrocarbon polymer. Attention has recently been called by the chemist Melvin Calvin, an expert on photosynthesis at the University of California at Berkeley, to shrubs which produce hydrocarbons similar to those predominant in petroleum. These shrubs, members of the genus *Euphorbia*, synthesize significant amounts of a hydrocarbon mixture which could probably be processed in existing petroleum refineries to produce gasoline and other hydrocarbon products. Experimental cultivation of these and related plants is now under way in an

437

14-13 HOW CAN THE INDIRECT EFFECTS OF THE SUN'S ENERGY BE USED AS SOURCES OF ENERGY?

$$\left[\begin{array}{c} CH_3 \\ | \\ -CH_2-C{=}CH-CH_2- \end{array} \right]_x$$

General formula for natural rubber produced by the rubber tree; x varies over a wide range in the component molecules, from about 1500 to 40,000 and beyond attempt to learn more about the practical aspects of using them as a source of fuel. Preliminary estimates suggest, however, that an area about the size of the state of Arizona would probably be required to a grow a crop sufficient to meet the current consumption of gasoline in the United States.

D. Waste Material as an Energy Source

Cellulose is the primary component of the solid carbon-containing waste materials which the United States generates each year. These have long been under consideration as possible sources of energy. Such a use would not only provide energy which is now being discarded but would also serve to dispose of some of the 2 billion tons of organic wastes produced each year. These waste products come primarily from agricultural crops, urban refuse, manure, wood processing, sewage, and industrial production. Unfortunately, only about 20 percent is readily collectable for processing.

There are three major methods of converting organic wastes into usable synthetic fuels. One process under development starts with urban refuse of all kinds. After shredding and sorting, the organic wastes are heated to 500°C in the absence of air (destructive distillation), and the gas and distillable liquids are collected. A ton of dry typical refuse yields about 42 gallons of an oil similar to heating oil, 140 pounds of iron, 120 pounds of glass, 160 pounds of a charry residue, and some combustible gas. In a second process organic waste is heated under pressure at 240 to 380°C in the presence of carbon monoxide and steam. The carbon monoxide is produced by partially burning some of the oil obtained in a previous run. It reacts with the steam to form hydrogen, the agent which converts the cellulose, $(C_6H_{10}O_5)_n$, and similar substances, into petroleumlike compounds containing less oxygen. Each ton of dry waste yields up to 50 gallons of petroleumlike heating oil. A third method of processing organic wastes involves the production of methane, CH_4, through the action of bacteria.

All three processes are now under development. Estimates indicate that if all the readily collectable organic wastes were processed into petroleumlike heating oil, the yield would be equivalent to about 3 percent of the annual consumption of petroleum. If the wastes were converted into methane, CH_4, the yield would be equivalent to about 5 or 6 percent of our annual consumption of natural gas. These yields would be useful supplements to our energy sources, but more importantly, the processes could be an excellent way of disposing of significant amounts of organic wastes.

CO

Carbon
monoxide

+

H_2O

Steam

↓

CO_2

Carbon
dioxide

+

H_2

Hydrogen

Reaction for generating hydrogen in one process for obtaining hydrocarbonlike fuels from organic wastes

E. Ocean Thermal Power Plants

This approach to using the energy of the sun would take advantage of the temperature difference between the sun-heated warm surface ocean water in tropical and semitropical regions and the cold water at depths of 1000 feet or more. Such temperature differences are in the range of 15 to 20°C. In one plan pressurized ammonia gas would be liquefied by the cold water from deep below the surface. The liquid ammonia would be evaporated by the warm surface water and used to drive a turbine to generate electricity. Such a power plant would probably be very large, one plan calling for 235,000 tons of concrete and 26,000 tons of steel in a floating platform the size of a supertanker (490 by 200 feet) (Fig. 14-16).

One of the advantages of such an energy source if it could be developed would be its continuous operation, since it would not be affected by the intermittent nature of sunshine. This would eliminate the necessity for energy storage. The technology of such a scheme is incomplete, however, and although costs are still uncertain, they appear to be high. The environmental impact of such offshore stations would be something with which to reckon, and there are also the problems of saltwater corrosion and transporting the energy back to shore.

439

14-13 HOW CAN THE INDIRECT EFFECTS OF THE SUN'S ENERGY BE USED AS SOURCES OF ENERGY?

FIGURE 14-16
Sketch of a proposed plant for generating electric energy using the temperature difference between surface and deep ocean waters. Pressurized ammonia gas is liquefied by the cool water from a depth of about 1000 feet and vaporized by the warmer water at the surface. The platform is about 200 feet wide and 490 feet long. (*Courtesy of the Applied Physics Laboratory of the The Johns Hopkins University.*)

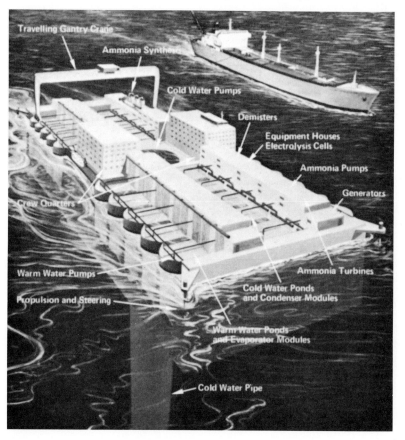

14-14 ENERGY FROM GEOTHERMAL SOURCES

Natural steam fields are being used in several parts of the world to generate electricity, and underground hot water is used in many countries to heat buildings. Considerable attention has recently been focused on the power stations using underground steam installed in New Zealand, Italy, and California (Fig. 14-17). Efforts are now under way to make further use of this special source of energy in those parts of the world where the resources are available and to improve the underlying technology, which is now at an early stage of development.

Geothermal energy is the energy derived from within the earth, mostly from underground sources of steam and hot water. While hot springs are found throughout the world, areas in which the temperatures are high enough for practical use are generally confined to regions of recent volcanic action or other activity in the earth's crust. Where geological conditions permit, cold surface water finds its way to hot rocks many miles below. The heated water then moves up toward the surface, dissolving chemicals and minerals along the way and following routes of least resistance. Cap-rock formations, supplemented by mineral deposits from the hot water itself, serve to prevent rapid escape of the water and the steam formed from it. Wells are dug, typically about a mile deep, to tap the

FIGURE 14-17
Part of the geothermal steam field at The Geysers, 90 miles northeast of San Francisco, where Pacific Gas and Electric Company generates 502,000 kilowatts of electricity from underground steam. On top of the hill are units which went into operation late in 1972. In the lower left are units which became operational in 1967 and 1968. Steam at the right and center comes from active geothermal wells. (*Courtesy of Pacific Gas and Electric Company.*)

supply of steam and hot water (Fig. 14-18). The steam is then used to generate electricity with the use of a conventional steam turbine, and the hot water is piped to nearby homes and other buildings for space heating. All the space-heating requirements in Reykjavik, the capital of Iceland, are met by a geothermal hot-water system.

The power installations are presently few. The total operating capacity throughout the world is about equal to that of one modern fossil-fuel power station, and about half this is at one geothermal location, The Geysers, in northern California (Fig. 14-19). While no fuel is required once the power station is in operation, the installation costs are much higher than those of conventional power plants.

Geothermal energy is not as trouble-free as one might imagine. An electrical generating station discharges huge amounts of hot water, which could cause harmful thermal pollution of local waterways. Underground hot water dissolves many substances and becomes an aqueous solution with an appreciable concentration of many salts. These form deposits, especially of calcium carbonate, $CaCO_3$, and silica, SiO_2, in conducting pipes and surface equipment; they also present serious corrosion problems. Hydrogen sulfide, H_2S, a foul-smelling, very toxic gas, often accompanies the geothermal water and steam, and amounts of troublesome compounds of boron, arsenic, and mercury may cause water-disposal problems. Moreover, geothermal production wells remove huge quantities of underground water, and this has led in some cases to local earth tremors and appreciable lowering of the ground level.

FIGURE 14-18
A geothermal field. Cold surface water penetrates down to hot rocks. The heated water rises through routes with the least resistance and reaches the top either as a hot spring or as a steam outlet called a *fumarole*. Wells are typically 1 mile deep.

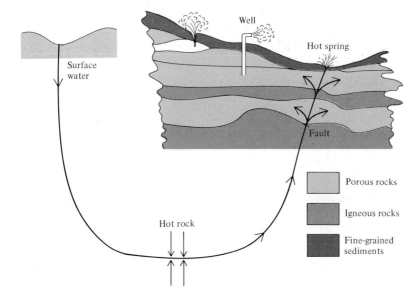

FIGURE 14-19
Electric-power-generating units at The Geysers geothermal power plant in northern California. The pipes in the foreground lead the steam from the wells to the generating plant. The steam on the extreme left comes from a natural surface steam vent. *(Courtesy of Pacific Gas and Electric Company.)*

14-15 ENERGY FROM OCEAN TIDES

Dams and waterfalls are not the only means by which the flow of water can be used as a source of energy. Tidal currents, arising from gravitational forces of the earth, moon, and sun can also be harnessed. The most favorable locations are V-shaped inlets where the tides are high and the volume of water flow very large. The world's first tidal-power dam, 7 miles long, was completed in 1966 across the Rance River on the French coast of the English Channel. The tides here average about 40 feet, and the turbine generators are designed to operate equally well whether the tide is rising or falling. The maximum power output is equivalent to about one-fourth the generating capacity of a modern conventional power plant. A similar power project to be located at Passamaquoddy Bay, on the boundary between Maine and Canada, has been under consideration for many years. The harnessing of tides could provide significant supplemental sources of energy, but the combined output is not expected to exceed a small fraction of the total energy needs.

The advantages and disadvantages of various energy sources are summarized in Table 14-2.

Table 14-2
Advantages and Disadvantages of Principal Present and Potential Sources of Energy

	Disadvantages	Advantages
Fossil fuels		
Natural gas, petroleum	Resources decreasing, currently importing about half crude petroleum and increasing amount of natural gas; consume current sources of raw materials of drugs, dyes, polymers, etc.; fire and explosion hazards; liquefied natural gas especially hazardous; tanker oil spills in waterways; offshore oilwell pollution; combustion produces carbon dioxide, potential greenhouse effect; use in motor vehicles produces toxic air pollutants	Products readily available and usable with existing technology; although costs increasing, still relatively low; natural gas burns cleanly; low-sulfur oil available for electric-energy generation and space heating
Coal	Hazardous to mine underground; strip surface mining disruptive; acid drainage polluting; combustion produces toxic sulfur oxides, nitrogen oxides, and particulate matter (removal of sulfur oxides under development); combustion produces carbon dioxide, potential greenhouse effect; expensive to transport; bulky solid nature inconvenient to handle; competes with source of raw materials for production of carbon-containing products	Abundant supplies in United States could last 200–300 years or more; no necessity for imports; costs relatively low; usable with existing technology; probable large-scale conversion to more convenient gas and liquid fuels–technology under development
Nuclear		
Uranium 235 reactor, breeder reactor	Long-range storage of radioactive waste products not adequately solved; major accidents with lethal radiation a possible hazard; shipping radioactive wastes hazardous; power plants and fuel reprocessing facilities vulnerable to sabotage; plutonium 239 possible source of terrorist bombs; fuel resources of uranium 235 ample but not unlimited; breeder-reactor technology not completely developed; uranium 235 reactors release abnormally high amount of waste heat	Replaces fossil fuels, conserving them for unique use in making drugs, dyes, polymers, etc.; no air pollution with oxides of sulfur, nitrogen, and carbon; avoids possible greenhouse effect; breeder-reactor fuel resources practically unlimited; reduces necessity of imports; breeder reactor avoids human hazards and land disruption of coal mining; fuel bulk extremely small; Ford Foundation study judges nuclear power using uranium 235 reactors to be about as hazardous as use of coal, or less so

(Continued on next page)

	Disadvantages	Advantages

Nuclear (*Continued*)

	Disadvantages	Advantages
Fusion	Technology not yet been developed and may not be for 50 years, if ever; many factors unknown	Unlimited supplies of fuels from seawater; radiation and pollution possibly low but largely undetermined

Direct solar

	Disadvantages	Advantages
Space heating and cooling, central power stations, and photovoltaic cells	Sunlight of low intensity; moderate-sized central power station could require from 7 to 10 square miles of collectors; sunlight intermittent; adequate storage technology not yet developed; regional variation in sunlight intensity (Southwest has greatest potential); technology for central power station or other large-scale use remains to be developed; costs for space heating and central power stations estimated to be from 2 to 4 times higher than present fossil-fuel sources; present photovoltaic cells inefficient, expensive, and of limited life	Unlimited supply of energy; free from environmental pollution; avoids necessity for imports; avoids problems of mining and transporting coal or drilling for and transporting oil; readily adaptable for industrial, small-scale use; technology for small-scale use already developed; photovoltaic cells in orbiting space stations could avoid intermittent availability of sunlight

Indirect solar

	Disadvantages	Advantages
Water power	Formation of lakes removes agricultural lands and wilderness areas; maximum potential for energy generation limited by stream flow; construction costs high; major dam failure could lead to major accident	Pollution-free; relatively safe once construction has been completed; dams can make land irrigation possible; operating costs low
Wind	Wind intermittent; adequate storage technology not yet developed; wind power has a low intensity; medium-sized central power station would require something like 10,000 towers 125 feet tall; windmill installation must withstand sudden gusts, vibration, and (in some areas) ice accumulation; technology not yet perfected; costs uncertain	Unlimited resources; pollution-free
Biomass	Competes for land with food production; large areas of land involved; large amounts of nitrogen fertilizer required for some crops; costs not yet competitive with fossil sources	Renewable; does not add to accumulation of carbon dioxide; potential production greatest in areas needing energy; possibility of producing convenient liquid fuels; absorbs solar energy in form which can be used continuously

Table 14-2
Advantages and Disadvantages of Principal Present and Potential Sources of Energy (*Continued*)

	Disadvantages	Advantages
Indirect solar (*Continued*)		
Waste	Capacity limited by available wastes; many waste products difficult to collect and sort; technology not yet perfected	Excellent method of waste disposal; potential source of convenient liquid and gaseous fuels
Ocean thermal	Technology not yet developed; very large off-shore plants required; corrosion problems at sea; necessity of transporting energy to shore; costs uncertain	Resources virtually unlimited; could operate continuously; no problem of energy storage; probably nonpolluting
Other		
Geothermal	Limited to regions of recent volcanic action or other activity in the earth's crust; discharge of huge amounts of hot water containing some harmful substances and polluting deposits; may cause lowering of ground level and earth tremors; equipment corrosion problems may be severe; resource capacity uncertain	Relatively clean; few hazards; technology incomplete but developing rapidly; costs at reasonably competitive level
Tidal	Limited to coastal regions with high tides and favorable geography; construction costs high	Technology developed and operating; pollution-free; operating costs low

14-16 HOW IS ENERGY USED IN THE UNITED STATES?

The United States is consuming enormous amounts of energy. In 1973 we used about 850 million tons of crude oil, about 460 million tons of natural gas, and about 540 million tons of coal. Refined oil products are being consumed at the rate of about 1100 gallons a year by each person. These quantities represent about 30 percent of all the crude oil consumed by the world as a whole, almost 50 percent of the natural gas, and about 17 percent of the coal.

The crude oil is used principally for motor gasoline (about 40 percent), heating oil (about 33 percent), jet-airplane fuel (about 7 percent), and lubricants (about 5 percent). (About 6 percent is used for industrial chemical synthesis and about 3 percent for asphalt.) The natural gas is used principally for residential heating and cooking (about 33 percent), electric-power generation (about 32 percent), commercial building heating (about 10 percent), industrial chemical synthesis, and metallurgical processes. The coal is used primarily for electric-energy generation (about 68 percent), production of coke and chemicals, mining, steel production, and manufacturing.

Table 14-3
Uses of energy in
the United States (1970)†

Classification	Percentage of total energy consumption
Transportation	24.7
Space heating (homes and commercial buildings)	17.7
Industrial:	
Process steam	16.4
Direct heat	11.0
Electric drive	8.1
Raw materials	5.6
Water heating	4.0
Air conditioning	2.9
Refrigeration	2.3
Cooking	1.2
Industrial electrolytic processes	1.2
Lighting, small appliances, and other uses	4.9

† Adapted from Eric Hirst and
John C. Moyers, Efficiency of En-
ergy Use in the United States,
Science, 179:1299 (1973).
Copyright 1973 by the American
Association for the Advancement
of Science.

The various end uses of energy in the United States are listed
in Table 14-3. Industrial production and other commercial activities consume about 40 percent of the total energy used. About
25 percent is used in the transportation of people and goods, the
largest single use. The space heating of homes and commercial
buildings accounts for another major portion of 18 percent.

14-17 PRESENT AND POTENTIAL FUTURE SOURCES OF ENERGY IN THE UNITED STATES

In 1800 about 99 percent of all the energy consumed in the
United States came from burning wood (Table 14-4), but 100
years later the combustion of coal was supplying about 70 percent of our energy needs. Although petroleum and natural gas
did not become significant sources of energy until about 1925,
the convenience and low cost of the products of these sources
led to a rapid expansion of their use. Today their combustion
accounts for about 75 percent of all of the energy generated.
Meanwhile, the contribution from coal has shrunk to about 18
percent. About 93 percent of our current energy supply is derived from fossil fuels (Table 14-5).

The great dependence on nonrenewable fossil fuels is a
matter of general concern. The problem is particularly acute in
regard to natural gas and petroleum. About 75 percent of our
current energy supplies depend on these sources, yet both
domestic and worldwide sources are limited (Table 14-5 and
Fig. 14-20). While estimations of reserves and undiscovered
resources are subject to considerable error, there is general
agreement that at current rates of consumption worldwide
sources will be near exhaustion in another 100 or 150 years.
Without the help of imports the sources in the United States
could become scarce in only about 40 years. The nation is
already importing almost half of the crude petroleum it is using,
and the imports of natural gas are growing.

Despite the great need for changing the patterns of energy
production, it will not be a simple matter. We have in place a
complex array of installations for obtaining, refining, and dis-
tributing the refined products of petroleum and natural gas:
wells (Fig. 14-21), refineries, pipelines and other transportation

Table 14-4
Historical Energy-Consumption Patterns in the United States

	Fuel as Percent of Total					
Year	Wood	Coal	Petroleum and natural gas	Water power	Nuclear	Total, 10^{16} kilocalories
1800	99	<1	0	<1	0	0.015
1860	84	16	<1	<1	0	0.078
1880	57	41	2	<1	0	0.13
1900	21	71	5	3	0	0.25
1920	8	73	16	4	0	0.54
1940	5	50	41	4	0	0.63
1960	<1	23	74	4	0	1.3
1976	<1	18	75	4	3	1.9

systems, gas stations and other distribution centers. Adjustments to accommodate other fuel sources will probably require large amounts of new equipment and considerable time for development, construction, and installation.

The data of Table 14-5 indicate the large resources of coal in the United States and in the world. Figure 14-4 shows the locations of the nation's main coal fields and mining areas. An increase in the use of coal would therefore appear to be a logical response to expected shortages in gas and oil supplies. But filling most of our energy needs by using coal would present many problems. Obtaining and using natural gas and petroleum causes pollution and other unfavorable environmental effects, but the use of coal involves many more (Table 14-2). It has been estimated that in order for coal production to meet 50 percent of our present energy needs, it would require putting into operation at least 200 new coal mines, each producing about 5 million tons a year. At current levels of mining efficiency, the total work force of the mining industry would have to be expanded to 260,000 mine workers.

Table 14-5
Principal Present and Potential Future Sources of Energy

	Percentage of current total energy sources in United States	Estimated lifetime, years[a]			
		Known reserves		Undiscovered resources	
		United States	World	United States	World
Fossil fuels:					
Natural gas	32	11	45	30–40?	100?
Petroleum[b]	43	11,[c] 6[d]	30	32,[c] 17[d]	50?
Coal	18	300	300?	5000?	3000?
Nuclear:					
Uranium 235 reactor	3	90?	90?	[e]	[e]
Breeder reactor	0	Unlimited			
Fusion	0	Unlimited			
Direct solar:					
Space heating and cooling	Small	Unlimited			
Central power station	0	Unlimited			
Photovoltaic cells	Negligible	Very large			
Indirect solar:					
Water power	4	Renewable			
Wind	Small	Unlimited			
Biomass	Small	Renewable			
Waste	Small	Renewable			
Ocean thermal	0	Unlimited			
Other:					
Geothermal	Small	Uncertain	Uncertain		
Tides	Small	Renewable			

[a] At current rates of consumption.

[b] Not including shale and tar sand oil.

[c] Assuming current import rate.

[d] Without imports.

[e] Not available.

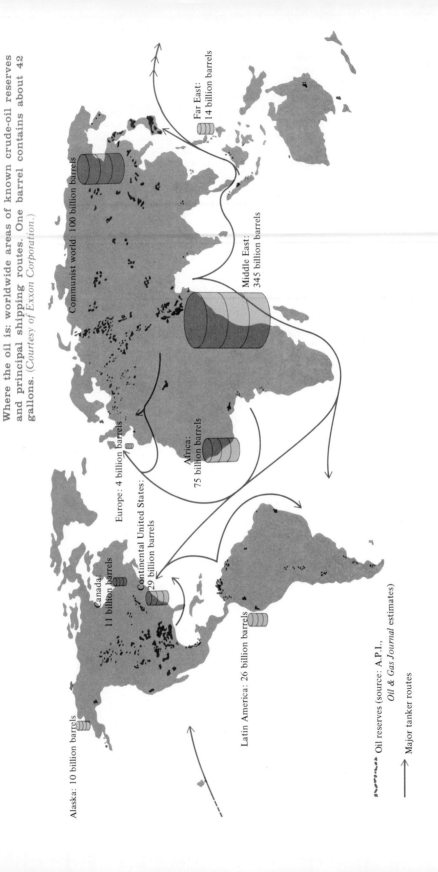

FIGURE 14-20
Where the oil is: worldwide areas of known crude-oil reserves and principal shipping routes. One barrel contains about 42 gallons. (*Courtesy of Exxon Corporation.*)

Far East:
14 billion barrels

Communist world: 100 billion barrels

Middle East:
345 billion barrels

Africa:
75 billion barrels

Europe: 4 billion barrels

Continental United States:
29 billion barrels

Canada:
11 billion barrels

Alaska: 10 billion barrels

Latin America: 26 billion barrels

Oil reserves (source: A.P.I.,
Oil & Gas Journal estimates)

Major tanker routes

For the expanded coal production to be used most conveniently, much of the coal would have to be converted into gaseous or liquid fuels. The technology for such conversion is at least partially developed (Sec. 14-5), but costs would be substantially above the levels of current petroleum-based gaseous and liquid fuels. Even if coal production can be suitably expanded, coal reserves, while ample, will not last forever. Sooner or later energy sources other than fossil fuels will have to be found. It is hoped that as a result of further research and development, significant amounts of energy will be derived in the future from a combination of water power, geothermal sources, windmills, tides, biomass conversion, waste conversion, and perhaps ocean thermal sources. Present indications suggest, however, that even with vigorous development these sources would collectively serve to meet only a portion of the total energy needs. Attention, therefore, turns to the possibilities of solar energy, nuclear fusion, and nuclear fission plants (especially breeder reactors).

Nuclear energy from fission has been under development for many years, and today 64 nuclear power plants are producing electricity at costs competitive with oil. Collectively they account for almost 12 percent of the total electrical generating capacity, about 3 percent of the nation's total energy production. The costs of nuclear-power-plant construction have been escalating, and although the public so far has been supporting nuclear energy when it has become an issue at the polls, considerable antinuclear sentiment has been expressed. Even if all the new nuclear plants now under construction or planned are carried through to production, their combined capacity is expected to be only about 20 percent of the total electric-energy generation by 1985, or 6 percent of the total energy being consumed. To what extent nuclear energy will be expanded beyond this point, with the possible introduction of breeder reactors (Sec. 14-8), is not yet clear.

The possibility of producing energy by controlled nuclear fusion has been under study for many years (Sec. 14-10). The problems are formidable, and progress has been slow. Although eventual success is not clearly indicated at the present stage of development, encouraging results have been reported. A period of 50 years is often estimated, however, as the time required for solving the production problems.

Solar energy, with its unlimited resources, has tremendous potential (the sun is expected to keep shining for at least several billion more years). The technology already exists for space and water heating by solar energy, and plans are under way to have 2 or 3 million homes equipped in about 10 years with one or the other. Even if this objective is met, however, the total energy collected will be less than 1 percent of our total needs.

An experimental central power station is now under construction in California's Mohave Desert to test the feasibility of solar-powered electrical generating plants. A field of over 5000 sun-following mirrors will be focused onto a water boiler atop a 280-foot tower. The steam will be used to generate electricity in the conventional manner. It is planned to have it generating energy for commercial distribution in 1981. The direct conversion of solar energy into electricity by photovoltaic cells has many attractive possibilities (Sec. 14-11C), but at present such cells are very expensive to make.

One of the principal uncertainties about the expanded use of solar energy concerns the costs of the various methods of using it. Small-scale individual solar units for residential heating and cooling are already commercially available. Further investigation is required to determine their economic practicality and that of large-scale central stations, but it seems fairly certain that solar energy will play an important role in energy generation in at least a supplemental way. Whether it can be expanded to furnish the major portion of our energy needs remains to be seen, but it is an objective well worth pursuing.

14-18 HOW CAN THE CONSUMPTION OF ENERGY BE REDUCED?

The use of energy is influenced by the size of the population, its level of financial prosperity, and the efficiency with which the energy is used. Energy in the United States has been very inexpensive and readily available for so long that there has been very little incentive to use it efficiently. Until recently our commercial and industrial practices and our social customs have been molded by the use of cheap energy. As efforts are made to extend energy resources and reduce environmental pollution in the face of pollution increases, however, the efficiency of energy use will become of paramount importance.

It is estimated that less than half of the energy currently consumed in the United States ends up as useful light, transportation, space heating, industrial production, and other applications. Automobiles on the average are only about 20 percent efficient. Less than 5 percent of the energy consumed by an incandescent light bulb is converted into visible light. Airplanes are fast but only about one-sixteenth as efficient as trains. Surveys have suggested that the consumption of energy in the United States could be reduced by 25 percent without affecting the general standard of living. Significant savings can be realized, for example, with more efficient heating and cooling systems, greater use of fluorescent lights, more efficient automobiles, and better insulation in buildings. About 11 percent of all energy consumed is used in residential space heating and 7 percent in commercial space heating. Yet more than one-third of this is unnecessarily lost because of poor insulation. Earl Cook of Texas A & M University has summarized diagrammatically the distribution of production, consumption, and waste of energy in the United States (Fig. 14-22).

The convenience and flexibility of electric energy has caused its use to rise faster in recent years than any other form of energy, until it now accounts for about 29 percent of all the energy consumed. The generation, storage, and distribution of electric energy, however, is much less efficient than many other forms of energy. The average existing power plant, for example, converts only 32 percent of the heat it generates into electricity; the remainder is passed on to a nearby river or to the atmosphere through a cooling tower. If electricity were to supply all of our

FIGURE 14-22
Graphical representation of the approximate flow of energy through the United States economy in 1976. Note the high proportion of energy that was unused. (*Courtesy of Earl Cook. Texas A & M University.*)

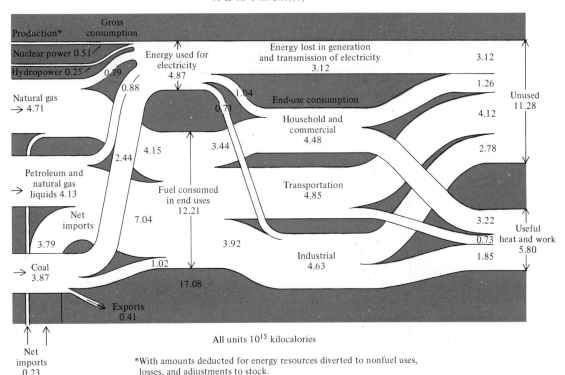

All units 10^{15} kilocalories

*With amounts deducted for energy resources diverted to nonfuel uses, losses, and adjustments to stock.

Table 14-6
Sources of Electric Energy†

Source	Percentage of total electricity generated, 1976
Coal	46.8
Oil	16.9
Natural gas	12.8
Hydroelectric	11.8
Nuclear	11.7

† From Earl V. Anderson, Nuclear Energy: A Key Role Despite Its Problems, **Chemical and Engineering News**, Mar. 7, 1977, p. 8.

energy needs, the cost of energy would increase fourfold. The current sources of electric energy are summarized in Table 14-6.

Transportation of freight and passengers by current means requires about 25 percent of all energy generated. Here, too, the availability of inexpensive energy has encouraged wasteful practices. Eric Hirst and John C. Moyers, of the Oak Ridge National Laboratory, have compared the rate of energy consumption by the various means of transporting freight between cities (Table 14-7). Railroads, waterways, and pipelines are very efficient. Although trucks are more flexible and faster than railroads, they consume energy 4 times as fast. Airplanes are extremely inefficient. Yet the federal government has been spending several billion dollars each year on highway and airport construction, while the nation's railroads have fallen into disrepair.

The same authors have also compared the energy-consumption rates of various means of transporting passengers (Table 14-7). Again the inefficiency of air travel stands out, as well as that of automobile travel within cities. The operation of the 100 million or so automobiles in the United States consumes about 60 percent of the total energy consumed for transportation and 15 percent of the energy used for all purposes. The effects of various factors on the gasoline mileage of automobiles have been estimated by Eric S. Cheney, of the University of Washington (Table 14-8).

An increase in the efficiency of energy use is in many ways the easiest method of relieving pressure on scarce energy resources and reducing environmental pollution due to the generation of energy. Clearly there must be more emphasis on the conservation of energy in the United States.

Table 14-7
Comparisons of Rate of Energy Consumption by
Alternative Means of Freight and Passenger Transport†

Freight, between cities		Passengers, between and within cities	
Means of transportation	Rate of energy consumption, Btu per ton-mile‡	Means of transportation	Rate of energy consumption (Btu per passenger-mile§)
Pipeline	450	Between cities:	
Railroad	670	Bus	1600
Waterway	680	Railroad	2900
Truck	2,800	Automobile	3400
Airplane	42,000	Airplane	8400
		Within cities:	
		Mass transit (~60% buses)	3800
		Automobile	8100

† From Eric Hirst and John C. Moyers, Efficiency of Energy Use in the United States, **Science, 179**:1299 (1973). Copyright 1973 by the American Association for the Advancement of Science.
‡ British thermal units (Sec. 14-2) consumed by transporting 1 ton of freight 1 mile.
§ British thermal units consumed by transporting one passenger 1 mile.

Table 14-8
Estimated Effects of Various Factors on Automobile Gasoline Mileage†

Factor	Percentage change
From a 2500- to a 5000-lb. car	− 65 to 100
Air conditioning	− 9 to 20
From 55 to 65 mph	− 11
Automatic transmission	− 5 to 6
Power steering	− 3 to 4
Steel-belted radial tires	Up to + 10

†Adapted from Eric S. Cheney, U.S. Energy Resources: Limits and Future Outlook, **American Scientist**, 62:14 (1974).

"I have a feeling it's too soon for fossil fuels around here."

© 1974 American Scientist; reprinted by permission of Sidney Harris.

KEY WORDS

1. Energy (Sec. 14-2)
2. Calorie (Sec. 14-2)
3. British thermal unit (Btu) (Sec. 14-2)
4. Power (Sec. 14-2)
5. Watt (Sec. 14-2)
6. Kilowatthour (Sec. 14-2)
7. Fossil fuel (Sec. 14-3)
8. Combustion (Sec. 14-3)
9. Gasoline (Sec. 14-3)
10. Anthracite coal (Sec. 14-3)
11. Bituminous coal (Sec. 14-3)
12. Complete combustion (Sec. 14-3)
13. Incomplete combustion (Sec. 14-3)
14. Greenhouse effect (Sec. 14-4)
15. Nuclear fission (Sec. 14-8)
16. Nuclear energy (Sec. 14-8)
17. Atomic energy (Sec. 14-8)
18. Nuclear chain reaction (Sec. 14-8)
19. Nuclear reactor (Sec. 14-8)
20. Breeder reactor (Sec. 14-8)
21. Nuclear fusion (Sec. 14-10)
22. Heliostat (Sec. 14-11B)
23. Photon (Sec. 14-11C)
24. Photovoltaic cell (Sec. 14-11C)
25. Photosynthesis (Sec. 14-13C)
26. Biomass (Sec. 14-13C)
27. Destructive distillation of wood (Sec. 14-13C)
28. Geothermal energy (Sec. 14-14)

SUMMARIZING QUESTIONS FOR SELF-STUDY

1. Q. About what percentage of the energy consumed in the world is being consumed in the United States?

A. About 35 percent.

2. Q. About what percent of the world's population lives in the United States?

A. About 6 percent.

3. Q. About what percent of our energy needs are being supplied by petroleum?

A. Over 40 percent.

4. Q. About what fraction of the petroleum used in the United States is imported?

A. Almost half.

Section 14-2

5. Q. What is one definition of energy?

A. The ability to do work.

6. Q. What is a calorie?

A. The amount of heat energy required to raise the temperature of 1 gram of water 1°C.

7. Q. What is a Btu?

A. A British thermal unit; the heat energy consumed in raising the temperature of 1 pound of water 1°F. It is equivalent to 252 calories.

8. Q. What is power?

A. The rate of expending energy.

9. Q. What is a watt?

A. A unit of power equivalent to the expenditure of about 860 calories of energy in an hour.

10. Q. What is a kilowatthour?

A. An amount of energy equivalent to what is consumed in an hour at the rate of 1000 watts.

Section 14-3

11. Q. What is a fossil fuel?

A. A fuel which is derived from the remains of living organisms, for the most part petroleum, natural gas, and coal.

12. Q. What is combustion?

A. Rapid reaction with oxygen which generates heat and light.

13. Q. From the viewpoint of interatomic bonds breaking and forming during a chemical reaction, why is energy evolved during the process of combustion?

A. Because the energy consumed in breaking the interatomic bonds of the reactants is less than the energy evolved in forming the interatomic bonds of the products.

14. Q. What is gasoline?

A. A mixture of hydrocarbon molecules containing from 5 to 12 carbon atoms.

15. Q. What is anthracite coal?

A. Coal which contains a relatively low proportion of hydrogen- and oxygen-containing substances.

16. Q. What is bituminous coal?

A. Coal which contains a substantial proportion of hydrogen- and oxygen-containing substances.

17. Q. What is meant by complete combustion?

A. A combustion reaction in which all the hydrogen atoms of the reactants are converted into water molecules and all the carbon atoms are converted into carbon dioxide molecules.

18. Q. What is meant by incomplete combustion?

A. Combustion which takes place with a limited supply of oxygen. Carbon monoxide is formed from the carbon as well as carbon dioxide.

Section 14-4

19. Q. Why is the carbon dioxide in the atmosphere essential to the life processes of plants?

A. Because together with water it is a raw material for the process of photosynthesis.

20. Q. What natural processes release carbon dioxide into the atmosphere?

A. The processes of respiration and the decay of animals and plants.

21. Q. Why doesn't the production of carbon dioxide from these sources cause the concentration of carbon dioxide in the atmosphere to increase?

A. Because the amount of carbon dioxide released is about equal to the amount removed by photosynthesis.

22. Q. What is the principal source of carbon dioxide in the atmosphere resulting from the generation of energy?

A. The combustion of natural gas, petroleum products, and coal.

23. Q. What is the greenhouse effect?

A. The insulating action of an atmospheric substance such as carbon dioxide; like the glass of a greenhouse, it permits the energy from the sun to pass through but prevents the heat from leaving.

24. Q. There is some evidence that the concentration of carbon dioxide in the atmosphere is increasing. If this is confirmed, why would it cause concern?

A. Because of the greenhouse effect it might cause the temperature of the earth's surface to increase, melting polar icecaps and flooding low-lying ocean coastal areas. A severe change in weather patterns might also occur.

25. Q. What factors other than a greenhouse effect due to carbon dioxide may influence the temperature of the earth's surface?

A. Turbidity in the atmosphere blocking the sun's rays (cooling); fluctuations in the intensity of the sun's radiation; greenhouse effect due to water vapor.

Section 14-5

26. Q. Gaseous and liquid hydrocarbon fuels can be obtained from coal. How is coal treated to produce such substances?

A. It is treated at a high temperature with a mixture of steam and oxygen under pressure to form (among other substances) carbon monoxide and hydrogen. These two substances then react to form the hydrocarbons.

27. Q. Give an equation for the reaction of carbon monoxide and hydrogen to form methane.

A. $CO + 3H_2 \rightarrow CH_4 + H_2O$.

28. Q. What is the name for the process of producing hydrocarbons from coal?

A. The Fischer-Tropsch synthesis, named for the German chemists who developed the original process.

29. Q. Why is this method of producing hydrocarbons not used more often?

A. Because it has been more expensive than obtaining hydrocarbons from petroleum and natural gas.

Sections 14-6 and 14-7

30. Q. What is probably the principal advantage in using petroleum and natural gas as sources of energy?

A. Choose from: relatively low cost, convenience, and established technology.

31. Q. What is probably the principal disadvantage of using petroleum and natural gas as sources of energy?

A. The limitations of supplies.

32. Q. What is the leading advantage of the use of coal as a source of energy?

A. The abundance of resources.

33. Q. What are three principal disadvantages in the use of coal as a source of energy?

A. Dangers of underground mining, disruption of surface mining, and the pollution caused by the combustion of coal.

34. Q. What is one of the substances used in the removal of sulfur oxides from the effluent gases of the combustion of coal?

A. Limestone, essentially $CaCO_3$.

35. Q. What is its function in the removal, for example, of sulfur trioxide?

A. In the heat of the combustion furnace the limestone is decomposed into calcium oxide and carbon dioxide. The calcium oxide subsequently reacts with the sulfur trioxide to form calcium sulfate:

$$CaCO_3 \quad \rightarrow \quad CaO \quad + \quad CO_2$$

Calcium carbonate, limestone   Calcium oxide, lime

$$CaO + SO_3 \rightarrow CaSO_4$$

Calcium sulfate

Sections 14-8 and 14-9

36. Q. What is nuclear fission?

A. A nuclear reaction in which atomic nuclei are split into two nuclei of substantial size, accompanied by the evolution of a very large amount of energy.

37. Q. What is a nuclear chain reaction?

A. A self-sustaining sequence of nuclear fission reactions made possible by the ability of the neutrons produced in one nuclear reaction to carry out a subsequent nuclear fission.

38. Q. What is a nuclear reactor?

A. An apparatus for carrying out a controlled self-sustaining nuclear fission reaction.

39. Q. What is a breeder reactor?

A. A nuclear reactor which produces more nuclear fuel than it consumes.

40. Q. What two substances that can be synthesized in breeder reactors can serve as nuclear fuels?

A. Plutonium 239 or uranium 233.

41. Q. What are two of the principal advantages of nuclear reactors as a source of energy?

A. Choose from: (1) breeder reactors are a potential source of an almost unlimited amount of energy, (2) they do not compete with the production of carbon-containing products for the available resources of petroleum, natural gas, and coal, and (3) they do not pollute the air with sulfur oxides, nitrogen oxides, and particles. No possible greenhouse effect.

42. Q. What is the most controversial aspect of use of nuclear energy?

A. Concern about the overall safety of the process.

Section 14-10

43. Q. What is nuclear fusion?

A. A type of nuclear change in which the nuclei of two atoms combine to form a more complex nucleus of higher atomic number, accompanied by the evolution of a large amount of energy.

44. Q. What would be the principal advantage of controlled nuclear fusion as a source of energy?

A. The practically unlimited supply of fuel (ocean water).

45. Q. What is the principal deterrent to the practical development of controlled nuclear fusion as an energy supply?

A. The extremely high temperature required (about 100 million degrees).

Sections 14-11 and 14-12

46. Q. What is the principal part of a solar heating system for a house?

A. The solar-energy collector, typically a large shallow tank with a black metal base to absorb energy, covered with transparent glass to prevent heat from escaping.

47. Q. What is the overall principle of a solar heating system for a house?

A. Water is heated in the collectors and circulated in pipes throughout the house.

48. Q. What is an heliostat?

A. A sun-following mirror used to reflect and focus the sun's energy.

49. Q. What is one plan under development for the use of solar energy in a central power station?

A. A field of heliostats is focused on a steam generator on top of a high tower. The steam is used to generate electricity in the conventional manner.

50. Q. Photovoltaic cells are devices for converting solar energy directly into electric energy. They have been used effectively in space vehicles. Why are they not more generally used?

A. At present they are extremely expensive.

51. Q. What are the two principal advantages to the use of solar energy as a source of energy?

A. Its huge supply and the low environmental pollution.

52. Q. What are the two principal disadvantages to the use of solar energy as a source of energy?

A. Its low intensity and its intermittent nature.

Section 14-13

53. Q. What are the principal indirect effects of sunlight which may be used as sources of energy?

A. Water power, wind power, biomass sources, waste material, and ocean thermal power.

54. Q. What is meant by a biomass source of energy?

A. A source derived from plant or animal life, such as burning wood.

55. Q. How can the fuel methyl alcohol, CH_3OH, be obtained from a biomass source?

A. From the destructive distillation of wood, i.e., heating wood in the absence of air.

56. Q. How can the fuel ethyl alcohol be obtained from a biomass source?

A. From the fermentation of natural sugars, e.g., those obtainable from sugar cane.

57. Q. What is one way of obtaining fuels from organic waste material?

A. Choose from (1) heating the waste in the absence of air to obtain a liquid fuel, like heating oil, (2) heating the waste with carbon monoxide and steam under pressure to obtain a liquid fuel, like heating oil, or (3) using the action of bacteria to produce methane.

Sections 14-14 and 14-15

58. Q. What is meant by geothermal energy?

A. Energy derived from within the earth, mostly from underground sources of steam and hot water.

59. Q. How is geothermal energy used?

A. Wells are dug to underground sources of hot water and steam. The hot water can be circulated through nearby buildings to heat them. The steam can be used to generate electricity in the conventional manner.

60. Q. What is the principal limitation to the use of geothermal energy?

A. Areas where the temperature of the underground water is high enough are largley confined to places of recent volcanic action or other activity of the earth's crust.

61. Q. How can the energy of ocean tides be harnessed?

A. By enclosing with a tidal-power dam the mouth of a V-shaped inlet where tides are unusually high. The tidal flow in and out of the enclosed basin is directed through turbine generators of electricity.

Sections 14-16 to 14-18

62. Q. About 25 percent of the energy consumed in the United States is used for transportation. What would be two approaches to reducing the amount of energy used for this purpose?

A. Producing more efficient motor vehicle and airplane engines and using the low-energy consuming means of travel such as railroads and waterways more and the high-energy-consuming means such as airplanes, trucks, and automobiles less.

63. Q. About 18 percent of the energy consumed in the United States is used for space heating. How could the energy used for this purpose be reduced?

A. By using more and more effective heat insulation in building construction.

64. Q. What unlimited or renewable sources of energy are available in principle without regard for their present availability on a practical basis?

A. Solar energy, water power, wind power, breeder reactors, fusion reactors, ocean thermal power, and tidal power.

65. Q. Of these, which probably have the greatest potential for supplying a major source of energy?

A. Solar energy, breeder reactors, and fusion reactors.

66. Q. What nonrenewable source of energy is in greatest supply?

A. Coal.

67. Q. How long do authorities estimate the known reserves of coal in the United States will last?

A. About 200 years or so, depending on rate of consumption, although such estimates are subject to a large margin of error.

68. Q. How long do authorities estimate the known reserves and undiscovered resources of petroleum and natural gas in the United States will last?

A. About 40 or 50 years at the current rate of consumption, although such estimates are subject to a large margin of error.

PRACTICE EXERCISES

1. Match each definition or other statement with the numbered term above with which it is most closely associated. Each numbered term may be used only once.

1. Power	2. Calorie
3. Bituminous coal	4. Combustion
5. Watt	6. Energy
7. Fossil fuels	8. Kilowatthour
9. Anthracite	10. Gasoline
11. Kilocalorie	

   (a) Fuels derived from petroleum, natural gas, and coal
   (b) An expenditure of energy at the rate of 860 calories in an hour
   (c) The ability to do work
   (d) A mixture of hydrocarbon molecules containing from 5 to 12 carbon atoms
   (e) The amount of heat energy required to raise the temperature of 1 gram of water 1°C
   (f) The energy consumed in an hour at the rate of 1000 watts
   (g) Coal with a relatively low proportion of hydrogen- and oxygen-containing substances
   (h) The rate of expending energy
   (i) The rapid reaction with oxygen which generates heat and light

2. What general process is involved in using fossil fuels as a source of energy?

3. Using methane as a representative fossil fuel, discuss the source of the energy evolved when it is used as a fuel.

4. What are the three principal products of the reactions involved in the use of fossil fuels as a source of energy?

5. What is the difference between complete combustion and incomplete combustion?

6. Match each definition or other statement with the numbered term above with which it is most closely associated. Each numbered term may be used only once.

1. Nuclear reactor	2. Cyclotron
3. Nuclear energy	4. Anthracite
5. Complete combustion	6. Nuclear fission
7. Bituminous coal	8. Breeder reactor
10. Incomplete combustion	9. Greenhouse effect
	11. Nuclear chain reaction

   (a) The insulating action of atmospheric carbon dioxide
   (b) A reaction in which two atomic nuclei are split into two nuclei of substantial size
   (c) Coal with a substantial proportion of hydrogen- and oxygen-containing substances
   (d) A reaction of hydrocarbons with oxygen in which the amount of oxygen is limited and insufficient for complete reaction
   (e) A self-sustaining sequence of nuclear fission reactions
   (f) A reaction of a hydrocarbon in which all the carbon atoms are converted into carbon dioxide and all the hydrogen atoms into water molecules
   (g) An apparatus for carrying out a controlled self-sustaining nuclear chain reaction
   (h) The energy released during the fission of the nucleus of an atom
   (i) A nuclear reactor which produces more nuclear fuel than it consumes

7. What is the greenhouse effect, and why is there concern about it?

8. (a) What is the Fischer-Tropsch synthesis? (b) What is the significance of this process?

9. Give equations for representative reactions involved in the Fischer-Tropsch synthesis.

10. Match each definition or other statement with the numbered term with which it is most closely associated. Each numbered term may be used only once.

1. Photovoltaic cell	2. Heliostat
3. Combustion	4. Biomass
5. Geothermal energy	6. Nuclear fusion
7. Photosynthesis	8. Destructive distillation
9. Photostat	10. Photon

    (a) The process whereby solar energy is converted into chemical energy as plants synthesize glucose from carbon dioxide and water
    (b) A type of nuclear change in which the nuclei of two atoms combine to form a more complex nucleus of higher atomic number
    (c) A device which converts light energy directly into an electric current
    (d) The process of heating wood in the absence of air
    (e) A sun-following mirror
    (f) A pulse of light energy
    (g) Energy derived from within the earth
    (h) The various forms of plant and animal life on earth

11. (a) List four advantages of using petroleum and natural gas as sources of energy. (b) List eight disadvantages of using petroleum and natural gas as sources of energy.

12. (a) List four advantages of using coal as a source of energy. (b) List eight disadvantages of using coal as a source of energy.

13. What is black lung disease?

14. (a) What is the source of acid mine drainage? (b) What harm does it do?

15. Using equations for reactions where appropriate, account for the formation of sulfur dioxide, SO_2, and sulfuric acid, H_2SO_4, in the use of coal as a source of energy.

16. What are the typical components of industrial smog?

17. Using equations for reactions where appropriate, discuss in overall terms how limestone can be used to reduce the amount of sulfur compounds expelled into the air as coal is used as a source of energy.

18. Explain (a) nuclear fission, (b) nuclear energy, (c) nuclear chain reaction, (d) nuclear reactor, and (e) breeder reactor.

19. Give the appropriate nuclear equations for the reactions which relate to one method of producing plutonium 239 in a breeder reactor.

20. Give the appropriate nuclear equations for reactions which relate to one method of producing uranium 233 in a breeder reactor.

21. (a) List six advantages to the use of nuclear reactors as a source of energy. (b) List seven disadvantages to the use of nuclear reactors as a source of energy.

22. (a) What is meant by nuclear fusion? (b) Give a nuclear equation for an example of a nuclear fusion reaction.

23. What are the major potential advantages and disadvantages in using nuclear fusion as a source of energy?

24. Describe in general terms how solar energy can be used to heat buildings.

25. Describe in general terms the principle of an experimental solar-powered station for the generation of electricity now under investigation.

26. Describe in general terms a method potentially available for the generation of electricity directly by light energy.

27. List (a) seven advantages and (b) six disadvantages in using direct solar energy as a source of energy.

28. Name five different methods of using indirect effects of the sun's energy as sources of energy.

29. List (a) four advantages and (b) four disadvantages in using water power as a source of energy.

30. What are the advantages and disadvantages of using wind power as a source of energy?

31. (a) What is meant by biomass and how can it be used as a source of energy? (b) What is the ultimate source of such energy?

32. (a) What is meant by photosynthesis? (b) Give a balanced equation for the overall reaction in photosynthesis. (c) What is the special significance of photosynthesis?

33. (a) What is the destructive distillation of wood? (b) What product obtained can be used as a source of energy? (c) How can this product be used as a source of energy?

34. (a) How may sugar cane be used to give a liquid product which can be used as a motor fuel? (b) Give equations for the reactions involved in the production of the liquid product.

35. How can waste material be used as a source of energy?

36. What is the operating principle of an ocean thermal power plant?

37. What are (a) five advantages and (b) three disadvantages to using biomass material for the generation of energy?

38. What are the advantages and disadvantages of the use of waste material as a source of energy?

39. What would be the potential advantages and disadvantages of the use of ocean thermal power stations as a source of energy?

40. (a) What is geothermal energy? (b) Explain in general terms how it is currently being used as a source of energy.

41. List (a) four advantages and (b) five disadvantages in using geothermal energy as an energy source.

42. How is energy obtained from ocean tides?

43. What are the advantages and the disadvantages of the current method of using ocean tides as a source of energy?

44. What are the four leading uses of energy in the United States?

45. Give one means of reducing the amount of energy used by each of the two most demanding uses of energy.

46. Why is there concern about the use of fossil fuels as a source of energy in the United States?

47. (a) What alternatives to the use of fossil fuels are available as sources of energy? (b) Of these alternatives which have the greatest potential for supplying the huge amounts of energy required by the United States?

SUGGESTIONS FOR FURTHER READING

The following articles are at an introductory level:

Anderson, Earl V.: Nuclear Energy: A Key Role Despite Problems, **Chemical and Engineering News**, Mar. 7, 1977, p. 8.

Atwood, Genevieve: The Strip-Mining of Western Coal, **Scientific American**, December 1975, p. 23.

Axtmann, Robert C.: Environmental Impact of a Geothermal Plant, **Science**, **187**:795 (1975).

Baes, C. F., Jr., H. E. Goeller, J. S. Olson, and R. M. Rotty: Carbon Dioxide and Climate: The

Uncontrolled Experiment, **American Scientist,** **65**:310 (1977).

Bailey, Maurice E.: The Chemistry of Coal and Its Constituents, **Journal of Chemical Education, 51**:446 (1974).

Bebbington, William P.: The Reprocessing of Nuclear Fuels, **Scientific American,** December 1976, p. 30

Bethe, H. A.: The Necessity of Fission Power, **Scientific American,** January 1976, p. 21.

Bozak, Richard E., and Manuel Garcia, Jr.: Chemistry in the Oil Shales, **Journal of Chemical Education, 53**:154 (1976.)

Carter, Luther J.: Radioactive Wastes: Some Urgent Unfinished Business, **Science, 195**:661 (1977).

Cheney, Eric S.: U.S. Energy Resources: Limits and Future Outlook, **American Scientist, 62**:14 (January-February 1974).

Christiansen, Bill, and Theodore H. Clack, Jr.: A Western Perspective on Energy: A Plea for Rational Energy Planning, **Science, 194**:578 (1976).

Cochran, Neal P.: Oil and Gas from Coal, **Scientific American,** May 1976, p. 24.

Cohen, Bernard L.: Impact of the Nuclear Energy Industry on Human Health and Safety, **American Scientist, 64**:550 (1976).

Cohen, Bernard L.: The Disposal of Radioactive Wastes from Fission Reactors, **Scientific American,** June 1977, p. 21.

Cook, C. Sharp: Energy: Planning for the Future, **American Scientist, 61**:61 (1973).

Cook, Earl: The Flow of Energy in an Industrial Society, **Scientific American,** September 1971, p. 135.

Damon, Paul E., and Stephen M. Kunen: Global Cooling?, **Science, 193**:447 (1976).

De Nevers, Noel: Tar Sands and Oil Shales, **Scientific American,** February 1966, p. 21.

Drake, Elizabeth, and Robert C. Reid: The Importation of Liquefied Natural Gas, **Scientific American,** April 1977, p. 22.

Duffie, John A., and William A. Backman: Solar Heating and Cooling, **Science, 191**:143 (1976).

Ellis, A. J.: Geothermal Systems and Power Development, **American Scientist, 63**:510 (1975).

Gough, William C., and Bernard J. Eastland: The Prospects of Fusion Power, **Scientific American,** February 1971, p. 50.

Gregory, Derek P.: The Hydrogen Economy, **Scientific American,** January 1973, p. 13.

Hammond, Allen L.: Solar Energy: The Largest Resource, **Science, 177**:1088 (1972).

Hammond, Allen L.: Fission: The Pros and Cons of Nuclear Power, **Science, 178**:147 (1972).

Hammond, Allen L.: Conservation of Energy: The Potential for More Efficient Use, **Science, 178**:1079 (1972).

Hammond, Allen L.: Individual Self-Sufficiency in Energy, **Science, 184**:278 (1974).

Hammond, Allen L.: Cleaning Up Coal: A New Entry in the Energy Sweepstakes, **Science, 189**:128 (1975).

Hammond, Allen L.: Solar Energy Reconsidered: ERDA Sees Bright Future, **Science, 189**:538 (1975).

Hammond, Allen L.: Photovoltaics: The Semiconductor Revolution Comes to Solar, **Science, 197**:445 (1977).

Hammond, Allen L.: Alcohol: A Brazilian Answer to the Energy Crisis, **Science, 195**:564 (1977).

Hammond, Allen L.: Photosynthetic Solar Energy: Rediscovering Biomass Fuels, **Science, 197**:745 (1977).

Hammond, R. Philip: Nuclear Power Risks, **American Scientist, 62**:155 (1974).

Harris, John R.: The Rise of Coal Technology, **Scientific American,** August 1974, p. 92.

Hildebrandt, Alvin F., and Lorin L. Vant-Hull: Power with Heliostats, **Science, 197**:1139 (1977).

Hirst, Eric, and John C. Moyers: Efficiency of Energy Use in the United States, **Science, 179**:1299 (1973).

Hohenemser, Christop, Roger Kasperson, and Robert Kates: The Distrust of Nuclear Power, **Science, 196**:25 (1977).

Hubbert, M. King: The Energy Resources of the Earth, **Scientific American,** September 1971, p. 60.

Krieger, James H.: Energy: The Squeeze Begins, **Chemical and Engineering News,** Nov., 13, 1972, p. 20.

Landsberg, H. H.: Low-Cost Abundant Energy: Paradise Lost?, **Science, 184**:247 (1974).

Lepkowski, Wil: Carbon Dioxide: A Problem of Producing Usable Data, **Chemical and Engineering News,** Oct. 17, 1977, p. 26.

Metz, William P.: Solar Thermal Energy: Bringing the Pieces Together, **Science, 197**:650 (1977).

Metz, William P.: Wind Energy: Large and Small Systems Compared, **Science, 197**:971 (1977).

Metz, William P.: Ocean Thermal Energy, **Science, 198**:178 (1977).

Perry, Harry: The Gasification of Coal, **Scientific American,** March 1974, p. 19.

Pierce, John R.: Fuel Consumption of Automobiles, **Scientific American,** January 1975, p. 34.

Plass, Gilbert N.: Carbon Dioxide and Climate, **Scientific American,** July 1959, p. 41.

Pollard, William G.: The Long-Range Prospects for Solar Energy, **American Scientist, 64**:424, (1976).

Rose, David J.: Energy Policy in the U.S., **Scientific American,** January 1974, p. 20.

Rose, David J.: The Long-Range Prospects for Solar-derived Fuels, **American Scientist, 64**:509 (1976).

Rose, David J., Patrick W. Walsh, and Larry L. Leskovjan: Nuclear Power: Compared to What?, **American Scientist, 64**:291 (1976).

Seltzer, Richard J.: Efforts to Tap Ocean Thermal Energy Gain, **Chemical and Engineering News,** Feb. 9, 1976, p. 19.

Squires, Arthur M.: Chemicals from Coal, **Science, 191**:689 (1976).

Starr, Chauncey: Energy and Power, **Scientific American,** September 1971, p. 37.

Walsh, John: Problems of Expanding Coal Production, **Science, 184**:336 (1974).

Walters, Edward A., and Eugene M. Wewerka: An Overview of the Energy Crisis: **Journal of Chemical Education, 52**:282 (1975).

Woodwell, George M.: The Energy Cycle of the Biosphere, **Scientific American,** September 1970, p. 64.

Woodwell, George M.: The Carbon Dioxide Question, **Scientific American,** January 1978, p. 34.

Yohe, G. R.: Coal, **Chemistry,** January 1967, p. 8.

Solar Thermal Electricity: Power Tower Dominates Research, Metz, William D., **Science, 197**:353 (1977).

The following articles are at a more advanced level:

Calvin, Melvin: Photosynthesis as a Resource for Energy and Materials, **American Scientist, 64**:270 (1976).

Chalmers, Bruce: The Photovoltaic Generation of Electricity, **Scientific American,** October 1976, p. 34.

de Marsily, G., E. Ledoux, A. Barbeau, and J. Margat: Nuclear Waste Disposal: Can the Geologist Guarantee Isolation? **Science, 197**:519 (1977).

Maugh, Thomas H., II: Hydrogen: Synthetic Fuel of the Future, **Science, 178**:849 (1972).

15

CHEMISTRY AND ENVIRONMENTAL PROBLEMS

15-1 THE LETHAL POTENTIAL OF AIR POLLUTION

In London, on the morning of December 4, 1952, there was no indication that by nightfall erratic weather conditions would bring on a deadly combination of fog and smoke which would hold the entire city and much of the surrounding area in its grip for several days and take the lives of thousands. Earlier air-pollution crises had warned the city how vulnerable it was to such disasters because of the soot, fly ash, and sulfur oxides produced by soft-coal combustion, its principal means of heating homes and generating energy for industry. It was generally known that as the sulfur-containing coal is burned, appreciable amounts of gaseous sulfur dioxide, SO_2, are formed. Much of this is further converted into gaseous sulfur trioxide, SO_3, which in turn reacts with the moisture of the air to produce the liquid sulfuric acid, H_2SO_4, usually in the form of an easily airborne mist. (See Sec. 15-2B for additional details.) In sum, the combustion of the coal produces a suffocating combination of sulfur dioxide, sulfuric acid mist, and smoke, which poses a greater hazard to health than any of the components separately, however harmful each one may be by itself.

Ordinarily, the air pollutants generated at ground level are swept away by the motion of the wind or carried upward as the warm air of lower altitudes rises toward the cooler air above. Normally the temperature of the air becomes cooler at higher altitudes. Warm air at the earth's surface, being more buoyant, tends to rise, and as it does, it carries pollutants with it. This upward motion of the air facilitates replacement with the cooler, less polluted air from higher altitudes. Such motion is especially important for the removal of lower-level contaminants when the wind is not blowing.

Under certain atmospheric circumstances the air at the ground level may be cooler than the air above it. In a **temperature inversion** a mass of warm air lies above a mass of cool air, a reverse of the normal arrangement. Under such circumstances there is very little motion of the air. The cool air close to the surface has no tendency to rise to the warm air above it (Fig. 15-1). In the absence of any wind, pollutants can concentrate in the calm air at ground level.

During the afternoon of December 4, a temperature inversion took place over much of the lower Thames valley, trapping most of greater London. By early evening a thick fog had formed in the still, cold air. The accumulation of coal fumes further decreased the visibility by the morning of the next day and made the air smelly and irritating. The smog became so thick that all forms of transportation were disrupted. Serious collisions occurred between trains and between ships. Thousands of automobiles were abandoned on the road as it became impossible to see beyond the front of a car.

FIGURE 15-1
(a) Under normal atmospheric conditions the buoyant warm air at ground level rises to the cooler air above, carrying pollutants with it, even when the wind is not blowing. (b) During a temperature inversion on a windless day there is no tendency for the air at the surface to rise, and the pollutants collect in the calm air at ground level.

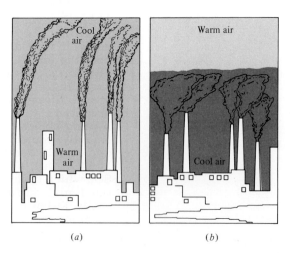

(a)   (b)

For 4 days the air remained still and freezing cold. Unusually large amounts of coal were burned to combat the chill, and the pollution intensified. For the most part the attention of the public was focused on the spectacular accidents, the traffic paralysis, and the general disruption of normal activities. Although physicians, health officials, and some others knew of the unusual number of deaths occurring, the population as a whole was unaware of the true dimensions of the disaster which had descended upon it.

The demand for hospital beds rose each day the smog held over the city, until it became 3 to 4 times the normal demand, and hospitals were compelled to restrict admissions to the most severely stricken. Emergency cases continued to appear at a high rate for many days after a wind swept the smog away. People with chronic bronchitis, asthma, and other respiratory ailments were the first to be affected. As breathing difficulties placed excessive strains on the body, heart attacks became more frequent. While the loss of life was greatest among the elderly, abnormal numbers of deaths occurred in all age groups, including infants. Weeks later, when the total impact of the 4-day smog was analyzed, conservative estimates placed the death toll at 4000 and other estimates ran as high as 8000. At least 100,000 others were made ill, many seriously so.

Although the 1952 air-pollution disaster in London was probably the worst that has occurred, it was by no means an isolated incident. Over 1000 people died, for example, in a similar episode in London in 1873. In 1930 about 60 were killed in Meuse, Belgium. The first recorded instance of lethal air pollution in the United States occurred in 1948 in western Pennsylvania, at Donora, where 17 died. All in all, a surprising number of air-pollution crises occurred before any concerted effort to minimize them was made.

15-2 WHAT ARE THE PRINCIPAL AIR POLLUTANTS?

The normal composition of clean dry air near sea level is given in Table 15-1. The atmosphere actually also contains a variable amount of water vapor. A **pollutant** is any form of matter which, when added to the surroundings in sufficient amount, produces a detectable injurious effect on people, animals, plants, or the environment in general. The principal pollutants in the atmosphere in the United States are listed in Table 15-2,

Table 15-1
Composition of Clean,
Dry Air at Sea Level†

Substance	Formula	Molecules in 1 million molecules of air
Nitrogen	N_2	780,840
Oxygen	O_2	209,480
Argon	Ar	9,340
Carbon dioxide	CO_2	314
Neon	Ne	18
Helium	He	5
Methane	CH_4	2
Krypton	Kr	1

† Plus traces of other substances. The atmosphere actually also contains a variable amount of water vapor.

Table 15-2
Principal Air Pollutants in the United States and Their Chief Sources,
Millions of Tons per Year (1970)

	Carbon monoxide	Sulfur oxides, SO_2 and SO_3	Nitrogen oxides, NO and NO_2	Hydrocarbons	Suspended particles
Automobiles	111	1	12	20	1
Electric-power plants and space heating	1	27	10	1	7
Industry	11	6	<1	6	13
Refuse disposal	7	<1	<1	2	1
Others	17	<1	<1	7	3
Total	147	34	22	36	25

together with their chief sources. These pollutants include carbon monoxide, CO; sulfur dioxide, SO_2; sulfur trioxide, SO_3; nitric oxide, NO; nitrogen dioxide, NO_2; such hydrocarbons as propane, $CH_3CH_2CH_3$, and benzene, C_6H_6; and suspended matter, e.g., fine particles of ammonium sulfate, $(NH_4)_2SO_4$, ammonium nitrate, NH_4NO_3, and carbonaceous soot. Carbon dioxide, CO_2, has not been included since it is a natural component of the atmosphere (Table 15-1) and its role as a pollutant has not been clearly determined. (The carbon dioxide in the atmosphere and its possible relationship to a greenhouse effect were discussed in Sec. 14-4.) The information in Table 15-2 indicates that at present the principal cause of air pollution is the gasoline-powered engine of the automobile.

A. The Sources and Effects of Carbon Monoxide

If the hydrocarbons of gasoline are burned with a sufficient supply of oxygen, the sole products are carbon dioxide and water (Sec.14-3), as the following typical reaction suggests:

$$CH_3\overset{\displaystyle |}{\underset{\displaystyle CH_3}{C}}HCH_2CH_2CH_2CH_3 + 11O_2 \rightarrow 7CO_2 + 8H_2O$$

2-Methylhexane,   Oxygen   Carbon   Water
a typical component   (ample   dioxide
of gasoline   supply)

**The reaction for the complete combustion of
a typical component of gasoline, which occurs when the
hydrocarbon is burned with a sufficient
supply of oxygen**

If there is less oxygen available, however, considerable amounts of carbon monoxide may be formed due to the incomplete combustion of the gasoline:

$$CH_3\overset{\displaystyle |}{\underset{\displaystyle CH_3}{C}}HCH_2CH_2CH_2CH_3 + 10O_2 \rightarrow 5CO_2 + 2CO + 8H_2O$$

2-Methylexane,   Oxygen   Carbon   Carbon   Water
a typical component   (limited   dioxide   monoxide
of gasoline   supply)

**A representative reaction for the incomplete
combustion of a typical component of gasoline, which
occurs when the hydrocarbon is burned with an
insufficient supply of oxygen; carbon monoxide
is among the products**

Note to Students

Many topics related to environmental problems have been discussed in previous chapters, including radiation hazards (Sec. 9-13), chemical insecticides (Sec. 11-10), food additives (Sec. 11-12), the safety of medicinal agents (Sec. 12-10), biodegradable detergents (Sec. 13-9), the greenhouse effect, and other environmental problems associated with the generation of energy (Chap. 14).

Carbon monoxide may also be produced when diesel fuel, heating oil, coal, or coke are burned in a limited supply of oxygen. It has been estimated that worldwide nearly 270 million tons of carbon monoxide are released each year from man-made sources. But evidence suggests that this is only about 7 percent of the total amount of carbon monoxide generated from natural and man-made sources combined.

The principal natural source of carbon monoxide results from the oxidation of methane. This hydrocarbon is produced in large quantitites from the decay of organic matter. It has been estimated, for example, that an average acre of swamp generates about 3000 pounds of methane each year. The methane is oxidized to carbon monoxide, principally in the atmosphere. Other sources of carbon monoxide include the decay of dying plants, the formation of chlorophyll, and natural processes in the oceans. It is estimated that the carbon monoxide from all natural sources amounts to some 3.5 billion tons a year, more than 90 percent of all the carbon monoxide released to the atmosphere.

How is it that this enormous release of carbon monoxide does not seriously affect the composition of the atmosphere? Carbon monoxide is apparently removed from the atmosphere at about the same rate it is released although the mechanisms whereby it is transformed into other substances are not thoroughly understood. It is known that the process of nature which converts methane into carbon monoxide in turn converts the carbon monoxide into carbon dioxide. Further, micro-organisms in the soil are capable of bringing about the same conversion. It has been estimated that the soil in the United States alone is capable of removing about 600 million tons of carbon monoxide each year.

Although natural processes apparently have an enormous capacity for removing carbon monoxide, the localized accumulation of the gas from man-made sources can be a serious health hazard. The principal toxic effect of carbon monoxide results from its ability to interfere with the transport of oxygen by the blood. Exposure for more than 1 hour to a concentration of 120 ppm (parts per million) is considered to be harmful, and a concentration of 240 ppm for 1 hour is dangerous. Prolonged exposure to higher concentrations can cause death. Concentrations as high as 100 ppm have been found in areas of heavy auto traffic. Concentrations of 20,000 to 40,000 ppm have been found in cigarette smoke, and the concentration of a poorly ventilated smoke-filled room usually amounts to several hundred parts per million.

B. The Sources and Effects of Sulfur Oxides

The chief source of sulfur dioxide and sulfur trioxide in the atmosphere is the combustion of fossil fuels for the generation of energy. The fundamental reactions, already described in Sec. 14-7, are restated here. Sulfur oxides are also expelled into the atmosphere from such sources as petroleum refining, papermaking, and the smelting of copper and many other metals. They also arise from natural sources.

Although healthy adults are able to tolerate sulfur oxide concentrations of about 1 ppm without apparent adverse effects,

S

Sulfur, present in coal, gasoline, and heating oil as sulfur compounds

+

O_2

Oxygen from air

$\downarrow$

SO_2

Sulfur dioxide

$2SO_2$

Sulfur dioxide

+

O_2

Oxygen from air

$\downarrow$

$2SO_3$

Sulfur trioxide

Reactions occurring during the combustion of sulfur-containing coal, gasoline, or heating oils; sulfur is present as compounds, and the products are sulfur dioxide and sulfur trioxide

SO_3

Sulfur
trioxide

+

H_2O

Water
from
atmosphere

↓

H_2SO_4

Sulfuric acid,
often as a mist

Reaction of sulfur trioxide
with water in the atmo-
sphere to form sulfuric acid,
which often forms a mist

H_2SO_4

Sulfuric
acid, formed
from polluting
sulfur trioxide

+

$2NH_3$

Ammonia,
from
natural
processes

↓

$(NH_4)_2SO_4$

Ammonium sulfate,
usually as very
fine particles

The reaction of sulfuric
acid with ammonia in the
atmosphere to form very fine
airborne particles of
ammonium sulfate

people with respiratory problems such as asthma and bronchitis are more susceptible. At higher concentrations (10 ppm) it is harmful to everyone. Most of the sulfur trioxide, SO_3, reacts with water in the atmosphere to form sulfuric acid, which may remain suspended as a mist. The acid formed contributes to the acidity of rain and snow, which has been increasing in parts of the United States and Europe. The consequences of acidic precipitation have not been completely determined, but there is suspicion that it may damage vegetation, kill off the fish in lakes and streams, and add to corrosion problems. Where ammonia, NH_3, is in the air, it reacts with the sulfuric acid to form solid ammonium sulfate, $(NH_4)_2SO_4$, which usually remains suspended as very fine particles (margin).

Industrial smog, the hazardous and sometimes deadly smog which is generated in highly industrialized areas, is made up chiefly of sulfur dioxide, sulfuric acid mist, and airborne particles of smoke, ash, and substances like ammonium sulfate. This type of air pollution has been known for some time and was responsible for the 1952 disaster in London (Sec. 15-1) and many similar crises. (The removal of sulfur oxides from coal combustion gases was discussed in Sec. 14-7B.) Buildings and sculpture made of stone containing calcium carbonate, $CaCO_3$, such as limestone and marble, are particularly vulnerable to the eroding effects of industrial smog because calcium carbonate reacts with sulfuric acid so easily (Fig. 15-2) in a proton-transfer reaction (Sec. 7-4):

$$CaCO_3 + H_2SO_4 \rightarrow CaSO_4 + CO_2 + H_2O$$

Calcium carbonate, found in limestone, marble, etc. — Sulfuric acid, formed from polluting sulfur trioxide — Calcium sulfate, slightly soluble in water — Carbon dioxide — Water

Reaction illustrating the effect of sulfuric acid in the atmosphere on stone containing calcium carbonate, $CaCO_3$; water can leach out the slightly soluble calcium sulfate formed

C. The Sources and Effects of Suspended Particles

The **particulates** which pollute the atmosphere are the finely divided particles of liquids and solids which remain suspended in the atmosphere for a long time. The smoke from coal combustion contains soot, which is mostly unreacted carbon, and fly ash, which is noncombustible inorganic material. Fine particles are also formed during cement manufacture, mining procedures, ore processing, the incineration of refuse, and a wide variety of other operations. Although the most readily noticed effects are the loss of atmospheric visibility and the soiling of clothes, house paints, and sometimes whole countrysides, the suspended particles can also be harmful to health. Workers who are exposed for a period of years to air laden with fine particles, for example, become susceptible to tuberculosis and lung cancer.

FIGURE 15-2
The effect of industrial smog on outdoor sculpture on a castle
built in 1702 in the Rhein-Ruhr section of Germany: (*a*) appearance in 1908; damage has just begun; (*b*) appearance in 1969;
very little of the original remains. (*Courtesy of Erhard M. Winkler
and Springer-Verlag New York, Inc.*)

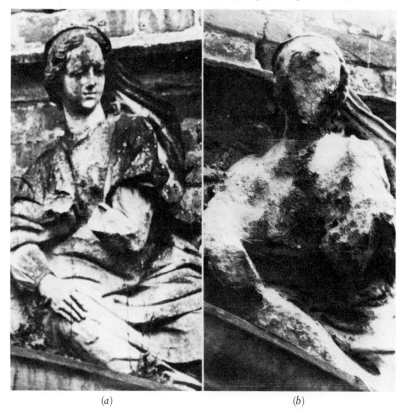

(*a*)   (*b*)

N₂

Nitrogen
from
air

+

O₂

Oxygen
from
air

high temp. | formed during
combustion

↓

2NO

Nitric
oxide

D. The Sources and Effects of Nitrogen Oxides

The oxides of nitrogen in the atmosphere which are of greatest
concern are the toxic air pollutants nitric oxide, NO, and
nitrogen dioxide, NO_2, sometimes represented collectively as
NO_x. They are produced by natural processes, e.g., bacterial
action in the soil and the passage of lightning through the air.
The principal source outside nature is high-temperature combustion processes, e.g., burning gasoline in an automobile
engine. The nitrogen and oxygen of the air do not react with
each other at ordinary temperatures, but at the high temperatures of an internal combustion engine (about 2000°C) they combine to form nitric oxide, as they also do in the presence of
lightning (margin).

Nitric oxide, NO, slowly reacts with the oxygen of the air to
form nitrogen dioxide, NO_2, a yellowish-brown gas. Nitric oxide
is converted into nitrogen dioxide much more rapidly by the

The combination of nitrogen
and oxygen to form nitric
oxide takes place at the high
temperatures of many
combustion processes

$$2NO \; + \; O_2 \; \xrightarrow{\text{slow}} \; 2NO_2$$

Nitric   Oxygen   Nitrogen
oxide,   from   dioxide,
colorless   air   yellowish-
gas   brown gas

The slow reaction between
nitric oxide and oxygen to
form nitrogen dioxide

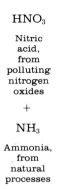

HNO₃ → HNO_3

Nitric
acid,
from
polluting
nitrogen
oxides

+

NH_3

Ammonia,
from
natural
processes

↓

NH_4NO_3

Ammonium
nitrate,
usually as very
fine particles

**The reaction of nitric acid
with ammonia in the
atmosphere to form very
fine airborne particles
of ammonium nitrate**

$$O_2 + O \rightarrow O_3$$

Oxygen　　Atomic　　Ozone
in the　　oxygen
atmosphere

**The production of ozone
from the action of atomic
oxygen on molecular
oxygen**

```
    H  H  H
    |  |  |
H—C=C—C=O
```

Acrolein

```
  H  O        O
  |  ||        ||
H—C—C—O—O—N—O
  |
  H
```

Peroxyacetyl nitrate (PAN)

**Eye irritants found in
photochemical smog**

action of ozone, O_3, which is also present in the atmosphere.

$$NO + O_3 \xrightarrow{\text{rapid}} NO_2 + O_2$$

Nitric　　Ozone　　　Nitrogen　　Oxygen
oxide　　in　　　　dioxide
atmosphere

**The rapid conversion of nitric oxide to nitrogen
dioxide by ozone in the atmosphere**

Nitrogen dioxide reacts with the water in the atmosphere to form nitric acid, HNO_3, which, along with sulfuric acid, is a principal cause of the acidity of rainfall.

$$2NO_2 + H_2O \rightarrow 2HNO_3 + NO$$

Nitrogen　　Water　　Nitric　　Nitric
dioxide　　from　　acid　　oxide
　　　atmosphere

**The simplest of the reactions of nitrogen dioxide with water in the
atmosphere to form nitric acid which contributes to the acidity
of rainfall**

Nitric acid may also react with ammonia, NH_3, in the atmosphere to form solid ammonium nitrate, NH_4NO_3, which, like ammonium sulfate, contributes to the airborne solid particles (margin). Nitrogen oxides also play a significant role in the formation of one type of smog, discussed in the next section.

E. The Sources and Effects of Hydrocarbons

The decay of organic matter produces large quantities of methane, most of which is apparently converted at high altitudes into carbon monoxide (Sec. 15-2A). The principal sources of hydrocarbons involved in atmospheric pollution are automobile exhausts, gasoline evaporation from fuel tanks, carburetor evaporation, crankcase evaporation, petroleum refineries, and supply depots. **Photochemical smog** is the irritating and toxic gaseous mixture found in the atmosphere when hydrocarbons and nitrogen oxides are irradiated by sunlight. This form of air pollution has been recognized only in the last 30 years or so, but occurs with annoying frequency in cities as widespread as Los Angeles, Tokyo, and Sydney. It reduces atmospheric visibility, causes eye irritation, and damages plant life. The complex sequence of reactions responsible for its formation have not been completely deciphered, but it is initiated by the action of the sunlight on nitrogen dioxide, appropriately called photochemical decomposition.

$$NO_2 \xrightarrow{\text{sunlight}} NO + O$$

Nitrogen　　　Nitric　　Atomic oxygen;
dioxide　　　oxide　　very reactive

**The decomposition action of sunlight on nitrogen dioxide in the
atmosphere, producing atomic oxygen, a very reactive species**

Most of the reactive oxygen atoms convert oxygen into ozone (margin). Both the ozone and the atomic oxygen react with hydrocarbons in the atmosphere to form the photochemical smog mixture. In additional to nitrogen oxides, ozone, and hydrocarbons, the mixture includes a variety of substances such as the eye irritants listed in the margin.

A. Motor-Vehicle Pollution

The 100 million motor vehicles in the United States are the principal source of air pollution (Table 15-2 and Figs. 15-3 and 15-4), although it is hoped that their contribution will be markedly reduced with the new pollution controls now being introduced. In 1970 Congress enacted the Clean Air Act, which, among many provisions aimed at reducing air pollution, established a set of maximum motor-vehicle emission standards for hydrocarbons, carbon monoxide, and nitrogen oxides (Table 15-3). The phasing in of these standards has been slower than the timetable of deadlines originally intended, but steps are being taken to achieve the goals originally established. Unfortunately, the standards apply only to new cars. In most states no regulations apply to cars already on the road.

FIGURE 15-3
The increase of smog in downtown Boston in the course of one day. (*Photograph by Allen Morgan, courtesy of the Massachusetts Audubon Society.*)

FIGURE 15-4
The 100 million motor vehicles in the United States are the principal source of air pollution. (*Blair Pittman, EPA-DOCUMERICA.*)

Several different approaches are available in principle for the reduction of polluting emissions from motor vehicles. The steps taken or considered by most automobile manufacturers involve (1) providing for the recyling of the volatile hydrocarbons in the cranckcase back to the combustion chamber, (2) carburetor and timing adjustments to increase the degree of fuel combustion and to reduce carbon monoxide and hydrocarbons in the exhaust, (3) recirculating part of the exhaust gases back through the engine to cool the fuel mixture and reduce the production of nitrogen oxides, (4) using catalytic converters in the exhaust systems to change carbon monoxide and hydrocarbons into carbon dioxide and water and nitrogen oxides into nitrogen.

The catalytic converters for reducing emissions of carbon monoxide and hydrocarbons use platinum and palladium catalysts which are easily deactivated by lead compounds (Fig. 15-5). For many years the performance of gasoline as a motor fuel has been improved by the addition of tetraethyllead, $Pb(CH_3CH_2)_4$, but cars equipped with catalytic converters must use lead-free gasoline. It is hoped that the steps used to maintain gasoline performance in the absence of tetraethyllead will not cause pollution problems in themselves.

It was originally feared that the ability of catalytic converters to change sulfur compounds in motor-vehicle exhaust to sulfur trioxide would serve to produce sulfuric acid in harmfully polluting amounts. Recent tests have shown, however, that in actual use the catalysts do not change as much of the sulfur compounds as predicted and that the sulfuric acid produced is readily dispersed. Further experience will be required, however, before the full impact of the catalytic converters is known.

B. Reduction of Suspended Particles

The suspended particles, or particulates as they are often called, which originate in combustion processes and a wide variety of other operations (Sec. 15-2C), may to a large extent be prevented from entering the atmosphere by several different processes. These include settling chambers, in which the heavier particles are removed by gravitation, various filtration processes, centrifugation units, and liquid scrubbing devices.

Table 15-3
Auto-Emission Standards in the United States Established by Clean Air Act, 1970, Grams per Mile

Type of emission	Typical values before controls	Air Act standards by 1976
Carbon monoxide	80	3.4
Nitrogen oxides	4	0.4
Hydrocarbons	18	0.4

FIGURE 15-5
Overall sketch of a catalytic-conversion installation in a motor vehicle.

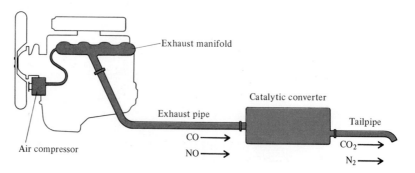

Electrostatic precipitation is an interesting method of removing fine particles by means of electric charges. The effluent gases containing the suspended particulates are led through a chamber in which a field of ions is generated. The particulates are subjected to the action of the ions and the influence of an electric field. They acquire electric charges by picking up ions. Under the influence of the electric field the charged particulates are attracted to electrodes, from which they fall into collecting hoppers (Fig. 15-6). The effectiveness of any of the methods of reducing particulates in effluent gases depends on the characteristics of specific situations, e.g., the rate of airflow, the size and mass of the particles, their chemical nature, and their concentration (Fig. 15-7).

15-4 LEAD COMPOUNDS AS POLLUTANTS

The severe toxic effects of lead and lead compounds include damage to the brain, coma, and death. Less poisonous doses lead to anemia, fatigue, and headaches. Children and infants are the most susceptible. A concentration in the bloodstream of

FIGURE 15-6
Diagram of the action of an electrostatic precipitator for reducing particulates in effluent gas. The gases are led into a chamber which gives the particles an electric charge and subjects them to the action of an electric field. The charged particulates are attracted to electrodes, from which they fall into collecting hoppers.

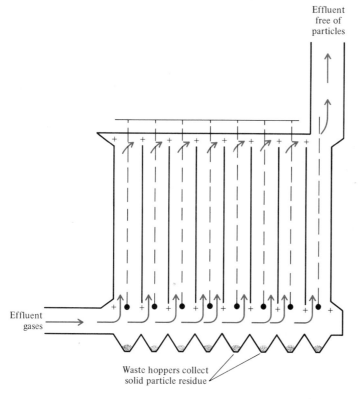

Effluent
free of
particles

Effluent
gases

Waste hoppers collect
solid particle residue

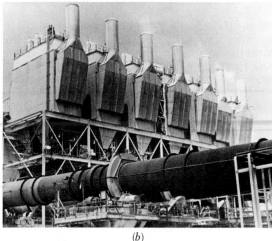

(a)   (b)

about 40 micrograms per 100 milliliters is considered to be acceptable (1 microgram is 1×10^{-6} gram), but concentrations above 80 micrograms per 100 milliliters may bring on mild symptoms, and concentrations above 120 micrograms per 100 milliliters are dangerous.

About 1.6 million tons of lead are used in the United States each year. Principal applications include storage batteries, metal alloys and products, paint pigments, and gasoline additives such as tetraethyllead, $Pb(CH_3CH_2)_4$. Lead compounds are found in the earth's crust, in the soil, in water, and in the atmosphere. One source of lead in the soil derives from the former use of lead arsenate as an insecticide. This finds its way, for example, into cigarette tobacco, where it is responsible for the 1 or 2 micrograms of lead found in the smoke of the average cigarette. Mining and other industrial operations can sometimes spill lead compounds into streams and lakes. Acidic beverages may leach lead from earthenware if it has lead-containing ceramic glaze which has not been properly fired. Lead compounds are widely used as paint pigments, but they are now restricted to exterior paints. Some young children have a tendency to eat such ordinarily inedible objects as paint chips. Old houses with lead-containing paints on their walls are a source of an unfortunate amount of lead poisoning in small children.

The major source of lead pollution in the atmosphere is the exhausts of automobiles using gasoline containing additives such as tetraethyllead. About 250,000 tons of lead were used in 1974 for this purpose (about 16 percent of the total lead consumed), resulting in an estimated 170,000 tons of lead being discharged into the atmosphere as lead compounds. The lead content of nonpolluted air in remote regions is about 0.02 microgram in a cubic meter. In downtown urban locations of many cities it is about 100 times greater. Studies of the lead content of ice in Greenland have shown that deep ice formed about 200 years ago contains very little lead, but newer ice, closer to the surface, contains an appreciable amount. Significantly, the beginning of a sharp increase in the lead content of the ice coincides with the widespread use of lead additives in gasoline.

The lead content of the human bloodstream is the principal method of monitoring the possibility of lead poisoning. Normal lead content appears to be somewhere around 10 to 20 micrograms per 100 milliliters of whole blood, with men tending, for some unknown reason, to have somewhat higher concentrations than women. One study found a sample of residents living near a busy California highway to have lead levels of about 23 micrograms per 100 milliliters (men) and 17 micrograms per 100 milliliters (women). Residents living away from the highway had an average of about 7 micrograms less. Samplings of traffic policemen, garage mechanics, and parking-lot attendants average lead blood levels in the range of 30 to 38 micrograms for each 100 milliliters of whole blood.

Fortunately, the use of leaded gasoline is declining because lead deactivates the catalytic converters installed in most automobiles made since 1975 to reduce pollutants (Sec. 15-3A). The Environmental Protection Agency has insisted since mid-1974 that all gasoline stations have leadfree gasoline available. The use of lead additives in gasoline is expected to end altogether in a few years.

15-5 THE HAZARDS OF FLUOROCARBONS

Before the early 1930s refrigeration units had to use compounds like ammonia and sulfur dioxide as the easily liquefiable gases which remove heat from enclosed spaces in the cooling cycle. Such substances were either highly toxic or flammable or both, and the possibility of accidental leaks presented considerable hazard. There was great enthusiasm, therefore, when nontoxic, nonflammable fluorocarbons were developed for refrigeration units in the early 1930s. **Fluorocarbons,** as the name implies, are substances containing both fluorine and carbon atoms. Two common ones are Freon-11 (trichlorofluoromethane, $CFCl_3$) and Freon-12 (dichlorodifluoromethane, CF_2Cl_2). Because of their inertness and safety they also found considerable use as propellants in aerosol spray cans.

In 1974, however, it was discovered that fluorocarbons in the atmosphere may pose a threat to the protective ozone found in the stratosphere. Paradoxically, ozone in the lower atmosphere can be an annoying and toxic air pollutant, especially as a component of photochemical smog (Sec. 15-2E), but at high altitudes it performs the very important function of absorbing the high-energy ultraviolet light of the sun, screening this harmful radiation from the people, plants, and animals of the earth. Indeed, it is thought that early forms of life were able to move from the sea to the land only after this protective blanket of ozone had been formed from the oxygen produced by photosynthesis.

$$Cl-\underset{\underset{\displaystyle Cl}{|}}{\overset{\overset{\displaystyle Cl}{|}}{C}}-F$$

Trichlorofluoromethane
(Freon-11)

$$Cl-\underset{\underset{\displaystyle F}{|}}{\overset{\overset{\displaystyle Cl}{|}}{C}}-F$$

Dichlorodifluoromethane
(Freon-12)

Two common fluorocarbons used as refrigerants and (formerly) as propellants in aerosol sprays

$$(1) \quad O_3 \xrightarrow[\text{radiation}]{\text{ultraviolet}} O_2 + O$$

Ozone   Oxygen   Atomic oxygen

The decomposition of ozone as it absorbs ultraviolet radiation in the stratosphere

$$(2) \quad O + O_2 \rightleftharpoons O_3$$

Atomic oxygen   Oxygen   Ozone

Dynamic equilibrium balance between the decomposition and formation of ozone in the stratosphere

$$Cl$$
Atomic chlorine
+
$$O_3$$
Ozone
$$(3) \quad \downarrow$$
$$ClO$$
Chlorine oxide
+
$$O_2$$
Oxygen

The action of atomic chlorine on ozone

$$ClO$$
Chlorine oxide
+
$$O$$
Atomic oxygen
$$(4) \quad \downarrow$$
$$Cl$$
Atomic chlorine
+
$$O_2$$
Oxygen

Reaction of chlorine oxide and atomic oxygen to form atomic chlorine and molecular oxygen; the atomic chlorine can then act on additional ozone, as in the reaction above

In order to understand the potential threat of fluorocarbons to the ozone concentration in the upper atmosphere we should be aware first of all that as ozone absorbs ultraviolet radiation, it is decomposed into molecular and atomic oxygen [margin (1)]. The atomic and molecular oxygen subsequently recombine to form ozone, and a sensitive equilibrium situation prevails [margin (2)]. The hazard presented by the infiltration of the stratosphere by fluorocarbons is due to the ease with which they are decomposed by ultraviolet radiation to form very reactive atomic chlorine:

$$Cl-\underset{\underset{Cl}{|}}{\overset{\overset{Cl}{|}}{C}}-F \xrightarrow[\text{radiation}]{\text{ultraviolet}} Cl-\underset{}{\overset{\overset{Cl}{|}}{C}}-F + Cl$$

Trichlorofluoromethane, a representative fluoro-carbon   Dichlorofluoro radical   Atomic chlorine

The decomposition of trichlorofluoromethane by ultraviolet radiation in the stratosphere

The atomic chlorine can subsequently act directly on ozone [margin (3)]. The chlorine oxide, ClO, can then react with atomic oxygen to form atomic chlorine [margin (4)]. The atomic chlorine is then available to act on more ozone.

These reactions could serve to reduce the amount of protective ozone and thus increase the amount of ultraviolet radiation reaching the earth. This, in turn, could lead to an increase in skin cancer in people and could be hazardous to plants and animals as well, although little is known about what the specific effects might be. The increase in ultraviolet radiation could also bring about unfavorble changes in the climate.

The threat of atmospheric fluorocarbons to the concentration of ozone in the stratosphere was originally suggested by interpretations of laboratory data. This led to an intensive investigation by a committee of the National Research Council. Measurements of key substances in the upper atmosphere were made by balloon-borne instruments and remote-sensing techniques. Although many important questions about the chemistry of the ozone layer remain unanswered, the committee found sufficient confirmation of the reasoning behind the original concern to recommend that all fluorocarbon use in aerosol sprays be banned. Moreover, they further recommended that potential fluorocarbon hazards be further studied with the idea that it might prove advisable to ban fluorocarbons in motor-vehicle air conditioners and industrial refrigeration units. Such investigations are now under way, along with attempts to determine the extent to which fluorocarbons may be removed by as yet undiscovered reactions in the atmosphere before they have an opportunity to reduce the ozone concentration.

Additional factors are currently suspected as possible causes of stratospheric ozone depletion. The high-altitude emissions of nitrogen oxides by supersonic transport planes could possibly lead to the destruction of some ozone. More ominous consequences are suggested by the discovery that nitrous oxide, N_2O, produced in the soil from nitrogen-containing fertilizers, is possibly converted in the stratosphere into ozone depleting nitric oxide, NO. Information about the nature of stratospheric ozone is incomplete, the chemistry involved is very complex, and much further investigation is needed.

15-6 THE PROBLEM OF SOLID WASTES

For many years our solid wastes have been disposed of by burying them, by burning them, or by dumping them in the ocean. But as the amount of waste has increased to alarming proportions, it has become clear that other methods must be developed. We are now discarding about 5 billion tons of solid waste each year, and we are spending over $6 billion to collect and dispose of it. Although interest in the waste problem is growing and methods of handling waste products are improving, disposal techniques for the most part are far from adequate. All too often waste products end up polluting the air and water, causing unsightly messes on land areas (Fig. 15-8),

FIGURE 15-8
A metals junkyard. (*McAllister, EPA-DOCUMERICA.*)

and contributing to the spread of disease. Moreover, many of us have a tendency to discard whatever we have been using in a casual and irresponsible way. The number of popcorn cups and candy wrappers on the floor of many movie houses is a phenomenon in itself.

A. The Increasing Demand for Raw Materials

At least some of the waste problem should be solved by the growing difficulties in obtaining energy supplies and raw materials for manufacturing processes. A technological society like ours in the United States requires enormous quantities of raw materials to prosper. The nation is currently consuming about 23 percent of the minerals used in the entire world, and our needs are growing rapidly. The collective annual consumption of many mineral products is presented in Table 15-4. S. Victor Radcliffe, a Fellow at Resources for the Future, has added up the soaring consumption of materials by each person in the United States. His data are presented in Table 15-5.

The United States has found it necessary to go to increasingly greater lengths to find the raw materials it is consuming. It currently imports most of its sources of over 20 important minerals, placing itself in the vulnerable position which accompanies such foreign sources. Both supply and costs are subject to sudden and unexpected changes. We are now importing, for example, 88 percent of our aluminum sources, 23 percent of our iron sources, and about 70 percent of our nickel, silver, and gold sources. It is estimated that in another 25 years we shall be importing more than half of our nonfuel mineral requirements.

If the composition of seawater and the earth's crust is examined, it would appear that mineral resources are practically inexhaustible. A cubic kilometer of the earth's crust is estimated to contain, for example, 2×10^8 tons of aluminum, 8×10^5 tons of zinc, and 2×10^5 tons of copper. A cubic mile of seawater contains, among other elements, 6×10^6 tons of magnesium, 1.8×10^6 tons of potassium, and 12 tons of tin. H. E. Goeller, of the Oak Ridge National Laboratory, and Alvin M. Weinberg, of the Institute for Energy Analysis, have estimated the world resources of a number of minerals (Table 15-6), and the composition of seawater is given in Table 15-7. (The details of the extraction of magnesium from seawater were discussed in Sec. 13-7.)

The problem is that most of these sources are in the form of low-grade ores with low concentrations, and the use of lower-grade sources requires the use of greater amounts of energy. Indeed, while future supplies of raw materials cannot be taken for granted, the problem is more one of energy sources than of other resources. But with questions about future energy supplies hovering over us, the dimensions of the overall problem are not reduced by this finding. Moreover, as the nation turns to more dilute mineral sources, land disruption and the pollution from mining and ore processing could intensify. Even the current mining of ores presents severe environmental problems. Learning how to use low-grade ores within stringent environmental constraints presents a great challenge.

Table 15-4
Approximate Annual Consumption of Mineral Products in the United States (1973)[†]

	Thousands of tons
Metals:	
Iron	102,000
Aluminum	6,500
Copper	2,400
Zinc	1,900
Lead	1,500
Manganese	700
Chromium	640
Nickel	198
Magnesium	116
Tin	84
Uranium	10.3
Silver	6.7
Mercury	2.1
Nonmetals:	
Crushed stone	1,060,000
Sand and gravel	980,000
Petroleum	580,000
Coal	540,000
Natural gas	460,000
Clays	64,800
Salt	46,500
Lime	21,100
Phosphate rock	14,000
Sulfur	10,200
Potash	5,600
Asbestos	880

[†] From Minerals Yearbook 1974, Bureau of Mines, United States Department of the Interior.

B. The Role of Recycling

As such problems increase it would appear eminently logical to extract what is economically reusable from the nation's overflowing trash heaps. The possibilities of using organic wastes to produce fuels have already been discussed (Sec. 14-13D). The retrieval of scrap materials also has great potential. Recovery of substantial amounts of paper and metals has been going on for some time. And a scrap industry has been in operation which facilitates getting the scrap in the hands of users. A summary of scrap metals currently being recycled is presented in Fig. 15-9; see also Fig. 15-10. In order for the recovery of scrap materials to be markedly expanded, however, several technical and organizational problems will have to be solved. Ways must be found to improve recovery methods and reduce costs, and markets for the scrap products must be developed. The time for such action is obviously at hand.

15-7 THE PROBLEM OF CANCER-CAUSING SUBSTANCES IN THE ENVIRONMENT

A. Carcinogenic Agents

Almost 20 percent of all deaths in the United States are caused by the various forms of cancer. Only diseases of the heart are responsible for greater loss of life. Although the elderly are the

Table 15-6
Estimated Ultimate Resources of the Compounds of 18 Elements
Some sources are given and in addition the maximum amount in the best resource is estimated in terms of the percent of total sources; in some cases the extraction technology has not yet been developed.[†]

Element	Resource	Maximum percentage in best resource	World resource, tons
C (oxidized)	Limestone	12	2×10^{15}
Si	Sand, sandstone	45	1.2×10^{16}
Ca	Limestone	40	5×10^{15}
H	Water	11	1.7×10^{17}
Fe	Basalt, laterite	10	1.8×10^{15}
N	Air	80	4.5×10^{15}
Na	Rock salt, seawater	39	1.6×10^{16}
O	Air	20	1.1×10^{15}
S	Gypsum, seawater	23	1.1×10^{15}
Cl	Rock salt, seawater	61	2.9×10^{16}
P	Phosphate rock	14	1.6×10^{10}
K	Sylvite, seawater	52	5.7×10^{14}
Al	Clay (kaolin)	21	1.7×10^{15}
Mg	Seawater	0.13	2×10^{15}
Mn	Sea floor nodules	30	1×10^{11}
Ar	Air	1	5×10^{13}
Br	Seawater		1×10^{14}
Ni	Peridotite	0.2	6×10^{11}

† Adapted from H. E. Goeller and Alvin M. Weinberg, The Age of Substitutability, **Science, 191**:685 (1976). Copyright 1976 by the American Association for the Advancement of Science.

Table 15-5
The Consumption of Materials by Each Person in the United States Each Year in Pounds Using Information for 1972[†]

Materials	Amount, pounds per person[‡]
Nonrenewable resources	
Nonmetallic minerals;	
Sand and gravel	9,000
Stone	8,500
Cement	800
Clays	600
Total	18,900
Metals:	
Iron and steel	1,200
Aluminum	50
Copper	25
Lead	15
Zinc	15
Other metals	35
Total	1,340
Polymers (synthetics, mostly from oil and natural gas):	
Plastics and resins	100
Synthetic rubber	26
Noncellulosic fibers	22
Total	148
Renewable resources	
Wood and wood products (1971):	
Lumber	1,141
Plywood and veneer	224
Pulp products	780
Other	77
Total	2,222
Natural rubber	8
Fibers:	
Cotton	19
Other plant	1
Animal (wool, silk)	1
Synthetic (cellulosic)	8
Total	29
Leather	14

† From S. Victor Radcliffe, World Changes and Chances: Some New Perspectives for Materials, **Science, 191**:701 (1976). Copyright 1976 by the American Association for the Advancement of Science.

‡ Data include imports.

FIGURE 15-9
A summary of scrap metals recycled annually in the United States
(as of 1972). Significant amounts have been recycled for some
time, and expansion is expected. [*From S. L. Blum, Tapping Resources
in Municipal Solid Waste, Science,* **191**:671 *(1976); copyright 1976 by the
American Association for the Advancement of Science.*]

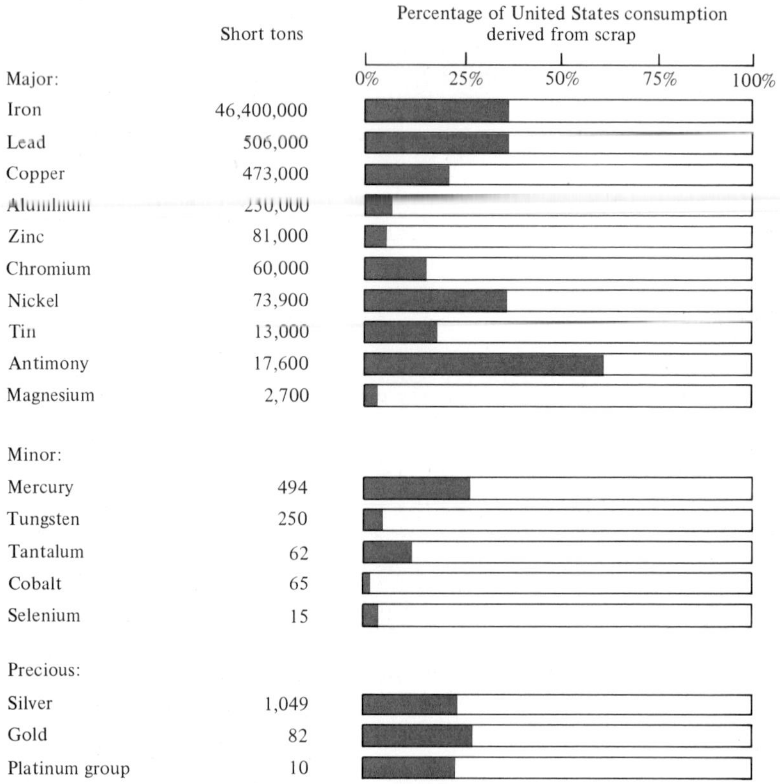

	Short tons	Percentage of United States consumption derived from scrap
		0% 25% 50% 75% 100%
Major:		
Iron	46,400,000	
Lead	506,000	
Copper	473,000	
Aluminum	230,000	
Zinc	81,000	
Chromium	60,000	
Nickel	73,900	
Tin	13,000	
Antimony	17,600	
Magnesium	2,700	
Minor:		
Mercury	494	
Tungsten	250	
Tantalum	62	
Cobalt	65	
Selenium	15	
Precious:		
Silver	1,049	
Gold	82	
Platinum group	10	

Table 15-7
Amount of Various Elements
in Seawater
The volume of seawater is
estimated to be about 350
million cubic miles.

Element	Thousand tons per cubic mile, approximate
Chlorine	90,000
Sodium	50,000
Magnesium	6,000
Sulfur	4,000
Calcium	1,900
Potassium	1,800
Bromine	300
Carbon	130
Strontium	40
Boron	22
Silicon	15
Fluorine	6
Argon	3
Nitrogen	2
Lithium	0.8
Rubidium	0.6
Phosphorus	0.3
Iodine	0.3
Barium	0.15
Indium	0.1
Zinc	0.045
Iron	0.045
Aluminum	0.045
Molybdenum	0.040
Selenium	0.020
Tin	0.012
Copper	0.012
Arsenic	0.012
Uranium†	0.012

† Plus smaller amounts of at
least 28 other elements including silver and gold.

most susceptible, young people are by no means immune. John
Cairns of the Imperial Cancer Research Fund of London has
compared the causes of death ranked according to the annual
loss of life with a ranking by annual loss of working life up to
the age of 65 (Fig. 15-11). The second ranking reflects the age
distribution of those who succomb to the various causes of
death. Among the forms of cancer the relative significance of
leukemia is increased, since the incidence of this type is high
among the young. Lung cancer, however, maintains its position
as the leading fatal form of cancer. The ranking according to
loss of working life also serves to emphasize the seriousness of
accidents as a cause of death, moving them into first place. But
cancer retains a leading role in the loss of life, in third position
after arterial diseases.

Opinions differ, but many authorities estimate that somewhere between 50 and 90 percent of human cancers may be
caused by factors in our personal environment. These include
diet, smoking, radiation exposure, and cancer-causing substances which find their way into the atmosphere, water sup-

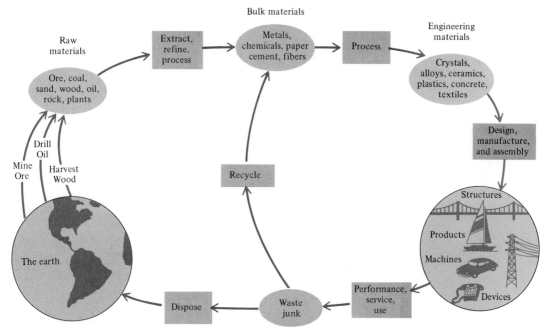

FIGURE 15-10
The cycle of materials from sources of the earth through various processes of extraction, refinement, and fabrication to forms of end uses and finally back to the earth. (*Adapted from "Mineral Resources and the Environment," p. 30, 1975, with the permission of the National Academy of Sciences.*)

plies, foods, cosmetics, and other products with which we come in contact one way or another. That specific substances may cause cancer was established many years ago. Following through on the discovery that workers exposed to coal tar were more apt to have skin cancer than most, Japanese scientists convincingly demonstrated that coal tar itself was responsible. When they painted the ears of rabbits with coal tar, the rabbits developed skin cancer. Ten years or so later English investigators identified a specific cancer-causing substance, benzo[a]-pyrene (Table 15-8) in coal tar. This hydrocarbon has recently been identified as one of the major cancer-causing agents in tobacco smoke.

A **carcinogenic agent** is any substance or material capable of causing cancer. The chemicals which are considered to be carcinogenic or are strongly suspected of being so include a wide variety of substances (Table 15-8). Many are synthetic carbon-containing compounds such as benzidine, 2-naphthylamine, and vinyl chloride, used (or formerly used) in the production of chemical products. Others are natural carbon-containing substances, such as aflatoxin B_1, found in a mold on peanuts and other plants, and safrole, found in sassafras root. Some are inorganic substances, such as cadmium sulfide and chromium trioxide. Still others are radioactive substances, like lead 210, which are carcinogenic because of the radiation they produce.

FIGURE 15-11
Causes of death in the United States ranked according to the annual loss of life (left) and by annual loss of working life up to the age of 65 (right). The second ranking increases the significance of leukemia among the forms of cancer and the seriousness of accidents among all causes of death. (*From John Cairns, The Cancer Problem. Copyright by Scientific American, Inc. All rights reserved.*)

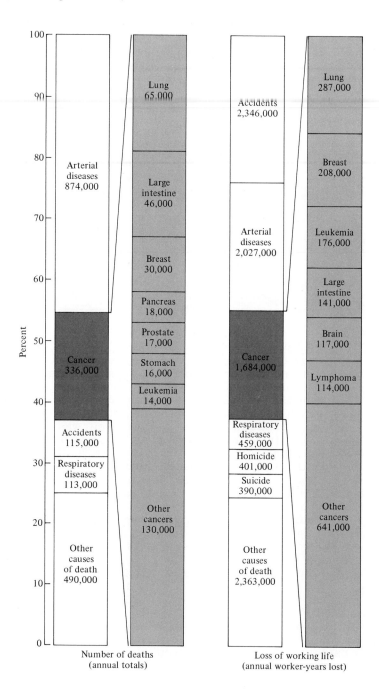

Table 15-8
Representative Substances Found to Cause Cancer in Small Animals

Substance	Use or source
Aflatoxin B₁	Found naturally in mold on peanut meal, corn, dried chili peppers, etc.
Benzidine	Synthetic, used to synthesize dyestuffs
Benzo[a]pyrene	Found naturally in coal tar and cigarette smoke
CdS — Cadmium sulfide	Synthetic, used as a coloring material for glass, textiles, rubber, paper, paint, etc.
CrO₃ — Chromium trioxide	Synthetic, used in chromium plating, photography, etc.
4-Dimethylaminoazobenzene	Synthetic, formerly used as a dairy-food coloring agent
$^{210}_{82}Pb$ — Lead 210 (radioactive)	Found in cigarette smoke
2-Naphthylamine	Synthetic, formerly used to synthesize dyes

(Continued on next page)

.e 15-8
presentative Substances Found to Cause Cancer in Small Animals (*Continued*)

Substance	Use or source
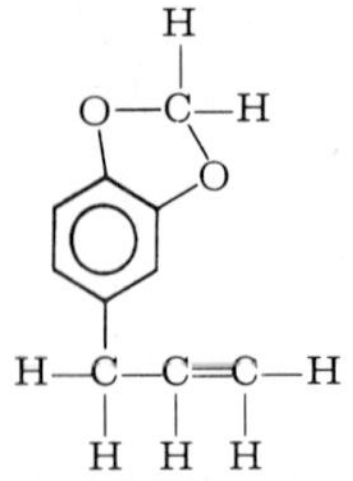 Safrole	Found in sassafras root, used in making sassafras tea
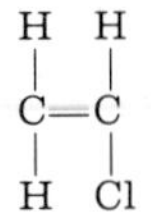 Vinyl chloride	Synthetic, used to synthesize polyvinyl chloride

B. The Nature of Cancerous Tumors

Tumors are growths composed of cells which lack the normal controls on growth and reproduction. They often grow rapidly. A **benign tumor** is one which remains within the tissue in which it develops. A **cancerous tumor** is a growth whose cells invade the surrounding tissue (Fig. 15-12). Many cancerous tumors undergo **metastasis;** i.e., they release malignant cells which travel through the body fluids and establish new cancerous growths in distant parts. Such spreading is their most destructive characteristic. (The treatment of cancer was discussed in Sec. 12-6.)

There is no general agreement about the mechanism responsible for the formation of cancerous cells. One of the most reasonable and interesting of the current hypotheses proposes that carcinogenic substances interact directly with the genetic material of a cell. This concept holds that cells are ordinarily restricted to normal growth and reproduction by several genes in the cell. A **gene** is a basic unit of heredity; it is a segment of the substances in cells, called nucleic acids, which preserve hereditary information and control the synthesis of cell materials. (The structure and function on nucleic acids were discussed in optional Sec. 10-8.) It has been established that at least some carcinogenic substances are capable of interfering with the normal behavior of a gene. They serve to inactivate it and remove the gene's growth-restraining ability. But since each of the controlling genes presumably acts independently, all of them must be inactivated before the constraints on the cell's growth and reproductive capacity are removed and the potential for tumor growth is formed (Fig. 15-13). The successive inactivation of all of the genes may extend over many years.

Once the inactivation of genes has transformed cells to a pre-tumor condition, they may form a benign tumor, they may regress to normal behavior, or they may form a cancerous tumor which invades neighboring tissues and metastasizes (Fig. 15-13). Such a mechanism could conceivably explain why ex-

FIGURE 15-12
A benign tumor (*a*) remains within the tissue in which it develops, but a cancerous tumor (*b*) invades the surrounding tissue and usually undergoes metastasis, releasing malignant cells which migrate through the body fluids to establish new cancerous growths in distant parts of the body. [*From I. J. Fidler and M. L. Kripke, Tumor Growth and Spread, Chemistry,* **50**:18 (1977). *Reprinted with permission of Chemistry. Copyright American Chemical Society.*]

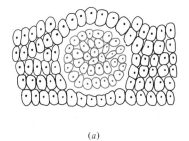

(*a*)

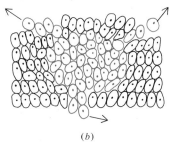

(*b*)

FIGURE 15-13
Diagram of a hypothesis of the origin of a cancerous tumor. Several successive exposures to carcinogens inactivate all the growth-restraining genes in a cell, producing a pretumor condition. This condition can lead to a regression to normal cells, a benign tumor, or a malignant tumor, which undergoes metastasis, releasing malignant cells to other sites. (*From John Cairns, The Cancer Problem. Copyright by Scientific American, Inc. All rights reserved.*)

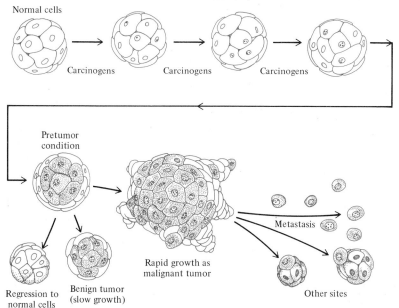

Normal cells

Carcinogens   Carcinogens   Carcinogens

Pretumor
condition

Regression to
normal cells

Benign tumor
(slow growth)

Rapid growth as
malignant tumor

Metastasis

Other sites

posure to carcinogens does not always produce a cancerous tumor (not all the genes have been inactivated). It could also account for the long period often observed between exposure and cancer appearance (required for all of the genes to be inactivated). Moreover, the hypothesis could explain the additive effects of successive carcinogen exposures which sometimes have been observed. It would also make it difficult to say how much cancer is due to a specific carcinogenic agent. Although this hypothesis has attractive features, it has by no means been firmly established.

C. The Difficulties of Identifying Cancer-causing Agents

The long latency period in people It is entirely possible that specific carcinogenic substances in the diet, in water supplies, in industrial workplaces, in cigarette smoke, or in the general environment are the most important causes of cancer. If this is indeed the case, it would be supremely reasonable to seek them out and either remove them or severely restrict the opportunity for exposure. Unfortunately, difficulties arise in determining whether or not a given substance is indeed a carcinogen in human beings. There are differences of opinion about the validity of some of the various kinds of evidence. What gives rise to these differences?

One of the difficulties in assessing the potential carcinogenic effects of a substance is caused by the 15- to 40-year latency period frequently observed between exposure and cancer appearance in people. This is demonstrated by the statistical relationship between cigarette smoking and lung cancer. The data presented in Fig. 15-14, obtained from populations in England and Wales, show how the increase in lung cancer followed the increase in smoking after about 25 or 30 years. The data are most complete for men because they began smoking cigarettes around the turn of the century. Women did not begin until about 1925, and lung cancer in women has only recently started to increase.

The reactions often required in the body before carcinogens are generated Many carcinogens are not cancer-causing themselves but are converted in the body to other substances which are the actual carcinogenic agents. This phenomenon further complicates the assessment of carcinogenic activity. Benzo[a]-pyrene, the powerful cancer-causing substance found in coal tar and cigarette smoke, is not carcinogenic itself, but in the body it is converted into an oxygen derivative which is the active cancer-causing substance (margin). This reaction is made possible by the presence in the body of one or more enzymes which normally catalyze reactions involved in the inactivation of foreign chemicals in the body. Evidence suggests, however, that the capacity of animals and humans to form the actual carbinogens is not uniform. Some are able to carry out the conversions more readily than others, which may explain why some people are more susceptible to certain cancer agents than others.

Benzo[a]pyrene, found in coal tar and cigarette smoke; not carcinogenic itself

enzymes | in body

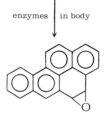

Oxygen derivative, the actual cancer-causing agent

The conversion, with the aid of enzymes in the body, of benzo[a]pyrene into the active carcinogen responsible for the cancer-causing action

FIGURE 15-14
Data from residents of England and Wales relating the increase
in lung cancer to an increase in cigarette smoking. The time interval between exposure and cancer appearance is striking. Men
began smoking cigarettes around the turn of the century, but
lung cancer in men did not start to rise until about 30 years later.
Women did not begin smoking until around 1925, and lung
cancer in women has only recently started to rise. (*From John
Cairns, The Cancer Problem. Copyright by Scientific American, Inc. All rights
reserved.*)

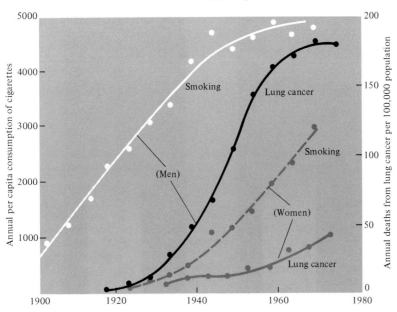

The interpretation of tests on animals Perhaps the most
troublesome part of determining which substances cause cancer
in human beings is related to disagreements about the significance of tests on animals. Mice and rats, the animals most often
used, live for only about 2 years. To make this time represent a
span of 50 or 60 years in humans the doses of test substances
given to the animals are greatly exaggerated. This is based on
the experience that larger doses accelerate the formation of
tumors. Some experienced investigators maintain that such
animal tests are valid indicators of the cancer risk in humans. If
large doses over a short time bring about cancer in animals, it
is held, then there is risk that cancer may be caused in humans
by low doses of the same agent over a much longer time.

Other experts in the field have some doubts about this view.
They are concerned that the excessive test doses may interfere
with the normal physiology of the animals and cause the cancer
results to be misleading. They are particularly suspicious of
animal experiments which use only one species of animal, in-
sisting that a test should use at least two different kinds, prefer-
ably several. Some also question the relevancy of tests on mice,
suggesting that tests on monkeys and other nonhuman pri-
mates might be more indicative of what can be expected in man.

Studies of the prevalence of cancer Warning of a cancer cause may also be obtained by studying patients' records, family histories, as well as the regional and occupational variations of cancer incidence. Such information was important, for example, in the discovery of the cancer-causing activity of vinyl chloride. This substance is the starting material in the production of polyvinyl chloride polymers (Table 13-3), millions of tons of which are used in the United States and elsewhere in the world to make such products as vinyl electrical insulation, phonograph records, house sidings, upholstery fabrics, and automobile top coverings. Vinyl chloride itself, a gas, has also been used as a propellant for aerosol sprays.

$$
3n\ \underset{\substack{|\\ \text{H}}}{\overset{\substack{\text{H}\\ |}}{\text{C}}}\!=\!\underset{\substack{|\\ \text{Cl}}}{\overset{\substack{\text{H}\\ |}}{\text{C}}}\ \xrightarrow{\text{cat.}}\ \left[\ \underset{\substack{|\\ \text{H}}}{\overset{\substack{\text{H}\\ |}}{-\text{C}}}\!-\!\underset{\substack{|\\ \text{Cl}}}{\overset{\substack{\text{H}\\ |}}{\text{C}}}\!-\!\underset{\substack{|\\ \text{H}}}{\overset{\substack{\text{H}\\ |}}{\text{C}}}\!-\!\underset{\substack{|\\ \text{Cl}}}{\overset{\substack{\text{H}\\ |}}{\text{C}}}\!-\!\underset{\substack{|\\ \text{H}}}{\overset{\substack{\text{H}\\ |}}{\text{C}}}\!-\!\underset{\substack{|\\ \text{Cl}}}{\overset{\substack{\text{H}\\ |}}{\text{C}}}\!-\ \right]_n
$$

Vinyl chloride Polyvinyl chloride polymer

The overall reaction for the formation of polyvinyl chloride from vinyl chloride; it is the gas, vinyl chloride, which causes cancer, not the polymer

Between 6000 and 7000 people in the United States are involved in the production of vinyl chloride and its conversion into polyvinyl chloride. Another 600,000 or 700,000 workers are involved in the fabrication of the various end products made from the polymer. Fortunately, however, the polymer has not been found to be carcinogenic.

In January 1974 one industrial user of vinyl chloride called the attention of federal and state health agencies to the deaths of three workers in one of their plants of liver cancer during the preceding 3 years. Since the form of cancer was rare, vinyl chloride, rather than a more common cause, was suspected as being responsible.

One of the difficulties of interpreting evidence from the statistical study of regional and occupational variations of cancer incidence is the frequent lack of a firmly established cause-and-effect relationship. The evidence is more convincing if the occurrence of cancer can be demonstrated to be actually caused by the suspected agent. In the case of vinyl chloride, it turned out that it had already been discovered in 1972 by an Italian investigator that in rats vinyl chloride causes liver cancer of the same type found in the vinyl chloride workers. About 350 rats had been exposed to more than 50 ppm of vinyl chloride gas for 12 months, and the same form of liver cancer appeared in about 10 percent of them.

A total of 32 cases of cancer were subsequently reported among the more than 6000 workers in the vinyl chloride plants. The use of vinyl chloride as a propellant in aerosol sprays has been banned,and the Occupational Safety and Health Administration has set a 1-ppm exposure limit over an 8-hour average period for all workers in the industry.

The average length of exposure for the cancer victims among the vinyl chloride workers was 19 years. Evidence suggests, however, that chronic exposure is not necessary in order for vinyl chloride to increase the risk of cancer, a single dose of a sufficient amount being all that is apparently necessary. It is hoped that the number of additional cancer victims will be held to a low level despite the great numbers of people who have been exposed to vinyl chloride in the many years it has been produced in high volume.

The expense and slowness of tests on animals There is a great need for a carcinogenicity test which will give strongly convincing indication of the cancer-causing potential of a substance in humans. Such a test should also be quick and inexpensive. Animal experiments, even with their uncertainties, take from 2 to 3 years and cost at least $100,000 for each substance tested. The work of Bruce Ames and his colleagues at the University of California at Berkeley has provided a test which is one of the most promising of several quick tests recently developed. The validity of the Ames test has not yet been completely settled, but it appears to be a significant step in the right direction.

The Ames test takes advantage of the connection between the cancer-causing effect of a substance and its ability to cause a mutation, i.e., alter a cell's heredity-controlling nucleic acids (Sec. 10-8). Ames uses a strain of bacteria which, as a result of a mutation, has lost its ability to produce the substance histi-

FIGURE 15-15

The Ames test for potential carcinogenicity: (A) without a potential carcinogen few bacteria colonies form; (B) 1 microgram of the Japanese food additive furylfuramide increases the number of bacteria colonies substantially; (C) the effect of 1 microgram of the mold carcinogen aflatoxin B_1; (D) the effect of 10 micrograms of 2-aminofluorene. [*Reprinted by permission, from B. N. Ames, J. McCann, and E. Yamasaki, Mutation Research,* **31**:347 *(1975), courtesy of Bruce N. Ames.*]

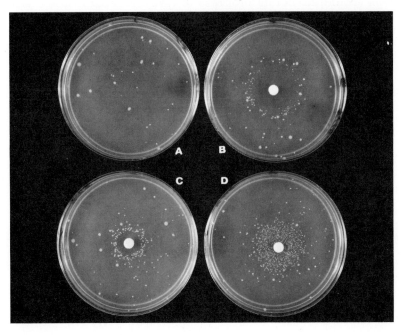

dine. (A **mutation** is an inheritable change in the genetic material of a living organism. An organism which has undergone a mutation is called a **mutant**.) As a consequence of the mutation, these bacteria need histidine in order to function normally, and they will not grow unless it is supplied to the nutrient culture medium in which their behavior is examined. Certain chemicals, however, can bring about a mutation which restores the bacteria's ability to produce histidine. The rehabilitated bacteria grow in the absence of an outside source of histidine (Fig. 15-15).

When a substance is tested for potential carcinogenic properties, it is added to a culture of the mutant bacteria containing all the nutrients for their growth and reproduction except histidine. If the bacteria do not grow, it follows that the test substance cannot bring about a mutation and is therefore considered to be without cancer-causing effect. On the other hand, if the bacteria grow and reproduce after the test chemical is added, the chemical has brought about a mutation of the bacteria and enabled them to produce histidine. The test chemical is then considered to be a potential carcinogen. Ames has shown that 90 percent of all of the known carcinogens tested to date bring about mutations and cause the bacteria to grow and all but a few of those substances considered to be noncarcinogenic do not bring about bacterial growth. Further work is required, however, to establish firmly that a substance which brings about a mutation of the bacteria is actually a carcinogen in humans. Significantly, the Ames test takes only 3 days and costs only about $500 for each substance.

More than 7000 scientists are currently working on research projects related to the problem of cancer. It is to be hoped that a clear method of assessing the risk associated with potential cancer-causing chemicals will soon be found. Such a development would accelerate progress toward the identification and the elimination of the environmental causes of cancer without unnecessarily restricting useful substances. All chemicals, however, should be handled cautiously and thoughtfully at all times.

KEY WORDS

1. Temperature inversion (Sec. 15-1)
2. Pollutant (Sec. 15-2)
3. Industrial smog (Sec. 15-2B)
4. Particulate (Sec. 15-2C)
5. Photochemical smog (Sec. 15-2E)
6. Electrostatic precipitation (Sec. 15-3B)
7. Fluorocarbon (Sec. 15-5)
8. Carcinogenic agent (Sec. 15-7A)
9. Tumor (Sec. 15-7B)
10. Benign tumor (Sec. 15-7B)
11. Cancerous tumor (Sec. 15-7B)
12. Metastasis (Sec. 15-7B)
13. Gene (Sec. 15-7B)
14. Mutation (Sec. 15-7C)
15. Mutant (Sec. 15-7C)

"You win a little and you lose a little. Yesterday the air
didn't look as good, but it smelled better."

SUMMARIZING QUESTIONS FOR SELF-STUDY

Sections 15-1 and 15-2

1. **Q.** Ordinarily how are ground-level air pollutants removed by natural processes of the atmosphere?

A. They are swept away by the motion of the wind or carried upward as the warm air of lower altitudes rises toward the cooler air above.

2. **Q.** What is a temperature inversion?

A. The atmospheric phenomenon in which a mass of warm air lies above a mass of cool air, a reversal of the normal arrangement.

3. **Q.** What is potentially hazardous about a temperature inversion in an industrial region?

A. Under such circumstances there is almost no motion of the air, and pollutants can concentrate at ground level.

4. **Q.** What is a pollutant?

A. Any form of matter which when added to the surroundings in sufficient amount produces a detectable injurious effect on people, animals, plants, or the environment in general.

5. **Q.** What is the principal cause of air pollution in the United States?

A. The gasoline-powered engine of the automobile.

6. **Q.** More than 90 percent of the carbon monox-

ide released to the atmosphere is generated by natural processes. What is the principal source?

A. The oxidation of methane, produced by the decay of organic matter.

7. Q. Why does the concentration of carbon monoxide in the atmosphere not increase as a result?

A. Carbon monoxide is apparently removed from the atmosphere at about the same rate it is formed.

8. Q. What is the largest individual man-made source of carbon monoxide in the United States (Table 15-2)?

A. Automobiles.

9. Q. Carbon monoxide is formed in an automobile engine during the incomplete combustion of the hydrocarbons in gasoline. What is meant by incomplete combustion?

A. Combustion in a limited supply of oxygen.

10. Q. With what would the carbon monoxide be replaced during a complete combustion of hydrocarbons?

A. Carbon dioxide.

11. Q. To what property is the principal toxic effect of carbon monoxide due?

A. Its ability to interfere with the transport of oxygen by the blood.

12. Q. What is the chief source of the air pollutants sulfur dioxide and sulfur trioxide?

A. The combustion of the fossil fuels: coal and petroleum products.

13. Q. How does the emission of sulfur oxides contribute to the acidity of rainfall?

A. Sulfur dioxide is converted in the atmosphere into sulfuric acid as follows:

$$2SO_2 + O_2 \rightarrow 2SO_3$$

$$SO_3 + H_2O \rightarrow H_2SO_4$$

14. Q. What is industrial smog?

A. A mixture of sulfur dioxide, sulfuric acid mist, and airborne particles of smoke, ash, and substances such as ammonium sulfate.

15. Q. How does industrial smog lead to the deterioration of buildings and sculpture made of stone containing calcium carbonate?

A. The sulfuric acid in the smog reacts with the calcium carbonate:

$$CaCO_3 + H_2SO_4 \rightarrow CaSO_4 + CO_2 + H_2O$$

The calcium sulfate, $CaSO_4$, being slightly soluble in water, is leached away by atmospheric moisture.

16. Q. Which oxides of nitrogen are of greatest concern as pollutants?

A. Nitric oxide, NO, and nitrogen dioxide, NO_2.

17. Q. What is the principal source of nitric oxide?

A. The combination of nitrogen and oxygen in the air at the high temperatures of combustion processes, like those in automobile engines:

$$N_2 + O_2 \rightarrow 2NO$$

18. Q. How does nitric oxide in the atmosphere contribute to the acidity of rainfall?

A. The nitric oxide is converted into nitric acid, HNO_3, in the atmosphere as follows:

$$2NO + O_2 \rightarrow 2NO_2$$

or

$$NO + O_3 \rightarrow NO_2 + O_2$$

$$2NO_2 + H_2O \rightarrow 2HNO_3 + NO$$

19. Q. What is photochemical smog?

A. A mixture of nitric oxide, nitrogen dioxide, ozone, hydrocarbons, and many other substances such as the eye irritant PAN.

20. Q. What is the source of the hydrocarbons?

A. Automobile fuel tanks, gasoline engines, engine exhaust, petroleum refineries, gasoline supply depots.

21. Q. How is photochemical smog formed?

A. By the action of sunlight on a mixture of nitric oxides and hydrocarbons in the atmosphere.

Section 15-3

22. Q. What steps are most frequently taken to reduce hydrocarbon emissions from automobiles?

A. Recycling the volatile hydrocarbons in the crankcase back to the combustion chamber, adjustments to increase the degree of fuel combustion, catalytic converters in the exhaust system to complete the combustion of unburned hydrocarbons to carbon dioxide and water.

23. Q. What steps are most frequently taken to reduce the emission of carbon monoxide from automobiles?

A. Adjustments to increase the degree of fuel combustion and catalytic converters in the exhaust system to complete the conversion of carbon monoxide to carbon dioxide.

24. Q. What steps are most frequently taken to reduce the emission of nitrogen oxides from automobiles?

A. Recirculating part of the exhaust gases back through the engine to reduce the temperature of the fuel mixture and catalytic converters to convert nitrogen oxides to nitrogen.

25. Q. What is an electrostatic precipitator?

A. It is a device for removing fine particles from the effluent gases of combustion and other processes by means of electric charges.

Section 15-4

26. Q. Why is paint containing lead compounds as pigments no longer permitted for indoor use?

A. Because children have a tendency to eat ordinarily inedible objects like paint chips and become poisoned by the lead.

27. Q. What is the principal source of lead pollution in the atmosphere?

A. The exhausts of automobiles using gasoline containing additives such as tetraethyllead, $Pb(CH_3CH_2)_4$.

28. Q. Why is tetraethyllead added to gasoline?

A. To increase the performance of gasoline as a motor fuel.

29. Q. Why is the use of leaded gasoline decreasing?

A. Because the catalytic converters used to reduce pollutants in automobile exhaust systems are deactivated by lead compounds.

Section 15-5

30. Q. What is a fluorocarbon?

A. A substance which contains both fluorine and carbon atoms.

31. Q. Give the formula of a popular fluorocarbon.

A.

Cl−C−F with Cl, Cl, Cl   or   Cl−C−F with Cl, Cl, F

Trichlorofluoromethane (Freon-11)   Dichlorodifluoromethane (Freon-12)

32. Q. Why was the availability of fluorocarbons welcomed enthusiastically in the early 1930s?

A. Because they provided a nonflammable, nontoxic replacement for the flammable, toxic sulfur dioxide and ammonia used as the liquefiable gases in refrigeration units.

33. Q. Why has the use of fluorocarbons as the propellants in aerosol spray cans been banned?

A. Because of the concern that fluorocarbons are a threat to the ozone in the stratosphere, nonessential uses have been banned.

34. Q. Why the interest in the amount of ozone in the stratosphere?

A. Because it absorbs the high-energy ultraviolet light of the sun, screening this harmful radiation from the earth's surface.

Section 15-6

35. Q. The United States is currently importing most of its sources of over 20 important minerals, and it has been estimated that in another 25 years we shall be importing more than half of our nonfuel mineral requirements. How could the necessity for these imports be reduced?

A. By more intensive recycling programs and by mining low-concentration ores.

36. Q. What are two of the principal disadvantages in mining low-concentration ores?

A. Such mining consumes greater amounts of energy for a given amount of useful substance produced and usually causes more disruption to the environment.

37. Q. The recycling of substantial amounts of paper and metals has been going on for some time. What are some of the improvements needed to encourage the expansion of recycling?

A. Improvements in recovery methods, reduction of recovery costs, and development of markets for the recovered products.

38. Q. From what point of view are mineral resources practically inexhaustible?

A. The earth's crust and seawater contain very large amounts of most of the materials now widely used.

39. Q. What is the problem of future mineral resources?

A. Generating the large amounts of energy required to use low-concentration ores and learning how to use low-grade ores within stringent environmental constraints.

Section 15-7

40. Q. What are the three leading causes of death in the United States in a ranking according to the relative loss of working life up to the age of 65 (Fig. 15-11)?

A. Accidents, arterial diseases, and cancer.

41. Q. What is the leading lethal form of cancer in the United States in a ranking according to the relative loss of working life up to the age of 65 (Fig. 15-11)?

A. Lung cancer.

42. Q. What is considered to be the principal cause of lung cancer?

A. Cigarette smoking.

43. Q. What is a carcinogenic agent?

A. Any substance or material capable of causing cancer.

44. Q. What is a tumor?

A. A growth composed of cells which lack the normal controls on growth and reproduction.

45. Q. What is the difference between a benign tumor and a cancerous tumor?

A. A benign tumor remains within the tissue in which it develops. A cancerous tumor invades the surrounding tissue.

46. Q. What is metastasis?

A. The process in which cancerous tumors release malignant cells which migrate through the body fluids and establish new cancerous growths in distant parts of the body.

47. Q. What is a gene?

A. A basic unit of heredity; a segment of the substances called nucleic acids which preserve the hereditary information of cells and control the synthesis of cell materials.

48. Q. One of the current hypotheses for the formation of cancerous cells proposes that carcinogenic substances interact directly with nucleic acids to inactivate the genes which control the cell's growth and reproduction. Why does the inactivation of one gene not produce a cancerous cell according to this hypothesis?

A. Because a cell's growth and reproduction are controlled by a group of genes and they all have to be inactivated before the constraints on growth and reproduction are removed and the potential for tumor growth is formed.

49. Q. What are the principal difficulties encountered in identifying cancer-causing agents?

A. The long latency period in people between exposure to carcinogenic agents and the formation of a cancer growth, the reactions often required in the body before the actual active carcinogens are generated, the difference of opinion about the interpretation of tests on animals, the frequent lack of a firm cause-and-effect relationship in evidence gathered by studies of cancer incidence, and the expense and slowness of tests for carcinogenicity in animals.

PRACTICE EXERCISES

1. Match each definition or other statement with the numbered term above with which it is most closely associated. Each numbered term may be used only once.

   1. Pollutant
   2. Thermal blanket
   3. Fluorocarbon
   4. Photochemical smog
   5. Particulates
   6. Temperature inversion
   7. Electrostatic precipitator
   8. Greenhouse effect
   9. Industrial smog

   (a) A substance which contains carbon and fluorine atoms
   (b) An air-pollution mixture made up principally of sulfur dioxide, sulfuric acid, and airborne particles of smoke, ash, and substances like ammonium sulfate
   (c) An atmospheric phenomenon in which a a mass of warm air lies above a mass of cool air
   (d) The removal of fine particles from smoke stacks by electric charges
   (e) A form of matter which when added to the surroundings in sufficient amount produces a detectable injurious effect on people, animals, and plants
   (f) An air-pollution mixture formed when hydrocarbons and nitrogen oxides are irradiated by sunlight
   (g) Finely divided particles of liquids and solids which remain suspended in the atmosphere for a long time

2. (a) Name seven different air pollutants in the United States. (b) Why is carbon dioxide usually not included?

3. (a) What is the principal source of air pollutants in the United States? (b) What pollutants derive from this source in appreciable amount?

4. What is the principal source of (a) sulfur-containing pollutants and (b) polluting suspended particles?

5. (a) What is the principal source of carbon monoxide in the atmosphere? (b) What is the principal man-made source of carbon monoxide in the atmosphere? (c) In general terms to what is the toxic effect of carbon monoxide due?

6. Give balanced equations for reactions which account for the formation of the following components of industrial smog:
   (a) Sulfur dioxide, SO_2

   (b) Sulfur trioxide, SO_3
   (c) Sulfuric acid, H_2SO_4
   (d) Ammonium sulfate, $(NH_4)_2SO_4$

7. (a) Give a balanced equation for a reaction which accounts in part for the eroding effects of industrial smog on limestone and marble. (b) To what type does the reaction belong?

8. Give a balanced equation for a reaction which accounts for the formation of each of the following air pollutants:
   (a) Nitric oxide, NO
   (b) Nitrogen dioxide, NO_2 (using oxygen, O_2)
   (c) Nitrogen dioxide, NO_2 (using ozone, O_3)
   (d) Nitric acid, HNO_3
   (e) Ammonium nitrate, NH_4NO_3

9. What are the principal sources of the hydrocarbons involved in atmospheric pollution?

10. What is meant by photochemical smog?

11. Give equations for the two reactions which are thought to account for the atmospheric pollutant ozone, O_3.

12. Give the names of two eye irritants found in photochemical smog.

13. What steps have been considered, or are being taken, to reduce the amount of (a) hydrocarbons, (b) carbon monoxide, and (c) nitrogen oxides emitted from automobiles?

14. Why must automobiles equipped with catalytic converters not use leaded gasoline?

15. (a) The production of what pollutant was feared might occur through the use of catalytic converters on automobiles? (b) From what would it have been produced? (c) Where does the matter stand?

16. In very general terms what is the principle of an electrostatic precipitator for removing suspended particles from the effluent gases of combustion processes?

17. (a) What is the major source of lead pollution in the atmosphere? (b) Why is the amount of lead pollution from this source decreasing?

18. Why are paint pigments using lead compounds no longer permitted in interior house paints?

19. (a) What are fluorocarbons, and what are their principal uses? (b) What are their advantages?

20. To what fundamental property of fluorocarbons is their potential hazard in the stratosphere due? Give the equation for an illustrative reaction.

21. What important function does ozone perform in the stratosphere?

22. In general terms why is the use of fluorocarbons in aerosol spray cans banned?

23. How could fluorocarbons serve to deplete the amount of ozone in the atmosphere?

24. Give equations for the reactions which could be involved as fluorocarbons possibly serve to interfere with the ozone content of the stratosphere?

25. What are the three methods of disposing of solid wastes used for many years?

26. (a) Why is there some concern about supplies of raw materials? (b) On a worldwide basis is there a shortage of raw materials for non-energy purposes? (c) What action would provide a partial solution to raw-material supply problems and at the same time make a positive contribution to the environment?

27. Match each definition or other statement with the numbered term above with which it is most closely associated. Each numbered term may be used only once.

    1. Benign tumor   2. Linus test
    3. Metastasis   4. Gene
    5. Ames test   6. Carcinogenic agent
    7. Metathesis   8. Cancerous tumor

    (a) A growth whose cells invade the tissue surrounding the tissue in which it originally developed
    (b) A substance or material capable of causing cancer
    (c) A segment of a nucleic acid which preserves hereditary information
    (d) The release of malignant cells, which

migrate through the body fluids to establish new cancerous growths elsewhere in the body
    (e) A tumor which remains within the tissue in which it develops
    (f) A test for potential carcinogenic properties

28. (a) What is a carcinogenic substance? (b) What was the first carcinogenic substance recognized, and how did it come to be recognized? (c) What are two places where the substance is found?

29. Give the name of (a) two synthetic substances which are carcinogenic and (b) a natural substance which is carcinogenic.

30. (a) What is a tumor? (b) What is the difference between a benign tumor and a cancerous tumor? (c) What is the process of metastasis?

31. (a) What is a gene? (b) In a current hypothesis what is the relationship between genes and the normal growth and reproduction of cells? (c) In the same hypothesis what is the relationship between carcinogenic agents and these genes? (d) How does this hypothesis account for the fact that exposure to carcinogens does not always produce a cancerous tumor?

32. List five difficulties encountered in identifying agents which may cause cancer in humans.

33. (a) What is the Ames test? (b) What are the advantages of the Ames test over comparable tests? (c) Does a positive Ames test indicate that a substance is carcinogenic in humans? Explain briefly. (d) What is the principle of the Ames test?

SUGGESTIONS FOR FURTHER READING

The following articles are of an introductory nature.

American Chemical Society: "Cleaning Our Environment: The Chemical Basis for Action," Washington, D.C., 1969. The report of a group of authorities covering a full range of environmental problems.

Anderson, Earl V.: World's Nations Scramble for Seas' Riches, **Chemical and Engineering News**, Mar. 4, 1974, p. 18.

Arnold, James R., and E. A. Martell: The Circulation of Radioactive Isotopes, **Scientific American**, September 1955, p. 84.

Carbon Monoxide: The Invisible Enemy, **Chemistry**, January 1973, p. 18.

Cairns, John: The Cancer Problem, **Scientific American**, November 1975, p. 64.

Carter, Luther J.: Cancer and the Environment, I: A Creaky System Grinds On, **Science, 186**:239 (1974).

Carter, Luther J.: Chemical Carcinogens: Industry Adopts Controversial "Quick" Tests, **Science, 192**:1215 (1976).

Chisolm, J. Julian, Jr.: Lead Poisoning, **Scientific American**, February 1971, p. 15.

Clark, John R.: Thermal Pollution and Aquatic Life, **Scientific American**, March 1969, p. 19.

Cook, Earl: Limits to Exploitation of Nonrenewable Resources, **Science, 191**:677 (1976).

Culliton, Barbara J.: Saccharin: A Chemical in Search of an Identity, **Science, 196**:1179 (1977).

de Nevers, Noel: Enforcing the Clean Air Act of 1970, **Scientific American**, June 1973, p. 14.

Dubos, Rene: Symbiosis between Earth and Humankind, **Science, 193**:459 (1976).

Fennelly, Paul F.: The Origin and Influence of Airborne Particulates, **American Scientist, 64**:46 (1976).

Ferguson, Lloyd N.: Cancer: How Can Chemists Help?, **Journal of Chemical Education, 52**:688 (1975).

Fox, Jeffrey L.: Ames Test Success Paves Way for Short-Term Cancer Testing, **Chemical and Engineering News,** Dec. 12, 1977, p. 34.

Gillette, Robert, Cancer and the Environment, II: Groping for New Remedies, **Science, 186:**242 (1974).

Goeller, H. E., and Alvin M. Weinberg: The Age of Substitutability, **Science, 191:**683 (1976).

Goldsmith, John R., and Stephen A. Landaw: Carbon Monoxide and Human Health, **Science, 162:**1352 (1968).

Goldwater, Leonard J.: Mercury in the Environment, **Scientific American,** May 1971, p. 15.

Gribble, Gordon W.: TCDD: A Deadly Molecule, **Chemistry,** February 1974, p. 15.

Haagen-Smit, A. J.: The Control of Air Pollution, **Scientific American,** January 1964, p. 24.

Hammond, Allen L.: Manganese Nodules, I: Mineral Resources on the Deep Sea Bed, **Science, 183:**502 (1974).

Haust, Philip L.: Noxious Trace Gases in the Air, **Chemistry,** January-February 1978, p. 8.

Hoffman, Alan J., Thomas C. Curran, Thomas B. McMullen, William M. Cox, and William F. Hunt, Jr.: EPA's Role in Ambient Air Quality Monitoring, **Science, 190:**243 (1975).

Holdren, John P., and Paul R. Ehrlich: Human Population and the Global Environment, **American Scientist, 62:**282 (1974).

How Accurate Are Environmental Data?, **Chemistry,** March 1974, p. 18.

Jensen, Clayton E., Dail W Brown, and John A. Mirabito, Earthwatch, **Science, 190:**432 (1975). Guidelines for implementing a global environmental assessment program.

Jones, Richard F.: Lead: A Case Study, **Chemistry,** March 1975, p. 12.

Keller, Eugenia: What Is Happening to Our Drinking Water?, **Chemistry,** February 1975, p. 16.

Kirby, Ralph C., and Andrew S. Prokopovitsh: Technological Insurance Against Shortages in Minerals and Metals, **Science, 191:**713 (1976).

Lepkowski, Wilbert C.: Saccharin Ban Goes Beyond Issue of Cancer, **Chemical and Engineering News,** Apr. 11, 1977, p. 17.

Likens, Gene E.: Acid Precipitation, **Chemical and Engineering News,** Nov. 22, 1976, p. 29.

Martell, Edward A.: Tobacco Radioactivity and Cancer in Smokers, **American Scientist, 63:**404 (1975).

Maugh, Thomas H., II: Carbon Monoxide: Natural Sources Dwarf Man's Output, **Science, 177:**338 (1972).

Maugh, Thomas H., II: Chemical Carcinogenesis: A Long-neglected Field Blossoms, **Science, 183:**940 (1974).

Maugh, Thomas H., II: Air Pollution: Where Do Hydrocarbons Come From? **Science, 189:**277 (1975).

Maugh, Thomas H., II: Sulfuric Acid from Cars: A Problem That Never Materialized, **Science, 198:**280 (1977).

McFarland, John H., and C. S. Benton: The Oxides of Nitrogen and Their Detection in Automotive Exhaust, **Journal of Chemical Education, 49:**21 (1972).

Moore, John W.: The Vinyl Chloride Story, **Chemistry,** June 1975, p. 12.

Moore, John W., and Elizabeth A. Moore: Resources in Environmental Chemistry, I: An Annotated Bibliography of the Chemistry of Pollution and Resources, **Journal of Chemical Education, 53:**167 (1976).

Moore, John W., and Elizabeth A. Moore: Resources in Environmental Chemistry, II: An Annotated Bibliography of Water, Life and Health, and Population Problems, **Journal of Chemical Education, 53:**240 (1976).

Newell, Reginald E.: The Global Circulation of Atmospheric Pollutants, **Scientific American,** January 1971, p. 32.

Neyman, Jerzy: Public Health Hazards from Electricity-producing Plants, **Science, 195:**754 (1977).

O'Sullivan, Dermot A.: Air Pollution, **Chemical and Engineering News,** June 8, 1970, p. 38.

Prud'homme, Robert K.: Automobile Emissions Abatement and Fuels Policy, **American Scientist, 62:**191 (1974).

Rawls, Rebecca, and Dermot A. O'Sullivan: Italy Seeks Answers Following Toxic Release [of TCDD], **Chemical and Engineering News,** Aug. 23, 1976, p. 27.

Should the Delaney Clause Be Changed?, **Chemical and Engineering News,** June 27, 1977, p. 24. A debate of food-additive safety, animal tests, and cancer.

Wade, Nicholas: Raw Materials: U.S. Grows More Vulnerable to Third World Cartels, **Science, 183:**185 (1974).

Wade, Nicholas: Control of Toxic Substances: An Idea Whose Time Has Nearly Come, **Science, 191:**541 (1976).

Walsh, John: Seveso [Italy]: The Questions Persist Where Dioxin Created a Wasteland, **Science, 197:**1064 (1977).

Weisburger, Elizabeth K.: Cancer-causing Chemicals, **Chemistry,** January-February 1977, p. 42.

Wenk, Edward, Jr.: The Physical Resources of the Ocean, **Scientific American,** September 1969, p. 166.

Wolman, Abel: Ecologic Dilemmas, **Science, 193:**740 (1976).

The following articles are at a more advanced level.

Chesick, John P.: Effects of Water and Nitrogen Dioxides on the Stratospheric Ozone Shield, **Journal of Chemical Education, 49:**722 (1972).

Finlayson, Barbara J., and James N. Pitts, Jr.: Photochemistry of the Polluted Troposphere, **Science, 192:**111 (1976).

SUPPLEMENTS

SUPPLEMENT 1

Expressing Very Large and Very Small Numbers

S1-1 THE REPRESENTATION OF VERY LARGE NUMBERS

In chemistry one often uses very large numbers. The number of molecules in 2.016 grams of hydrogen, for example, is

$$602,200,000,000,000,000,000,000$$

It is much more convenient to express this number by using an **exponent,** a number written as a superscript which indicates the number of times a number is multiplied by itself. Thus 10^3 is equivalent to $10 \times 10 \times 10 = 1000$, and 2^6 is equivalent to $2 \times 2 \times 2 \times 2 \times 2 \times 2 = 64$. The exponent of a number is also known as the power of a number. The **exponential form of a number** expresses a number as a product of two numbers: the first (called the coefficient) is a number between 1 and 10 (for example, 1.3, 6.78, or 8.4), and the second is 10 raised to some power expressed as an exponent. The exponential form of 543,000,000, for example, is 5.43×10^8, which is equivalent to $5.43 \times 10 \times 10 \times 10 \times 10 \times 10 \times 10 \times 10 \times 10$.

Converting a large number into exponential notation involves (1) writing the coefficient as a number between 1 and 10 and (2) writing 10 raised to a power the value of which is the same as the number of places that the decimal point is moved to the left in going from the original form of the number to the coefficient. Thus 216.4 expressed in exponential form is 2.164×10^2 since in deriving the coefficient 2.164 from the original form of the number 216.4 the decimal point is moved two places to the left. Consider the examples in Table S1-1.

Table S1-1
Some Large Numbers and Their Equivalents in Exponential Notation

Number	Exponential notation
10.82	$1.082 \times 10^1 = 1.082 \times 10$
108.2	$1.082 \times 10^2 = 1.082 \times 100$
1,082	$1.082 \times 10^3 = 1.082 \times 1000$
10,820	$1.082 \times 10^4 = 1.082 \times 10,000$
108,200	$1.082 \times 10^5 = 1.082 \times 100,000$
108,200,000	$1.082 \times 10^8 = 1.082 \times 100,000,000$

S1-2 THE REPRESENTATION OF VERY SMALL NUMBERS

We also have to express very small numbers in chemistry. The mass of a single sulfur atom, for example, is

$$0.00000000000000000000005324 \text{ gram}$$

It is much more convenient to express this number in the exponential form 5.324×10^{-23}. The exponential form of very small numbers uses negative exponents. Converting a very small number into exponential notation involves (1) writing the coefficient as a number between 1 and 10, and (2) writing 10 raised to a negative power the value of which is the same as the number of places that the decimal point is moved to the right in going from the original number to the coefficient. Thus 0.06391 expressed in exponential form is 6.391×10^{-2} since in deriving the coefficient 6.391 from the original form of the number 0.06391 the decimal point is moved two places to the right. Consider the examples in Table S1-2.

Table S1-2
Some Small Numbers and Their Equivalents in Exponential Notation

Number	Exponential notation
0.7628	$7.628 \times 10^{-1} = 7.628 \times 0.1$
0.07628	$7.628 \times 10^{-2} = 7.628 \times 0.01$
0.007628	$7.628 \times 10^{-3} = 7.628 \times 0.001$
0.0007628	$7.628 \times 10^{-4} = 7.628 \times 0.0001$
0.0000007628	$7.628 \times 10^{-7} = 7.628 \times 0.0000001$

S1-3 MULTIPLYING AND DIVIDING NUMBERS EXPRESSED IN EXPONENTIAL NOTATION

To multiply numbers expressed in exponential notation the exponents are added:

$$10^2 \times 10^2 = 10^4$$
$$10^3 \times 10^6 = 10^9$$
$$10^5 \times 10^{-2} = 10^3$$
$$10^{-4} \times 10^{-2} = 10^{-6}$$
$$10^3 \times 10^{-4} \times 10^5 = 10^4$$

The coefficients and the powers of 10 are multiplied separately.

$$(2.0 \times 10^2)(2.0 \times 10^3) = (2.0 \times 2.0)(10^2 \times 10^3) = 4.0 \times 10^5$$
$$(2.0 \times 10^5)(2.0 \times 10^{-3}) = (2.0 \times 2.0)(10^5 \times 10^{-3}) = 4.0 \times 10^2$$
$$(3.0 \times 10^4)(4.0 \times 10^{-6}) = (3.0 \times 4.0)(10^4 \times 10^{-6}) = 12 \times 10^{-2} = 1.2 \times 10^{-1}$$
$$(1.63 \times 10^5)(2.86 \times 10^{-2}) = (1.63 \times 2.86)(10^5 \times 10^{-2}) = 4.66 \times 10^3$$

To divide numbers expressed in exponential notation the exponents are subtracted:

$$\frac{10^4}{10^2} = 10^2 \qquad \frac{10^9}{10^3} = 10^6 \qquad \frac{10^{-3}}{10^2} = 10^{-5} \qquad \frac{10^{-5}}{10^{-2}} = 10^{-3}$$

The coefficients and the powers of 10 are divided separately:

$$\frac{4.0 \times 10^4}{2.0 \times 10^2} = \frac{4.0}{2.0} \times \frac{10^4}{10^{-2}} = 2.0 \times 10^2$$

$$\frac{9.0 \times 10^3}{3.0 \times 10^4} = \frac{9.0}{3.0} \times \frac{10^3}{10^4} = 3.0 \times 10^{-1}$$

$$\frac{3.73 \times 10^2}{1.29 \times 10^1} = \frac{3.73}{1.29} \times \frac{10^2}{10^1} = 2.89 \times 10^3$$

It is much more convenient to multiply and divide very large or very small numbers when they are expressed in exponential notation. For example,

$$602,000 \times 0.0000018 = 1.08$$

becomes

$$(6.02 \times 10^5)(1.8 \times 10^{-6}) = 10.8 \times 10^{-1} = 1.08$$

And

$$\frac{0.00068}{5128} = 0.00000013$$

becomes

$$\frac{6.8 \times 10^{-4}}{5.128 \times 10^3} = 1.3 \times 10^{-7}$$

The convenience of the exponential notation is particularly evident when multiplication and division are combined:

$$\frac{3860 \times 0.0963}{61,100}$$

becomes

$$\frac{(3.86 \times 10^3)(9.63 \times 10^{-2})}{6.11 \times 10^4} = \frac{3.86 \times 9.63}{6.11} \frac{10^3 \times 10^{-2}}{10^4} = 6.08 \times 10^{-3}$$

It is customary to express the exponential form of a number using a coefficient which is between 1 and 10. For example, 47,360 is expressed as 4.736×10^4, not as 47.36×10^3; and 0.00215 is expressed as 2.15×10^{-3}, not as 0.215×10^{-2}. When the operations of multiplication or division yield answers with coefficients not between 1 and 10 a conversion should be made to obtain such a coefficient:

$$(5.83 \times 10^3)(8.29 \times 10^2) = 48.3 \times 10^5 = 4.83 \times 10^6$$

$$\frac{1.17 \times 10^{-1}}{9.62 \times 10^3} = 0.122 \times 10^{-4} = 1.22 \times 10^{-5}$$

KEY WORDS

1. Exponent (Sec. S1-1)   2. Exponential form of a number (Sec. S1-1)

PRACTICE EXERCISES

1. Give the exponential form of:
   (a) 68,200,000   (b) 1,300
   (c) 0.00559   (d) 123,000,000,000
   (e) 0.176   (f) 0.000008332

2. Express each of the following numbers in the conventional (nonexponential) form:
   (a) 7.72×10^{-2} (b) 9.16×10^9
   (c) 3.81×10^{-5} (d) 4.47×10^2
   (e) 5.83×10^{-8} (f) 2.22×10^6

3. Carry out the following computations, making use of exponential notation:

   (a) 0.0000735×1289 (b) $\dfrac{867,736}{293}$

   (c) $\dfrac{0.05344}{28,700}$ (d) $\dfrac{8,720,000}{41,700 \times 0.00395}$

   (e) $\dfrac{5119 \times 0.2781}{0.0003377}$

SUPPLEMENT 2

Experimental Measurements and Approximate Calculations†

S2-1 THE ACCURACY AND PRECISION OF EXPERIMENTAL MEASUREMENTS

Experiments in which quantities are measured provide the foundations of the facts, ideas, and theories of chemistry. Experimental measurements must be determined, expressed, and used with care, however, in order to keep the conclusions derived from them free from error. Chemists are very concerned with the accuracy of experimental measurements. **Accuracy** refers to the closeness with which the average of a series of repeated measurements approaches the true or accepted value of the quantity being determined. The agreement between a series of measurements is also of great interest. **Precision** refers to the closeness of approach of a series of repeated measurements to the average value of the measurements. For example, two series of measurements of the diameter of a quarter coin with a ruler result in the following values:

First series		Second series	
Measurement	Centimeters	Measurement	Centimeters
1	2.34	1	2.24
2	2.36	2	2.41
3	2.37	3	2.30
4	2.35	4	2.39
Average	2.36	Average	2.34

It is rather easy to see by inspection of the data that the agreement between the measurements in the first series is closer than the agreement in the second series of measurements. The precision in the first series of measurements is therefore greater.

The attainment of good precision generates confidence in the consistency of a series of measurements, but it does not in itself prove that the average value is of high accuracy. Constant errors may affect a series of measurements uniformly. If all measurements of the diameter of a quarter coin were made using the same ruler and the ruler was inaccurate, the agreement in the measurements could be close but the accuracy of the determination would be poor. If reliable procedures are followed carefully, however, a good degree of precision increases the probability of the measurements' being accurate.

S2-2 SIGNIFICANT FIGURES

Experimental measurements should be expressed by numbers in a way which reflects their reliability. If two measurements, for example, of the diameter of a quarter coin with a ruler

† Should be preceded by Supplement 1.

are found to be 2.34 and 2.35 centimeters, the average value may be expressed as 2.35 centimeters. The average value should not be expressed as 2.345 centimeters, however, since this would imply that the value is known to 0.001 centimeter, which it is not. Each of the measurements was determined only to 0.01 centimeter, and the average value should reflect this fact. If, on the other hand, the measurements were made with a caliper equipped to read to 0.001 centimeter and the values obtained were 2.340 and 2.350 centimeters, then it would be appropriate to express the average value as 2.345 centimeters. The number implies that the value is known to 0.001 centimeter, which with the caliper measurements is true.

Care must be especially taken in expressing the numerical results of calculations based on experimental measurements to ensure that the accuracy implied by the result does not exceed the accuracy of the measurements on which the calculations are based. Consider, for example, an analytical procedure carried out to determine the silver content of an aqueous solution. The final step of this procedure involves weighing a precipitate of silver chloride, AgCl. Suppose that the weighing is carried out on a balance capable of weighing accurately only to 0.01 gram, and the weight of silver chloride is found to be 1.53 grams. To calculate the weight of silver in the precipitate, the very accurately known values for the atomic weight of silver, Ag (107.868), and molecular weight of silver chloride, AgCl (143.321), are used as follows:

$$1.53 \text{ grams AgCl} \times \frac{107.868}{143.321} = 1.15 \text{ grams Ag}$$

The weight of the silver, Ag, should be expressed with only three figures because the weight of the silver chloride, AgCl, was determined to only three figures. Although the values of the atomic and molecular weights are known to six figures, the result should be expressed only in three figures. Electronic calculators, with many digits in the display, can be particularly misleading about the number of useful figures in a result. A calculator with an eight-digit display will give a value of 1.1515272 for the calculation, but obviously using all eight figures would be ridiculous because it implies an accuracy not present in the original numbers in the calculation.

A **significant figure** is a digit in a number that is reasonably reliable. In expressing the value of measurements standard practice calls for including only one digit about which there is any uncertainty. If the usual 30-centimeter ruler is used to measure a distance of between 1 and 2 centimeters, the distance can be determined, for example, to be 1.78 centimeters. In reading the ruler calibrations there is no doubt about the 1.7, but the last digit is estimated with some uncertainty. Approved practice calls for including one such uncertain digit among the significant figures but only one. The position of the decimal point does not affect the number of significant figures. To express a distance of 3.82 centimeters as 0.0382 meter does not affect the number of significant figures, which is three in both cases.

In expressing the results of calculations such as addition, subtraction, multiplication, and division, it is very important that numerical answers not use more significant figures than the number of significant figures in the least accurately known measurement.

A. Addition

In addition the number of decimal places in the answer should be the same as the number of decimal places in the measurement with the least number of decimal places.

The weights of three objects are known to be as follows:

Object	Weight, grams
1	37.425
2	215.6
3	21.1134
Sum	274.1

The sum is expressed to one decimal place since the weight of object 2 is expressed by a number with only one decimal place.

B. Subtraction

In subtraction the number of decimal places in the answer should be the same as the number of decimal places in the measurement with the least number of decimal places.

A container is weighed with and without its contents on two different balances to determine the weight of the contents.

	Weight, grams
With contents	268.4308
Without contents	53.22
Contents	215.21

The answer is expressed to two decimal places since the weight of the container without the contents is expressed to two decimal places.

C. Multiplication

A determination of the silver, Ag, content of an aqueous solution yields 2.13 grams of silver chloride, AgCl. The calculation of the silver which corresponds to this weight uses the following atomic and molecular weights:

Atomic weight of silver $= 107.868$

Molecular weight of silver chloride $= 143.321$

The calculation is carried out as follows:

$$2.13 \text{ grams AgCl} \times \frac{107.868}{143.321} = 1.60 \text{ grams Ag}$$

The result may be expressed in only three significant figures because the weight of silver chloride is expressed in only three significant figures. If the weight of silver chloride were determined to more significant figures (as it is in good laboratory work), more significant figures could be used in the result. For example,

$$\text{Experimental weight of silver chloride} = 2.1343 \text{ grams}$$

Calculation: $2.1343 \text{ grams AgCl} \times \dfrac{107.868}{143.321} = 1.6063 \text{ grams Ag}$

D. Division

A sample of sodium chloride, NaCl, is weighed on a rough scale and found to weigh 3.52 grams. The amount can be converted into ounces by dividing by the number of grams in an ounce, 28.35:

$$\frac{3.52 \text{ grams NaCl}}{28.35 \text{ grams/ounce}} = 0.124 \text{ ounce NaCl}$$

The answer may be expressed in only three significant figures because the mass of the sodium chloride was expressed in only three significant figures.

S2-3 ROUNDING OFF NUMBERS

Restricting the expression of the result of a calculation to the correct number of significant figures frequently involves rounding off numbers. We can state the conventions for carrying out this operation sufficient for our purposes as follows:

1. When the number dropped is 5 or greater, the last remaining digit is increased by 1, for example,

	Answers
Round off 23.246 to 4 digits	23.25
Round off 6.3647 to 2 digits	6.4
Round off 0.00753 to 1 digit	0.008

2. When the number dropped is less than 5, the retained digits are not changed, for example,

	Answers
Round off 23.243 to 4 digits	23.24
Round off 6.3442 to 2 digits	6.3
Round off 0.00733 to 1 digit	0.007

S2-4 USING APPROXIMATE CALCULATIONS

The approximate answers to many calculations can quickly be estimated by using rounded numbers and exponential notation (Supplement 1). Such approximate answers are of great value in avoiding careless errors in calculation. They are particularly helpful in checking an answer from an electronic calculator.

If it is necessary, for example, to carry out the multiplication

$$588,476 \times 39,628 =$$

the approximate answer can be obtained very quickly by rounding off the numbers, expressing them in exponential form, and using mental arithmetic:

$$6.0 \times 10^5 \times 4.0 \times 10^4 = 24 \times 10^9 = 2.4 \times 10^{10}$$

The value obtained from an electronic calculator is 2.3320×10^{10}. The quick estimate was close. Additional examples are:

Operations with answers obtained with calculator	Aproximate calculations using exponential notation, rounded numbers, and mental arithmetic
$4398 \times 0.0207 = 91.0$	$4.4 \times 10^3 \times 2.1 \times 10^{-2} = 9 \times 10^1 = 90$
$\dfrac{384,968}{213} = 1.81 \times 10^3$	$\dfrac{3.8 \times 10^5}{2 \times 10^2} = 1.9 \times 10^3$
$\dfrac{0.00817 \times 9079 \times 0.0112}{5903 \times 6.29} = 0.0000224$	$\dfrac{(8 \times 10^{-3})(9 \times 10^3)(1 \times 10^{-2})}{(6.0 \times 10^3)(6)} = \dfrac{72 \times 10^{-2}}{36 \times 10^3} = 2 \times 10^{-5}$

KEY WORDS

1. Accuracy (Sec. S2-1)   2. Precision (Sec. S2-1)   3. Significant figure (Sec. S2-2)

SUMMARIZING QUESTIONS FOR SELF-STUDY

1. Q. What is meant by the accuracy of a series of repeated measurements?

A. The closeness with which the average approaches the true or accepted value of the quantity being determined.

2. Q. What is meant by the precision of a series of repeated measurements?

A. The closeness of approach to the average value of the measurements.

3. Q. Is a series of measurements with good precision necessarily accurate?

A. No. (But the good precision may make you more confident about the accuracy.)

4. Q. What is meant by a significant figure?

A. A digit in a number that is reasonably reliable.

5. Q. According to the prevailing custom, how many uncertain digits are included in a number which expresses the result of a measurement?

A. One.

6. Q. In expressing the result of calculations based on experimental measurements what limits the number of significant figures which may be used in the answer?

A. In multiplication and division the limiting factor is the number of significant figures in the least precisely known measurement. In addition and subtraction the limiting factor is the number of decimal places in the measurement expressed with the least number of decimal places.

7. Q. An electronic calculator gives 197.6179 as the sum of the following experimentally derived numbers.

$$
\begin{array}{r}
13.2679 \\
21.35 \\
\underline{163} \\
\end{array}
$$

How should the result be expressed?

A. 198 (because one of the numbers has no decimal places).

8. Q. An electronic calculator gives 25.424791 as the result of the operation

$$16.1 \times \dfrac{283.458}{179.497}$$

How should the result be expressed?

A. 25.4, because one of the numbers has only three significant figures.

9. Q. Round off (a) 16.7385 to three figures, (b) 0.001197 to two figures, (c) 243.78 to four figures, (d) 0.0006666 to three figures.

A. (*a*) 16.7, (*b*) 0.0012, (*c*) 243.8, (*d*) 0.000667.

10. Q. An electronic calculator gives 34.464974 as the result of the operation

$$\frac{73.9963}{2.147}$$

How should the result be expressed?

A. 34.46 (because one of the numbers has only four significant figures).

11. Q. Using rounded numbers and exponential notation, give an approximate answer to the following calculation. Show your method.

$$0.0691 \times 5116 =$$

A. $(7 \times 10^{-2})(5 \times 10^{3}) = 35 \times 10^{1} = 350.$

PRACTICE EXERCISES

1. Round off the following numbers.
   (*a*) 1713 to two figures
   (*b*) 68,962 to three figures
   (*c*) 0.0033487 to three figures
   (*d*) 3333.3 to four figures
   (*e*) 0.048076 to three figures

2. Carry out the following operations and express the answers in the correct number of significant figures, rounding off properly.
   (*a*) Add 396.2 + 84.229 + 27.6633 + 100.17
   (*b*) Subtract 493.557 from 29.76.
   (*c*) $\dfrac{368.117}{453.229} \times 4722$

   (*d*) $\dfrac{28.899}{1.7}$

   (*e*) $\dfrac{0.006377}{0.0003}$

3. Using exponential notation, rounded numbers, mental arithmetic (but no calculator), give approximate answers for the following:
   (*a*) 6789 × 0.07113
   (*b*) $\dfrac{732,622}{176}$
   (*c*) $\dfrac{5402 \times 0.00993 \times 0.216}{671 \times 206}$

SUPPLEMENT 3

Units and Unit Conversions†

S3-1 UNITS OF LENGTH, VOLUME, AND MASS

A. The Metric System

The metric system of units was first used in France in 1790. Since it is based on tens, hundreds, thousands, and other multiples of 10, it is a convenient system. It has been used by scientists for many years. It has also been used in commerce in all countries throughout the world except the United States and the United Kingdom, which held on to the English system of feet, pounds, and quarts. Now these countries are in the process of converting to the metric system, and the exchange of manufactured goods will be facilitated. The metric system was revised in 1960, and the version now in use is known as the International Metric System or Système International (SI). The term **SI unit** is applied to the units of the revised system.

B. Prefixes Used to Designate Multiple Numbers of Units

A series of prefixes is used to designate multiples of basic units. The prefix *kilo-*, for example, signifies 1000 times the basic unit: a kilogram is 1000 grams; and a kilometer is 1000 meters. The prefixes are summarized in Table S3-1.

Table S3-1
Prefixes for the Metric System of Units

Prefix	Abbreviation	Meaning of multiple		Exponential notation
pico	p	0.000000000001		10^{-12}
nano	n	0.000000001		10^{-9}
micro	μ	0.000001		10^{-6}
milli	m	0.001		10^{-3}
centi	c	0.01		10^{-2}
kilo	k	1000		10^{3}
mega	M	1,000,000	(1 million)	10^{6}
giga	G	1,000,000,000	(1 billion)	10^{9}
tera	T	1,000,000,000,000	(1 trillion)	10^{12}

C. Metric Units of Length, Volume, and Mass

The **meter** (m) is the basic unit of length, originally defined as the distance between etched lines on a platinum-iridium bar kept as an international standard in France. The meter is now defined in terms of a certain wavelength of light.

The **cubic meter** (m^3) is the basic unit of volume, and the **kilogram** (kg) is the basic unit of mass. Table S3-2 summarizes some basic metric and English system units. Table S3-3 gives some equivalents.

† Should be preceded by Supplements 1 and 2.

Table S3-2
Basic Metric and English System
Units of Length, Volume, and Mass

Metric system	English system
Length	
1 kilometer (km) = 10^3 m	1 mile (mi) = 5280 ft
1 centimeter (cm) = 10^{-2} m	1 yard (yd) = 3 ft
1 millimeter (mm) = 10^{-3} m	1 foot (ft) = 12 in
1 angstrom (Å) = 10^{-10} m	
Volume	
1 cubic meter (m^3) = 10^3 L	1 gallon (gal) = 4 qt
1 liter (L) = 10^{-3} m^3	1 quart (qt) = 2 pt
1 milliliter (mL) = 10^{-3} L	
Mass	
1 metric ton = 10^3 kg	1 ton = 2000 lb
1 kilogram (kg) = 10^3 g	1 pound (lb) = 16 oz
1 milligram (mg) = 10^{-3} g	
1 microgram (μg) = 10^{-6} g	

Table S3-3
Some Conversion Equivalents Between the Metric and English Systems†

English to metric	Conversion factor	Metric to English	Conversion factor
Length			
1 mi = 1.609 km	1.609 km/mi	1 km = 0.621 mi	0.621 mi/km
1 ft = 30.48 cm	30.48 cm/ft	1 m = 1.094 yd	1.094 yd/m
1 in = 2.54 cm	2.54 cm/in	1 m = 39.37 in	39.37 in/m
Volume			
1 gal = 3.785 L	3.785 L/gal	1 L = 0.2642 gal	0.2642 gal/L
1 qt = 0.9464 L	0.9464 L/qt	1 L = 1.057 qt	1.057 qt/L
Mass			
1 ton = 0.907 metric tons	0.907 t/ton	1 ton = 1.102 tons	1.102 tons/t
1 lb = 453.6 g	453.6 g/lb	1 kg = 2.205 lb	2.205 lb/kg
1 oz = 28.35 g	28.35 g/oz	1 g = 0.03527 oz	0.03527 oz/g

† Some of these numbers have been rounded off.

The following examples will illustrate how conversions can be carried out between the metric and English scales.

EXAMPLE 1   The highway distance between San Francisco and New York is 3120 mi. What is this distance in kilometers?

Solution
(a) Information given: 3120 mi.
(b) Information sought: the same distance expressed in kilometers.

(c) The conversion factor which applies (from Table S3-3) is 1.609 km/mi.

(d) The solution can be set up as follows, keeping in mind that the units should cancel properly:

$$3120 \text{ mi} \times 1.609 \text{ km/mi} = 5020 \text{ km} \qquad (Ans.)$$

EXAMPLE 2 A person is 5.50 ft tall. What is this height in meters?

Solution

(a) Information given: 5.50 ft.

(b) Information sought: the same height expressed in meters.

(c) The first step will be to express the height in centimeters. The conversion factor which applies (from Table S3-3) is 30.48 cm/ft.

(d) The solution to the first step can be set up as follows, keep-in mind that the units should cancel properly:

$$5.50 \text{ ft} \times 30.48 \text{ cm/ft} = 168 \text{ cm}$$

(e) The height in centimeters can be converted into the height in meters by keeping in mind that there are 100 cm in 1 m:

$$\frac{168 \text{ cm}}{100 \text{ cm/m}} = 1.68 \text{ m} \qquad (Ans.)$$

EXAMPLE 3 The price of gasoline in many European cities in 1977 was $1.76 per gallon. How many dollars did a liter cost?

Solution

(a) Information given: $1.76 per gallon.

(b) Information sought: the cost in dollars per liter.

(c) Conversion factor which applies (from Table S3-3):

$$0.2642 \text{ gal/L}$$

(d) The solution can be set up as follows, keeping in mind that the units should cancel properly:

$$\frac{\$1.76}{\text{gal}} \times \frac{0.2642 \text{ gal}}{\text{L}} = \$0.465/\text{L} \qquad (Ans.)$$

EXAMPLE 4 If a car travels 20 mi on a gallon of gasoline, how many kilometers will it travel on a liter?

Solution

(a) Information given: 20 mi/gal.

(b) Information sought: the distance the car will travel on a liter.

(c) The first step will be to determine how many kilometers the car will travel on a gallon. The conversion factor which applies (from Table S3-3) is 1.609 km/mi.

(d) The solution to the first step can be set up as follows, keep-in mind that the units should cancel properly:

$$\frac{20 \text{ mi}}{\text{gal}} \times \frac{1.609 \text{ km}}{\text{mi}} = 32 \text{ km/gal}$$

(e) The final step will be to determine how many kilometers the car will travel on a liter. The conversion factor which applies (from Table S3-3) is 0.2642 gal/L.

(f) The final step can be set up as follows, keeping in mind that the units should cancel properly:

$$32 \frac{km}{\cancel{gal}} \times \frac{0.2642 \; \cancel{gal}}{L} = 8.5 \; km/L \qquad (Ans.)$$

EXAMPLE 5 A person weighs 160 lb. What is this weight in kilograms?

Solution
(a) Information given: 160 lb.
(b) Information sought: the weight expressed in kilograms.
(c) The first step will be to determine the weight expressed in grams. The conversion factor which applies (from Table S3-3) is 453.6 g/lb.
(d) The solution to the first step can be set up as follows, keeping in mind that the units should cancel properly:

$$160 \; \cancel{lb} \times 453.6 \; g/\cancel{lb} = 72,600 \; g$$

(e) The final step will be to express the weight in kilograms, keeping in mind that there are 1000 g in 1 kg.
(f) The final step can be set up as follows, keeping in mind that the units should cancel properly:

$$72,600 \; \cancel{g} \times \frac{1 \; kg}{1000 \; \cancel{g}} = 72.6 \; kg \qquad (Ans.)$$

EXAMPLE 6 An automobile weighs 1280 kg. What is its weight in tons?

Solution
(a) Information given: 1280 kg.
(b) Information sought: the weight expressed in tons.
(c) The first step will be to determine the weight expressed in pounds. The conversion factor which applies (from Table S3-3) is 2.205 lb/kg.
(d) The solution to the first step can be set up as follows, keeping in mind that the units should cancel properly:

$$1280 \; \cancel{kg} \times \frac{2.205 \; lb}{\cancel{kg}} = 2822 \; lb$$

(e) The final step will be to express the weight in tons, keeping in mind that there are 2000 lb/ton.
(f) The final step can be set up as follows, keeping in mind that the units should cancel properly:

$$2822 \; \cancel{lb} \times \frac{1 \; ton}{2000 \; \cancel{lb}} = 1.411 \; tons \qquad (Ans.)$$

S3-2 SOME UNITS OF ENERGY AND POWER

A **calorie** (cal) is a small metric unit of energy equal to the amount of heat energy required to raise the temperature of 1 g of water 1°C. A kilocalorie (the Calorie of nutritionists) contains 1000 cal. The **British thermal unit** (Btu) is a unit of energy in the English system; one such unit contains the amount of heat energy required to raise the temperature of 1 lb of water 1°F. A **watt** (W) is a unit of power (the rate of expending energy) equivalent to the expenditure of about 860 cal of energy in an hour. A **kilowatthour** (kWh) is the amount of electric energy consumed in 1 hour at the rate of 1000 W. Table S3-4 gives some conversion equivalents of energy units.

Table S3-4
Some Conversion
Equivalents of Energy Units

	Conversion factor
1 Btu = 252 cal	252 cal/Btu
1 cal = 0.00397 Btu	0.00397 Btu/ cal
1 kWh = 859.2 kcal	859.2 kcal/kWh
1 kWh = 3410 Btu	3410 Btu/kWh

The following examples illustrate how conversions can be carried out between the various energy units.

EXAMPLE 1   A 100-W light bulb consumes 4.80 kWh of electricity when lighted continuously for 48.0 hours. What is the equivalent number of kilocalories?

Solution
(*a*) Information given: 4.80 kWh.
(*b*) Information sought: an expression of the amount in kilocalories.
(*c*) Conversion factor which applies (from Table S3-4): 859.2 kcal/kWh.
(*d*) The solution can be set up as follows, keeping in mind that the units should cancel properly:

$$4.80 \, \cancel{kWh} \times 859.2 \, kcal/\cancel{kWh} = 4120 \, kcal \qquad (Ans.)$$

EXAMPLE 2   The combustion of 16.04 g of methane (natural gas) produces 213 kcal of energy. What is the equivalent number of Btu?

Solution
(*a*) Information given: 213 kcal.
(*b*) Information sought: an expression of the amount in Btu.
(*c*) The first step will be to convert the amount of energy from kilocalories to calories, keeping in mind that there are 1000 cal in a kilocalorie.
(*d*) The solution to the first step can be set up as follows, keeping in mind that the units should cancel properly:

$$213 \, \cancel{kcal} \times \frac{1000 \, cal}{\cancel{kcal}} = 213,000 \, cal$$

(*e*) The final step will be to express the amount of energy in Btu. The conversion factor which applies (from Table S3-4) is 0.00397 Btu/cal.
(*f*) The final step can be set up as follows, keeping in mind that the units should cancel properly:

$$213,000 \, \cancel{cal} \times 0.00397 \, \frac{Btu}{\cancel{cal}} = 846 \, Btu \qquad (Ans.)$$

S3-3 SOME UNITS OF PRESSURE

A **millimeter of mercury** (mmHg) is the pressure exerted by a column of mercury 1 mm in height. It is also called a **torr.** An **atmosphere** is a pressure equivalent to that exerted by a column of mercury 760 mm (29.92 in) high. Table S3-5 lists a few conversion equivalents.

Table S3-5
Some Conversion
Equivalents of Pressure Units

	Conversion factor
1 atm = 760 mmHg	760 mmHg/atm
1 atm = 29.92 inHg	29.92 inHg/atm
1 torr = 1 mmHg	1 mmHg/torr

S3-4 SOME UNITS OF TIME

See Table S3-6.

Table S3-6
Some Conversion
Equivalents of Time Units

	Conversion factor
1 day = 24 hours	24 hours/day
1 hour = 60 minutes	60 minutes/hour
1 minute = 60 seconds	60 seconds/minute

S3-5 TEMPERATURE SCALES

The scale of temperature favored by chemists is the Kelvin scale, also called the absolute scale. The lowest temperature on this scale is called **absolute zero,** the temperature below which it is impossible to cool matter. The freezing point of water is taken to be 273 kelvins (K). (The temperature in careful work is actually 273.16 K.) The boiling point of water is at 373 K, and the difference between the freezing point and the boiling point of water is therefore 100 K (or absolute degrees).

The Celsius (formerly centigrade) temperature scale has the same size degrees as the Kelvin scale. The freezing point of water, however, is labeled 0°C. Temperatures below the freezing point of water are given a minus sign. Absolute zero on the Celsius scale is at −273°C (or more accurately −273.16°C). A temperature reading in degrees Celsius can be converted into kelvins very simply:

$$K = °C + 273$$

A reading in kelvins can be converted into degrees Celsius by a corresponding relationship.

$$°C = K - 273$$

The Fahrenheit temperature scale has degrees which are 1.8 times larger than the degrees on the Celsius and Kelvin scales. The freezing point of water is labeled 32°F. The boiling point of water is labeled 212°F. Temperatures below 0°F are given a minus sign. Absolute zero on the Fahrenheit scale is at $-459.7°$F. To convert a temperature on the Celsius scale into the corresponding temperature on the Fahrenheit scale we use the relationship

$$°F = °C \times \frac{1.8°F}{1.0°C} + 32°F$$

To convert a temperature on the Fahrenheit scale to the corresponding temperature on the Celsius scale we use the relationship

$$°C = \frac{1.0°C}{1.8°F}(°F - 32°F)$$

The Kelvin, Celsius, and Fahrenheit scales are compared in Fig. S3-1. The following examples illustrate conversions between the three scales.

EXAMPLE 1 Temperatures as high as 100°F occur from time to time in many parts of the United States. What is the corresponding temperature on the Celsius scale?

Solution The fundamental relationship is

$$°C = \frac{1.0°C}{1.8°F}(°F - 32°F)$$

In this case

$$°C = \frac{1.0°C}{1.8°F}(100°F - 32°F) = \frac{1.0°C}{1.8°F}(68°F) = 38°C \qquad (Ans.)$$

EXAMPLE 2 To what temperature does 100°F correspond on the Kelvin scale?

Solution
(a) Converting °F to °C as in Example 1 gives 38°C.
(b) The fundamental relationship is K = °C + 273. In this case

$$K = 38°C + 273 = 311 \text{ K} \qquad (Ans.)$$

EXAMPLE 3 What temperatures on the Fahrenheit scale and the Kelvin scale correspond to $-20°$C?

Solution The fundamental relationships are

$$°F = °C \times \frac{1.8°F}{1.0°C} + 32°F \qquad K = °C + 273$$

In this case

$$°F = -20°C \times \frac{1.8°F}{1.0°C} + 32°F = -36° + 32°F = -4°F \qquad (Ans.)$$

$$K = -20°C + 273 = 253 \text{ K} \qquad (Ans.)$$

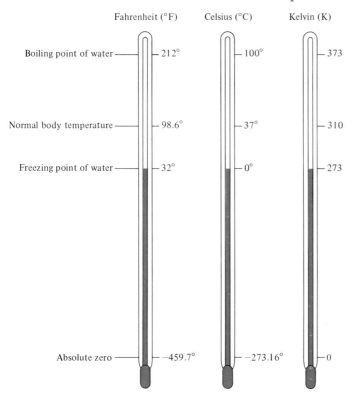

FIGURE S3-1
Comparison of the Fahrenheit, Celsius (formerly centigrade), and Kelvin temperature scales.

Fahrenheit (°F)   Celsius (°C)   Kelvin (K)

Boiling point of water —— 212°   —— 100°   —— 373

Normal body temperature —— 98.6°   —— 37°   —— 310

Freezing point of water —— 32°   —— 0°   —— 273

Absolute zero —— −459.7°   —— −273.16°   —— 0

© *United Feature Syndicate, Inc.*

KEY WORDS

1. SI unit (Sec. S3-1A)
2. Meter (Sec. S3-1C)
3. Cubic meter (Sec. S3-1C)
4. Kilogram (Sec. S3-1C)
5. Calorie (Sec. S3-2)
6. British thermal unit (Sec. S3-2)
7. Watt (Sec. S3-2)
8. Kilowatthour (Sec. S3-2)
9. Millimeter of mercury (Sec. S3-3)
10. Torr (Sec. S3-3)
11. Atmosphere (Sec. S3-3)
12. Absolute zero (Sec. S3-5)

PRACTICE EXERCISES

1. The highway distance between Chicago and Miami is 1330 mi. What is the distance in kilometers?

2. A person is 6.00 ft tall. What is this height in meters?

3. The price of gasoline in many European cities in 1970 was $1.53 per gallon. How many dollars did a liter cost?

4. If a car travels 16 mi on a gallon of gasoline, how many kilometers will it travel on 2.000 L?

5. A person weighs 120 lb. What is this weight in kilograms?

6. A truck weighs 9670 kg. What is its weight in tons?

7. If milk costs 47 cents a quart, how many cents will a liter cost?

8. A barrel weighs 80.0 kg. What is this weight in pounds?

9. A 100-W light bulb consumes 14.4 kWh of electricity when lighted continuously for 6.00 days. What is the equivalent number of kilocalories?

10. How many Btu are equivalent to 4673 kcal?

11. How many kilocalories are equivalent to 10,000 Btu?

12. To what temperature on the Celsius scale does 70°F correspond?

13. To what temperature does 70°F correspond on the Kelvin scale?

14. What temperatures on the Fahrenheit scale and the Kelvin scale correspond to $-26°C$?

15. To what temperature on the Celsius scale does $-20°F$ correspond?

SUPPLEMENT 4

The Weight Relationships of Chemical Reactions†

S4-1 THE IMPORTANCE OF QUANTITATIVE RELATIONSHIPS

Although the discussions of this text have only hinted at it, the facts and theories which are the foundation of all aspects of chemistry rest on a firm basis of quantitative measurements and relationships. Indeed, it was only after questions like: How much? entered laboratory experiments that chemistry became a highly productive form of inquiry. In this supplement we discuss a few of the quantitative aspects of chemical reactions.

S4-2 ATOMIC WEIGHTS, MOLECULAR WEIGHTS, AND EQUATION BALANCING

A chemical equation cannot provide useful information about the weight relationships of a chemical reaction unless the equation is balanced. The mechanics of bringing an equation into balance were given in Sec. 6-4. To recapitulate, let us balance the equation

$$NO_2 + H_2O \rightarrow HNO_3 + NO$$

Nitrogen Water Nitric Nitric
dioxide   acid   oxide

Unbalanced chemical equation; starting point of an exercise in equation balancing

To bring the equation into balance using a trial-and-error approach we must devise numbers of molecules for each reactant and product, i.e., the numbers in front of the formulas, so that the number of each kind of atom represented among the reactants is equal to the number among the products. Both atoms and ions count as atoms in this tally. The balanced form of the equation under consideration is

$$3NO_2 + H_2O \rightarrow 2HNO_3 + NO$$

Nitrogen Water Nitric Nitric
dioxide   acid   oxide

Balanced form of the equation in the equation balancing exercise

EXAMPLE 1 Balance

$$SO_2 + O_2 \rightarrow SO_3$$

Solution

$$2SO_2 + O_2 \rightarrow 2SO_3$$

---

† Prerequisite, Chaps. 1 to 6; recommended, Supplement 2.

EXAMPLE 2   Balance

$$Al + O_2 \rightarrow Al_2O_3$$

Solution

$$4Al + 3O_2 \rightarrow 2Al_2O_3$$

In calculating the weight relationships of a chemical reaction we need to know how the information about the relative weights of atoms and molecules is expressed. The reason for using relative weights rather than absolute weights was discussed in Sec. 6-6. The scale has been chosen in such a way that the weight of even the smallest atom (of the hydrogen 1 atom) is greater than 1. It is also useful to recall that the relative weight used for an atom reflects the distribution of isotopes as they occur in nature (the nature of isotopes is discussed in Sec. 2-5; their role in relative atomic weights is explained in Sec. 6-6). A complete listing of atomic weights is given in Table S4-1. For our present purposes the use of atomic weights expressed to only four significant figures† will be sufficient (see Table S4-2).

The relative weights of molecules are expressed on the same scale as the relative weights of atoms. As discussed in Sec. 6-6, the relative molecular weight of a substance can be obtained by adding the relative weights of the component atoms.

EXAMPLE 3   What is the relative molecular weight of methyl alcohol, CH_3OH?

Solution

Relative weight of carbon atom (Table S4-2)	12.01
Relative weight of oxygen atom (Table S4-2)	16.00
Relative weight of four hydrogen atoms (Table S4-2)	
4×1.008	4.032
Relative weight of methyl alcohol molecule	32.04

Ordinarily relative atomic and molecular weights are referred to simply as atomic and molecular weights, it being understood that they are relative. The value 32.04, for example, is ordinarily called the molecular weight of methyl alcohol; further, since it is a relative value it has no units.

S4-3 WHAT IS MEANT BY A MOLE IN THE CHEMICAL SENSE?

It was mentioned in Sec. 6-6 that the actual weights of atoms and molecules are very small. The actual weight of the hydrogen 1 isotope, for example, is 1.673×10^{-24} g. To avoid using such awkwardly small numbers it is convenient to work in terms of the weight of a standard reference number of atoms. The standard reference number has the value 6.022×10^{23}, called Avogadro's number, in honor of the Italian chemist Amedeo Avogadro, who made significant contributions to the develop-

---

† The method of rounding off numbers is discussed in Supplement 2, Sec. S2-3.

Table S4-1
Atomic Weights of Atoms in Order of Name
on a Scale Where the Carbon 12 Isotope Has a Value of 12.0000†

	Symbol	Atomic number	Atomic weight		Symbol	Atomic number	Atomic weight
Actinium	Ac	89	(227)	Mercury	Hg	80	200.59
Aluminum	Al	13	26.98154	Molybdenum	Mo	42	95.94
Americium	Am	95	(243)	Neodymium	Nd	60	144.24
Antimony	Sb	51	121.75	Neon	Ne	10	20.179
Argon	Ar	18	39.948	Neptunium	Np	93	(237)
Arsenic	As	33	74.9216	Nickel	Ni	28	58.70
Astatine	At	85	(210)	Niobium	Nb	41	92.9064
Barium	Ba	56	137.34	Nitrogen	N	7	14.0067
Berkelium	Bk	97	(247)	Nobelium	No	102	(254)
Beryllium	Be	4	9.01218	Osmium	Os	76	190.2
Bismuth	Bi	83	208.9804	Oxygen	O	8	15.9994
Boron	B	5	10.81	Palladium	Pd	46	106.4
Bromine	Br	35	79.904	Phosphorus	P	15	30.97376
Cadmium	Cd	48	112.40	Platinum	Pt	78	195.09
Calcium	Ca	20	40.08	Plutonium	Pu	94	(242)
Californium	Cf	98	(251)	Polonium	Po	84	(210)
Carbon	C	6	12.011	Potassium	K	19	39.098
Cerium	Ce	58	140.12	Praseodymium	Pr	59	140.9077
Cesium	Cs	55	132.9054	Promethium	Pm	61	(147)
Chlorine	Cl	17	35.453	Protactinium	Pa	91	(231)
Chromium	Cr	24	51.996	Radium	Ra	88	(226)
Cobalt	Co	27	58.9332	Radon	Rn	86	(222)
Copper	Cu	29	63.546	Rhenium	Re	75	186.207
Curium	Cm	96	(247)	Rhodium	Rh	45	102.9055
Dysprosium	Dy	66	162.50	Rubidium	Rb	37	85.4678
Einsteinium	Es	99	(254)	Ruthenium	Ru	44	101.07
Erbium	Er	68	167.26	Samarium	Sm	62	150.4
Europium	Eu	63	151.96	Scandium	Sc	21	44.9559
Fermium	Fm	100	(253)	Selenium	Se	34	78.96
Fluorine	F	9	18.99840	Silicon	Si	14	28.086
Francium	Fr	87	(223)	Silver	Ag	47	107.868
Gadolinium	Gd	64	157.25	Sodium	Na	11	22.98977
Gallium	Ga	31	69.72	Strontium	Sr	38	87.62
Germanium	Ge	32	72.59	Sulfur	S	16	32.06
Gold	Au	79	196.9665	Tantalum	Ta	73	180.9479
Hafnium	Hf	72	178.49	Technetium	Tc	43	(99)
Helium	He	2	4.00260	Tellurium	Te	52	127.60
Holmium	Ho	67	164.9304	Terbium	Tb	65	158.9254
Hydrogen	H	1	1.0079	Thallium	Tl	81	204.37
Indium	In	49	114.82	Thorium	Th	90	232.0381
Iodine	I	53	126.9045	Thulium	Tm	69	168.9342
Iridium	Ir	77	192.22	Tin	Sn	50	118.69
Iron	Fe	26	55.847	Titanium	Ti	22	47.90
Krypton	Kr	36	83.80	Tungsten	W	74	183.85
Lanthanum	La	57	138.9055	Uranium	U	92	238.029
Lawrencium	Lr	103	(257)	Vanadium	V	23	50.9414
Lead	Pb	82	207.2	Xenon	Xe	54	131.30
Lithium	Li	3	6.941	Ytterbium	Yb	70	173.04
Lutetium	Lu	71	174.97	Yttrium	Y	39	88.9059
Magnesium	Mg	12	24.305	Zinc	Zn	30	65.38
Manganese	Mn	25	54.9380	Zirconium	Zr	40	91.22
Mendelevium	Md	101	(256)				

† Values in parentheses are mass numbers of isotopes best known or of longest half-life.

Table S4-2
Approximate Atomic Weights of Common Elements

Aluminum	Al	26.98	Hydrogen	H	1.008	Radium	Ra	226.0
Argon	Ar	39.95	Iodine	I	126.9	Silicon	Si	28.09
Arsenic	As	74.92	Iron	Fe	55.85	Silver	Ag	107.9
Barium	Ba	137.3	Lanthanum	La	138.9	Sodium	Na	22.99
Bismuth	Bi	209.0	Lead	Pb	207.2	Strontium	Sr	87.62
Boron	B	10.81	Lithium	Li	6.941	Sulfur	S	32.06
Bromine	Br	79.90	Magnesium	Mg	24.31	Thorium	Th	232.0
Cadmium	Cd	112.4	Manganese	Mn	54.94	Tin	Sn	118.7
Calcium	Ca	40.08	Mercury	Hg	200.6	Titanium	Ti	47.90
Carbon	C	12.01	Molybdenum	Mo	95.94	Tungsten	W	183.9
Chlorine	Cl	35.45	Neon	Ne	20.18	Uranium	U	238.0
Chromium	Cr	52.00	Nickel	Ni	58.70	Vanadium	V	50.94
Cobalt	Co	58.93	Nitrogen	N	14.01	Xenon	Xe	131.3
Copper	Cu	63.55	Oxygen	O	16.00	Zinc	Zn	65.38
Fluorine	F	19.00	Phosphorus	P	30.97	Zirconium	Zr	91.22
Gold	Au	197.0	Platinum	Pt	195.1			
Helium	He	4.003	Potassium	K	39.10			

ment of the atomic theory. The specific value assigned to the standard reference number has been chosen for two reasons: (1) the weight of this number of the smallest atoms has a value greater than 1 and (2) the weight of this number of any kind of atom is the relative atomic weight expressed as grams. The relative atomic weight of the hydrogen 1 isotope, for example, is 1.0078. The actual weight of 6.022×10^{23} hydrogen 1 atoms is 1.0078 g.

A special name is given to an Avogadro number of atoms; it is called a mole. A **mole** (mol) is an Avogadro number (6.022×10^{23}) of atoms or of anything else. A mole of hydrogen 1 atoms contains 6.022×10^{23} atoms and weighs 1.0078 g. A mole of carbon 12 atoms contains 6.022×10^{23} atoms and weighs 12.0000 g. A mole of marbles contains 6.022×10^{23} marbles and if each marble weighs 1.000 g, the mole of marbles weighs 6.022×10^{23} g.

A mole of carbon atoms with the distribution of isotopes occurring in nature weighs 12.011 g. A mole of hydrogen atoms with a natural distribution of isotopes weighs 1.0079 g. A mole of hydrogen molecules with a natural distribution of isotopes weighs 2.0158 g. This value is obtained from the relative weight of hydrogen molecules, in turn calculated from the relative weight of hydrogen atoms.

In order to become familiar with the meaning of the quantity called a mole let us consider the following illustrative examples.

EXAMPLE 1 What does a mole of oxygen atoms weigh in grams?

Solution The weight of an Avogadro number (6.022×10^{23}) of oxygen atoms, which is the same as the atomic weight expressed in grams, or 16.00 grams (using Table S4-2).

EXAMPLE 2 What does a mole of oxygen molecules, O_2, weigh in grams?

Solution The weight of an Avogadro number (6.022×10^{23}) of oxygen molecules, which is the same as the molecular weight expressed in grams, or 32.00 grams.

EXAMPLE 3 What does 1 mole of methyl alcohol, CH_3OH, weigh in grams?

Solution The weight of 6.022×10^{23} methyl alcohol molecules, which is the same as the molecular weight expressed in grams: 32.04 g (obtained from adding the atomic-weight values from Table S4-2).

EXAMPLE 4 How many moles of methyl alcohol are contained in 64.08 g?

Solution The weight of a mole of methyl alcohol is 32.04 g/mol. In 64.08 g

$$\frac{64.08 \text{ g}}{32.04 \text{ g/mol}} = 2.000 \text{ mol} \qquad (Ans.)$$

Note that the units cancel out, so that they are the same on both sides of the equation:

$$\frac{\cancel{g}}{\cancel{g}/\text{mol}} = \text{mol}$$

EXAMPLE 5 How many moles of methyl alcohol are contained in 100.00 g?

Solution The weight of a mole of methyl alcohol is 32.04 g/mol. In 100.00 g

$$\frac{100.00 \cancel{g}}{32.04 \cancel{g}/\text{mol}} = 3.121 \text{ mol} \qquad (Ans.)$$

EXAMPLE 6 How many grams does 3.000 mol of methyl alcohol weigh?

Solution The weight of a mole of methyl alcohol is 32.042 g/mol. Then

$$3.000 \cancel{\text{mol}} \times 32.04 \text{ g/}\cancel{\text{mol}} = 96.12 \text{ g} \qquad (Ans.)$$

EXAMPLE 7 How many grams does 24.30 mol of water weigh?

Solution The weight of 1 mol of water is 18.0152 g/mol (obtained by adding the atomic weight values from Table S4-2). Then

$$24.30 \cancel{\text{mol}} \times 18.02 \text{ g/}\cancel{\text{mol}} = 437.9 \text{ g} \qquad (Ans.)$$

EXAMPLE 8 How many moles of water are contained in 200.00 g?

Solution The weight of a mole of water is 18.02 g/mol. Then

$$\frac{200.00 \cancel{g}}{18.02 \cancel{g}/\text{mol}} = 11.10 \text{ mol} \qquad (Ans.)$$

EXAMPLE 9 How many hydrogen molecules are contained in 4.000 mol?

Solution Since 1 mol of hydrogen molecules contains 6.022 × 10^{23} molecules per mol, in 4.000 mol

$$(4.000 \text{ mol})(6.022 \times 10^{23} \text{ molecules/mol}) = 2.409 \times 10^{24} \text{ molecules} \quad (Ans.)$$

EXAMPLE 10 How many molecules of methyl alcohol are in 43.68 g?

Solution Since 1 mol of methyl alcohol weighs 32.04 g/mol (Example 3) and 1 mol contains 6.022 × 10^{23} molecules, in 43.68 g

$$\frac{43.68 \text{ g}}{32.04 \text{ g/mol}} = 1.363 \text{ mol}$$

$$1.363 \text{ mol} \times 6.022 \times 10^{23} \text{ molecules/mol} = 8.208 \times 10^{23} \text{ molecules} \quad (Ans.)$$

S4-4 THE WEIGHT RELATIONSHIPS IN A CHEMICAL REACTION

$$2H_2 \;+\; O_2 \;\rightarrow\; 2H_2O$$

Hydrogen   Oxygen   Water

Balanced equation for the reaction between hydrogen and oxygen to form water

Let us consider the information in a chemical equation which tells us the quantities of reactants which undergo reaction and the quantities of products obtained. As a sample reaction let us turn to the combination of hydrogen and oxygen to form water, a highly exothermic chemical change used to propel rockets. The statement made by this balanced equation tells us that the following relative amounts of reactants and product will be involved:

	$2H_2$	$+$	O_2	$\rightarrow$	$2H_2O$
Molecules:	2		1		2
Avogadro units of molecules:	$2 \times 6.022 \times 10^{23}$		6.022×10^{23}		$2 \times 6.022 \times 10^{23}$
Moles:	2		1		2
Grams:	2×2.016		32.00		2×18.02

Relative amounts of reactants and product of a reaction

It should be noted carefully that an equation gives only the relative amounts of substances involved in a reaction. This equation indicates for example, that if 2 mol of hydrogen and 1 mol of oxygen are available, 2 mol of water will be formed. The amount of water formed in a particular case depends on the specific amounts of hydrogen and oxygen used.

Moreover, an equation indicates that the reactants react on a molecular basis and products are formed on a molecular basis. In the reaction under consideration, two molecules of hydrogen react with one molecule of oxygen to produce two molecules of water. Or the weight relationships can be stated on a molar basis: 2 mol of hydrogen react with 1 mol of oxygen to form 2 mol of water. If the information about a weight of reactant or product is given in grams, the amount must be converted into moles before the information provided by the equation can be used.

An equation, however, provides the key information which permits calculation of the specific amounts of products formed from a given amount of the reactants or a calculation of how much of a reactant is required to produce a given amount of product. Moreover, an equation provides the information which permits the calculation of what weight of oxygen will be required to react with a specific weight of hydrogen and vice versa. The following examples will illustrate.

EXAMPLE 1 In the reaction between hydrogen and oxygen to form water, how many moles of oxygen are required to react with 8.000 mol of hydrogen?

Solution

$$8.000 \text{ mol H}_2 \times \quad \frac{1 \text{ mol O}_2}{2 \text{ mol H}_2} \quad = 4.000 \text{ mol O}_2 \quad (Ans.)$$

Ratio from
inspection of
balanced equation

EXAMPLE 2 In the reaction between hydrogen and oxygen to form water, how many moles of oxygen are required to react with 8.064 g of hydrogen?

Solution
(*a*) Convert grams of hydrogen into moles of hydrogen:

$$\frac{8.064 \text{ g}}{2.016 \text{ g/mol}} = 4.000 \text{ mol H}_2$$

(*b*) Use information from the equation:

$$4.000 \text{ mol H}_2 \times \quad \frac{1 \text{ mol O}_2}{2 \text{ mol H}_2} \quad = 2.000 \text{ mol O}_2 \quad (Ans.)$$

Ratio from
inspection of
balanced equation

EXAMPLE 3 In the reaction between hydrogen and oxygen to form water, how many grams of hydrogen are required to react with 64.00 g of oxygen?

Solution
(*a*) Convert grams of oxygen into moles of oxygen:

$$\frac{64.00 \text{ g}}{32.00 \text{ g/mol}} = 2.000 \text{ mol O}_2$$

(*b*) Use information from the equation:

$$2.000 \text{ mol O}_2 \times \quad \frac{2 \text{ mol H}_2}{1 \text{ mol O}_2} \quad = 4.000 \text{ mol H}_2$$

Ratio from
inspection of
balanced equation

(c) Convert 4.000 mol of hydrogen into grams:

$$4.000 \, \text{mol} \, H_2 \times 2.016 \, \text{g/mol} = 8.064 \, \text{g} \, H_2 \qquad (Ans.)$$

Note that only step (b) requires information from the equation; steps (a) and (c) have nothing to do with the equation.

The equation for the reaction between hydrogen and oxygen predicts that when the two substances react, 2 mol of hydrogen react with 1 mol of oxygen. In order to react, however, they do not have to be brought together in exactly this molar ratio. If, for example, 2 mol of hydrogen and 2 mol of oxygen are brought together, one of the moles of oxygen will react with the 2 mol of hydrogen, as the reaction predicts, and 1 mol of oxygen will remain unreacted. The oxygen in such a case is said to be "in excess," i.e., in excess of the amount which will react with the hydrogen available. Or if 4 mol of hydrogen and 1 mol of oxygen are mixed, the hydrogen will be in excess, and only 2 mol will react and 2 mol of hydrogen will remain unreacted.

In practice, it often happens that the exact amount of one reactant is known, but the amount of the other reactant is not known exactly, although it is known to be in excess. Problems dealing with such situations say something like "10.000 g of hydrogen reacts with an excess of oxygen to form . . . ," or "10.000 g of hydrogen reacts with all the oxygen required to form. . . ." Two examples will illustrate.

EXAMPLE 1   All the oxygen required to react with 3.600 g of hydrogen in the reaction $H_2 + O_2 \rightarrow H_2O$ is available. How many grams of water will be formed?

Solution
(a) Balance the equation:

$$2H_2 + O_2 \rightarrow 2H_2O$$

(b) Convert 3.600 g of hydrogen into moles of hydrogen:

$$\frac{3.600 \, \text{g}}{2.016 \, \text{g/mol}} = 1.786 \, \text{mol} \, H_2$$

(c) Using information from the equation, calculate the number of moles of water formed from 1.786 mol of hydrogen and all the oxygen required:

$$1.786 \, \text{mol} \, H_2 \times \frac{2 \, \text{mol} \, H_2O}{2 \, \text{mol} \, H_2} = 1.786 \, \text{mol} \, H_2O$$

Ratio from
inspection of
balanced equation

(d) Convert 1.786 mol of water into grams of water:

$$1.786 \, \text{mol} \, H_2O \times 18.02 \, \text{g/mol} = 32.18 \, \text{g} \, H_2O \qquad (Ans.)$$

Note that only step (c) requires information from the equation. Steps (b) and (d) have nothing to do with the equation. In particular, the numbers in front of the formulas in the balanced equation have nothing to do with steps (b) and (d). In order to use the information from the equation the grams of hydrogen given in the question must first be converted into moles of hydrogen [step (b)].

EXAMPLE 2 An excess of hydrogen undergoes reaction with 30.00 g of oxygen in the reaction $H_2 + O_2 \rightarrow H_2O$. How many grams of water will be formed?

Solution
(a) Balance the equation:

$$2H_2 + O_2 \rightarrow 2H_2O$$

(b) Convert 30.00 g of oxygen into moles of oxygen:

$$\frac{30.00 \text{ g}}{32.00 \text{ g/mol}} = 0.9375 \text{ mol } O_2$$

(c) Using information from the equation, calculate the number of moles of water formed from 0.9375 mol of oxygen and all the hydrogen required:

$$0.9375 \text{ mol } O_2 \times \frac{2 \text{ mol } H_2O}{1 \text{ mol } O_2} = 1.875 \text{ mol } H_2O$$

Ratio from
inspection of
balanced equation

(d) Convert 1.875 mol of water into grams of water:

$$1.875 \text{ mol } H_2O \times 18.02 \text{ g/mol} = 33.79 \text{ g } H_2O \quad (Ans.)$$

Note that the information from the balanced equation is used only in step (c). In order to use the information from the equation the grams of oxygen given in the question must first be converted into moles of oxygen [step (b)].

S4-5 SUMMARY OF METHOD OF DETERMINING WEIGHT RELATIONSHIPS OF A REACTION FROM AN EQUATION

(a) Complete and balance the equation which represents the reaction.
(b) Determine the weights of substances involved in the reaction according to the following sequence:

Ratio from
balanced
equation

$$\underset{\substack{\text{Substance about which} \\ \text{data are known}}}{\text{Grams} \rightarrow \text{moles}} \xrightarrow{\hspace{2cm}} \underset{\substack{\text{Substance about which} \\ \text{data are sought}}}{\text{moles} \rightarrow \text{grams}}$$

This method of determining weight relationships from chemical equations assumes that the reactions take place completely. Actually this does not often happen. Most of the time the reaction will stop before all the available reactants are consumed. While such cases have been thoroughly studied, further discussion is omitted here.

The following examples will illustrate the implementation of the suggested sequence of steps in the solution.

EXAMPLE 1   How many grams of magnesium oxide, MgO, will be obtained from 40.00 g of magnesium in the reaction $Mg + O_2 \rightarrow MgO$ assuming that all of the necessary oxygen is available?

Solution

(a) Balance the equation:

$$2Mg + O_2 \rightarrow 2MgO$$

(b) Convert 40.00 g of magnesium into moles of magnesium:

$$\frac{40.00 \text{ g}}{24.31 \text{ g/mol}} = 1.645 \text{ mol Mg}$$

(c) Using information from the equation, calculate the number of moles of magnesium oxide formed from 1.645 mol of magnesium:

$$1.645 \text{ mol Mg} \times \frac{1 \text{ mol MgO}}{1 \text{ mol Mg}} = 1.645 \text{ mol MgO}$$

(d) Convert 1.645 mol of magnesium oxide into grams of magnesium oxide (the molecular weight of magnesium oxide is calculated from the atomic weights of Table S4-2):

$$1.645 \text{ mol MgO} \times 40.31 \text{ g/mol} = 66.31 \text{ g MgO} \quad (Ans.)$$

EXAMPLE 2   If 50.00 g of hydrogen is treated with an excess of nitrogen to form ammonia according to the reaction $N_2 + H_2 \rightarrow NH_3$, how many grams of ammonia will be formed?

Solution

(a) Balance the equation:

$$N_2 + 3H_2 \rightarrow 2NH_3$$

(b) Convert 50.00 g of hydrogen into moles of hydrogen:

$$\frac{50.00 \text{ g}}{2.016 \text{ g/mol}} = 24.80 \text{ mol H}_2$$

(c) Using information from the equation, calculate the number of moles of ammonia formed from 24.80 mol of hydrogen.

$$24.80 \text{ mol H}_2 \times \frac{2 \text{ mol NH}_3}{3 \text{ mol H}_2} = 16.53 \text{ mol NH}_3$$

(d) Convert moles of ammonia into grams of ammonia (the molecular weight of ammonia is calculated from the atomic weights of Table S4-2):

$$16.53 \text{ mol NH}_3 \times 17.03 \text{ g/mol} = 281.5 \text{ g NH}_3 \quad (Ans.)$$

1. Mole (Sec. S4-3)

SUMMARIZING QUESTIONS FOR SELF-STUDY

1. Q. How many grams does 1.000 mol of propane, $CH_3CH_2CH_3$, weigh?

   A. Using atomic weights of Table S4-2, we get

   $$3C = 3 \times 12.01 = \quad 36.03$$
   $$8H = 8 \times 1.008 = \quad \underline{8.064}$$
   $$\text{Weight 1.000 mol} \quad 44.09 \text{ g} \qquad (Ans.)$$

2. Q. How many grams does 3.600 mol of methane, CH_4, weigh?

   A. Using atomic weights of Table S4-2, we have

   $$3.600 \text{ mol} \times 16.04 \text{ g/mol} = 57.74 \text{ g} \quad (Ans.)$$

3. Q. How many moles of propane, $CH_3CH_2CH_3$, are contained in 100.0 g?

   A. Using the answer to Example 1, we have

   $$\frac{100.0 \text{ g}}{44.09 \text{ g/mol}} = 2.268 \text{ mol } CH_3CH_2CH_3 \quad (Ans.)$$

4. Q. How many molecules of ethane, CH_3CH_3, are contained in 2.80 mol?

   A. $2.80 \text{ mol} \times 6.022 \times 10^{23} \text{ molecules/mol} = 1.69 \times 10^{24} \text{ molecules}$ (Ans.)

5. Q. How many molecules of methyl alcohol, CH_3OH, are contained in 50.00 g?

   A. $\dfrac{50.00 \text{ g}}{32.04 \text{ g/mol}} = 1.561 \text{ mol } CH_3OH$

   $1.561 \text{ mol} \times 6.022 \times 10^{23} \text{ molecules/mol} = 9.400 \times 10^{23} \text{ molecules}$ (Ans.)

6. Q. How many grams of carbon dioxide, CO_2, will be obtained from 40.00 g of methane and the necessary oxygen in the reaction $CH_4 + O_2 \rightarrow CO_2 + H_2O$?

   A. Balanced equation:

   $$CH_4 + 2O_2 \rightarrow CO_2 + 2H_2O$$
   $$\frac{40.00 \text{ g}}{16.04 \text{ g/mol}} = 2.494 \text{ mol } CH_4$$

$$2.494 \text{ mol } CH_4 \times \frac{1 \text{ mol } CO_2}{1 \text{ mol } CH_4} = 2.494 \text{ mol } CO_2$$

$$2.494 \text{ mol } CO_2 \times 44.01 \text{ g/mol} =$$
$$109.8 \text{ g } CO_2 \quad (Ans.)$$

7. Q. Sodium hydroxide, NaOH, and sulfuric acid, H_2SO_4, undergo the reaction: NaOH + $H_2SO_4 \rightarrow Na_2SO_4 + H_2O$. How many grams of sodium sulfate, Na_2SO_4, are formed from 60.00 g of sodium hydroxide, NaOH, and an excess of sulfuric acid, H_2SO_4?

   A. Balanced equation:

   $$2NaOH + H_2SO_4 \rightarrow Na_2SO_4 + 2H_2O$$

   $$\frac{60.00 \text{ g}}{40.00 \text{ g/mol}} = 1.500 \text{ mol NaOH}$$

   $$1.500 \text{ mol} \times \frac{1 \text{ mol } Na_2SO_4}{2 \text{ mol NaOH}} = 0.7500 \text{ mol } Na_2SO_4$$

   $$0.7500 \text{ mol } Na_2SO_4 \times 142.0 \text{ g/mol} = 106.5 \text{ g } Na_2SO_4 \quad (Ans.)$$

8. Q. Phosphorus pentoxide, P_4O_{10}, reacts with water to form phosphoric acid, H_3PO_4, according to the reaction $P_4O_{10} + H_2O \rightarrow H_3PO_4$. How many grams of phosphoric acid, H_3PO_4, are formed from 100.00 g of phosphorus pentoxide, P_4O_{10}, and the necessary water?

   A. Balanced equation:

   $$P_4O_{10} + 6H_2O \rightarrow 4H_3PO_4$$

   $$\frac{100.00 \text{ g}}{283.9 \text{ g/mol}} = 0.3522 \text{ mol } P_4O_{10}$$

   $$0.3522 \text{ mol } P_4O_{10} \times \frac{4 \text{ mol } H_3PO_4}{1 \text{ mol } P_4O_{10}} = 1.409 \text{ mol } H_3PO_4$$

   $$1.409 \text{ mol } H_3PO_4 \times 97.99 \text{ g/mol} = 138.1 \text{ g } H_3PO_4 \quad (Ans.)$$

PRACTICE EXERCISES

1. What does a mole of nitrogen atoms weigh in grams?

2. What does a mole of nitrogen molecules, N_2, weigh in grams?

3. What does a mole of acetic acid, CH_3COOH, weigh in grams?

4. How many moles of acetic acid, CH_3COOH, are contained in 10.00 g?

5. How many grams does 2.30 mol of acetic acid, CH_3COOH, weigh?

6. How many acetic acid molecules are contained in 200.00 g?

7. In the reaction between the rocket fuel hydrazine, N_2H_4, and oxygen, O_2, to form nitrogen, N_2, and water, H_2O,

   $$N_2H_4 + O_2 \rightarrow N_2 + H_2O$$

   how many moles of water are formed from 4.000 mol of hydrazine?

8. In the reaction between methane, CH_4, and oxygen, O_2, to form carbon dioxide, CO_2, and water, H_2O,

   $$CH_4 + O_2 \rightarrow CO_2 + H_2O$$

   how many moles of oxygen are required to react with 50.00 g of methane?

9. In the reaction of Question 8 how many grams of methane are required to react with 100.00 g of oxygen?

10. How many grams of water are formed from 140.00 g of methane in the reaction of Question 8?

11. If 78.00 g of methane is treated with an excess of oxygen, how many grams of carbon dioxide will be formed according to the reaction of Question 8?

12. How many grams of phosphorus pentoxide, P_4O_{10}, are required to produce 200.00 g of phosphoric acid, H_3PO_4, assuming an excess of water in the reaction?

$$P_4O_{10} + H_2O \rightarrow H_3PO_4$$

SUPPLEMENT 5

The Concentration of Solutions†

S5-1 PERCENT BY WEIGHT

The concentration of a solution is the amount of the solute (substance dissolved) in a given amount of the solution (Sec. 5-1). One of the most common methods of expressing the concentration of solutions is by percent by weight. The **percent by weight** of a solution is the number of grams of solute in each 100.00 g of solution:

$$\text{Percent by weight (wt \%) of solute} = \frac{\text{weight solute}}{\text{weight solution}} \times 100$$

The following examples will illustrate how the value is determined and used.

EXAMPLE 1 What is the percent by weight of sodium chloride, NaCl, in an aqueous solution if 10.00 g is dissolved in 48.00 g of solution?

Solution
(*a*) Amount of solute: 10.00 g (directly from question).
(*b*) Amount of solution: 48.00 g (directly from question).

(*c*)
$$\frac{10.00 \text{ g NaCl}}{48.00 \text{ g solution}} \times 100 = 20.83 \text{ wt \%} \qquad (Ans.)$$

Grams of NaCl in
1.00 g solution

Grams of NaCl in
100.00 g solution

EXAMPLE 2. What is the percent by weight of sodium chloride, NaCl, in an aqueous solution prepared by dissolving 12.00 g in 50.00 g of water?

Solution
(*a*) Amount of solute: 12.00 g (directly from question).
(*b*) Amount of solution:

$$12.00 \text{ g NaCl} + 50.00 \text{ g water} = 62.00 \text{ g}$$

(*c*)
$$\frac{12.00 \text{ g NaCl}}{62.00 \text{ g solution}} \times 100 = 19.35 \text{ wt \%} \qquad (Ans.)$$

Grams of NaCl in
1.00 g solution

Grams of NaCl in
100.00 g solution

---

† An expansion of Chap. 5; should be preceded by Secs. S4-1 to S4-3.

EXAMPLE 3 How many grams of magnesium bromide, $MgBr_2$, are contained in 72.00 g of a solution which is 11.20 percent by weight?

Solution

(*a*) Determine the grams of $MgBr_2$ in 1.00 g of solution (incorporated into solution below).

(*b*) Determine the grams of $MgBr_2$ in 72.00 g of solution.

(*c*) For both steps:

$$\underbrace{\frac{11.20 \text{ g } MgBr_2}{100.00 \text{ g solution}}}_{\substack{\text{Grams of } MgBr_2 \text{ in} \\ 1.00 \text{ g solution}}} \times 72.00 \text{ g solution} - 8.064 \text{ g} \quad (Ans.)$$

S5-2 PARTS PER MILLION PARTS BY WEIGHT

The percent by weight of a solution is, by definition, the parts per hundred parts by weight of a solution. When concentrations are very dilute, it is often convenient to express them in parts per million parts by weight (ppm). The **parts per million parts by weight** of a solution is the number of grams of solute in each million (1.00×10^6) g of solution:

$$\text{ppm} = \frac{\text{weight solute}}{\text{weight solution}} \times 1.00 \times 10^6$$

The following examples will illustrate how the value is obtained and used.

Note to Students

It has been calculated that 1 ppm is equivalent to 1 inch in 16 miles or 1 minute in 2 years.

EXAMPLE 1 What is the concentration of an aqueous solution of potassium chloride, KCl, in parts per million parts by weight if 0.100 g is dissolved in 1206 g of solution?

Solution

(*a*) Amount of solute: 0.100 g (directly from question).

(*b*) Amount of solution: 1206 g (directly from question).

(*c*)
$$\underbrace{\frac{0.100 \text{ g } KCl}{1206 \text{ g solution}}}_{\substack{\text{Grams of KCl in} \\ 1 \text{ g of solution}}} \times 1.00 \times 10^6 = 82.9 \text{ ppm} \quad (Ans.)$$

$$\underbrace{\phantom{\frac{0.100 \text{ g } KCl}{1206 \text{ g solution}}}}_{\substack{\text{Grams of KCl in } 1 \times 10^6 \text{ g} \\ \text{solution}}}$$

EXAMPLE 2 How many grams of calcium chloride, $CaCl_2$, are contained in 18,000 (1.8000×10^4) g of a solution which has a concentration of 126 ppm by weight?

Solution

(a) Determine the grams of $CaCl_2$ in 1.00 g of solution (incorporated into solution below).

(b) Determine the grams of $CaCl_2$ in 1.8000×10^4 g of solution.

(c) For both steps:

$$\frac{126 \text{ g } CaCl_2}{1.00 \times 10^6 \text{ g solution}} \times 1.8000 \times 10^4 \text{ g solution} = 2.27 \text{ g} \quad (Ans.)$$

Grams of $CaCl_2$ in
1.00 g solution

EXAMPLE 3   What is the concentration of magnesium ion, Mg^{2+}, in parts per million parts by weight if 4.80 mg of magnesium ion is contained in 40.00 g of solution?

Solution

(a) Amount of solute: $4.80 \text{ mg} = 4.80 \times 10^{-3}$ g.

(b) Amount of solution: $40.00 \text{ g} = 4.000 \times 10^1$ g.

(c)
$$\frac{4.80 \times 10^{-3} \text{ g } Mg^{2+}}{4.000 \times 10^1 \text{ g solution}} \times 1.00 \times 10^6 = 120 \text{ ppm} \quad (Ans.)$$

Grams of Mg^{2+} in 1.00 g
solution

S5-3 MOLARITY

A mole of a substance has been defined as the amount which contains 6.02×10^{23} molecules (or other fundamental unit of matter) (Sec. S4-3). The **molarity** M of a solution is the number of moles of the solute contained in 1.000 L of solution:

$$M = \frac{\text{moles of solute}}{\text{liters of solution}}$$

The following examples will illustrate how the molarity of a solution is calculated and used.

EXAMPLE 1   If 24.00 g of sodium hydroxide, NaOH, is contained in 0.100 L of a solution, what is the molarity of the solution?

Solution

(a) Determine amount of solute in moles (the conversion of an amount in grams into an amount in moles was discussed in Sec. S4-3):

$$\frac{24.00 \text{ g NaOH}}{40.00 \text{ g/mol}} = 0.6000 \text{ mol NaOH}$$

(b) Determine molarity M:

$$M = \frac{0.6000 \text{ mol NaOH}}{0.100 \text{ L}} = 6.00 \text{ M} \quad (Ans.)$$

EXAMPLE 2   If 100.00 g of ethyl alcohol, CH_3CH_2OH, is contained in 600 mL of solution, what is the molarity of the solution?

Solution

(*a*) Determine amount of solute in moles:

$$\frac{100.00 \text{ g CH}_3\text{CH}_2\text{OH}}{46.07 \text{ g/mol}} = 2.171 \text{ mol CH}_3\text{CH}_2\text{OH}$$

(*b*) Determine amount of solution in liters:

$$600 \text{ mL} \times \frac{1.00 \text{ L}}{1000 \text{ mL}} = 0.600 \text{ L}$$

(*c*) Determine molarity M:

$$M = \frac{2.171 \text{ mol CH}_3\text{CH}_2\text{OH}}{0.600 \text{ L}} = 3.62 \text{ mol/L} \qquad (Ans.)$$

EXAMPLE 3 How many moles of sodium bromide, NaBr, are contained in 400 mL of a 3.000 M solution?

Solution

(*a*) Since

$$M = \frac{\text{moles solute}}{\text{liters of solution}}$$

Moles solute = M (mol/L) × volume solution (L)

(*b*) Determine volume of solution in liters:

$$400 \text{ mL} \times \frac{1.000 \text{ L}}{1000 \text{ mL}} = 0.400 \text{ L}$$

(*c*) Determine moles of solute:

Moles solute = 3.000 mol/L × 0.400 L = 1.20 mol   (*Ans.*)

EXAMPLE 4 How many grams of methyl alcohol, CH$_3$OH, are contained in 1200 mL of a 2.100 M solution?

Solution

(*a*) Since

$$M = \frac{\text{moles solute}}{\text{liters of solution}}$$

Moles solute = M (mol/L) × volume solution (L)

(*b*) Determine volume of solution in liters:

$$1200 \text{ mL} \times \frac{1.000 \text{ L}}{1000 \text{ mL}} = 1.200 \text{ L}$$

(*c*) Determine moles of solute:

Moles of solute = 2.100 mol/L × 1.200 L = 2.520 mol

(*d*) Determine grams of solute:

Grams of solute = 2.520 mol × 32.04 g/mol = 80.74 g (*Ans.*)

PRACTICE EXERCISES

1. What is the percent by weight of potassium bromide, KBr, in an aqueous solution if 20.00 g is dissolved in 62.00 g of solution?

2. What is the percent by weight of potassium bromide, KBr, in an aqueous solution prepared by dissolving 16.00 g in 60.00 g of water?

3. How many grams of calcium chloride, $CaCl_2$, are contained in 84.00 g of a solution which is 14.80 percent by weight?

4. What is the concentration of an aqueous solution of lithium chloride, LiCl, in parts per million parts by weight if 0.150 g is dissolved in 1870 g of solution?

5. How many grams of sodium iodide, NaI, are contained in 21,000 (2.100×10^4) g of a solution which has a concentration of 118 ppm by weight?

6. What is the concentration of calcium ion, Ca^{2+}, in parts per million parts by weight if 5.40 mg of calcium ion is contained in 30.00 g of solution?

7. If 16.00 g of potassium hydroxide, KOH, is contained in 0.200 L of a solution, what is the molarity of the solution?

8. If 80.00 g of acetone, CH_3COCH_3, is contained in 700 mL of solution, what is the molarity of the solution?

9. How many moles of sodium chloride, NaCl, are contained in 600 mL of a 6.200 M solution?

10. How many grams of ethyl alcohol, CH_3CH_2OH, are contained in 1800 mL of a 1.700 M solution?

11. An amount of methyl alcohol, CH_3OH, is dissolved in water and diluted with water until the total volume of solution is 2.000 L. How many grams of methyl alcohol should be used to obtain a solution which is 2.500 M?

12. An amount of sodium hydroxide, NaOH, is dissolved in water and diluted with water until the total volume of solution is 500 mL. How many grams of sodium hydroxide should be used to obtain a solution which is 0.500 M?

SUPPLEMENT 6

The Names and Formulas of Some Common Salts and Acids†

S6-1 THE NAMES AND FORMULAS OF SOME SIMPLE IONS

Much of the information needed to name many common salts and acids can be condensed into a small reference table. As a first step in the construction of such a table, recall that ions have been described in Sec. 3-4 as particles with one or more positive or negative electric charges. The simplest ions are closely related to the structures of individual atoms. Sodium ions, for example, are sodium atoms with the outermost electron missing. Chloride ions are chlorine atoms with an additional electron. Consider the comparisons in Table S6-1.

Table S6-1

	Symbol	Electronic symbol	Simplified electron configuration
Sodium atom	Na	Na·	2, 8, 1
Sodium ion	Na$^+$	[Na]$^+$	2, 8
Chlorine atom	Cl	:C̈l·	2, 8, 7
Chloride ion	Cl$^-$	[:C̈l:]$^-$	2, 8, 8

As a second step in the construction of the reference table, consider Table S6-2, which gives the names and symbols of some commonly occurring ions whose structures, like those of sodium ions and chloride ions, are closely related to individual atoms. Note particularly the relationship between the charge on the ions and the number of valence electrons in the electronic symbols of the related atoms.

S6-2 THE NAMES AND FORMULAS OF SOME COMMON ACIDS

The names and formulas of some common acids are given in Table S6-3. Sometimes the name of an aqueous solution of an acid differs from the name of the acid itself. Hydrogen chloride, for example, is the name of the pure acid. The name of an aqueous solution of hydrogen chloride is hydrochloric acid. This acid is generally used as an aqueous solution.

S6-3 THE NAMES AND FORMULAS OF SOME IONIC RADICALS

As a further step in the construction of a reference table let us consider the names and formulas of some ionic radicals. An **ionic radical** is an ion which contains more than one atom. The structures of many common ionic radicals are related to the

† An expansion of Chap. 3.

Table S6-2
Some Ions with Structures Closely Related to Individual Atoms

Name of ion	Symbol of ion	Name of atom with structural relationship	Electronic symbol of atom
Positive Ions			
Lithium	Li^+	Lithium	Li·
Sodium	Na^+	Sodium	Na·
Potassium	K^+	Potassium	K·
Silver	Ag^+	Silver	Ag·
Magnesium	Mg^{2+}	Magnesium	·Mg·
Calcium	Ca^{2+}	Calcium	·Ca·
Strontium	Sr^{2+}	Strontium	·Sr·
Barium	Ba^{2+}	Barium	·Ba·
Radium	Ra^{2+}	Radium	·Ra·
Aluminum	Al^{3+}	Aluminum	·Al·
Negative Ions			
Fluoride	F^-	Fluorine	:Ḟ·
Chloride	Cl^-	Chlorine	:Ċl·
Bromide	Br^-	Bromine	:Ḃr·
Iodide	I^-	Iodine	:İ·
Oxide	O^{2-}	Oxygen	:Ö·
Sulfide	S^{2-}	Sulfur	:Ṡ·

Table S6-3
The Names of Some Representative Acids

Formula	Name	Name of aqueous solution
HF	Hydrogen fluoride	Hydrofluoric acid
HCl	Hydrogen chloride	Hydrochloric acid
HBr	Hydrogen bromide	Hydrobromic acid
HI	Hydrogen iodide	Hydriodic acid
H_2S	Hydrogen sulfide	Hydrosulfuric acid
HCN	Hydrogen cyanide	Hydrocyanic acid
HNO_3	Nitric acid	Nitric acid
H_2SO_4	Sulfuric acid	Sulfuric acid
H_3PO_4	Phosphoric acid	Phosphoric acid
H_2CO_3	Carbonic acid	Carbonic acid
CH_3COOH	Acetic acid	Acetic acid

structures of oxy acids. An **oxy acid,** as the name implies, is an acid containing oxygen. If the oxy acid contains only one hydrogen atom in its molecule, only one ionic radical is associated with it. Nitric acid, HNO_3, is such an oxy acid, and the related ionic radical is the nitrate radical, NO_3^-. If the oxy acid contains more than one hydrogen atom, more than one ionic radical is associated with it. Sulfuric acid, H_2SO_4, for example, has two hydrogen atoms. The two related ionic radicals are the hydrogen sulfate ion, HSO_4^-, and the sulfate ion, SO_4^{2-}. (Note that ionic radicals may contain hydrogen atoms.) The formulas of five common oxy acids are given in Table S6-4 together with the ionic radicals associated with them. Table S6-4 also includes the names and formulas of four ionic radicals not related to oxy acids.

Table S6-4
The Names and Formulas of Some Common Ionic Radicals

Name	Acid formula	Related ionic radical	Radical name
Nitric acid	HNO_3	NO_3^-	Nitrate ion
Sulfuric acid	H_2SO_4	HSO_4^-	Hydrogen sulfate ion
		SO_4^{2-}	Sulfate ion
Carbonic acid	H_2CO_3	HCO_3^-	Hydrogen carbonate ion†
		CO_3^{2-}	Carbonate ion
Phosphoric acid	H_3PO_4	$H_2PO_4^-$	Dihydrogen phosphate ion
		HPO_4^{2-}	Hydrogen phosphate ion
		PO_4^{3-}	Phosphate ion
Acetic acid‡	CH_3COOH	CH_3COO^-	Acetate ion

Ionic radicals not related to oxy acids			
Water	HOH	OH^-	Hydroxide ion
Hydrogen sulfide	HSH	SH^-	Hydrosulfide ion
Hydrogen cyanide	HCN	CN^-	Cyanide ion
Ammonia	NH_3	NH_4^+	Ammonium ion§

† Also commonly called the bicarbonate ion.

‡ Only one of the hydrogen atoms in acetic acid ionizes in the molecule.

§ The ammonium ion results from the *addition* of a hydrogen ion to a molecule of ammonia.

S6-4 THE NAMES AND FORMULAS OF SALTS

Illustrations of the Naming of Salts

The names are derived from the names of the component ions, with the positive ion named first

Formula	Name
Li_2CO_3	Lithium carbonate
$NaHCO_3$	Sodium hydrogen carbonate (sodium bicarbonate)
$(NH_4)_2SO_4$	Ammonium sulfate
$Ca(NO_3)_2$	Calcium nitrate

As an additional step toward constructing a reference table let us consider the names of salts. A salt has been defined in Sec. 3-6 as a substance made up of ions. The name of a salt is derived from the names of its component ions, with the positive ion named first. Thus the name of the salt containing potassium ions, K^+, and cyanide ions, CN^-, is potassium cyanide. Additional examples are given in the margin. The formulas of salts can easily be derived from the formulas of the component ions, keeping in mind that the numbers of positive and negative ionic charges in each formula must be equal. The formula of the common fertilizer ammonium nitrate is NH_4NO_3 since the ammonium ion, NH_4^+, has a single positive charge, and the nitrate ion, NO_3^-, has a single negative charge. The formula for potassium sulfate is K_2SO_4 since the sulfate ion, SO_4^{2-}, has two negative charges and each potassium ion, K^+, has one positive charge. Similarly, the formula for aluminum fluoride is AlF_3, since the aluminum ion is Al^{3+} and the fluoride ion is F^-.

The electronic structures of many atoms are such that more than one ion is formed from them. There are, for example, two common ions of iron, one named the iron(II) ion, or ferrous ion, Fe^{2+}, and the other the iron(III) ion, or ferric ion, Fe^{3+}. The relationships between such ions and the atoms with which they are associated are important, but the chemistry is too complex to be presented here. The names and atomic symbols of some atoms which have more than one ion associated with them are given in Table S6-5. The table also includes the formulas and names of the associated ions.

The examples in Table S6-6 illustrate how salts containing these ions are named.

Table S6-5
The Symbols and Principal Ions of a
Group of Atoms with More than One Associated Ion†

Atom	Atomic number	Associated ion
Chromium, Cr	24	Cr^{2+}, chromium(II) or chromous; Cr^{3+}, chromium(III) or chromic
Manganese, Mn	25	Mn^{2+}, manganese(II) or manganous
Iron, Fe	26	Fe^{2+}, iron(II) or ferrous; Fe^{3+}, iron(III) or ferric
Cobalt, Co	27	Co^{2+}, cobalt(II) or cobaltous
Nickel, Ni	28	Ni^{2+}, nickel
Copper, Cu	29	Cu^+, copper(I) or cuprous; Cu^{2+}, copper(II) or cupric
Zinc, Zn	30	Zn^{2+}, zinc
Tin, Sn	50	Sn^{2+}, tin(II) or stannous; Sn^{4+}, tin(IV) or stannic
Mercury, Hg	80	Hg_2^{2+}, mercury(I) or mercurous‡; Hg^{2+}, mercury(II) or mercuric
Lead, Pb	82	Pb^{2+}, lead

† For simplicity in some cases only one of the ions is considered here.

‡ The structure of the mercury(I) ion is unusual. It consists of two mercury atoms bonded by a single covalent bond, which collectively have two positive ionic charges: $[Hg\!-\!Hg]^{2+}$

Table S6-6
Illustrations of the Naming
of Salts Containing Ions Listed in Table S6-5

As is usual in the naming of salts, the names of the salts are derived from the names of the component ions, with the positive ion named first

Formula	Name	Alternate (older) name
$FeCl_2$	Iron(II) chloride	Ferrous chloride
$Fe_2(SO_4)_3$	Iron(III) sulfate	Ferric sulfate
CrS	Chromium(II) sulfide	Chromous sulfide
CrF_3	Chromium(III) fluoride	Chromic fluoride
Hg_2O	Mercury(I) oxide	Mercurous oxide
$Hg(NO_3)_2$	Mercury(II) nitrate	Mercuric nitrate

S6-5 REFERENCE TABLES FOR NAMING ACIDS AND SALTS

Now that we have considered the names and formulas of some common acids and salts we can examine Tables S6-7 and S6-8. Table S6-7 contains the formulas and names of some common ions and acids listed alphabetically by formula. This arrangement is particularly useful in naming of salts and acids if the formula is known. Table S6-8 contains the same information listed alphabetically by name. This arrangement is very useful

Table S6-7

Selected Ions and Acids for Naming Salts and Deriving Their Formulas
Arranged alphabetically by formula, condensed notation

Ag^+	Silver ion	HPO_4^{2-}	Hydrogen phosphate ion
Al^{3+}	Aluminum ion	$H_2PO_4^-$	Dihydrogen phosphate ion
Ba^{2+}	Barium ion	H_3PO_4	Phosphoric acid
Br^-	Bromide ion	HS^-	Hydrogen sulfide ion
Ca^{2+}	Calcium ion	H_2S	Hydrogen sulfide
CH_3COO^-	Acetate ion	HSO_4^-	Hydrogen sulfate ion
CH_3COOH	Acetic acid	H_2SO_4	Sulfuric acid
Cl^-	Chloride ion	I^-	Iodide ion
CN^-	Cyanide ion	K^+	Potassium ion
Co^{2+}	Cobalt(II) (cobaltous) ion	Li^+	Lithium ion
CO_3^{2-}	Carbonate ion	Mg^{2+}	Magnesium ion
Cr^{2+}	Chromium(II) (chromous) ion	Mn^{2+}	Manganese ion
Cr^{3+}	Chromium(III) (chromic) ion	Na^+	Sodium ion
Cu^+	Copper(I), (cuprous) ion	NH_4^+	Ammonium ion
Cu^{2+}	Copper(II) (cupric) ion	Ni^{2+}	Nickel ion
F^-	Fluoride ion	NO_3^-	Nitrate ion
Fe^{2+}	Iron(II) (ferrous) ion	O^{2-}	Oxide ion
Fe^{3+}	Iron(III) (ferric) ion	OH^-	Hydroxide ion
HBr	Hydrogen bromide	Pb^{2+}	Lead ion
HCl	Hydrogen chloride	PO_4^{3-}	Phosphate ion
HCN	Hydrogen cyanide	Ra^{2+}	Radium ion
HCO_3^-	Hydrogen carbonate ion	S^{2-}	Sulfide ion
H_2CO_3	Carbonic acid	Sn^{2+}	Tin(II) (stannous) ion
HF	Hydrogen fluoride	Sn^{4+}	Tin(IV) (stannic) ion
Hg^{2+}	Mercury(II) (mercuric) ion	SO_4^{2-}	Sulfate ion
Hg_2^{2+}	Mercury(I) (mercurous) ion	Sr^{2+}	Strontium ion
HI	Hydrogen iodide	Zn^{2+}	Zinc ion
HNO_3	Nitric acid		

in deriving a formula from the name of a compound. To illustrate the use of these reference tables, let us consider the following examples.

EXAMPLE 1   The name of a compound is calcium chloride. What is its formula?

Solution   From Table S6-8 we find that the formula for the calcium ion is Ca^{2+} and that the formula for the chloride ion is Cl^-. In order for the formula of the compound to represent an equal number of positive and negative ionic charges, the formula must be $CaCl_2$.

EXAMPLE 2   The name of a compound is copper(II) nitrate. What is its formula?

Solution   From Table S6-8 we find that the formula for the copper(II) ion is Cu^{2+} and that the formula for the nitrate ion is NO_3^-. In order for the formula of the compound to represent an equal number of positive and negative ionic charges, the formula must be $Cu(NO_3)_2$. (Note use of parentheses in representing the two nitrate ions.)

EXAMPLE 3   The formula of a compound is $Mg(OH)_2$. What is its name?

Table S6-8
Selected Ions and Acids for Naming Salts and Deriving Their Formulas
Arranged alphabetically by name, condensed notation

Acetate ion	CH_3COO^-	Hydrogen sulfate ion	HSO_4^-
Acetic acid	CH_3COOH	Hydrogen sulfide	H_2S
Aluminum ion	Al^{3+}	Hydrogen phosphate ion	HPO_4^{2-}
Ammonium ion	NH_4^+	Hydrosulfide ion	HS^-
Barium ion	Ba^{2+}	Hydrosulfuric acid	H_2S
Bicarbonate ion	HCO_3^-	Hydroxide ion	OH^-
Bromide ion	Br^-	Iodide ion	I^-
Calcium ion	Ca^{2+}	Iron(II) ion	Fe^{2+}
Carbonate ion	CO_3^{2-}	Iron(III) ion	Fe^{3+}
Carbonic acid	H_2CO_3	Lead ion	Pb^{2+}
Chloride ion	Cl^-	Lithium ion	Li^+
Chromium(II) ion	Cr^{2+}	Magnesium ion	Mg^{2+}
Chromium(III) ion	Cr^{3+}	Manganese(II) ion	Mn^{2+}
Chromic ion	Cr^{3+}	Mercury(I) ion	Hg_2^{2+}
Chromous ion	Cr^{2+}	Mercury(II) ion	Hg^{2+}
Cobalt(II) ion	Co^{2+}	Mercuric ion	Hg^{2+}
Cobaltous ion	Co^{2+}	Mercurous ion	Hg_2^{2+}
Copper(I) ion	Cu^+	Nickel ion	Ni^{2+}
Copper(II) ion	Cu^{2+}	Nitrate ion	NO_3^-
Cupric ion	Cu^{2+}	Nitric acid	HNO_3
Cuprous ion	Cu^+	Oxide ion	O^{2-}
Dihydrogen phosphate ion	$H_2PO_4^-$	Phosphate ion	PO_4^{2-}
Ferric ion	Fe^{3+}	Phosphoric acid	H_3PO_4
Ferrous ion	Fe^{2+}	Potassium ion	K^+
Fluoride ion	F^-	Radium ion	Ra^{2+}
Hydriodic acid	HI	Silver ion	Ag^+
Hydrobromic acid	HBr	Sodium ion	Na^+
Hydrochloric acid	HCl	Stannic ion	Sn^{4+}
Hydrocyanic acid	HCN	Stannous ion	Sn^{2+}
Hydrofluoric acid	HF	Strontium ion	Sr^{2+}
Hydrogen bromide	HBr	Sulfide ion	S^{2-}
Hydrogen carbonate ion	HCO_3^-	Sulfate ion	SO_4^{2-}
Hydrogen chloride	HCl	Sulfuric acid	H_2SO_4
Hydrogen cyanide	HCN	Tin(II) ion	Sn^{2+}
Hydrogen fluoride	HF	Tin(IV) ion	Sn^{4+}
Hydrogen iodide	HI	Zinc ion	Zn^{2+}

Solution From Table S6-7 we see that the name of the Mg^{2+} ion is magnesium and the name of the OH^- ion is hydroxide. The name of the compound is therefore magnesium hydroxide.

EXAMPLE 4 The formula of a compound is HCN. What is its name?

Solution We see directly from Table S6-7 the name of the compound is hydrogen cyanide.

EXAMPLE 5 The formula of a compound is $NiSO_4$. What is its name?

Solution From Table S6-8 we see that the name of the Ni^{2+} ion is nickel and the name of the SO_4^{2-} ion is sulfate. The name of the compound is therefore nickel sulfate.

EXAMPLE 6 The name of a compound is magnesium phosphate. What is its formula?

Solution From Table S6-8 we find that the formula for the magnesium ion is Mg^{2+} and the formula for the phosphate ion is PO_4^{3-}. In order for the formula of the compound to represent an equal number of positive and negative ionic charges, the formula must be $Mg_3(PO_4)_2$.

KEY WORDS

1. Oxy acid (Sec. S6-3)   2. Ionic radical (Sec. S6-3)

SUMMARIZING QUESTIONS FOR SELF-STUDY

1. Q. Using Table S6-7 name the following compounds: (a) NaI, (b) KCN, (c) H_2SO_4, (d) $MgBr_2$, (e) $NiSO_4$, (f) $AlPO_4$, (g) Cr_2S_3, (h) Li_2CO_3.

    A. (a) Sodium iodide, (b) potassium cyanide, (c) sulfuric acid, (d) magnesium bromide, (e) nickel sulfate, (f) aluminum phosphate, (g) chromium(III) sulfide, (h) lithium carbonate.

2. Q. Using Table S6-8 give formulas for (a) ammonium nitrate, (b) hydrogen bromide, (c) zinc hydroxide, (d) tin(II) sulfate, (e) strontium carbonate, (f) iron(II) fluoride, (g) mercury(II) chloride, (h) radium iodide.

    A. (a) NH_4NO_3, (b) HBr, (c) $Zn(OH)_2$, (d) $SnSO_4$, (e) $SrCO_3$, (f) FeF_2, (g) $HgCl_2$, (h) RaI_2.

PRACTICE EXERCISES

1. Using Table S6-7, name
   - (a) $NaHCO_3$
   - (b) $Ba(CH_3COO)_2$
   - (c) $Co(CN)_2$
   - (d) Fe_2O_3
   - (e) PbS
   - (f) K_2HPO_4
   - (g) CuCl
   - (h) H_2CO_3
2. Using Table S6-8 give formulas for
   - (a) Radium sulfide
   - (b) Phosphoric acid
   - (c) Ammonium hydrogen sulfate
   - (d) Silver sulfate
   - (e) Manganese cyanide
   - (f) Strontium hydroxide
   - (g) Hydrogen cyanide
   - (h) Calcium acetate

SUPPLEMENT 7

Important Amino Acids†

Table S7-1
Important Amino Acids

NH_2CH_2COOH

Glycine

$CH_3CHCOOH$
|
NH_2

Alanine

$(CH_3)_2CHCHCOOH$
|
NH_2

Valine

$(CH_3)_2CHCH_2CHCOOH$
|
NH_2

Leucine

$CH_3CH_2CH{-}CHCOOH$
| |
CH_3 NH_2

Isoleucine

$C_6H_5CH_2CHCOOH$
|
NH_2

Phenylalanine

$HOCH_2CHCOOH$
|
NH_2

Serine

$CH_3CH{-}CHCOOH$
| |
OH NH_2

Threonine

$NH_2(CH_2)_4CHCOOH$
|
NH_2

Lysine

$NH_2CH_2CH(CH_2)_2CHCOOH$
| |
OH NH_2

δ-Hydroxylysine

NH
‖
$C{-}NH(CH_2)_3CHCOOH$
NH_2 NH_2

Arginine

$HOOCCH_2CHCOOH$
|
NH_2

Aspartic acid

$NH_2COCH_2CHCOOH$
|
NH_2

Asparagine

$HOOC(CH_2)_2CHCOOH$
|
NH_2

Glutamic acid

$NH_2CO(CH_2)_2CHCOOH$
|
NH_2

Glutamine

$HSCH_2CHCOOH$
|
NH_2

Cysteine

$S{-}CH_2CHCOOH$
|
NH_2
$S{-}CH_2CHCOOH$
|
NH_2

Cystine

$CH_3S(CH_2)_2CHCOOH$
|
NH_2

Methionine

Continued on next page

† An expansion of Chap. 10.

Table S7-1
Important Amino Acids (*Continued*)

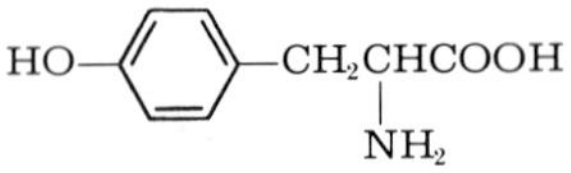

HO—⟨benzene⟩—CH$_2$CHCOOH
|
NH$_2$

Tyrosine

HOCH—CH$_2$
CH$_2$ CHCOOH
N
H

Hydroxyproline

HO—⟨benzene with I, I⟩—O—⟨benzene with I, I⟩—CH$_2$CHCOOH
|
NH$_2$

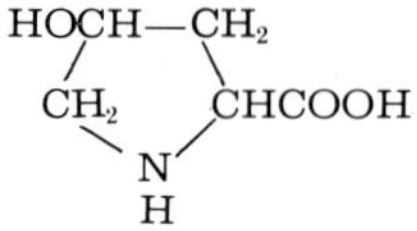

Thyroxine

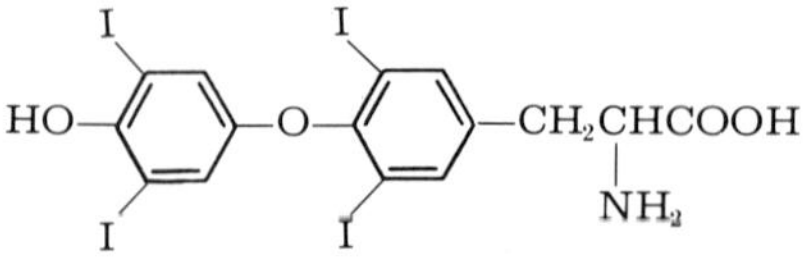

⟨indole ring⟩—CH$_2$CHCOOH
|
NH$_2$
N
H

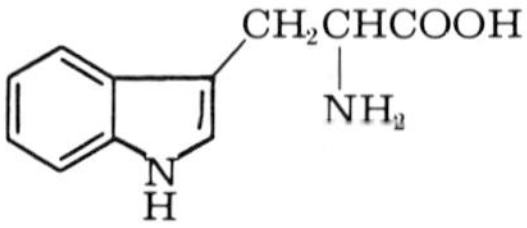

Tryptophan

CH$_2$—CH$_2$
CH$_2$ CHCOOH
H
N

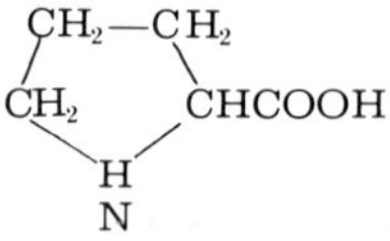

Proline

CH=CCH$_2$CHCOOH
N   NH   NH$_2$
CH

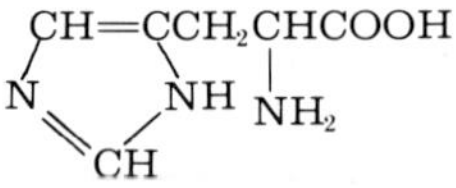

Histidine

GLOSSARY

Section numbers are given for terms defined in the text; for convenience other common terms have been added.

Absolute zero The temperature below which it is impossible to cool matter, that is, 0 K and $-273.17°C$ (Sec. S3-5).

Accuracy Closeness with which a measurement or the average of a series of measurements approaches the true or accepted value of the quantity being determined (Sec. S2-1).

Acid A substance capable of releasing one or more protons (Sec. 7-5).

Acid-base reaction Another name for a proton-transfer reaction: a reaction which involves the transfer of a proton, H^+, from one molecular species to another (Sec. 7-4).

Acidic solution An aqueous solution in which the concentration of hydrogen ions exceeds the concentration of hydroxide ions.

Activation energy Minimum energy which chemical reactants in a reaction must have for a given chemical reaction to occur.

Active site of an enzyme The specific region on the surface of an enzyme to which the substrate is bonded in an enzyme-substrate complex (Sec. 10-5).

Addition reaction A reaction which involves the overall combination of molecules; most commonly, the addition of substances such as hydrogen, halogens and halogen acids, to alkenes, $-\overset{|}{C}=\overset{|}{C}-$, or alkynes, $-C\equiv C-$.

Adipose tissue Deposits of reserve fat in the body (Sec. 11-6B).

Alcohol A carbon-containing compound with an alcohol functional group, $-\overset{|}{\underset{|}{C}}-O-H$; general formula ROH (Sec. 8-5).

Aldehyde A carbon-containing compound with an aldehyde functional group, $-\overset{O}{\overset{||}{C}}-H$; general formula, RCHO (Sec. 8-5).

Alicyclic substance A carbon-containing subtance whose molecular structure includes one or more rings of carbon atoms which are not aromatic (Sec. 8-6).

Aliphatic substance. A carbon-containing substance whose molecular structure has a fundamental skeleton of atoms with an open structure, containing no rings (Sec. 8-6).

Alkali metal Any of the metals of group IA of the periodic table, e.g., lithium, sodium, and potassium.

Alkaline earth metal Any of the elements of group IIA of the periodic table, e.g., magnesium, calcium, and barium.

Alkaline solution Another name for a basic solution, i.e., an aqueous solution in which the hydroxide-ion concentration exceeds the hydrogen-ion concentration.

Alkane A carbon-containing compound composed of hydrogen and carbon atoms with no functional groups; all the bonds are single covalent bonds (Sec. 8-4).

Alkene A compound containing an alkene function group, $-\overset{|}{C}=\overset{|}{C}-$; general formula (Sec. 8-5)

$$R-\overset{R}{\overset{|}{C}}=\overset{R}{\overset{|}{C}}-R$$

Alkyl group A hydrocarbon group like methyl CH_3-, ethyl CH_3CH_2-, or n-propyl, $CH_3CH_2CH_2-$.

Alkyne A compound containing an alkyne functional group, $-C\equiv C-$; general formula $R-C\equiv C-R$ (Sec. 8-5).

Allotrope An elemental substance made up of the same kind of atoms found in another elemental substance; oxygen, O_2, and ozone, O_3, are allotropes, and diamond and graphite are two allotropes of carbon (Chap. 2).

Alloy A solution of one metal in another.

Alpha particle The nucleus of a helium atom; alpha rays are made up of high-speed alpha particles; symbol α (Sec. 9-1).

Alpha ray Penetrating stream of high-speed alpha particles (helium nuclei) (Sec. 9-1).

Amalgam A solution of a metal in mercury.

Amide A carbon-containing compound with an amide functional group, $-\overset{|}{\underset{O}{\overset{||}{C}}}-\overset{|}{N}-$ (Sec. 8-2);

general formulas

$$R-\overset{}{\underset{O}{\overset{||}{C}}}-\overset{}{\underset{H}{\overset{|}{N}}}-H \quad R-\overset{}{\underset{O}{\overset{||}{C}}}-\overset{}{\underset{H}{\overset{|}{N}}}-R \quad R-\overset{}{\underset{O}{\overset{||}{C}}}-\overset{}{\underset{R}{\overset{|}{N}}}-R$$

Amine A carbon-containing compound with an amine functional group, $-\overset{|}{N}-$; general formulas, RNH_2, $RNHR$, and R_3N (Sec. 8-5).

Amino acid A carbon-containing compound which contains an amine functional group and a carboxylic acid functional group (Secs. 10-4A and 11-2C).

Amino acid residue The part of an amino acid molecule actually found in a protein molecule (Sec. 10-4A); general formula $-\overset{}{\underset{H}{\overset{R}{\overset{|}{N}}}}-\overset{}{\underset{H}{\overset{|}{C}}}-\overset{O}{\overset{||}{C}}-$

Amorphous solid A noncrystalline solid whose fundamental particles are in a more random arrangement than the fixed pattern of a crystalline solid (Sec. 4-7).

Amphoteric Capable of reacting with both hydrogen ions, H^+, and hydroxide ions, OH^-.

Analgesic A pain-relieving agent (Sec. 12-4).

Anesthetic An agent which relieves pain (*a*) over the entire body with loss of consciousness (general anesthetic) or (*b*) in a specific region of the body without loss of consciousness (local anesthetic) (Sec. 12-4).

Anhydride A substance derived from one or more other substances by the removal of water.

Anion Another name for a negative ion (Sec. 3-4).

Anthracite Coal which contains a relatively low proportion of hydrogen- and oxygen-containing substances, also called hard coal (Sec. 14-3).

Antibiotic A substance produced by a microorganism capable of destroying or inhibiting the growth of other microorganisms (Sec. 12-2B).

Anti-juvenile hormone A hormone antagonist which interferes with the production of juvenile hormones by insects (Sec. 11-10B).

Antimetabolite A substance with a structure similar to a normal metabolite which is capable of interfering with the reactions of the normal metabolite (a metabolite is a substance which is converted into other substances by the reactions of living organisms) (Sec. 12-5).

Antimetabolite theory of drug action It proposes that a substance may serve as a chemotherapeutic agent if it has a structure similar to that of a normal metabolite of a disease-causing microorganism (Sec. 12-5).

Aromatic substance A carbon-containing substance with a molecular structure in which the fundamental skeleton of atoms contains one or more rings of six carbon atoms linked together by a distinctive covalent bond associated with the presence of multiple $C=C$ double bonds (Sec. 8-6).

Artificial transmutation A nuclear change which does not occur in nature but is deliberately brought about by the bombardment of nuclei with high-speed particles (Sec. 9-3).

Atmosphere A unit of pressure equivalent to that exerted by a column of mercury 760 millimeters high (Sec. S3-3).

Atom The smallest conceivable unit of an element. (Sec. 2-10).

Atomic energy Energy generated by a nuclear reaction (Secs. 9-7 and 14-8).

Atomic fission A nuclear change in which an atomic nucleus is split into two nuclei of substantial size (Sec. 9-4).

Atomic mass unit (amu) A unit of mass equivalent to one-twelfth the mass of an atom of carbon 12.

Atomic number The number of positive charges on the nucleus of an atom; it corresponds to the number of protons in the nucleus (Sec. 2-4).

Atomic orbital The volume within which an electron is most probably to be found in an atom (Sec. 2-4).

Atomic weight The weight of an average atom of an element, relative to the weight of a carbon 12 isotope taken as 12.0000 units (Sec. 6-6A).

Atom in combined state An atom as it is found in a compound, i.e., in a molecule that contains atoms of more than one atomic number (Sec. 2-10).

Atom in free state An atom in the molecule of an element (Sec. 2-10).

Aufbau principle The arrangement of electrons in an atom is the one requiring the least amount of energy to attain; in more specific terms electrons ordinarily group themselves in orbitals with the lowest possible principal quantum numbers.

Avogadro's law Equal volumes of gases at the same temperature and pressure contain the same number of molecules.

Avogadro's number 6.022×10^{23}; the number of units in a mole (Sec. S4-3).

Baking soda A common name for sodium hydrogen carbonate, $NaHCO_3$, also called sodium bicarbonate.

Barbiturate One of a class of substances derived from barbituric acid used as sedatives, sleep producers, and anesthetics (Sec. 12-9).

Base A substance capable of combining with one or more protons (Sec. 7-5).

Basic solution An aqueous solution in which the hydroxide-ion concentration exceeds the hydrogen-ion concentration.

Benign tumor A tumor which remains within the tissue in which it develops (Sec. 15-7B).

Beta particle A high-speed electron ejected from a radioactive nucleus; symbol β (Sec. 9-1).

Beta ray A penetrating stream of beta particles: high-speed electrons ejected from radioactive nuclei (Sec. 9-1).

Biochemistry The study of the chemistry of plants and animals (Sec. 10-1).

Biodegradable A term applied to large molecules or the substances comprising them capable of being degraded by the biological or chemical changes of natural processes into small molecules readily assimilated by the environment (Sec. 13-9).

Biomass The various forms of plant and animal life on the earth (Sec. 14-13C).

Biopolymer A polymeric substance found in living organisms, e.g., proteins, carbohydrates, and nucleic acids.

Biosynthesis The synthesis of substances in a living organism, usually involving a sequence of reactions.

Bituminous coal Coal which contains a substantial proportion of hydrogen- and oxygen-containing substances (also called soft coal), burns with a very smoky flame (Sec. 14-3).

Boiling The conversion of a liquid into a gas within the body of the liquid (Sec. 4-8).

Boiling point The temperature required to boil a liquid (Sec. 4-8).

Bond energy The amount of energy required to break an interatomic bond (Sec. 6-7A).

Bonding pair of electrons A pair of electrons shared by two atoms in a covalent bond (Sec. 2-6).

Bond length The average distance between two nuclei joined by a chemical bond; also called an interatomic bond distance.

Boyle's law The volume of a given amount of gas varies inversely with the pressure at constant temperature.

Breeder reactor A nuclear reactor which produces more nuclear fuel than it consumes (Secs. 9-9 and 14-8).

Brine A solution of a salt in water, usually an aqueous solution of sodium chloride, NaCl.

British thermal unit (Btu) An amount of energy sufficient to raise the temperature of 1 pound of water $1°F$; equivalent to 252 calories (Secs. 14-2 and S3-2).

Brönsted-Lowry acid Another name for a proton donor in a proton-transfer reaction (Sec. 7-5).

Brönsted-Lowry base Another name for a proton acceptor in a proton-transfer reaction (Sec. 7-5).

Brown-air pollution Another name for photochemical smog (Sec. 15-2E), derived from the light-brownish color of the smog due principally to the presence of nitrogen oxides.

Buffer An aqueous solution whose hydrogen-ion concentration and pH do not change significantly on the addition of a reasonable amount of either an acid or a base, e.g., a solution of a weak acid, such as acetic acid, CH_3COOH, and one of its salts, such as sodium acetate, CH_3COONa.

Calorie An amount of energy equivalent to the heat energy required to raise the temperature of 1 gram of water $1°C$. The Calorie used by nutritionists is a kilocalorie (Secs. 11-11, 14-2, and S3-2).

Cancerous tumor A tumor growth whose cells invade the tissues surrounding the tissue in which it develops (Sec. 15-7B).

Carbohydrate One of a group of substances composed of carbon, hydrogen, and oxygen atoms which include sugars, starch, and cellulose; the molecular formulas of most can be reduced to the general formula $C_x(H_2O)_y$, although they are not actually hydrates (Sec. 11-2A). More specifically, they are substances with polyhydroxy molecules containing aldehyde groups, modified aldehyde groups, ketone groups, or modified ketone groups or molecules containing a sequence of such structural units from two to thousands (Sec. 10-2G).

Carboxylic acid A carbon-containing compound with a carboxylic acid functional group,
$$\overset{O}{\overset{\|}{-C}}-O-H,\ \text{general formula } R\overset{O}{\overset{\|}{C}}OH\ (\text{Sec. 8-5}).$$

Carboxylic ester A carbon-containing compound with a carboxylic ester functional group,
$$\overset{O}{\overset{\|}{-C}}-O-\overset{|}{\underset{|}{C}}-\ (\text{Sec. 8-5}),\ \text{general formula}$$
$$R-\overset{O}{\overset{\|}{C}}-O-R$$

Carcinogenic agent An agent capable of causing cancer (Sec. 15-7A).

Catalyst A substance which changes the reaction rate without itself being permanently altered in the process (Sec. 6-9).

Cation Another name for a positive ion (Sec. 3-4).

Caustic soda Another name for sodium hydroxide, NaOH.

Centrally acting analgesic A substance which relieves most forms of pain throughout the body, usually without loss of consciousness, e.g., morphine and aspirin (Sec. 12-4).

Chain reaction A type of reaction which proceeds by a mechanism involving a repeating sequence of steps; each step generates a reactive molecular species which brings about the succeeding step. It may be an ordinary chemical reaction or a nuclear reaction (Sec. 6-10).

Characterization The determination of the distinctive properties of a substance and the structure of its fundamental components; usually results in a structural formula of the molecule (Sec. 1-5).

Charles' law The volume of a given amount of gas varies directly with the absolute temperature at constant pressure.

Chemical equation The representation of a reaction in formula symbols which specifies each reactant, each product, and the relative number of molecules of each (Sec. 6-3).

Chemical equilibrium A condition of dynamic balance between two competing processes which take place at equal rates (Sec. 5-2E).

Chemical intermediate A substance which is derived by chemical reactions from raw materials or other chemical intermediates and which undergoes further chemical treatment along the pathway toward the production of an end product (Sec. 13-1E).

Chemical reaction A process in which substances are converted into other substances with different sets of properties (Sec. 6-2).

Chemistry That branch of science which is concerned with the structure and behavior of the various forms of matter; a molecular approach to understanding the world about us (Sec. 1-1).

Chemotherapeutic agent In a strict sense, a substance used to treat diseases of microbial origin without significant harm to the animal or person afflicted; in a broad sense, a chemical agent used for an ailment (Sec. 12-2).

Chemotherapy In a strict sense, the use of a substance to treat diseases of microbial origin without significant harm to the animal or person afflicted; in a broad sense, the use of a chemical agent in treating ailments.

Chlorinated hydrocarbon A carbon-containing compound which also contains chlorine atoms; frequently used to refer to certain insecticides (DDT, chlordane) but not restricted to them.

Coal A black or dark-brown combustible mineral composed of carbon with varying amounts of hydrogen- and oxygen-containing carbon compounds (Sec. 13-1B).

Coal gasification Any of several processes of obtaining combustible gaseous fuels from coal (Sec. 14-5).

Coke The carbonaceous solid residue obtained from the destructive distillation of coal (Sec. 13-1B).

Colligative property A property of a solution which depends on the concentration of the solute particles but not on their structure.

Combustion A rapid reaction with oxygen which generates heat and light (Sec. 14-3).

Complete combustion A combustion reaction in which all the carbon atoms of the principal reactant are converted into carbon dioxide molecules and all the hydrogen atoms into water molecules (Sec. 14-3).

Compound A substance which can be decomposed into simpler substances (Sec. 2-3).

Concentration of a reactant The relative amount of a reactant in a reaction mixture (Sec. 6-9).

Condensation The conversion of a gas into a liquid; also called liquefaction (Sec. 4-8).

Condensed formula A condensation of a structural formula with some indication of the atom sequence but without bonds specifically represented, e.g., hydrazine, H_2NNH_2 (Sec. 3-14).

Conjugated protein A protein which contains a (nonpolypeptide) prosthetic group in addition to a polypeptide structure (Sec. 10-4C and 11-2C).

Control rod A control element of a nuclear reactor which absorbs neutrons in order to control the rate of the nuclear chain reaction or to interrupt it completely if a shutdown is necessary (Sec. 9-8C).

Coordinate covalent bond A single covalent bond in which the bonding electrons are furnished by only one of the bonded atoms.

Covalent bond A link between two atoms resulting from their mutual attraction for electrons which they share between them (Sec. 3-7).

Covalent-network solid A type of crystalline solid composed of an indefinitely extended network of atoms joined by covalent bonds, e.g., quartz and diamond (Sec. 4-7).

Critical mass Another term for the critical size (Sec. 9-10).

Critical size The size of a sample of fissionable material for which the neutrons lost by surface escape or nonproductive capture are balanced by the neutrons produced in the fission reaction; also called critical mass (Sec. 9-10).

Crystal A sample of matter in which the component molecules, atoms or ions, are arranged in an orderly three-dimensional repeating pattern.

Crystalline solid A solid whose component particles are arranged in an orderly three-dimensional repeating structural pattern (Sec. 4-7).

Cubic meter The basic unit of volume in the metric system, a cube whose edges are each 1 meter in length (Sec. S3-1C).

Daughter element The principal product of a radioactive disintegration (Sec. 9-14).

Decomposition reaction A reaction in which a substance is decomposed into two or more other substances; e.g., the decomposition (on heating) of ammonium nitrate into nitrous oxide and water (Sec. 7-8).

Deficiency disease An ailment caused not by poisons or invading parasites but by the absence of an essential substance in the diet (Sec. 11-7).

Dehydration A chemical process in which water is removed.

Denaturation of a protein A change in the structure of a protein brought about by heat, chemical changes, and other changes in the protein's environment; it is thought that the only forces affected are those holding the amino acid residue sequence in a definite shape (Sec. 10-9).

Density Mass of a substance in a unit of volume.

Desalination The removal of salt; most commonly the removal of salt from ocean water to produce potable water.

Desiccant A drying agent.

Destructive distillation of coal The decomposition of bituminous (soft) coal in high-temperature ovens sealed from air (Sec. 13-1B).

Destructive distillation of wood Heating wood in the absence of air (Sec. 14-13C).

Deuterium One of the three isotopes of hydrogen, symbol ^2_1H or ^2_1D (Sec. 2-5).

Dibasic acid An acid capable of furnishing two hydrogen ions per molecule.

Diffusion A process by which a substance gradually mixes with another substance, due to the motion of its particles.

Digestion The series of changes whereby the complex substances of foods are converted into substances of simpler structure which can be absorbed by the body (Sec. 11-5).

Dilute Having a low concentration of solute.

Dipeptide A molecule containing only two amino acid residues (Sec. 10-4B).

Dipolar ion An ion which contains both positive and negative ionic charges (Sec. 10-10).

Dipole An object in which there is a separation of the centers of positive and negative charge (Sec. 4-11).

Dipole forces The cohesive force operating between polar molecules, resulting from the mutual attraction of regions of excess positive and excess negative charge (Sec. 4-11).

Disaccharide A carbohydrate which contains two structural units, e.g., sucrose and maltose (Secs. 10-2G and 11-2A).

Dissociation The separation of a molecule into two or more molecular species; most commonly applied to the separation of an electrolyte into ions (Sec. 5-2G).

Distillation The boiling and subsequent condensation of a liquid, a means of separating and purifying liquids (Secs. 4-8 and 5-5).

DNA Deoxyribonucleic acids, the nucleic acids which are responsible for preserving the information which guides the synthesis of proteins (Sec. 10-8F).

Double covalent bond A covalent bond which involves two pairs of electrons; each atom donates two electrons (Sec. 3-8).

Double helix The double-stranded spiral structure of deoxyribonucleic acids (DNA) (Sec. 10-8E).

Drug Any chemical agent which affects the processes of living systems (Sec. 12-2).

Drug abuse The use of a drug, usually by self-administration, in a way which departs from the prevailing medical or social conventions; generally implies the compulsive use of a substance capable of causing harm (Sec. 12-9).

Drug addiction A condition characterized by a physiological and psychological dependence on a drug which results in disagreeable and dangerous symptoms when use of the drug is stopped (Sec. 12-9).

Ecology The study of the relationships between living organisms and between them and their environment.

Electric charge A quantity of electricity (Sec. 2-3).

Electric current A flow of particles with electric charges, e.g., the flow of electrons along a conducting metal wire or the flow of ions in an aqueous solution (Sec. 2-3).

Electricity A kind of force involving positive and negative charges (Sec. 2-3).

Electrolysis A chemical reaction carried out with the use of electric energy (Sec. 6-7B).

Electrolyte A substance which is capable of furnishing ions when dissolved in water.

Electrolytic cell An apparatus for carrying out an electrolysis: a chemical reaction carried out with use of electric energy (Sec. 7-14).

Electron A tiny bit of exceedingly small mass and negative electric charge (Sec. 2-2).

Electron cloud A region of negative charge around the nucleus of an atom due to the electrons, related to an atomic orbital (Sec. 2-4).

Electron configuration The arrangement of electrons in an atom (Sec. 2-8).

Electronegativity A measure of the relative ability of an atom to attract the electrons of a covalent bond; the greater the electronegativity of an atom the greater its ability to attract the electrons of a covalent bond.

Electronic symbol A representation of an atom in which (a) the letter atomic symbol represents the nucleus of the atom and all the electrons except those in the outermost orbit and (b) the electrons in the outermost orbit (or valence shell) are represented by dots, one dot for each electron e.g., hydrogen, H·, chlorine, $:\ddot{C}l·$, oxygen, $:\ddot{O}·$ (Sec. 2-9).

Electron-pair repulsion theory Another name for the valence-shell electron-pair repulsion theory.

Electron spin A property of an electron in an atom which in mechanical terms is best described as a spin.

Electron-transfer reaction A reaction which involves the transfer of one or more electrons from one molecular species to another (Sec. 7-10).

Electrostatic forces The forces of attraction and repulsion between particles due to their electric charges (Sec. 3-6).

Electrostatic precipitation Removing fine particles from effluent gases by giving the particles electric charges and then subjecting them to the influence of an electric field (Sec. 15-3B).

Element A substance not readily decomposable into simpler substances; a substance composed of atoms all having the same atomic number (Secs. 2-3 and 2-10).

Empirical formula A formula for a substance which indicates only the ratio of the kinds of atoms in the molecule, e.g., hydrazine, NH_2, sodium chloride, $NaCl$, sulfur, S (Sec. 3-14).

Emulsifier A substance used to disperse particles of an oily, fatty, or other substance in a liquid.

Endothermic reaction A reaction which absorbs heat as it takes place (Sec. 6-7B).

End point The point in a titration when an indicator signals the equivalence point.

Energy The capacity to do work, e.g., the capacity of a force to set a body of matter in motion (Secs. 2-3 and 14-2).

Entropy A measure of a system related to its degree of disorder. A system of high disorder has a high entropy. Indicated by the symbol S.

Enzyme A substance which catalyzes a reaction of a living system (Secs. 10-5 and 11-2C).

Enzyme-substrate complex The weak association between enzyme and substrate, of central importance to the mechanism whereby enzymes exert their catalytic effect (Sec. 10-5).

Equation, chemical See Chemical equation.

Equivalence point The point, usually in a titration, when equivalent amounts of the reactants have reacted.

Equivalent amount (a) That amount of a substance which will be formed by a mole (6.022×10^{23}) of electrons in an electrolysis or that amount which will release a mole of electrons in an electrolysis. (b) That amount of an acid in a proton-transfer reaction which will react with a mole (6.022×10^{23}) of hydroxide ions, OH^-; that amount of a base in a proton-transfer reaction which will react with a mole of hydrogen ions, H^+; also called an equivalent.

Essential amino acid An amino acid needed by the body which cannot synthesize it or cannot synthesize it fast enough to meet bodily needs and which the body must obtain directly in food (Sec. 11-6C).

Essential fatty acid A fatty acid which the human body must obtain directly from food since the body is unable to synthesize it from other substances or synthesize it fast enough to meet bodily needs (Sec. 11-6B).

Ester See Carboxylic ester and Phosphate ester.

Ether A carbon-containing compound with an ether functional group $-\overset{|}{\underset{|}{C}}-O-\overset{|}{\underset{|}{C}}-$; general formula ROR; also a common name for ethyl ether, $CH_3CH_2OCH_2CH_3$, one of the original surgical anesthetics (Sec. 8-5).

Ethical drug A drug which can be bought only on a physician's prescription, e.g., penicillin and morphine (Sec. 12-10).

Eutrophication A condition of a body of water characterized by high nutrient content capable of supporting a large amount of aquatic plant life, leading in turn to the depletion of the oxygen content of the water.

Evaporation The conversion of a liquid into the gaseous state at the surface (Sec. 4-8).

Exclusion principle See Pauli exclusion principle.

Exothermic reaction A reaction which evolves heat as it occurs (Sec. 6-7B).

Exponent A number written as a superscript which indicates the number of times a number is multiplied by itself, for example, $10^4 = 10 \times 10 \times 10 \times 10$ (Sec. S1-1).

Exponential form of a number An expression of a number as a product of two numbers: the first (called the coefficient) is a number between 1 and 10, and the second is 10 raised to a power expressed as an exponent, for example, $23,400 = 2.34 \times 10^4$ (Sec. S1-1).

Fahrenheit degree A degree on the Fahrenheit temperature scale, where water freezes at 32°F and boils at 212°F (Sec. S3-5).

Fast neutron A neutron with a relatively great amount of energy (Sec. 9-4).

Fat A carboxylic ester of the alcohol glycerol and three long-chain carboxylic acids (Secs. 10-3 and 11-2B).

Fatty acid The name for a carboxylic acid obtained from the hydrolysis of a fat (Secs. 10-3 and 11-2B).

Fermentation The conversion of an organic (carbon-containing) substance into substances of simpler structure by reactions catalyzed by enzymes (Sec. 14-13C).

Fertile material A substance, such as thorium 232, which although not fissile it itself can be converted into a fissile material (Sec. 9-9).

First law of thermodynamics See Law of conservation of energy.

Fissile nuclide An isotope which is readily fissionable (Sec. 9-8).

Fission See Nuclear fission.

Fixation of nitrogen See Nitrogen fixation.

Fluorocarbon A substance which contains both fluorine and carbon atoms; e.g.; trichlorofluoromethane (Freon-11), $CFCl_3$, used in refrigeration and air-conditioning units (Sec. 15-5).

Food additive Any nutritive or nonnutritive substance added to food other than a basic foodstuff (Sec. 11-12).

Force That which is exerted on a motionless body of matter to put it in motion or to change the velocity of a body if it is already in motion (Sec. 2-3).

Fossil fuel A fuel derived from materials formed over a very long period of time from the remains of living organisms such as petroleum, natural gas, and coal (Sec. 14-3).

Fractional distillation A process involving successive distillations; used to separate components of a mixture according to their boiling points (Sec. 5-5).

Free energy A property of a chemical system which indicates its capacity to do useful work.

Free radical A highly reactive molecular species which contains a single, unpaired electron, e.g., the methyl radical $CH_3\cdot$.

Freezing The solidification of liquids (Sec. 4-8).

Freezing point The temperature at which freezing occurs (Sec. 4-8).

Fuel cell A variation of a galvanic cell which can provide electric energy continuously (Sec. 7-13).

Functional group A distinctive arrangement of atoms in a molecule which imparts characteristic chemical properties (Sec. 8-2).

Functional group principle A principle of organizing and predicting reactions which assumes that the chemical properties of a substance are determined by the functional groups present in its molecular structure (Sec. 8-2).

Fusion The melting of a solid to a liquid (Sec. 4-8); see also Nuclear fusion.

Galvanic cell An apparatus for using an electron-transfer reaction to generate the flow of electrons constituting an electric current in a metal conductor (Sec. 7-12).

Gamma ray A form of highly penetrating, invisible light rays, emitted during radioactive disintegrations; symbol γ (Sec. 9-1).

Gasoline A blended mixture of hydrocarbon molecules containing from 5 to 12 carbon atoms in greatest concentration, often containing nonhydrocarbon additives (Secs. 13-10C and 14-3).

Gene A basic unit of heredity; a segment of a deoxyribonucleic acid (DNA) molecule (Secs. 10-6 and 15-7B).

General anesthetic A substance which relieves pain throughout the body but usually brings about loss of consciousness, e.g., ethyl ether and nitrous oxide (Sec. 12-4).

Generic name for a drug The name of a drug which usually designates a chemical relationship and does not depend on the company supplying it (Sec. 12-10).

Genetic code The relationship between the base-unit sequence in ribonucleic acid molecules (RNA) or their deoxyribonucleic acid (DNA) parents and the sequence of amino acid residues in proteins (Sec. 10-8G).

Geothermal energy Energy derived from within the earth, mostly from underground sources of steam and hot water (Sec. 14-14).

Glucose A simple sugar (a monosaccharide), a white crystalline solid (Sec. 10-2A).

Glycogen A form of starch found in animals; made within the animal from α-glucose units, it serves as a reserve supply of carbohydrates (Sec. 10-2E).

Gram atomic weight An atomic weight expressed in grams; the weight in grams of a mole, i.e., an Avogadro number (6.022×10^{23}) of the atoms of a given element (Sec. S4-3).

Gram molecular volume The volume of a mole of any gas at 0°C and 760 mmHg; actually close to 22.4 liters; also called molar volume.

Gram molecular weight A molecular weight expressed in grams; the weight in grams of a mole, i.e., an Avogadro number (6.022×10^{23}) of the molecules of a given substance (Sec. S4-3).

GRAS list The official Food and Drug Administration list of food additives which are generally recognized as safe (Sec. 11-12).

Gravimetric A procedure which involves the measurement of mass; most commonly applied to analytical procedures.

Greenhouse effect The insulating action of an atmospheric substance, e.g., carbon dioxide; like the glass of a greenhouse, it permits the energy of the sun to pass through to the earth but prevents the heat from leaving (Sec. 14-4).

Half-life The time it takes for one-half the nuclei in a given sample of a specific radioactive isotope to decompose (Sec. 9-15).

Hallucinogen A substance which when ingested causes hallucinations (Table 12-7).

Halogen An element of periodic group VII: fluorine, chlorine, bromine, iodine, astatine.

Hard water Water with appreciable concentrations of such ions as calcium, magnesium, and iron (Sec. 13-9).

Heat of reaction The heat evolved or absorbed as a reaction occurs (Sec. 6-7).

Heliostat A sun-following mirror used in concentrating solar energy (Sec. 14-11B).

Herbicide A substance or other agent used to kill weeds.

Heterocyclic base A basic compound whose molecule contains a heterocyclic ring (Sec. 10-8B).

Heterocyclic compound A substance whose molecules contain heterocyclic rings (Sec. 10-8B).

Heterocyclic ring A structural-ring unit of a molecule which contains carbon atoms and other atoms (such as nitrogen, oxygen, or sulfur) (Sec. 10-8B).

Hexose A sugar whose molecules contain six carbon atoms, e.g., glucose and fructose.

Hormone A substance secreted by specific body organs which exerts important controlling influences on body chemistry, e.g., insulin, secreted by the pancreas, which regulates the reactions involving the use of glucose (Sec. 12-7).

Hund's rule of maximum multiplicity Electrons in an incompletely filled subshell occupy different orbitals insofar as possible.

Hydrate A substance containing water in chemically combined form, e.g., sodium sulfate decahydrate, $Na_2SO_4 \cdot 10H_2O$.

Hydration The special attraction between an ion and one or more water molecules in an aqueous solution due to ion-dipole forces (Sec. 5-2D).

Hydrocarbon A compound of carbon which contains only carbon and hydrogen atoms (Sec. 8-4).

Hydrogen bond The relatively weak but significant attractive force which arises between hydrogen atoms in certain molecules and nonbonding pairs of electrons in certain neighboring molecules (Sec. 4-12).

Hydrolysis A chemical reaction in which water is a reactant. (Sec. 8-2).

Hydronium ion A hydrated hydrogen ion, H_3O^+ (Sec. 5-2F).

Incomplete combustion Combustion with a limited amount of oxygen; in the incomplete combustion of hydrocarbons carbon monoxide, CO, is among the products as well as carbon dioxide, CO_2, and water (Sec. 14-3).

Industrial smog A hazardous and sometimes deadly smog which is generated in highly industrialized areas; made up for the most part of sulfur dioxide, SO_2, sulfuric acid mist, H_2SO_4, and airborne particles of smoke, ash, and substances such as ammonium sulfate, $(NH_4)_2SO_4$ (Sec. 15-2B).

Inert gas A name formerly applied to the noble gases, the elements of periodic group 0, because for many years they were considered unable to react chemically.

Inertia The tendency of a body of matter to remain at rest if it is at rest or to keep moving with a constant velocity if it is moving (Sec. 2-3).

Inorganic chemistry The study of the structure and behavior of substances composed of all the elements other than carbon (Sec. 1-15).

Insect hormone A substance secreted in extremely small amounts by insects which regulates the growth and developmental changes at the various stages of the insect's life (Sec. 11-10).

Insecticide A chemical agent used to kill insects (Sec. 11-10).

Ion A charged particle formed from an atom (or a group of atoms) through the gain or loss of one or more electrons (Sec. 3-4).

Ion combination reaction which forms a precipitate Combining ions in excess of their solubility in water and the resulting formation of a precipitate (Sec. 7-2).

Ion-dipole force The cohesive force between an ion and a polar molecule such as the water molecule (Sec. 5-2D).

Ionic bond An interatomic bond which arises from the forces of attraction between positive and negative ions, known as electrostatic attraction (Sec. 3-6).

Ionic compound A substance made up of positive and negative ions, also called a salt (Sec. 3-6).

Ionic crystal A crystalline solid composed of a three-dimensional pattern of positive and negative ions (Sec. 4-7).

Ionic radical An ion which contains more than one atom, e.g., nitrate ion, NO_3^-, sulfate ion, SO_4^{2-} (Sec. S6-3).

Ionization The process of ion formation (Sec. 3-4).

Ionization equilibrium The equilibrium condition prevailing in an aqueous solution of a substance only some of whose molecules are dissociated into ions (Sec. 5-2G).

Ion-product constant of water The value of the product of the molar concentration of hydrogen ions and hydroxide ions of water or of an aqueous solution:

$$[H^+][OH^-] = 1 \times 10^{-14} \text{ at } 25°C$$

Isomer A substance which has the same molecular formula as one or more other substances; i.e., all isomers of a given kind contain the same number of each variety of atom in their molecules but the arrangement in space of the atoms is different.

Isotope An atom which has the same atomic number as another atom but which differs from it in mass number (Sec. 2-5).

Juvenile hormone A hormone secreted by insects which regulates the metamorphic changes involved in the development of insects from egg to larva to pupa to adult (Sec. 11-10B).

Kelvin temperature The temperature on a scale where absolute zero is taken as zero degrees, same as the absolute scale (Sec. S3-5).

Ketone A carbon-containing compound which contains a ketone group

$$\begin{matrix} & & O & \\ | & & \| & | \\ -C- & C & -C- \\ | & & & | \end{matrix}$$

general formula $R-\overset{\overset{\displaystyle O}{\|}}{C}-R$ (Sec. 10-2B).

Kilogram The basic unit of mass in the metric system, equal to 1000 grams or 2.205 pounds (Sec. S3-1C).

Kilowatthour An amount of electric energy consumed in an hour at the rate of 1000 watts (Secs. 14-2 and S3-2).

Kinetic energy The energy of a body by virtue of its motion.

Kinetic molecular theory Another name for the kinetic theory.

Kinetic theory A group of explanatory concepts which have been elaborated to account for the behavior of gases, liquids, and solids; a fundamental part of this theory postulates that molecules and ions are constantly in motion (Sec. 4-5).

Law of conservation of energy Energy is conserved in energy transformations; it can neither be created nor destroyed but only changed in form (Sec. 6-8).

Law of conservation of mass During a chemical reaction the amount of mass remains constant (Sec. 6-5).

Law of constant composition The constituent elements of a compound are always present in the same fixed proportion by weight; also called the law of definite proportions.

Law of definite proportions Another name for the law of constant composition.

Le Châtelier's law If a change in condition is imposed on a system in equilibrium, the point of equilibrium tends to shift so as to minimize the effect of the change.

Lewis acid A molecular species which accepts a pair of electrons in the course of a reaction.

Lewis base A molecular species which donates a pair of electrons in the course of a reaction.

Lipid A member of a class of substances found in living organisms which includes fats, oils, and waxes.

Local anesthetic A substance capable of relieving pain in specific parts of the body without loss of consciousness, e.g., novocain (Sec. 12-4).

Magnetic quantum number One of the four quantum numbers; related, roughly, to the orientation of an atomic orbital in space; symbol m_l or m.

Mass A quantity of matter (Sec. 2-3).

Mass number The sum of the protons and neutrons of an atom (Sec. 2-4).

Matter A general term for anything which occupies space and has inertia; the stuff of the physical universe (Sec. 2-3).

Mechanism of reaction The actual sequence of steps which occurs during a chemical reaction (Sec. 6-10).

Melting The conversion of a solid into a liquid (Sec. 4-8).

Melting point The temperature at which a solid melts (Sec. 4-8).

Metabolism The chemical processes which food substances undergo in living organisms (Sec. 11-6).

Metabolite A substance which is converted into another substance by a reaction of a living organism (Sec. 12-5).

Metal A substance like copper and silver which has a characteristic luster and the ability to conduct electricity and heat (Sec. 4-13).

Metallic bond The net forces of attraction operating between the valence electrons and the positive ions of metals (Sec. 4-13).

Metalloid An element which has properties intermediate between those of a metal and a nonmetal.

Metallurgy The processes of obtaining metals from their ores.

Metastasis The process whereby cancerous tumors release malignant cells capable of migrating through the body fluids to establish new cancerous growths elsewhere in the body (Sec. 15-7B).

Meter The basic unit of length in the metric system, abbreviated m; defined in terms of a wavelength of light; equal to 39.37 inches (Sec. S3-1C).

Mineral A substance obtained by a mining operation; also a substance containing any of the atoms found in the substances of living systems other than carbon, hydrogen, oxygen, and nitrogen (Sec. 11-8).

Mixture A combination of two or more substances which are not linked chemically (Sec. 2-3).

Moderator A substance such as graphite or water which is capable of slowing down fast neutrons (Sec. 9-4).

Molarity A measure of the concentration of a solution; the number of moles of the solute contained in 1.000 liter of solution; abbreviated M (Sec. S5-3).

Molar volume The volume of 1 mole of a gas at $0°C$ and 760 mmHg.

Mole An Avogadro number (6.022×10^{23}) of atoms or of anything else (Sec. S4-3).

Molecular analog A synthetic variation of a known molecular structure; also called a structural variant (Sec. 1-8).

Molecular crystal A crystal whose fundamental particles are molecules (Sec. 4-7).

Molecular formula A formula for a substance which indicates only the numbers of each kind of atom in the molecule; e.g., ethyl alcohol, C_2H_6O (Sec. 3-14).

Molecular orbital The volume within which an electron is most probably to be found within a molecule (Sec. 2-6).

Molecular-orbital description A method of expressing the structure of a molecule which uses molecular orbitals (Sec. 2-6).

Molecular species A general term which refers to an atom, molecule, or ion (Sec. 2-10).

Molecule The smallest particle of a substance as it ordinarily exists, whether the substance is an element or a compound (Secs. 1-3 and 2-10).

Monomer A substance which is the starting compound in the preparation of a polymer.

Monosaccharide A carbohydrate such as glucose and fructose which contains only one structural unit (Secs. 10-2G and 11-2A).

Mutagen An agent capable of causing an alteration in basic genetic material and thereby causing a mutation in a living system.

Mutant An organism which has undergone a mutation (Sec. 15-7C).

Mutation A change in the genetic material of a living organism which can be inherited (Sec. 9-13).

Narcotic analgesic Morphine or a centrally acting, pain-relieving substance which acts like morphine; also called an opiate (Sec. 12-4A).

Narcotic antagonist A substance which is able to block the effects of a narcotic (Sec. 1-9).

Natural gas Mixture of gaseous hydrocarbons (primarily methane) found in underground rock formations (Sec. 14-3).

Negative ion An ion with one or more negative charges (Sec. 3-4).

Net ionic equation A chemical equation in which only the molecular species involved in the reaction are represented (Sec. 7-2).

Neutralization A proton-transfer reaction.

Neutral solution An aqueous solution in which there is an equal concentration of hydrogen ions and hydroxide ions.

Neutron A fundamental atomic particle which has a mass about the same as a proton but has no electric charge; found in the nucleus of an atom (Sec. 2-2).

Nitrogen cycle A chemical cycle in nature in which nitrogen as an element or in compounds circulates in various natural processes.

Nitrogen fixation A process of converting elemental nitrogen into a nitrogen-containing compound (Secs. 11-4 and 13-1A).

Noble gas An element of periodic group 0.

Nonbonding electron A valence electron in a molecule which is not involved in an interatomic bond (Sec. 3-7).

Nonelectrolyte A substance incapable of furnishing a concentration of ions to an aqueous solution

Nonmetal An element such as oxygen and chlorine which tends to share electrons to form a covalent bond or to gain electrons and form the negative ion of an ionic bond.

Nonpolar molecule A molecule in which the center of positive charge and the center of negative charge coincide (Sec. 4-11).

Normal boiling point The temperature at which a liquid boils at 1 atmosphere pressure (760 mmHg or 760 torr).

Nuclear chain reaction A self-sustaining sequence of nuclear fission reactions (Sec. 9-4).

Nuclear energy Energy generated by a nuclear reaction (Secs. 9-7 and 14-8).

Nuclear equation A representation of a nuclear reaction in which the nucleus of each reactant and each product is represented by an isotope symbol (Sec. 9-14).

Nuclear fission A nuclear change in which an atomic nucleus is split into two nuclei of substantial size; also called atomic fission (Sec. 9-4).

Nuclear fusion A nuclear reaction in which two nuclei combine to form a nucleus of higher atomic number (Secs. 9-6 and 14-10).

Nuclear radiation Alpha particles, beta particles, gamma rays, neutrons, and other particles associated with radioactive materials and nuclear processes (Sec. 9-13).

Nuclear reaction A change in which the structures of atomic nuclei are altered (Sec. 9-1).

Nuclear reactor An apparatus for carrying out a self-sustaining nuclear-fission chain reaction in a controlled manner (Secs. 9-8 and 14-8).

Nucleic acids A group of substances whose molecules contain long-chain polyester structures to which are attached characteristic sequences of heterocyclic bases; they contain information which serves to guide the synthesis of proteins, and they possess the means of preserving that information (Secs. 10-6 and 10-8E).

Nucleon A term applied to protons and neutrons collectively (Sec. 9-14).

Nucleotide A unit of a nucleic acid molecule containing a heterocyclic base, a five-carbon sugar, and a phosphate ester group.

Nucleus A very small, positively charged tightly packed arrangement of protons and neutrons at the center of an atom. Contains most of the mass of an atom (Sec. 2-4).

Octane number A number usually between 0 and 100 which indicates the relative performance of a fuel in an internal combustion engine, on a scale where the performance of isoctane

$$CH_3$$
$$|$$
$$CH_3CCH_2CHCH_3$$
$$|\quad\quad|$$
$$CH_3\quad CH_3$$

has been arbitrarily assigned an octane number of 100 and the performance of n-heptane,

$$CH_3CH_2CH_2CH_2CH_2CH_2CH_3$$

has arbitrarily been assigned an octane number of 0.

Octet rule Atoms with fewer than eight valence electrons have a strong tendency to achieve a total of eight valence electrons (or two for very simple atoms); this is a useful generalization in organizing the combining behavior of atoms (Sec. 3-5).

Oil shale Shale rock which contains hydrocarbon oils. The hydrocarbons can be obtained by crushing the rock and volatilizing the hydrocarbons.

Olefin Another name for an alkene; a hydrocarbon containing a carbon-carbon double bond (Sec. 8-5).

Opiate Morphine or a centrally acting analgesic like morphine; also called a narcotic analgesic (Sec. 12-4A).

Orbital The volume around an atomic nucleus in which an electron is most probably to be found (Sec. 2-4F).

Orbital quantum number One of the four quantum numbers, symbol l, which characterizes the shape of the atomic orbital within which the electron is most probably to be found.

Ore A mineral which serves as the commercial source of a raw material.

Organic chemistry That branch of chemistry which is concerned with the compounds of carbon (Sec. 8-1).

Over-the-counter drug A drug sold freely without a prescription, e.g., aspirin and tincture of iodine (Sec. 12-10).

Oxidation (a) A reaction involving oxygen as a reactant; (b) the loss of electrons in an electron-transfer reaction (Sec. 7-10).

Oxidation-reduction reaction Another name for an electron-transfer reaction; also called a redox reaction (Sec. 7-10).

Oxidizing agent The molecular species which accepts one or more electrons in an electron-transfer reaction (Sec. 7-10).

Oxy acid An acid which contains oxygen (Sec. S6-3).

Ozone layer A layer of ozone, O_3, in the upper atmosphere which absorbs the high-energy ultraviolet light of the sun, screening this harmful radiation from the people, plants, and animals of the earth (Sec. 15-5).

PAN Abbreviation for peroxyacetyl nitrate, one of a group of irritating compounds present in photochemical smog; formed by the interaction of hydrocarbons, nitrogen oxides, ozone, and sunlight (Sec. 15-2D).

Parent element The principal reactant in a radioactive disintegration (Sec. 9-14).

Particulate One of the finely divided polluting particles of liquids and solids which remain suspended in the atmosphere for a long time (Sec. 15-2C).

Parts per million parts by weight A method of expressing very low concentrations, e.g., the number of grams of solute in each million (1.00 $\times$ 10^6) grams of solution (Sec. S5-2).

Pauli exclusion principle All the electrons in any atom are distinguishable from each another; no two electrons in a given atom may have all four quantum numbers the same.

Peat An early stage of the formation of coal from plant material; used as a fuel.

Pentose A five-carbon sugar.

Peptide bond A special name given to the amide bonds of protein molecules (Secs. 10-4B and 11-2C).

Percent by weight A method of expressing concentrations, e.g., the number of grams of solute in each 100.00 grams of solution (Sec. S5-1).

Period A horizontal row of elements in the periodic table (Sec. 3-3).

Periodic law The properties of the elements vary periodically if the elements are arranged in order of their atomic numbers (Sec. 3-3).

Periodic table A tabular organization of the properties of elements in horizontal rows (periods) and vertical columns (groups) and overall according to increasing atomic number (Sec. 3-3).

Pesticide A chemical agent used to kill or control insects, rodents, weeds, or other pests.

Petrochemical A chemical intermediate derived from petroleum or natural gas (Sec. 13-11B).

pH The logarithm of the reciprocal of the molar concentration of hydrogen ion in an aqueous solution; pH = $\log (1/[H^+])$; a measure of the acidity of a solution; the lower the pH value the greater the acidity.

Pharmacology A branch of science which is concerned with the development of drugs, their use, and the mechanism of their action (Sec. 12-2).

Pheromone A substance secreted by one insect which affects the behavior of another insect of the same species (Sec. 11-10C).

Phosphate ester The class of substance formed when an alcohol reacts with phosphoric acid (Sec. 10-8A); general formulas

```
     O              O              O
     ‖              ‖              ‖
H—O—P—O—R    R—O—P—O—R    R—O—P—O—R
     |              |              |
     O              O              O
     |              |              |
     H              H              R
```

Photochemical smog The irritating and toxic gaseous mixture formed in the atmosphere when hydrocarbons and nitrogen oxides are irradiated by sunlight (Sec. 15-2E).

Photon A pulse of light energy (Sec. 14-11C).

Photosynthesis The process whereby solar energy is converted into chemical energy as plants synthesize the carbohydrate glucose from carbon dioxide and water (Secs. 11-3 and 14-13C).

Photovoltaic cell A device which converts light energy directly into an electric current (Sec. 14-11C).

Physical change A change in the state or form of a sample of matter which does not involve a change in chemical composition, e.g., freezing water and boiling water.

Plasma A state of matter at extremely high temperatures (several million degrees) containing a mixture of atomic nuclei and electrons.

Plastic A polymeric material capable of being molded into such objects as eyeglass frames and buttons (Sec. 13-3).

Polar molecule A molecule in which the centers of positive and negative charge are separated from each other (Sec. 4-11).

Pollutant Any form of matter which when added to the surroundings in sufficient amount produces a detectable injurious effect on people, animals, plants, or the environment in general (Sec. 15-2).

Polymer A substance whose molecules are made up of many structural units held together by interatomic bonds (Secs. 7-9 and 13-3).

Polymerization reaction A reaction in which many molecules of one or a small number of substances of relatively simple structure are linked together to make a large molecule (Secs. 7-9 and 13-4).

Polypeptide A molecule containing several amino acid residues (Sec. 10-4B).

Polypeptide structure The polyamino acid residue structure of a protein (Secs. 10-4B and 11-2C).

Polysaccharide A carbohydrate like starch and cellulose which contains many structural units (Secs. 10-2G and 11-2A).

Polyunsaturated fat Another name for an unsaturated fat, i.e., a fat which contains a substantial proportion of unsaturated fatty acid residues (residues with alkene groups) (Sec. 10-3).

Positive ion An ion with one or more positive charges (Sec. 3-4).

Potential energy The energy of a body by virtue of its position.

Power The rate of expending energy (Sec. 14-2).

Precipitate The deposit of a solid from a solution (Sec. 7-2).

Precision The closeness of approach of a series of repeated measurements to the average value of the measurements (Sec. S2-1).

Pressurized water reactor (PWR) A type of nuclear reactor in which the water used to remove heat from the reactor is kept under pressure to prevent formation of steam.

Primary level of protein structure The fundamental atom-by-atom sequence in the polypeptide chain and any prosthetic groups which may be present in a protein (Sec. 10-9).

Principal quantum number One of the four quantum numbers, symbol n; characterizes the average distance of an electron from the nucleus of an atom. A rough measure of the size of the orbital within which the electron is most probably to be found.

Product A substance formed in a chemical reaction (Sec. 6-2).

Prosthetic group The nonpolypeptide structural unit in a protein (Secs. 10-4C and 11-2C).

Protein A substance with a large molecule which contains a long chain of amino acid residues joined by peptide bonds and sometimes in addition a prosthetic (nonpolypeptide) structural unit (Sec. 10-4D).

Protium A name of the hydrogen isotope ^1_1H; one of the three isotopes of hydrogen; more commonly called hydrogen or hydrogen 1 (Sec. 2-5).

Proton A fundamental atomic particle with a mass close to that of a hydrogen 1 atom and a positive charge of the same magnitude as the negative charge on an electron; found in the nucleus of an atom; same as a hydrogen ion (Sec. 2-2).

Proton-transfer reaction A reaction which involves the transfer of one or more protons from one molecular species to another (Sec. 7-4).

Pure substance A substance free of any other substance; in short another word for a substance.

Quantum A small, separate packet of energy.

Quantum number One of four numbers assigned to an electron in an atom which designate the characteristics of the electron (Sec. 2-8).

Radical See Free radical and Ionic radical.

Radioactivity The spontaneous decomposition of atomic nuclei (Sec. 9-1).

Radiocarbon dating A method of determining the age of archeological specimens made from the materials of living organisms through a comparison of their carbon radioactivity with the carbon radioactivity of current living organisms (Sec. 9-16).

Radioisotope An isotope which is radioactive.

Rate of a reaction The amount of product formed (or reactant consumed) in a chemical reaction in a given unit of time (Sec. 6-9).

Reactant A starting substance in a chemical reaction (Sec. 6-2).

Reaction A chemical reaction; a process in which substances are converted into other substances with different sets of properties (Sec. 6-2).

Reactor fuel The substance which undergoes nuclear fission in a nuclear reactor (Sec. 9-8).

Recrystallization A process of purifying solids which takes advantage of the varying solubilities of substances and the extent to which these solubilities change with temperature (Sec. 5-4).

Redox reaction Another name for an electron-transfer (oxidation-reduction) reaction (Sec. 7-10).

Reducing agent The molecular species which donates one or more electrons in an electron-transfer reaction (Sec. 7-10).

Reduction The process of gaining electrons in an electron-transfer reaction (Sec. 7-10).

Reforming A petroleum-refinery process involving rearrangements of molecular structures of hydrocarbons which increase the branching of the carbon chains to make better motor fuels (Sec. 13-11A).

Resonance The discrepancy between the actual bonding in a molecule (or ionic radical) and the formulas which are used to represent that bonding (Sec. 3-16).

Resonance structure (resonance approximation, resonance form) Any approximate structural formula used as one of a series to represent an actual molecular structure (Sec. 3-16).

Resonance hybrid A molecular species whose structure must be represented by more than one approximate structural formula if it is to be represented by structural formulas at all (Sec. 3-16).

Reversible reaction A reaction which can take place in either direction depending on such conditions as concentration, temperature, and pressure.

RNA Ribonucleic acids, the nucleic acids which guide the synthesis of proteins (Sec. 10-8G).

Salt Another name for an ionic crystal; a substance composed of a three-dimensional lattice of positive and negative ions, e.g., sodium chloride, $NaCl$, and lithium carbonate, Li_2CO_3 (Sec. 3-6).

Saponification The hydrolysis of a carboxylic ester catalyzed by sodium or potassium hydroxide; specifically the hydrolysis of fats catalyzed by sodium or potassium hydroxide to form glycerol and a soap (Sec. 13-8).

Saturated fat A fat which contains fatty acid residues which are all, or almost all, saturated; i.e., the fat contains few if any alkene groups (Sec. 10-3).

Saturated fatty acid A carboxylic acid obtained from the hydrolysis of a fat which contains no alkene groups (Sec. 10-3).

Saturated hydrocarbon A hydrocarbon in which all the bonds are single covalent.

Saturated solution A solution in which the maximum amount of solute is dissolved (Sec. 5-1).

Secondary level of protein structure How chains of amino acid residues are arranged in three dimensions to form coils, sheets, or more compact arrangements (Sec. 10-9).

Shell The region of an atom which will accommodate electrons with the same principal quantum number.

Significant figure A digit in a number expressing a measurement that is reasonably reliable (Sec. S2-2).

Simple protein A protein with no prosthetic group; it contains only a polypeptide structure (Secs. 10-4C and 11-2C).

Single covalent bond A covalent bond in which one pair of electrons is shared by the bonded atoms (Sec. 3-8).

SI unit A unit in the latest revision of the metric system, the Système International (Sec. S3-1A).

Slow neutron A neutron with relatively low energy, which is effective in bringing about a nuclear fission (Sec. 9-4).

Smog See industrial smog or photochemical smog.

Soap For the most part, a sodium salt of a carboxylic acid containing between 12 and 18 carbon atoms (Sec. 13-8).

Solar cell Another name for a photovoltaic cell (Sec. 14-11C).

Solubility The maximum amount of solute which will dissolve in a solvent under specified conditions of temperature and pressure (Sec. 5-1).

Solubility equilibrium The specific equilibrium condition prevailing when a saturated solution is in contact with undissolved solute (Sec. 5-2E).

Solute The material which is dissolved in a solution (Sec. 5-1).

Solution A homogeneous mixture of two or more substances dispersed in each other at the molecular level (Sec. 5-1).

Solvent The dissolving medium in a solution (Sec. 5-1).

Spin quantum number One of the four quantum numbers, symbol m_s or s; it designates a property of an electron which in mechanical terms is best described as a spin.

Stereochemistry The part of chemistry concerned with the spatial arrangement of the structural units of matter (Sec. 3-17).

Stoichiometry The part of chemistry concerned with the amounts of substances involved in chemical reactions.

Straight-run gasoline The fraction of crude petroleum directly obtainable as gasoline by distillation (Sec. 13-11A).

Strong electrolyte A substance which when dissolved in water is capable of furnishing an appreciable concentration of ions.

Structural formula A formula representation of a molecule or ion in which each bonding electron pair of an electronic formula is replaced by a dash; individual atoms are represented by their letter symbols, and their sequence is indicated by the use of dashes (Secs. 1-6 and 3-10).

Structural variant A synthetic variation of a known molecular structure; also called a molecular analog (Sec. 1-8).

Sublimation The evaporation of a solid directly to a gas and subsequent conversion back to the solid state (Sec. 4-8).

Subshell The region of an atom which will accommodate electrons with the same principal and orbital quantum numbers, e.g., $1s$ subshell, $2p$ subshell.

Substance A form of matter with a definite composition and a consistent set of properties (Secs. 1-2 and 2-3).

Substrate A reactant whose participation in a specific reaction is catalyzed by an enzyme (Sec. 10-5).

Sugar A carbohydrate of simple structure such as glucose and sucrose (Secs. 10-2G and 11-2A).

Supersaturated A solution which contains more solute than equilibrium conditions allow.

Surface tension The property of liquids caused by the unbalanced distribution of the forces acting on the molecules at the surface; this results in an inward pull, which contracts the surface of the liquid and causes it to behave as though it were covered by a thin skin (Sec. 4-6).

Suspension A dispersion of a liquid or a solid in a liquid in which the dispersed substance does not dissolve.

Symbiotic relationship An interdependence of two species of organisms living together which is advantageous to both (Sec. 11-4).

Synthetic detergent Any of the synthetic substances (other than soaps) capable of soaplike cleaning action (Sec. 13-9).

Temperature A property of matter which is directly dependent on the amount of motion of its component particles.

Temperature inversion A natural atmospheric phenomenon in which a mass of warm air lies above a mass of cool air in the atmosphere, in a reversal of the normal arrangement (Sec. 15-1).

Titration A method of determining the concentration of a solution by determining the volume which will react in a known reaction with a given volume of another solution of known concentration.

Torr A unit of pressure equal to the pressure exerted by a column of mercury 1 millimeter high (Sec. S3-3).

Trade name of a drug A name given to a drug or a specific drug preparation by a specific manufacturer (Sec. 12-10).

Transuranium element An element of atomic number greater than uranium (Sec. 9-5).

Triple covalent bond A covalent bond in which three pairs of electrons are shared between the bonded atoms (Sec. 3-8).

Tritium A name for one of the three isotopes of hydrogen, ${}^3_1H$; also called hydrogen 3 (Sec. 2-5).

Tumor A growth in a living organism composed of cells which lack the normal controls on growth and reproduction (Sec. 15-7B).

Unsaturated fat A fat which contains a substantial proportion of unsaturated fatty acid residues (residues with alkene groups) (Sec. 10-3).

Unsaturated fatty acid A fatty acid which contains one or more alkene groups (Sec. 10-3).

Vacuum A region in space which contains no matter (Sec. 2-3).

Valence The capacity of an atom to form bonds with other atoms (Sec. 3-2).

Valence-bond resonance representation A multiple-structural-formula approach to the representation of molecular structure (Sec. 3-16).

Valence electron An electron in the outermost shell of an atom; of great influence on the combining behavior of an atom (Sec. 3-2).

Van der Waals forces Short-range electrical forces which are the weakest cohesive forces between molecules (Sec. 4-10).

Vapor A gas obtained by evaporating a liquid or a solid or boiling a liquid (Sec. 4-8).

Vapor pressure The pressure exerted by the escape of the fastest-moving particles from the surface of a liquid or a solid (Sec. 4-9).

Vegetable oil A fat of vegetable origin which is a liquid at room temperature (Sec. 10-3).

Vitamin One of a group of about 20 carbon-containing substances known to be necessary in the diet for normal health in addition to the usual carbohydrates, fats, and proteins (Sec. 11-7).

Voltaic cell Another name for a galvanic cell.

Vulcanization The process of heating raw rubber with elemental sulfur; at the molecular level the principal change is the formation of covalently bonded sulfur-containing cross-links between chains of isoprene residues; the rubber becomes less tacky, more elastic, and more resistant to abrasion (Sec. 13-5).

Washing soda A common name for sodium carbonate, Na_2CO_3.

Water-gas process A process in which steam and hot coke react to produce carbon monoxide and hydrogen (Sec. 13-10B).

Water softening A process for removing ions from water, especially calcium ions, Ca^{2+}, and magnesium ions, Mg^{2+}.

Watt A unit of power equivalent to the expenditure of 860 calories of energy in an hour (Secs. S3-2 and 14-2).

Weak electrolyte A substance which can furnish only a relatively small concentration of ions in an aqueous solution.

Weight The force exerted on a quantity of matter by the gravity of the earth (Sec. 2-3).

X-rays Invisible light rays having a wavelength from 0.1 to 10 angstrom units (Sec. 4-4).

INDEX

Page numbers in *italic* indicate tables.

British thermal unit, defined, 415, 509
Bromine:
    combining behavior, 67, 68
    uses, 35
Brown-air pollution, 543
Brownian motion, 82, 83
Btu (British thermal unit), 415, 509
Bubonic plague, 331, 332, 340
Buffer, defined, 543
Butadiene:
    production of, 384
    and synthetic rubber, 379
    and synthetic rubber production, 384–385
    use, 186
Butter fat, 241
Butylated hydroxyanisole (BHA) as food preservative, 313, 314
Butylated hydroxytoluene (BHT) as food preservative, 313, 314

Cade, John F. J., 156
Cadmium as nuclear control element, 207
Cadmium sulfide as carcinogenic agent, 481
Caffeine and drug abuse, 359
Calcium:
    combining behavior, 67, 68
    in human body, 291, 292
Calcium carbide and acetylene production, 397
Calcium carbonate:
    and acetylene production, 397
    as gastric antacid, 128, 160–161
    and magnesium production, 389, 392
    reaction with hydrochloric acid, 128–129, 160–161
    and sulfur oxide removal, 426
Calcium chloride, formula, 71
Calcium hydroxide and production of magnesium, 156–157, 389
Calcium oxide in sulfur oxide removal, 426
Calcium propionate as food preservative, 313, 314
Calorie:
    defined, 307, 415, 509
    of nutritionists, 307, 415
Calorie food intake, average, 307
    significance, 307
Calvin, Melvin, 437
Cancer:
    as cause of death, 478
        graph, 480
    and cigarette smoking, 484, 485
    environmental causes, 478–479
    nature of, 352
    treatment, 352–353
    radiation, 209
    types of lethal, graph, 480
Cancer-causing agents [see Carcinogenic agent(s)]
Cancer formation, latency period, 484
Cancerous tumor(s) [see Tumor(s), cancerous]

Cannizzaro, Stanislao, 21
Carbohydrates, 229–237, 275
    classes, 236
    defined, 235
    digestion, 284
    digestion products, metabolism, 285–286
    examples, 236–237
    in human body, 253
    as "hydrates of carbon," 275
    from photosynthesis, 280
Carbon, 35
    combining behavior, 67, 68
    in human body, 291
Carbon compounds:
    families, 185, 188
    and organic chemistry, 174–175
    why so many? 175
Carbon dioxide:
    in atmosphere, 416–419
    electronic formula, 54
    as fire extinguisher, 145
    and greenhouse effect, 416–419
    nonpolar character, 97
    in photosynthesis, 280, 417
Carbon 14:
    production, 208–209
    radiation, 216
    and radiocarbon dating, 220–221
    as tracer isotope, 208–209
Carbon monoxide:
    as air pollutant, sources, 464
    and gasoline production, 400
    hazard in fires, 254
    hazards, 465
    from incomplete combustion, 416, 464–465
    mechanism of poisoning, 254
    natural sources, 465
    toxicity, 254
Carbon tetrachloride:
    formula, 109
    solution with hexane, 109
Carbon 12 as relative atomic weight standard, 134
Carbonic acid, formation in proton-transfer reaction, 160–161
Carboxylic acid(s), 178
    from ester hydrolysis, 238–239
    examples, 184, 186–187
    functional group, 178, 187
        recognition, 190–192
    general formula, 187
Carboxylic acid residue, defined, 240
Carboxylic ester(s):
    and fat structure, 393
    in fats, 238–239
    functional group, 187
        recognition, 190, 192, 193
    general formula, 187
    hydrolysis, 238
Carcinogenic agent(s):
    activation in body, 484
    defined, 479
    in environment, 477–488
    identifying: Ames test, 487–488
        animal tests, 485, 487
        from cancer incidence, 486–487
        difficulties, 484–488
        latency period, 484, 485

Carcinogenic agent(s):
    in nature, 479, 481
    representative, 481–482
Carothers, Wallace, 388
Catalyst(s):
    action, mechanism of, 143–145
    defined, 143
    enzymes, 247–251
Catalytic converters:
    automobile, 470
    and lead tetraethyl, 470, 473
Cation, 543
Caustic soda, 543
Caventou, J. B., 342
Cellulose:
    presence in nature, 234
    structure, 234–235
Celsius temperature scale, 511–513
Centi, metric system meaning, 506
Centigrade temperature scale, 511–513
Cesium, combining behavior, 67, 68
Chain, Ernest, 335
Chain reaction(s), nuclear, 201–202
Chain-reaction mechanism, 146–147
Characterization, defined, 5
Charles' law, 543
Chemical changes (see Reactions)
Chemical equation(s) [see Equation(s), chemical]
Chemical equilibrium, defined, 114
Chemical industry:
    anonymous nature, 375
    distinguishing features, 369–376
    flexibility, 375
    hazards, 375–376
    pattern of production, 376–377
    products, 377
    scope, 376
Chemical intermediate(s):
    defined, 374
    examples, 377
Chemistry:
    applications, 3
    general aims and accomplishments, 15
    scope, 1
    special viewpoint, 3
    ultimate aim, 1
    unfinished nature, 15
Chemists, work of, 3
Chemotherapeutic agent(s), 334–339
    defined, 334
Chemotherapy, 334–338
    in treatment of cancer, 352–353
Cheney, Eric S., 452
Chlordane as insecticide, 298–299
Chlorinated hydrocarbon, 543
Chlorine, 35
    combining behavior, 67, 68
    disinfectant in water, 168, 181, 331
    in iodine production, 162–163
    in magnesium production, 390, 392
Chloroform as anesthetic, 176
Chlorophyll, role in photosynthesis, 280